SPARKNOTES™

SAT II Physics

2003–2004 Edition

Series Editor Ben Florman

Book Editor David Egan

Contributors Elizabeth Waterhouse, Elizabeth Shack

Production Porter Mason

Technology Tammy Hepps

This edition published by Spark Publishing

Spark Publishing
A Division of SparkNotes LLC
120 Fifth Avenue, 8th Floor
New York, NY 10011

02 03 04 05 SN 9 8 7 6 5 4 3 2 1

Please send all comments and questions or report errors to feedback@sparknotes.com.

Library of Congress information available upon request

Printed and bound in Canada

ISBN 1-58663-433-X

Welcome to SparkNotes™ Test Preparation

So you're planning to take the SAT II Physics test. Not only that, you're planning to get a great score on the SAT II Physics test. You've come to the right book. In order to help you reach your goal of extreme SAT II Physics success, the SparkNotes Guide to the SAT II Physics test provides:

The precise physics topics you'll need to know to do well on the test. This book is not designed to teach you all of physics. It's designed to teach you the physics you need to know to do well on the SAT II Physics test. That way, you won't waste your study time working on the wrong topics. Yet while the physics we teach is tailored to the test, we don't simply feed you formulas and tell you to memorize them, as other books do. The SAT II Physics test demands flexibility and understanding, not rigid memorization. We teach the physics you need to know so that you truly comprehend it, and will be able to deal with whatever the SAT II Physics throws at you.

Critical-thinking skills and specific SAT II Physics test-taking strategies. Understanding physics is the most important ingredient for success on the SAT II Physics. But only the proper test-taking strategies can make that knowledge really shine. SparkNotes explains all the strategies, from general approaches to taking the test to specific ways to answer questions on optics.

Five full-length practice tests, and a study method that will maximize their usefulness and teach you to learn from your mistakes. Practice tests can and should be an extremely important part of your studying for any standardized test. Practice tests help you to hone your test-taking skills, familiarize you with the format and time limits of the test, and show you your progress. In addition, if you follow our methods for studying the practice tests you take, the tests can be a powerful study tool that will help you to identify and eliminate your weaknesses.

General information about SAT II Subject Tests. Beyond teaching you what you need to know to do well on a particular SAT II test, we think it's also important to discuss the SAT IIs in general. This first chapter of this book is dedicated to helping you figure out how the SAT II tests are used by colleges, which SAT II tests are right for you, when to take the tests, and how to register for them.

Contents

ORIENTATION

Introduction to the SAT II Tests 3

Scoring the SAT II Subject Tests 4
SAT II Physics Score Conversion Table 6
Colleges and the SAT II Subject Tests. 7
Which SAT II Subject Tests to Take 8
When to Take an SAT II Subject Test 10
Registering for SAT II Tests 10

Introduction to the SAT II Physics 11

Content of the SAT II Physics 11
Format of the SAT II Physics. 13
How Your Knowledge Will Be Tested 16

Strategies for Taking the SAT II Physics 17

General Test-Taking Strategies 17
Strategies for Taking the SAT II Physics. . 20

SAT II PHYSICS REVIEW

Vectors 29

What's a Vector? 29
Vector Addition. 30
Vector Subtraction 33
Multiplication by a Scalar 34
Vector Components 34
Vector Multiplication 37
Key Formulas 41
Practice Questions 41
Explanations. 43

Kinematics 45

Displacement 45
Speed, Velocity, and Acceleration. 47
Kinematics with Graphs 50
One-Dimensional Motion
with Uniform Acceleration 58
Two-Dimensional Motion
with Uniform Acceleration 61
Key Formulas 62
Practice Questions 62
Explanations. 64

Dynamics 67

What Are Forces?. 67
Newton's Laws 68
Problem Solving with Newton's Laws 72
Types of Forces 74
Key Formulas 80
Practice Questions 80
Explanations. 83

Work, Energy, and Power 85

Work 85
Energy 91
Forms of Energy 92
Power 97
Key Formulas 98
Practice Questions 99
Explanations 101

Special Problems in Mechanics 103

The Three-Step Approach to Problem Solving 103
Pulleys 104
Inclined Planes 110
Springs 116
Pendulums 125
Key Formulas 129
Practice Questions 129
Explanations 132

Linear Momentum 135

What is Linear Momentum? 135
Impulse 136
Conservation of Momentum 138
Collisions 141
Center of Mass 145
Key Formulas 149
Practice Questions 150
Explanations 153

Rotational Motion 157

Important Definitions 158
Rotational Kinematics 160

Frequency and Period 162
Rotational Dynamics 166
Kinetic Energy 173
Angular Momentum 174
Key Formulas 177
Practice Questions 178
Explanations 181

Circular Motion and Gravitation 185

Uniform Circular Motion 185
Newton's Law of Universal Gravitation . . 189
Gravitational Potential Energy 193
Weightlessness 194
Kepler's Laws 196
Key Formulas 197
Practice Questions 198
Explanations 200

Thermal Physics 203

Heat and Temperature 204
The Kinetic Theory of Gases
& the Ideal Gas Law 211
The Laws of Thermodynamics 216
Heat Engines 219
Key Formulas 221
Practice Questions 222
Explanations 224

Electric Forces, Fields, and Potential 227

Electric Charge 228
Electric Force 229
Electric Field 232
Electric Potential 234
Conductors and Insulators 236
Key Formulas 237

Practice Questions 237
Explanations. 240

DC Circuits 243

Voltage. 243
Current. 244
Resistance 245
Energy, Power, and Heat 247
Circuits 249
Capacitors 262
Key Formulas 265
Practice Questions 266
Explanations. 268

Magnetic Force and Fields 271

Permanent Magnets. 271
Magnetic Force on Charges 273
Magnetic Force on
Current-Carrying Wires 278
The Magnetic Field Due to a Current . . . 279
Key Formulas 281
Practice Questions 282
Explanations. 286

Electromagnetic Induction 289

Motional Emf 289
Faraday's Law 291
Applications. 296
Key Formulas 297
Practice Questions 297
Explanations. 299

Wave Phenomena 301

Periodic Motion 301
Wave Motion 302
Transverse Waves and
Longitudinal Waves 305
Superposition 308
Standing Waves and Resonance 310
The Doppler Effect 313
Key Formulas 314
Practice Questions 315
Explanations 317

Optics 321

The Electromagnetic Spectrum 321
Classical Optics 322
Optical Instruments 327
Wave Optics 335
Key Formulas 341
Practice Questions 342
Explanations 344

Modern Physics 347

Special Relativity 348
The Discovery of the Atom 354
Quantum Physics 356
Nuclear Physics 365
Key Formulas 374
Practice Questions 375
Explanations 377

PRACTICE TESTS

Practice Tests Are Your Best Friends 381

Taking Advantage of the Test's Regularity 381
Taking a Practice Test 383
Scoring Your Practice Test 384
Studying Your... No, Wait, Go Take a Break 384
Studying Your Practice Test 384
The Secret Weapon: Talking to Yourself 386

Practice Test I 387
Explanations 403

Practice Test II 419
Explanations 437

Practice Test III 453
Explanations 469

ORIENTATION

Introduction to the SAT II

Chapter Contents

Scoring 4
Colleges and the SAT II Subject Tests 6
Which SAT II Subject Tests to Take 7
When to Take an SAT II Subject Test 9
Registering for SAT II Tests 9

THE SAT II SUBJECT TESTS ARE CREATED and administered by the College Board and the Educational Testing Service (ETS), the two organizations responsible for the dreaded SAT I, which most people just call the SAT. The SAT II Subject Tests were created to act as complements to the SAT I. Whereas the SAT I is a three-hour test that assesses your critical-thinking skills by asking math and verbal questions, the SAT II Subject Tests are hour-long affairs that test the specific skills and knowledge that you would have picked up in standard American high school courses. There are Subject Tests on writing, literature, history, math, science, and a host of languages from Korean to Hebrew.

In our opinion, the SAT II Subject Tests are better tests than the SAT I because they cover a clearly defined topic rather than an ambiguous set of critical-thinking skills. However, just because the SAT II Subject Tests do a better job of testing your actual knowledge, that doesn't mean they aren't necessarily easier or demand less study. A "better" test isn't necessarily "better for you" in terms of how easy it will be. It just means that your score will probably reflect your knowledge of that subject better than your SAT I verbal result will reflect your reading abilities.

In comparison to taking the SAT I, there are good things and bad things about taking an SAT II Subject Test.

The Good

- Because SAT II Subject Tests cover actual topics like physics and biology, you can very effectively study for them. If you don't know a topic in physics, such as how to deal with an inclined plane problem, you can look it up and learn it. The SAT II tests are therefore safe, predictable tests: if you know your stuff, you will do well on them.

- Often, the classes you've taken in school have already prepared you for the test. If you took a course in physics and did well, you probably covered most of the topics that are tested on the SAT II Physics test. All you need is some refreshing.

- In studying for the SAT II Biology, History, or Physics tests, you really are learning biology, history, and physics. In other words, you are learning valuable, interesting knowledge. If learning is something you enjoy, you might actually find the process of studying for an SAT II test to be worthwhile and gratifying. You probably can't say that about studying for the SAT I.

The Bad

- Because SAT II Subject Tests quiz you on specific knowledge, it is much harder to "beat" or "outsmart" an SAT II test than it is to outsmart the SAT I. For the SAT I, you can use all sorts of tricks or strategies to figure out an answer. There are far fewer strategies to help you on the SAT II. Don't get us wrong: having test-taking skills will help you on an SAT II test, but knowing the subject will help you much, much more. In other words, to do well on the SAT II, you can't just rely on your quick thinking and intelligence. You need to study!

Scoring

All SAT II tests are scored on a scale that ranges from 200 to 800. The ETS determines this scaled score by taking a "raw score" and using a mathematical formula to convert it to a "scaled score" that accords to a curve. This scaled score can then be converted to a "percentile" rank that tells you what percent of test takers got a lower score than you did.

The Raw Score

Your raw score on the SAT II Physics test is based on a few simple rules:

- You earn 1 point for every correct answer.

- You lose ¼ of a point for each incorrect answer.

- You receive 0 points for each question left blank.

Suppose, for example, that of the 75 questions on the test, you answered 52 questions correctly, 18 questions incorrectly, and left five blank:

$$(52 \times 1) - (18 \times \tfrac{1}{4}) = 52 - 4.5 = 47.5$$

The raw score is then rounded to the nearest whole number. In this case, your raw score would be 48.

The Scaled Score

Using a formula, the raw score is then coverted into the scaled score between 200 and 800 that you've probably heard so much about.

The curve to convert raw scores to scaled scores varies from test to test. For example, a raw score of 33 on the Math IC might scale to a 600, while the same raw score on the Math IIC might scale to a 700. In fact, the scaled score can even vary on different editions of the *same* test. A raw score of 33 on the February 2002 Math IIC might scale to a 710, while a 33 in June 2002 might scale to a 690. These differences in scaled scores reflect the differences in the difficulty level of the test from edition to edition.

You'll find a sample of a raw-score-to-scaled-score conversion chart at the back of the book alongside the practice tests. But to give you a sense of what sorts of performances convert into what sorts of scores, you could score:

800 if you answered 68 right, 7 wrong, and left 0 blank
750 if you answered 58 right, 12 wrong, and left 5 blank
700 if you answered 51 right, 13 wrong, and left 11 blank
650 if you answered 43 right, 16 wrong, and left 16 blank
600 if you answered 36 right, 19 wrong, and left 20 blank

This chart should show you that your score doesn't plummet with every question you can't answer confidently. You can do very well on this test without knowing or answering everything.

The Percentile Score

Along with the scaled score you receive from the ETS, you will also get a percentile score. Your percentile is based on the percentage of the total test takers who received a lower raw score than you did. For example, if Joan took the SAT II Physics test and got a score that placed her in the 37th percentile, that means she scored better on that test than did 36% of the other students who took the same test. It also means that 63% of the students taking that test got the same score or scored better than she did.

SAT II Physics Score Conversion Table

Scaled Score	Average Raw Score	Scaled Score	Average Raw Score	Scaled Score	Average Raw Score
800	75	680	43	480	11
800	74	670	42	480	10
800	73	670	41	470	9
800	72	660	40	470	8
800	71	650	39	460	7
800	70	640	38	450	6
800	69	640	37	450	5
800	68	630	36	440	4
800	67	620	35	440	3
800	66	610	34	430	2
790	65	610	33	430	1
790	64	600	32	420	0
790	63	600	31	410	–1
780	62	590	30	410	–2
780	61	590	29	400	–3
780	60	580	28	400	–4
770	59	580	27	390	–5
770	58	570	26	380	–6
760	57	560	25	380	–7
760	56	560	24	370	–8
750	55	550	23	360	–9
740	54	540	22	360	–10
740	53	540	21	360	–11
730	52	530	20	350	–12
720	51	530	19	350	–13
720	50	520	18	340	–14
710	49	520	17	340	–15
700	48	510	16	330	–16
690	47	510	15	320	–17
690	46	500	14	310	–18
680	45	490	13	310	–19
680	44	490	12		

Colleges and the SAT II Subject Tests

Why should you take an SAT II Subject Test? Is it to prove to yourself how much you've learned in the school year? That seems unlikely. Is it to prove to your teacher how much you've learned? No, you've got finals for that. Is it to win you a new car? You wish. No, there's only one reason to take an SAT II Subject Test: colleges want you to, and sometimes require you to do so.

Colleges care about SAT II Subject Tests for two related reasons:

1. Your performance on the tests demonstrates your interest, knowledge, and skill in a broad range of topics.
2. Because SAT II tests are standardized, they show your abilities in these topics *in relation to the entire country.* The grades you get in high school don't offer such a measurement to colleges: some high schools are more difficult than others, meaning that students of equal ability might receive different grades, even in physics classes that have basically the same curriculum. By contrast, SAT II tests provide colleges with a definite yardstick against which they can measure your, and every other applicant's, knowledge and skills.

When it comes down to it, colleges like the SAT II tests because they make the college's job easier. The tests are tools that allow the colleges to compare students easily.

Because the SAT II tests provide an objective standard by which colleges can judge all their applicants, they provide an ideal opportunity for you to shine. If you got a 93 in a physics class and some other kid got a 91, colleges won't necessarily know what to make of that difference. But if you get a 720 on the SAT II Physics and that other kid gets a 670, colleges will be far more convinced.

That said, the SAT II plays a relatively small role in the admissions process. High school grades, extracurricular activities, and SAT or ACT scores are all more important to colleges than your scores on SAT II tests. Still, a strong score shows that you stand out on a national level and are well prepared for college-level courses in that subject. Think of it not as a make-or-break test, but as an opportunity to add another feather to your cap.

Further, colleges occasionally use SAT II tests to determine placement. For example, if you do very well on the Writing SAT II, you might be exempted from a basic expository writing class. Many colleges also have a language requirement that can be waived by a sufficiently high score on a language SAT II. You should find out whether the colleges to which you are applying use the SAT II tests for placement.

Which SAT II Subject Tests to Take

There are three types of SAT II tests: those you must take, those you should take, and those you shouldn't take.

- The SAT II tests you must take are those that are required by the colleges in which you are interested.
- The SAT II tests you should take are those that aren't required, but that you'll do well on, thereby impressing the colleges looking at your application.
- The SAT II tests you shouldn't take are those that aren't required and that cover a subject about which you don't feel confident.

Determining Which SAT II Tests Are Required

To find out if the colleges to which you are applying require that you take a particular SAT II test, you'll need to do a bit of research. Call the schools you're interested in, look at their web pages online, or talk to your guidance counselor. Often, colleges request that you take the following SAT II tests:

- The Writing SAT II test
- One of the two Math SAT II tests (either Math IC or Math IIC)
- Another SAT II in some other subject of your choice

Not all colleges follow these guidelines, however, so you should take the time to research which tests you need to take in order to apply to the colleges that interest you.

Colleges do not usually require you to take the SAT II Physics, but taking it and doing well can show a liberal arts college that you are well-rounded, or show a technical-oriented college that you are serious about science. In general, it is a good idea to take one science-related SAT II, whether Biology, Chemistry, or Physics.

Determining Whether You Should Take a Particular SAT II

There are two rules of thumb for deciding which additional test to take beyond the Writing and Math tests:

1. **Go with what you know.** If history is your field, a strong score on the American History test will impress admissions officers far more than a bold but mediocre effort on the Physics test.

2. **Try to show breadth.** Scoring well on Math tests, Physics, and Chemistry will not be as impressive as respectable scores in Math, Writing, World History, and Physics. It's good to take a few practice tests in a number of different subjects to get a sense of how best to make your choices.

Of course, the bottom line is that you need to get a good score. Two factors should help you make this decision:

1. What a good score on that SAT II test is
2. Whether you can get that score

Below, we have included a list of the most commonly taken SAT II tests and the average scaled score on each. For most schools, a score that is 50 points above this average will significantly boost your college application. If you are applying to an elite school, you may need to aim closer to 100 points above the average. A little research in this area can get you a long way. Call the schools you're interested in and/or talk to a guidance counselor.

Test	Average Score
Writing	590–600
Literature	590–600
American History	580–590
World History	570–580
Math IC	580–590
Math IIC	655–665
Biology E/M	590–600
Chemistry	605–615
Physics	635–645

AP vs. SAT II

It's a good idea to take three tests that cover a range of subjects, such as one Math SAT II, one humanities SAT II (History or Writing), and one science SAT II. However, there's no real reason to take *more* than three SAT II tests. Once you've taken the SAT II tests you need to take, the best way to set yourself apart from other students is to take AP courses and tests.

AP tests are harder than the SAT II tests, but as a result they carry quite a bit more distinction. AP tests show that you're excited and able to take on responsibility, and that you want to challenge yourself. SAT II tests show your skill against the rest of the students your age. AP tests offer you the chance to show that you're the best of the best of the best.

When to Take an SAT II Subject Test

The best time to take an SAT II Subject Test is, naturally, right after you've finished a yearlong course in that subject. If, for example, you take a physics class in tenth grade, then you should take the SAT II Physics near the end of that year, when you've learned all the material and it's still fresh in your mind. This rule does not apply for the Writing, Literature, and Foreign Language SAT II tests: it's best to take those after you've had as much study in the area as possible.

You should plan to have taken all your SAT II Subject Tests by the winter of your senior year of high school. Unless the colleges to which you are applying use the SAT II for placement purposes, there is no point in taking SAT II tests after the winter of your senior year, since you won't get your scores back from ETS until after the college application deadline has passed.

ETS usually sets testing dates for SAT II Subject Tests in October, November, December, January, May, and June. However, not every Subject Test is administered in each of these months. To check when the test you want to take is being offered, visit the College Board website at www.collegeboard.com or do some research in your school's guidance office.

Registering for SAT II Tests

To register for the SAT II tests of your choice, you have to fill out some forms and pay a registration fee. We know—it's ridiculous that *you* have to pay for a test that colleges require you to take in order to make *their* jobs easier, but, sadly, there isn't anything we, or you, can do about it. It is acceptable for you to grumble here about the unfairness of the world.

After grumbling, of course, you still have to register. There are two ways to register: online or by mail. To register online, go to www.collegeboard.com. To register by mail, fill out and send in the forms enclosed in the *Registration Bulletin,* which should be available in your high school's guidance office. You can also request a copy of the *Bulletin* by calling the College Board at (609) 771-7600, or writing to:

College Board SAT Program
P.O. Box 6200
Princeton, NJ 08541-6200

You can register to take up to three SAT II tests on any given testing day. Unfortunately, even if you decide to take three tests in one day, you'll still have to pay a separate registration fee for each.

Introduction to the SAT II Physics

Chapter Contents

Content of the SAT II Physics *11*
Format of the SAT II Physics *13*
How Your Knowledge Will Be Tested *16*

THE BEST WAY TO DO WELL ON THE SAT II Physics is to be really good at physics. For that, there is no substitute. But the physics whiz who spends the week before taking the SAT II cramming Lagrangian mechanics and Dirac notation will probably not fare any better than the average student who reviews this book carefully. Why? Because SAT II Physics tests (and first-year university courses) do not cover Lagrangian mechanics or Dirac notation. Take this moment to sigh with relief.

Happy? Good. This chapter will tell you precisely what the SAT II Physics *will* test you on, how the test breaks down, and what format the questions will take. Take this information to heart, and base your study plan around it. There's no use spending hours on end studying for stuff that's not really important. Knowing nothing about electromagnetic induction will hurt you on the test, but nowhere near as much as knowing nothing about optics.

Content of the SAT II Physics

Math and physics go hand in hand, right? You might be surprised, then, to learn that you aren't allowed to use a calculator on the SAT II Physics. The math required of you never goes beyond simple arithmetic and manipulation of equations. You have, on average, 48 seconds to answer each question, and the people at ETS realize that isn't enough time to delve into problems involving simultaneous equations or complex trigonometry. They're more interested in testing your grasp of the basic concepts of physics. If you've grasped these concepts, your weakness in math problem solving isn't going to hurt you.

ETS breaks down the concepts you need to know for the test into six categories:

Topic	Percentage of the Test
Mechanics	34–38%
Electricity and Magnetism	22–26%
Waves	15–19%
Heat, Kinetic Theory, and Thermodynamics	8–12%
Modern Physics	8–12%
Miscellaneous	2–4%

This outline is a rough guide, and it isn't specific enough to be truly valuable. You may be a whiz with waves but a loser with lenses, and want to know how much of the waves portion of the test will be devoted to optics. To help you out, we've broken the test down even further so that you'll know exactly where to expect to feel the squeeze. These figures are only approximations, and may vary from test to test.

Topic	% of the Test	Number of Questions
Mechanics	**34–38%**	**25–29**
Vectors	2%	1-2
Kinematics	6%	4-5
Dynamics	10%	7-8
Work, Energy, and Power	6%	4-5
Special Problems in Mechanics	5%	3-4
Linear Momentum	2%	1-2
Rotational Motion	1%	0-1
Circular Motion and Gravitation	4%	2-4
Thermal Physics	**8–12%**	**6–10**
Heat and Temperature	4%	2-4
Kinetic Theory and Ideal Gas Laws	2-3%	1-2
Laws of Thermodynamics	1%	0-2
Heat Engines	2-3%	1-2
Electricity & Magnetism	**22–26%**	**16–20**
Electric Fields, Forces, Potential	10%	7-8
Magnetic Fields and Forces	6%	4-5
Electromagnetic Induction	1%	1
Circuits and Circuit Elements	6%	4-5

Waves	15–19%	11–15
Waves	10%	7-8
Optics	7%	5-6
Modern Physics	**8–12%**	**6–9**
Special Relativity	1-2%	1-2
Atomic Models	3%	2-3
Quantum Physics	2%	1-2
Nuclear Physics	3%	2-3
Miscellaneous	**2–4%**	**1–3**
Graph Analysis	1-2%	0-2
Equation Manipulation	0–1%	0–1
Significant Digits and Lab Skills	0–1%	0–1

The chapters of this book have been organized according to these divisions. Any of the material covered on these topics in this book is fair game for the SAT II. If it's not in this book, you don't need to know it. There are a few other things you might not be sure whether or not you'll need to know. So we'll tell you.

- **You need to know:** the formulas expressing physical relationships (such as $F = ma$), how to manipulate equations, how to read a graph
- **You don't need to know:** trig identities, calculus, three-dimensional vectors and graphs, physical constants (such as $G = 6.67 \times 10^{-11}\ \text{N}\cdot\text{m}^2/\text{kg}^2$)

Format of the SAT II Physics

The SAT II Physics is an hour-long test composed of 75 questions and divided into two parts. You can answer questions in any order you like, though you're less likely to leave a question out accidentally if you go through them in the order in which they're written. Part A—classification questions—takes up the first 12 or 13 questions of the test, while Part B—five-choice completion questions—takes up the remaining 62 or 63 questions. What are classification and five-choice completion questions? Funny you should ask: we were just about to tell you.

Part A: Classification Questions

Classification questions are the reverse of your normal multiple-choice question: they give you the answers first and the questions second. You'll be presented with five possible answer choices, and then a string of two to four questions to which those answer choices apply. The answer choices are usually either graphs or the names of five related laws or concepts. Because they allow for several questions on the same topic, classification questions will ask you to exhibit a fuller understanding of the topic at hand.

The level of difficulty within any set of questions is generally pretty random: you can't expect the first question in a set to be easier than the last. However, each set of classification questions is generally a bit harder than the one that came before. You should expect questions 11 to 13 to be harder than questions 1 to 4.

Classification Question Example

Directions: Each set of lettered choices below refers to the numbered questions immediately following it. Select the one lettered choice that best answers each question and then blacken the corresponding space on the answer sheet. A choice may be used once, more than once, or not at all in each set.

Questions 1–3

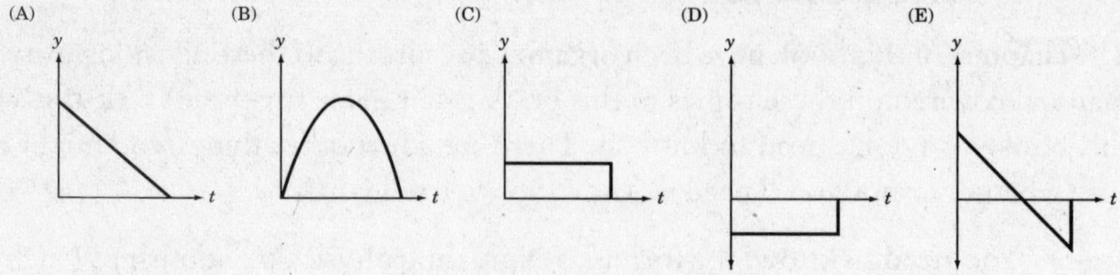

A boy throws a ball straight up in the air and then catches it again.

1. Which of the above graphs best represents the ball's position with respect to time?
2. Which of the above graphs best represents the ball's velocity with respect to time?
3. Which of the above graphs best represents the ball's acceleration with respect to time?

Explanation

You can usually answer classification questions a bit more quickly than the standard five-choice completion questions, since you only need to review one set of answer choices to answer a series of questions.

The answer to question 1 is (B). The ball's position with respect to time can be expressed by the equation $y = -\frac{1}{2} gt^2$, where g is the downward, and hence negative, acceleration due to gravity. As we can see, the graph of y against t is an upside-down parabola. In more intuitive terms, we know that, over time, a ball thrown in the air will rise, slow down, stop, and then descend.

The answer to question 2 is (E). The acceleration due to gravity means that the velocity of the ball will decrease at a steady rate. On the downward half of the ball's trajectory, the velocity will be negative, so (E), and not (A), is the correct graph.

The answer to question 3 is (D). The acceleration due to gravity is constant throughout the ball's trajectory, and since it is in a downward direction, its value is negative.

Don't worry if the question confused you and the explanations didn't help. This material and more will be covered in some detail in Chapter 2: Kinematics. This was just an exercise to show you how a classification question is formatted.

Part B: Five-Choice Completion Questions

These are the multiple-choice questions we all know and love, and the lifeblood of any multiple-choice exam. You know the drill: they ask a question, give you five possible answer choices, and you pick the best one. Got it? Good. There's an example below.

While you'll often find two or three questions in a row that all deal with the same topic in physics, there is no pattern. You might find a question on modern physics followed by a question on dynamics followed by a question on optics. However, there is a general tendency for the questions to become more difficult as you progress.

Five-Choice Completion Question Example

Directions: Each of the questions of incomplete statements below is followed by five suggested answers or completions. Select the one that is best in each case and then fill in the corresponding oval on the answer sheet.

1. A gas in a closed container is steadily heated over a period of time. Which of the following statements is true of this process?

(A) The average kinetic energy of the gas molecules decreases
(B) The mass of the container increases
(C) The pressure exerted by the gas on the walls of the container increases
(D) The gas changes phase into a liquid
(E) The specific heat of the gas decreases

Explanation

The answer to this question is (C). The key lies in remembering the ideal gas law: $PV = nRT$. According to this formula, an increase in temperature is accompanied by an increase in pressure. (A) is wrong, since the average kinetic energy of gas molecules corresponds to their temperature: if the temperature increases, so does the average kinetic energy of the molecules. (B) is wrong because we're dealing with a closed container: the mass cannot either increase or decrease. (D) is wrong because a gas must be cooled, not heated, to change phase into a liquid. Finally, (E) is wrong because the specific heat of any substance is a constant, and not subject to change. We'll touch on all this and more in Chapter 9: Thermal Physics.

How Your Knowledge Will Be Tested

ETS tests you on three different "levels of concept application." That is, there are three different levels on which your understanding of physics may be tested. While questions on kinematics often require that you make use of some of the formulas for kinematic motion, questions on quantum physics or atomic structure may often ask just that you remember the name of a particular concept. Knowing the different ways in which your knowledge may be tested should help you better prepare yourself for the exam.

Recall (20–33% of the test): These are questions of the either-you-know-it-or-you-don't variety. They test your understanding of the basic concepts of physics. No equations or calculations are necessary for these questions. They're simply a matter of knowing your stuff.

Single-Concept Problem (40–53% of the test): These questions expect you to recall, and make use of, one physical relationship, formula, or equation. This might involve plugging numbers into a kinematic equation of motion, or it might involve recalling the equation $E = hf$ and solving for E or f. These questions test to see if you know important formulas and how to apply them.

Multiple-Concept Problem (20–33% of the test): These questions expect you to bring together two or more different relationships, formulas, or equations. This could involve bringing together two formulas from the same subject—for instance, a problem in linear momentum that requires you to calculate the momentum of an object before a collision so that you can calculate its velocity after the collision—or it may bring together formulas from two different subjects—for instance, a problem that involves an electric point charge moving in circular motion in a magnetic field. These questions test not only your knowledge of physical relationships, but also your ability to integrate more than one in a complex problem.

You're probably thinking that the recall questions are the easiest, and the multiple-concept problems are the hardest. This isn't necessarily true. Most people have an easier time bringing together two simple principles of mechanics than recalling the significance of the Rutherford experiment. You'll find all three types of questions all over the test, and at all different levels of difficulty. Ultimately, they all test the very same thing: whether you've grasped the basic principles of physics.

Strategies for Taking the SAT II Physics

Strategies

Chapter Contents

General Test-Taking Strategies *17*
SAT II Physics Test-Taking Strategies . . . *21*

A MACHINE, NOT A PERSON, WILL SCORE your SAT II test. The tabulating machine sees only the filled-in ovals on your answer sheet, and doesn't care how you came to these answers; it cares only whether your answers are correct. A lucky guess counts in your favor just as much as an answer you give confidently. By the same token, if you accidentally fill in (B) where you meant (C), you won't get any credit for having known what the answer was. Think of the multiple-choice test as a message to you from the ETS: "We care only about your answers, and not about any of the work behind them."

So you should give ETS correct answers—as many as possible, using whatever means possible. It's obvious that the SAT II Physics test allows you to show off your knowledge of physics, but the test gives you the same opportunity to show off your foxlike cunning by figuring out what strategies will best allow you to display that knowledge. This chapter will first cover some general principles of test taking that apply equally to this test and any other SAT test you might come upon, and then will discuss a few strategies that are particularly useful to the SAT II Physics.

General Test-Taking Strategies

Most of these "strategies" are common sense; many of them you already know. The funny thing about high-pressure situations, though, is that common sense often goes out the window. If you review anything in the minutes before taking the test, review these strategies. Of course, that doesn't mean you should skip this section *now.* It's full of juicy hints, some of which might be new to you.

General Hint 1: Be Calm

The best way to do poorly on a test is to psych yourself out. Physics in particular calls for cool, systematic thinking: if your mind starts thrashing about wildly, it will have a hard time settling on the right answers. There are a number of preventative measures you can take, beginning weeks, or even months, before you take the test. Buying this book was a good start: it's reassuring to see all the information you'll need to ace the test in a compact, manageable form. But there are a number of other things you ought to keep in mind:

Study in advance. If you've studied at regular intervals leading up to the test, and don't do all your cramming the night before, the information will sit more easily in your mind.

Be well rested. Get a good night's sleep on the two nights leading up to the test. If you're frazzled or wired, you're going to have a harder time buckling down and concentrating when it really counts.

Come up for air. Don't assume that the best way to take an hour-long test is to spend the full hour nose-to-nose with the test questions. If you lift your head occasionally, look about you, and take a deep breath, you'll return to the test with a clearer mind. You'll lose maybe ten seconds of your total test-taking time, but you'll be all the more focused for the other fifty-nine minutes and fifty seconds.

General Hint 2: Grid Your Answers Carefully

No kidding. People make mistakes doing this and it can cost them big-time. This slip up occurs most frequently after you skip a question. If you left question 43 blank, and then unthinkingly put the answer to question 44 into row 43, you could be starting a long, painful chain of wrong answers. Don't do this.

Some test prep books advise that you fill in your answer sheet five questions at a time rather than one at a time. Some suggest that you fill out each oval as you answer the question. We think you should fill out the answer sheet in whatever way feels most natural to you, but make sure you're careful while doing it. In our opinion, the best way to ensure that you're being careful is to talk to yourself: as you figure out an answer in the test booklet and transfer it over to the answer sheet ovals, say to yourself: "Number 23, B. Number 24, E, Number 25, A."

General Hint 3: Pace Yourself

At the very least, aim to look at every question on the test. You can't afford to lose points because you didn't find the time to look at a question you could have easily answered. You can spend an average of forty-eight seconds on each question, though you'll probably breeze through some in ten seconds and dwell on others for two minutes. Knowing how to pace yourself is a critical skill—and these three guidelines should help:

Don't dwell on any one question for too long: If you've spent a couple minutes laboring over the question, you might just want to make a note of it and move on. If you feel the answer is on the tip of your tongue, it might come more easily if you just let it rest and come back to it later. Not only is it demoralizing to spend five minutes on a single question, but it also eats up precious time in which you might have answered a number of easier questions.

Nail the easy questions: As we said in the previous chapter, the test questions get progressively harder as you go along. Nonetheless, there will be some tough ones thrown in right at the start, and you'll be finding gimmes right up until the end. One of the reasons you don't want to dwell too long on tough questions is that you should ensure that you get to look at all the questions and snatch up the easy ones. Remember: you get as many points for answering an easy question as a difficult one. You get a lot more points for five quickly answered easy questions than for one hard-earned victory.

Skip the unfamiliar: If you encounter a question you can't make heads or tails of, just skip it. Don't sweat too hard trying to sort out what's going on. If you have time at the end, you can come back to it and see if you can make an educated guess. Your first priority should be to get all the easy questions, and your second priority should be to work through the questions you can solve with difficulty. Unfamiliar material should be at the bottom of your list of priorities.

General Hint 4: Set a Target Score

You can make the job of pacing yourself much easier if you go into the test knowing how many questions you have to answer correctly in order to earn the score you want. What score do you want to get? Obviously, you should strive for the best score possible, but be realistic: consider how much you know about physics and how well you do, generally, on these sorts of tests. You should also do a little research and find out what counts as a good score for the colleges to which you're applying: is it a 620? 680? Talk to the admissions offices of the colleges you might want to attend, do a little research in college guidebooks, or talk to your guidance counselor. Find out the average score of a student admitted to the schools of your choice, and set your target score above it (you want to be above average, right?). Then take a look at the chart we showed you before. You can score:

800 if you answered 68 right, 7 wrong, and left 0 blank
750 if you answered 58 right, 12 wrong, and left 5 blank
700 if you answered 51 right, 13 wrong, and left 11 blank
650 if you answered 43 right, 16 wrong, and left 16 blank
600 if you answered 36 right, 19 wrong, and left 20 blank

Suppose the average score on the SAT II Physics for the school you're interested in is 650. Set your target at about 700. To get that score, you need to get 51 questions right, while leaving yourself room to get 13 wrong and leave 11 blank. In other words, you can leave a number of tough questions blank, get a bunch more wrong, and still get the score you want. As long as you have some idea of how many questions you need to answer, bearing in mind that you'll likely get some questions wrong, you can pace yourself accordingly. Taking practice tests is the best way to work on your pacing. See how many questions you can leave blank and still get the score you want, and you'll have a better sense of what to aim for on the big day.

If you find yourself effortlessly hitting your target score when you take the practice tests, don't just pat yourself on the back. Set a higher target score and start gunning for that one. The purpose of buying this book and studying for the test is to improve your score as much as possible, so be sure to push your limits.

General Hint 5: Know What You're Being Asked

You can't know the answer until you know the question. This might sound obvious, but many a point has been lost by the careless student who scans the answer choices hastily before properly understanding the question. Take the following example:

> Two positively charged particles, one twice as massive as the other, are moving in the same circular orbit in a magnetic field. Which law explains to us why the less massive particle moves at twice the speed of the more massive particle?
>
> (A) Coulomb's Law
> (B) Conservation of angular momentum
> (C) Hooke's Law
> (D) The ideal gas law
> (E) Heisenberg's uncertainty principle

The hasty student will notice that the question is about charged particles, and see "Coulomb's Law" as the first answer choice. Without further ado, the student answers (A) and loses a quarter of a point.

A more careful student will not just read the question, but will take a moment to *understand* the question before glancing at the answer choices. This student will realize that the question ultimately deals with particles moving in circular orbits, and the relative speeds of these particles. Whether or not these particles are charged is irrelevant: you're facing a problem of rotational motion, not of electric forces. Once you've recognized what you're dealing with, you will have little trouble in correctly answering (B).

General Hint 6: Know How To Guess

ETS doesn't take off ¼ of a point for each wrong answer in order to punish you for guessing. They do it so as not to reward you for blind guessing. Suppose that, without looking at the questions at all, you just randomly entered responses in the first 20 spaces on your answer sheet. Because there's a 20% chance of guessing correctly on any given question, odds are you would guess right for four questions and wrong for 16 questions. Your raw score for those 20 questions would then be:

$$(4 \times 1) - (16 \times \tfrac{1}{4}) = 0$$

You would be no better off and no worse off than if you'd left those twenty spaces blank.

Now suppose in each of the first 20 questions you are able to eliminate just one possible answer choice, so that you guess with a 25% chance of being right. Odds are, you'd get five questions right and 15 questions wrong, giving you a raw score of:

$$(5 \times 1) - (15 \times \tfrac{1}{4}) = 1.25$$

All of a sudden, you're over a point up. It's not much, but every little bit helps.

The lesson to be learned here is that blind guessing doesn't help, but educated guessing does. If you can eliminate even one of the five possible answer choices, you should guess. We'll discuss how to eliminate answer choices on certain special kinds of questions in Physics Hint 5: Eliminate Wrong Answers.

Guessing As Partial Credit

Some students feel that guessing is like cheating—that guessing correctly means getting credit where none is due. But instead of looking at guessing as an attempt to gain undeserved points, you should look at it as a form of partial credit. Suppose you're stumped on the question we looked at earlier regarding the charged particle moving in circular motion in a magnetic field. Though you don't know the correct answer, you may know the answer isn't the ideal gas law, because the question doesn't deal with gases in any way. Suppose you also know that the answer isn't Hooke's Law, because Hooke's Law deals with force exerted by a spring, and there are no springs in this question. Don't you deserve something for that extra knowledge? Well, you do get something: when you look at this question, you can throw out (C) and (D) as answer choices, leaving you with a one in three chance of getting the question right if you guess. Your extra knowledge gives you better odds of getting this question right, exactly as extra knowledge should.

SAT II Physics Test-Taking Strategies

All the strategies discussed above can be applied equally to the SAT II Physics and the SAT II Modern Hebrew. That's why they're called "general hints." However, as you may have noticed in the past, there are a number of dissimilarities between the study of physics and the study of modern Hebrew. You read physics formulas from left to right, for instance. Because physics is unlike modern Hebrew, and even unlike math and chemistry, there are a number of strategies that apply uniquely to the SAT II Physics. Some of these strategies will help you out in physics generally, while some are suited to the unique idiosyncrasies of the SAT II format.

Physics Hint 1: Know Those Formulas!

You aren't allowed to bring a calculator into the SAT II, nor are you allowed to bring in a sheet of paper with useful information on it. That means that if you haven't memorized formulas like $F = ma$ and $F = k(q_1)(q_2)/r^2$, you're going to lose a lot of points. As we said earlier, 67–80% of the test requires that you know your formulas.

This doesn't mean you have to do a lot of rote memorization. In fact, if you just remember equations involving a jumble of letters, you'll probably lack the understanding necessary to apply them flexibly. You'll find that as you become more familiar with the principles of physics, the equations that express these principles will become increasingly intuitive. You'll find patterns: for instance, the force exerted at any point in a field, be it a gravitational field or an electric field, is inversely proportional to r^2. That's why Coulomb's Law and Newton's Law of Gravitation look similar. Knowing your physics will help you know your formulas.

A lot of people feel burdened coming into an exam with lots of formulas and equations in their head. It can feel like your mind is "full," and there's no room for the problem solving at hand. If you have trouble remembering formulas, you might want to look them over carefully in the minutes before the test, and then, before you even look at the first question, write down the formulas you have a hard time remembering on the back of the question booklet. That way, you can refer back to them without any painful effort of recollection.

Physics Hint 2: Estimate

This hint goes hand in hand with General Hint 5: Know What You're Being Asked. Don't dive blindly into five possible answer choices until you know what you're looking for. The first way to know what you're looking for is to understand the question properly—we've discussed that already. Once you understand the question, get a rough sense of what the correct answer should look like.

Estimation, obviously, is only useful in questions involving calculation: you can't "estimate" which Law of Thermodynamics states that the world tends toward increasing disorder. In questions involving a calculation, though, it may save you from foolish errors if you have a sense of the correct order of magnitude. If you're being asked to calculate the mass of a charging elephant, you can be pretty confident that the answer won't be 2 kg, which would be far too small, or 2×10^6, far too big. Estimation is a good way to eliminate some wrong answers when you're making an educated guess.

Physics Hint 3: Put It on Paper

Don't be afraid to write and draw compulsively. The first thing you should do once you've made sure you understand the question is to draw a diagram of what you're dealing with. Draw in force vectors, velocity vectors, field lines, ray tracing, or whatever else may be appropriate. Not only will a visual representation relieve some of the pressure on your beleaguered mind, it may also help the solution jump right off the page at you.

Drawing graphs can also make a solution appear out of thin air. Even if a problem doesn't ask you to express anything in graphic terms, you might find that a rough sketch of, say, the velocity of a particle with respect to time will give you a much clearer sense of what you're dealing with.

And don't forget to write down those equations! Writing down all the equations you can think of may lead you to a correct answer even if you don't really understand the question. Suppose you know the problem deals with an electric circuit, and you're given values for current and electric potential. Write down equations like $V = IR$ and $P = IV$, plug in values, fiddle around a little, and see if you can come up with an answer that looks right.

Physics Hint 4: Answers Are Not Convoluted

Remember, on the SAT II Physics you're not allowed to use a calculator, and you're only given, on average, 48 seconds to answer each question. If you're working on a problem and find yourself writing out lines and lines of simultaneous equations, trying to figure out $\sqrt{34.956}$ or to recall your trig identities, you're probably on the wrong track. These questions are designed in such a way that, if you understand what you're being asked, you will need at most a couple simple calculations to get the right answer.

Physics Hint 5: Eliminate Wrong Answers

In General Hint 6: Know How To Guess, we explained the virtues of eliminating answers you know to be wrong and taking a guess. On most questions, there will be at least one or two answer choices you can eliminate. There are also certain styles of questions that lend themselves to particular process-of-elimination methods.

Classification Questions

Questions 1–3 relate to the following quantities:

(A) Frequency
(B) Amplitude
(C) Period
(D) Wavelength
(E) Kinetic Energy

1. Which is measured in hertz?

2. For a mass on a spring, which is maximized when the displacement of the mass from its equilibrium position is zero?

3. Which quantity is not applied to pendulum motion?

Strategies

The weakness of classification questions is that the same five answer choices apply to several questions. Invariably, some of these answer choices will be tempting for some questions but not for others. For instance, you can be pretty sure that kinetic energy isn't measured in hertz: (E) may be a tempting answer choice for other questions but not for that one, so you can eliminate it.

Another point that may help you guess in a pinch is that you'll rarely find that the same answer choice is correct for two different questions. The directions for classification questions explicitly state that an answer choice "may be used once, more than once, or not at all," but on the whole, the ETS people shy away from the "more than once" possibility. This is by no means a sure bet, but if you're trying to eliminate answers, you might want to eliminate those choices that you've already used on other questions in the same set.

If you're wondering, the answers to the above questions are 1 (A), 2 (E), and 3 (D).

"EXCEPT" Questions

All of the following are true about an α-particle EXCEPT

(A) It has an atomic mass of 4
(B) It carries a positive charge
(C) It is identical to the nucleus of a helium atom
(D) It will always pass right through a thin sheet of gold foil
(E) It contains two neutrons

Questions of the "EXCEPT" variety contain a bunch of right answers and one wrong answer, and it's generally possible to spot one or two right answers. Even if you can't answer the question confidently, you might remember that alpha particles have a positive charge and that they are identical to the nucleus of a helium atom. Already, you've eliminated two possible answers, and can make a pretty good guess from there.

If you're interested, the answer is (D): Rutherford's gold foil experiment showed that alpha particles would occasionally deflect off the gold foil at extreme angles, thus proving that atoms have nuclei.

"I, II, and III" Questions

For which of the following is $f > 0$:

I. Concave mirror
II. Convex mirror
III. Converging lens

(A) I only
(B) II only
(C) I and III only
(D) II and III only
(E) I, II, and III

In this style of multiple-choice question, the "I, II, and III" questions provide you with three possible answers, and the five answer choices list different combinations of those three. There's an upside and a downside to questions like this. Suppose you know that a concave mirror has $f > 0$ and a convex mirror doesn't, but you're not sure about a converging lens. The downside is that you can't get the right answer for sure. The upside is that you can eliminate (B), (D), and (E), and have a 50% chance of guessing the right answer. As long as you're not afraid to guess—and you should never be afraid to guess if you've eliminated an answer—these questions shouldn't be daunting.

The value of f for a converging lens is positive, so the answer is (C).

Physics Hint 6: Be Flexible

Knowing your physics formulas is a must, but they're useless if you don't know how to apply them. You will probably never be asked to calculate the force acting on an object given its mass and acceleration. Far more likely, you will be asked for the acceleration given its mass and the force acting on it. Knowing that $F = ma$ is useless unless you can also sort out that $a = F/m$.

The ETS people don't want to test your ability to memorize formulas; they want to test your understanding of formulas and your ability to use formulas. To this end, they will word questions in unfamiliar ways and expect you to manipulate familiar equations in order to get the right answer. Let's look at an example.

A satellite orbits the Earth at a speed of 1000 m/s. Given that the mass of the Earth is 6.0×10^{24} kg and the universal gravitational constant is $6.67 \times 10^{-11}\ \text{N} \cdot \text{m}^2/\text{kg}^2$, what is the best approximation for the radius of the satellite's orbit?

(A) 4×10^{-2} m
(B) 4×10^{8} m
(C) 4×10^{12} m
(D) 2.5×10^{4} m
(E) 2.5×10^{10} m

What's the universal gravitational constant? Some people will know that this is the G in the equation for Newton's Law of Gravitation: $F = G(m_1)(m_2)/r^2$. Other people won't know that G is called the "universal gravitational constant," and ETS will have successfully separated the wheat from the chaff. It's not good enough to know some formulas: you have to know what they mean as well.

Given that we know what the universal gravitational constant is, how do we solve this problem? Well, we know the satellite is moving in a circular orbit, and we know that the force holding it in this circular orbit is the force of gravity. If we not only know our formulas, but also *understand* them, we will know that the gravitational force must be equated with the formula for centripetal force, $F = mv^2/r$. If we know to equate these two formulas, it's a simple matter of plugging in numbers and solving for r.

Knowing formulas, however, is a small part of getting the right answer. More important, you need to know how to put these two equations equal to one another and solve for r. On their own, without understanding how to use them, the equations are useless.

But there are two slightly underhanded ways of getting close to an answer without knowing any physics equations. These aren't foolproof methods, but they can help in a pinch.

Slightly Underhanded Way #1: Elimination through Logic

By scanning the possible answer choices, you can see that the answer will begin either with a 4 or a 2.5. There are three options beginning with 4 and only two beginning with 2.5. Odds are, the correct answer begins with 4. The test makers want to give answer choices that are close to the correct answer so that, even if you're on the right track, you might still get caught in a miscalculation.

Second, make a rough estimate. At what sorts of distances might a satellite orbit? We can eliminate (A) immediately: that answer has our satellite orbiting at 4 cm from the center of the Earth! That leaves us with a choice between (B) and (C). Those aren't bad odds for guessing.

Strategies

Slightly Underhanded Way #2: Work with the Letters

This is a method for those of you who like manipulating equations. From looking at the answer choices, you know the answer will be in meters. You've been given three quantities, one expressed in m/s, one expressed in kg, and one expressed in $N \cdot m^2/kg^2$. These are the only three quantities you'll be asked to draw upon in order to get your answer. Because $F = ma$, you know you can substitute $kg \cdot m/s^2$ for N. So a quantity expressed in $N \cdot m^2/kg^2$ can equally be expressed in $m^3/kg \cdot s^2$.

The trick, then, is to combine a quantity expressed in these terms with a quantity expressed in meters per second and a quantity expressed in kilograms, and wind up with a quantity expressed solely in meters. To do that, we need to get rid of the "kg" and the "s" by canceling them out. Let's start by canceling out the "kg":

$$(6.67 \times 10^{-11} \text{ N} \cdot \text{m}^2/\text{kg}^2)(6.0 \times 10^{24} \text{ kg}) \approx 10^{14} \text{ m}^3/\text{s}^2$$

Now we need to cancel out the "s^2" in the denominator. Let's divide by the square of our "m/s" quantity:

$$\frac{10^{14} \text{ m}^3/\text{s}^2}{(1000 \text{ m/s})^2} = 10^8 \text{m}$$

There you have it. We didn't need to use a single formula to get the answer. We just had to be aware of the terms in which our answer needed to be expressed, and manipulate the quantities we were given in the question.

Word to the wise: don't use this method unless you're absolutely stumped. It can backfire, and is of course no substitute for careful reasoning.

SAT II PHYSICS REVIEW

Vectors

VECTORS ARE USUALLY THE FIRST THING you learn in a physics class, and they're the first thing you'll learn here. Vectors are one of the fundamental mathematical tools the physicist uses, and one that is frequently misunderstood or misapplied by students. There aren't usually more than one or two questions on the SAT II that test your knowledge of vectors directly, but there are a host of problems—particularly in mechanics—where arriving at the right solution demands a solid grasp of how to apply and manipulate vectors. Even if you feel confident with vectors, we urge you to review this chapter and be absolutely sure you won't get tripped up on what would otherwise be some easy questions.

Chapter Contents

What's a Vector? **29**
Vector Addition **30**
Vector Subtraction **33**
Multiplication by a Scalar **34**
Vector Components **34**
Vector Multiplication **37**
Key Formulas **41**
Practice Questions **41**

What's a Vector?

A **vector** is a mathematical object possessing, and fully described by, a **magnitude** and a **direction**. It's possible to talk about vectors simply in terms of numbers, but it's often a lot easier to represent them graphically as arrows. The vector's magnitude is equal to the length of the arrow, and its direction corresponds to where the arrow is pointing. Physicists commonly refer to the point of a vector as its **tip** and the base as its **tail**.

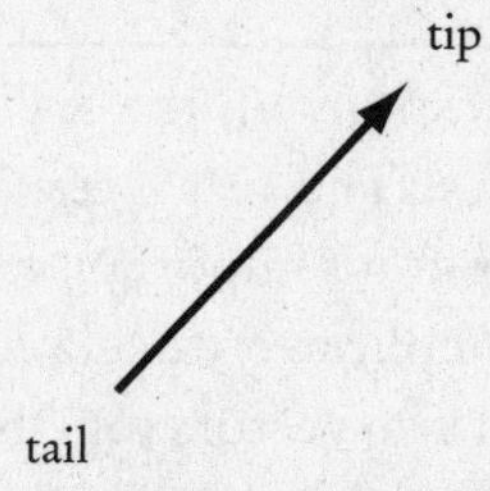

There are a number of ways to label vectors. You may have seen vectors labeled $\vec{A}$ or A. This book will follow the convention you'll find on the SAT II: vectors are written in boldface and vector magnitudes in plain script. For example, vector $\mathbf{A}$ has magnitude A.

Vectors vs. Scalars

In contrast to a vector quantity, a **scalar** quantity does not have a direction; it is fully described by just a magnitude. Examples of scalar quantities include the number of words in this sentence and the mass of the Hubble Space Telescope. Vector quantities you'll likely come across quite frequently in physics include displacement, $\boldsymbol{s}$; velocity, $\boldsymbol{v}$; acceleration, $\boldsymbol{a}$; force, $\boldsymbol{F}$; momentum, $\boldsymbol{p}$; electric field, $\boldsymbol{E}$; and magnetic field, $\boldsymbol{B}$.

When in doubt, ask yourself if a certain quantity comes with a direction. If it does, it's a vector. If it doesn't, it's a scalar.

Example

Which of the following sentences deal with vector quantities?

I. "I used to drive a 10-ton truck."
II. "You'll find a gas station if you follow this road 20 miles due north."
III. "The 10-volt battery is the one on your left."

(A) I only
(B) II only
(C) III ony
(D) II and III
(E) I, II, and III

"I used to drive a 10-ton truck" deals with mass, which is a scalar quantity. When we know that a truck weighs 10 tons, we don't need to ask, "in what direction?" "You'll find a gas station if you follow this road 20 miles due north" deals with the vector quantity of displacement. When asking directions to a gas station, you don't simply want to know how far it is from where you are, but also in what direction you need to go. "The 10-volt battery is the one on your left" is slightly tricky: volts are a scalar quantity—you don't ask in what direction the batteries volts are going. However, you might be deceived by the mention of "on your left." However, "on your left" is a reference to the battery, not to the volts. The magnitude "10 volts" doesn't have a direction, so that quantity is a scalar. The answer is (B).

Vector Addition

There are bound to be several questions on the SAT II Physics that involve vector addition, particularly in mechanics. The test doesn't demand a very sophisticated understanding of vector addition, but it's important that you grasp the principle. That is, you won't be asked to make complicated calculations, but you will be expected to know what happens when you add two vectors together.

The easiest way to learn how vector addition works is to look at it graphically. There are two equivalent ways to add vectors graphically: the **tip-to-tail method** and the **parallelogram method**. Both will get you to the same result, but one or the other is more convenient depending on the circumstances.

Tip-to-Tail Method

We can add any two vectors, ***A*** and ***B***, by placing the tail of ***B*** so that it meets the tip of ***A***. The sum, ***A*** + ***B***, is the vector from the tail of ***A*** to the tip of ***B***.

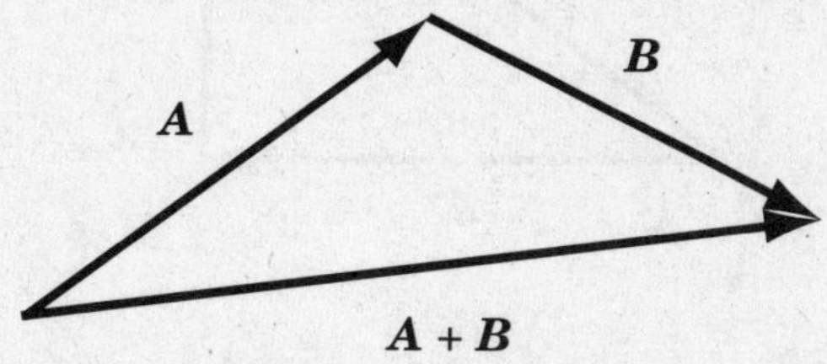

Note that you'll get the same vector if you place the tip of ***B*** against the tail of ***A***. In other words, ***A*** + ***B*** and ***B*** + ***A*** are equivalent.

Parallelogram Method

To add ***A*** and ***B*** using the parallelogram method, place the tail of ***B*** so that it meets the tail of ***A***. Take these two vectors to be the first two adjacent sides of a parallelogram, and draw in the remaining two sides. The vector sum, ***A*** + ***B***, extends from the tails of ***A*** and ***B*** across the diagonal to the opposite corner of the parallelogram. If the vectors are perpendicular and unequal in magnitude, the parallelogram will be a rectangle. If the vectors are perpendicular and equal in magnitude, the parallelogram will be a square.

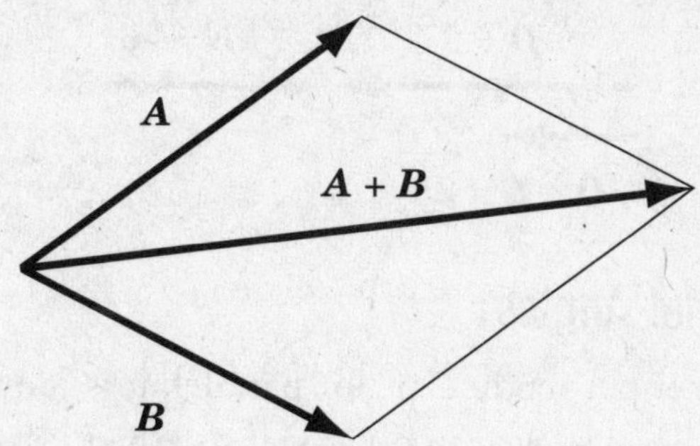

Adding Vector Magnitudes

Of course, knowing what the sum of two vectors looks like is often not enough. Sometimes you'll need to know the magnitude of the resultant vector. This, of course, depends not only on the magnitude of the two vectors you're adding, but also on the angle between the two vectors.

Adding Perpendicular Vectors

Suppose vector ***A*** has a magnitude of 8, and vector ***B*** is perpendicular to ***A*** with a magnitude of 6. What is the magnitude of ***A*** + ***B***? Since vectors ***A*** and ***B*** are perpendicular, the triangle formed by ***A***, ***B***, and ***A*** + ***B*** is a right triangle. We can use the Pythagorean Theorem to calculate the magnitude of ***A*** + ***B***, which is $\sqrt{8^2 + 6^2} = \sqrt{100} = 10$.

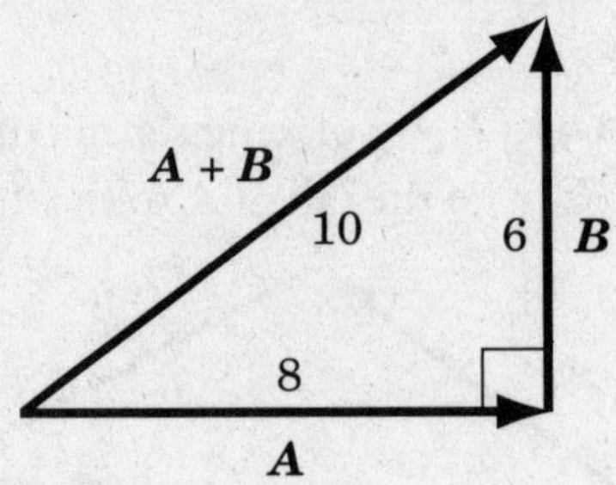

Adding Parallel Vectors

If the vectors you want to add are in the same direction, they can be added using simple arithmetic. For example, if you get in your car and drive eight miles east, stop for a break, and then drive six miles east, you will be 8 + 6 = 14 miles east of your origin. If you drive eight miles east and then six miles west, you will end up 8 – 6 = 2 miles east of your origin.

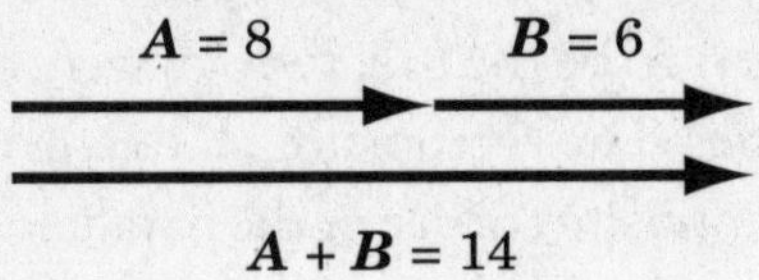

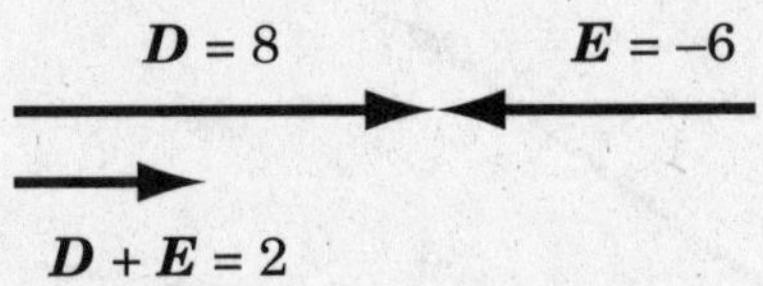

Adding Vectors at Other Angles

When ***A*** and ***B*** are neither perpendicular nor parallel, it is more difficult to calculate the magnitude of ***A*** + ***B*** because we can no longer use the Pythagorean Theorem. It is possible to calculate this sum using trigonometry, but the SAT II will never ask you to do this. For the most part, the SAT II will want you to show graphically what the sum will look like, following the tip-to-tail or parallelogram methods. On the rare occasions that you need to calculate the sum of vectors that are not perpendicular, you will be able to use the component method of vector addition, explained later in this chapter.

Example

> Vector ***A*** has a magnitude of 9 and points due north, vector ***B*** has a magnitude of 3 and points due north, and vector ***C*** has a magnitude of 5 and points due west. What is the magnitude of the resultant vector, ***A*** + ***B*** + ***C***?

First, add the two parallel vectors, ***A*** and ***B***. Because they are parallel, this is a simple matter of straightforward addition: 9 + 3 = 12. So the vector ***A*** + ***B*** has a magnitude of 12 and points due north. Next, add ***A*** + ***B*** to ***C***. These two vectors are perpendicular, so apply the Pythagorean Theorem:

$$\sqrt{12^2 + 5^2} = 13$$

The sum of the three vectors has a magnitude of 13. Though a little more time-consuming, adding three vectors is just as simple as adding two.

Vector Subtraction

You probably know that subtraction is the same thing as adding a negative. That is, 8 – 5 is the same thing as 8 + (–5). The easiest way to think about vector subtraction is in terms of adding a negative vector. What's a negative vector? It's the same vector as its positive counterpart, only pointing in the opposite direction.

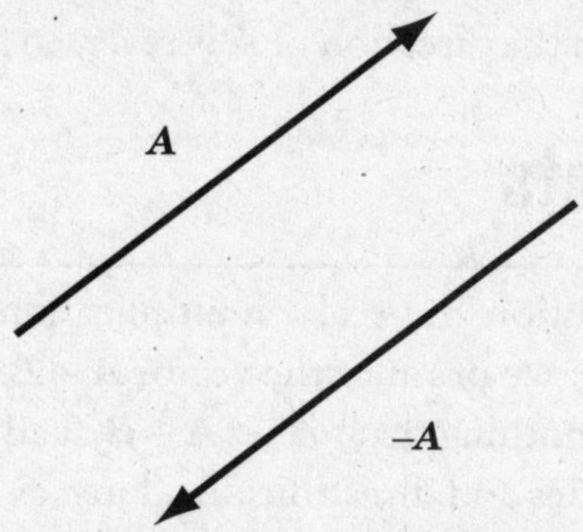

A – ***B***, then, is the same thing as ***A*** + (–***B***). For instance, let's take the two vectors ***A*** and ***B***:

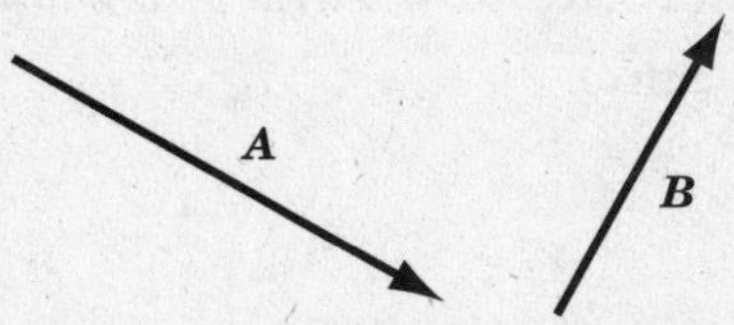

To subtract ***B*** from ***A***, take a vector of the same magnitude as ***B***, but pointing in the opposite direction, and add that vector to ***A***, using either the tip-to-tail method or the parallelogram method.

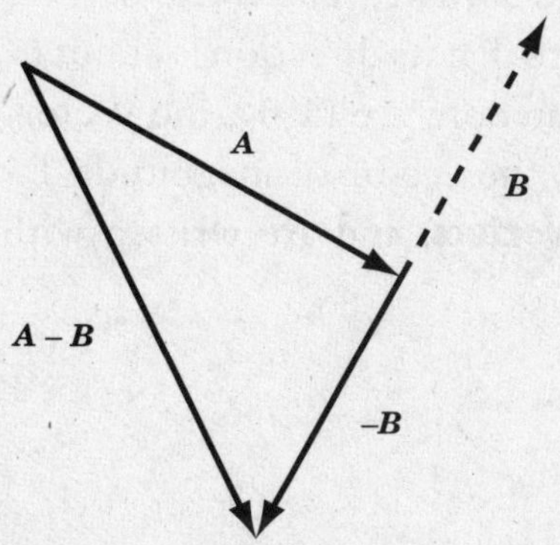

Multiplication by a Scalar

Multiplication is like repeated addition. Multiplying 4 by 3 means adding four three times: $3 \times 4 = 4 + 4 + 4 = 12$. The multiplication of a vector times a scalar works in the same way. Multiplying the vector $\boldsymbol{A}$ by the positive scalar c is equivalent to adding together c copies of the vector $\boldsymbol{A}$. Thus $3\boldsymbol{A} = \boldsymbol{A} + \boldsymbol{A} + \boldsymbol{A}$. Multiplying a vector by a scalar will get you a vector with the same direction, but different magnitude, as the original.

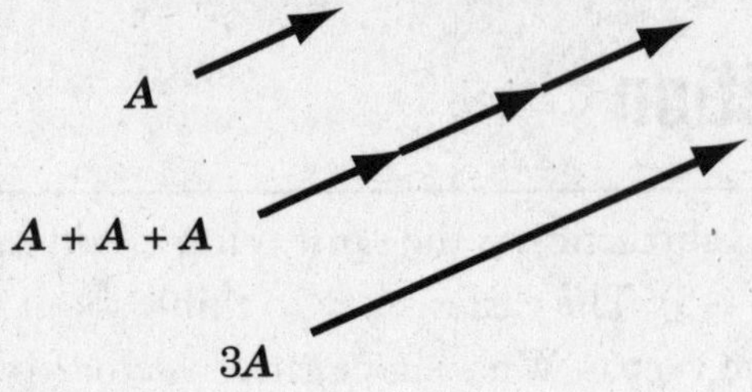

The result of multiplying $\boldsymbol{A}$ by c is a vector in the same direction as $\boldsymbol{A}$, with a magnitude of $c \times \boldsymbol{A}$. If c is negative, then the direction of $\boldsymbol{A}$ is reversed by scalar multiplication.

Vector Components

As we have seen, vector addition and scalar multiplication can produce new vectors out of old ones. For instance, we produce the vector $\boldsymbol{A} + \boldsymbol{B}$ by adding the two vectors $\boldsymbol{A}$ and $\boldsymbol{B}$. Of course, there is nothing that makes $\boldsymbol{A} + \boldsymbol{B}$ at all distinct as a vector from $\boldsymbol{A}$ or $\boldsymbol{B}$: all three have magnitudes and directions. And just as $\boldsymbol{A} + \boldsymbol{B}$ can be construed as the sum of two other vectors, so can $\boldsymbol{A}$ and $\boldsymbol{B}$. In problems involving vector addition, it's often convenient to break a vector down into two **components**; that is, two vectors whose sum is the vector in question.

Basis Vectors

We often graph vectors in an *xy*-coordinate system, where we can talk about vectors in purely numerical terms. For instance, the vector (3,4) is the vector whose tail is at the origin and whose tip is at the point (3,4) on the coordinate plane. From this coordinate, you can use the Pythagoean Theorem to calculate that the vector's magnitude is 5 and trigonometry to calculate that its direction is about 53.1° above the *x*-axis.

Two vectors of particular note are (1,0), the vector of magnitude 1 that points along the *x*-axis, and (0,1), the vector of magnitude 1 that points along the *y*-axis. These are called the **basis vectors**, and are written with the special hat notation: $\boldsymbol{x}$ and $\boldsymbol{y}$ respectively.

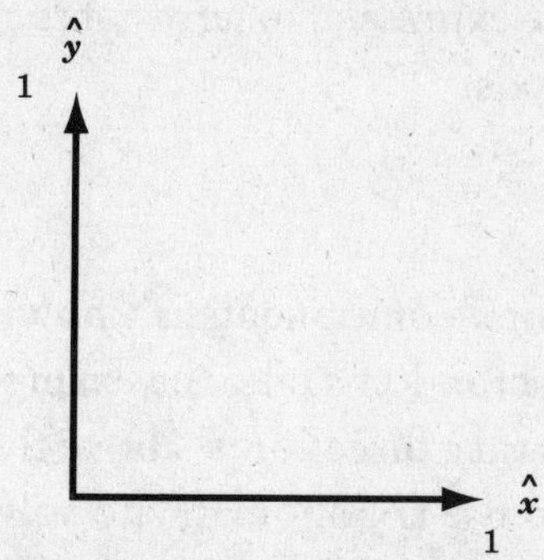

The basis vectors are important because you can express any other vectors in terms of the sum of multiples of the two basis vectors. For instance, the vector (3,4) that we discussed above—call it $\boldsymbol{A}$—can be expressed as the vector sum $3\boldsymbol{x} + 4\boldsymbol{y}$.

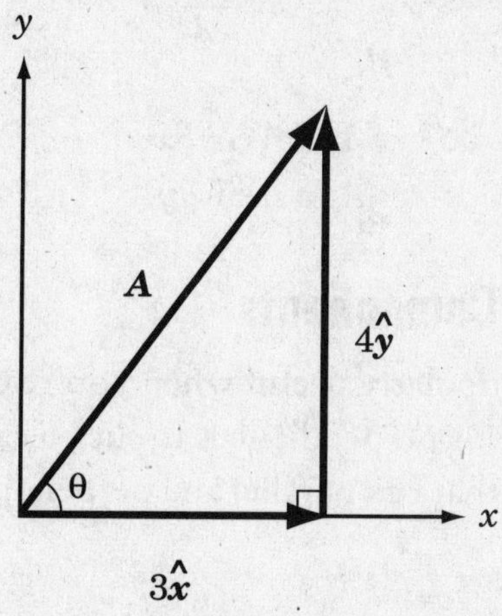

The vector $3\boldsymbol{x}$ is called the "x-component" of $\boldsymbol{A}$ and the $4\boldsymbol{y}$ is called the "y-component" of $\boldsymbol{A}$. In this book, we will use subscripts to denote vector components. For example, the x-component of $\boldsymbol{A}$ is $\boldsymbol{A}_x$ and the y-component of vector $\boldsymbol{A}$ is $\boldsymbol{A}_y$.

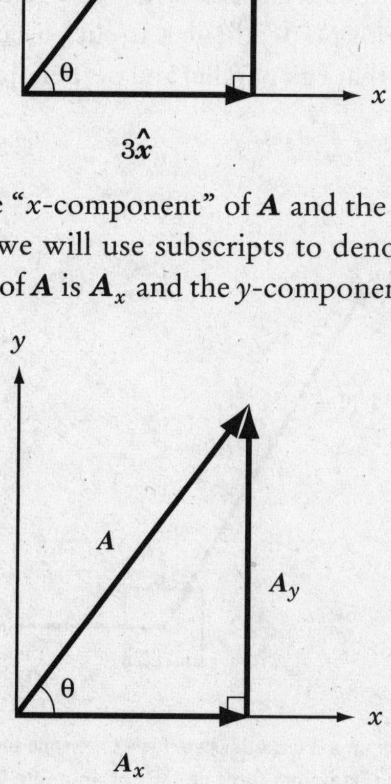

The direction of a vector can be expressed in terms of the angle θ by which it is rotated counterclockwise from the x-axis.

Vector Decomposition

The process of finding a vector's components is known as "resolving," "decomposing," or "breaking down" a vector. Let's take the example, illustrated above, of a vector, $\boldsymbol{A}$, with a magnitude of A and a direction θ above the x-axis. Because $\boldsymbol{A}_x$, $\boldsymbol{A}_y$, and $\boldsymbol{A}$ form a right triangle, we can use trigonometry to solve this problem. Applying the trigonometric definitions of cosine and sine,

$$\cos\theta = \frac{A_x}{A}$$

$$\sin\theta = \frac{A_y}{A}$$

we find:

$$\boldsymbol{A}_x = A\cos\theta\,\hat{\boldsymbol{x}}$$

$$\boldsymbol{A}_y = A\sin\theta\,\hat{\boldsymbol{y}}$$

Vector Addition Using Components

Vector decomposition is particularly useful when you're called upon to add two vectors that are neither parallel nor perpendicular. In such a case, you will want to resolve one vector into components that run parallel and perpendicular to the other vector.

Example

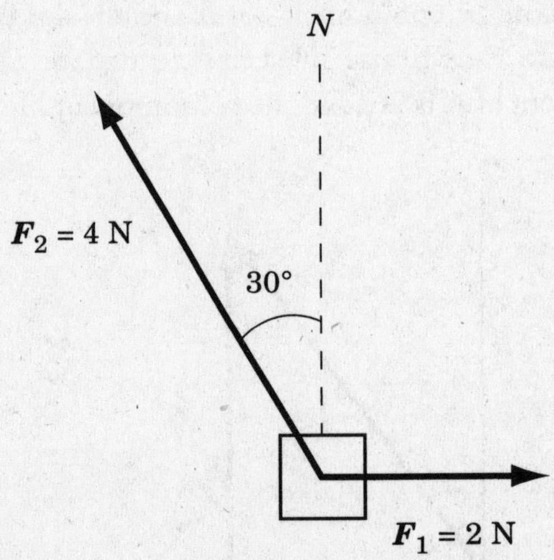

Two ropes are tied to a box on a frictionless surface. One rope pulls due east with a force of 2.0N. The second rope pulls with a force of 4.0N at an angle 30° west of north, as shown in the diagram. What is the total force acting on the box?

To solve this problem, we need to resolve the force on the second rope into its northward and westward components.

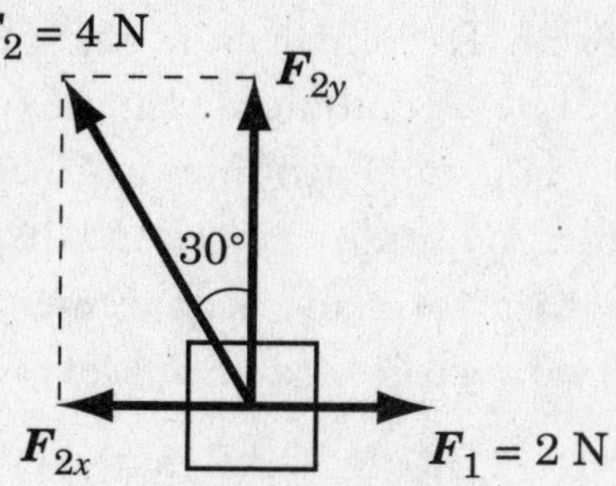

Because the force is directed 30° west of north, its northward component is

$$4.0 \cos 30° \approx 4.0 \times 0.86 = 3.4$$

and its westward component is

$$4.0 \sin 30° = 4.0 \times \frac{1}{2} = 2.0$$

Since the eastward component is also 2.0 N, the eastward and westward components cancel one another out. The resultant force is directed due north, with a force of approximately 3.4 N.

You can justify this answer by using the parallelogram method. If you fill out the half-completed parallelogram formed by the two vectors in the diagram above, you will find that the opposite corner of the parallelogram is directly above the corner made by the tails of those two vectors.

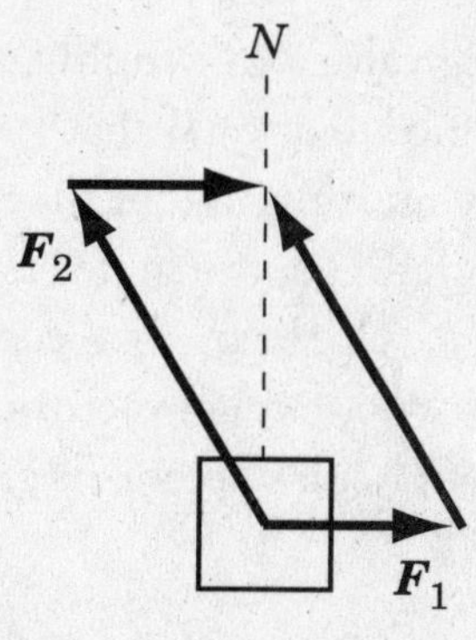

Vector Multiplication

When you multiply a vector by another vector, do you get a vector or a scalar? Well, that depends on the kind of multiplication you choose. There are two forms of vector multiplication: one results in a scalar, and one results in a vector.

Vectors

Dot Product

The **dot product**, also called the scalar product, takes two vectors, "multiplies" them together, and produces a scalar. The smaller the angle between the two vectors, the greater their dot product will be. A common example of the dot product in action is the formula for work, which you will encounter in Chapter 4. Work is a scalar quantity, but it is measured by the magnitude of force and displacement, both vector quantities, and the degree to which the force and displacement are parallel to one another.

The dot product of any two vectors, $\boldsymbol{A}$ and $\boldsymbol{B}$, is expressed by the equation:

$$\boldsymbol{A} \cdot \boldsymbol{B} = AB\cos\theta$$

where θ is the angle made by A and B when they are placed tail to tail.

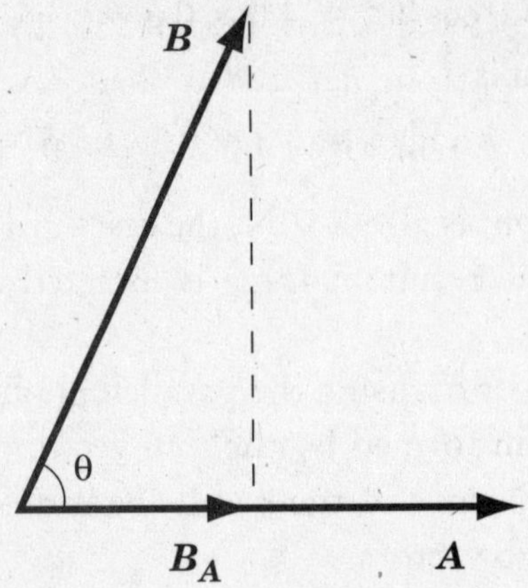

The dot product of $\boldsymbol{A}$ and $\boldsymbol{B}$ is the value you would get by multiplying the magnitude of $\boldsymbol{A}$ by the magnitude of the component of $\boldsymbol{B}$ that runs parallel to $\boldsymbol{A}$. Looking at the figure above, you can get $\boldsymbol{A} \cdot \boldsymbol{B}$ by multiplying the magnitude of $\boldsymbol{A}$ by the magnitude of $\boldsymbol{B_A}$, which equals $B\cos\theta$. You would get the same result if you multiplied the magnitude of $\boldsymbol{B}$ by the magnitude of $\boldsymbol{A_B}$, which equals $A\cos\theta$.

Note that the dot product of two identical vectors is their magnitude squared, and that the dot product of two perpendicular vectors is zero.

Example

Suppose the hands on a clock are vectors, where the hour hand has a length of 2 and the minute hand has a length of 4. What is the dot product of these two vectors when the clock reads 2 o'clock?

The angle between the hour hand and the minute hand at 2 o'clock is 60°. With this information, we can simply plug the numbers we have into the formula for the dot product:

$$\text{minute hand} \cdot \text{hour hand} = (2)(4)\cos 60^\circ = 4$$

Vectors

The Cross Product

The **cross product**, also called the vector product, "multiplies" two vectors together to produce a third vector, which is perpendicular to both of the original vectors. The closer the angle between the two vectors is to the perpendicular, the greater the cross product will be. We encounter the cross product a great deal in our discussions of magnetic fields. Magnetic force acts perpendicular both to the magnetic field that produces the force, and to the charged particles experiencing the force.

The cross product can be a bit tricky, because you have to think in three dimensions. The cross product of two vectors, $\boldsymbol{A}$ and $\boldsymbol{B}$, is defined by the equation:

$$\boldsymbol{A} \times \boldsymbol{B} = AB \sin \theta \boldsymbol{n}$$

where $\boldsymbol{n}$ is a unit vector perpendicular to both $\boldsymbol{A}$ and $\boldsymbol{B}$. The magnitude of the cross product vector is equal to the area made by a parallelogram of $\boldsymbol{A}$ and $\boldsymbol{B}$. In other words, the greater the area of the parallelogram, the longer the cross product vector.

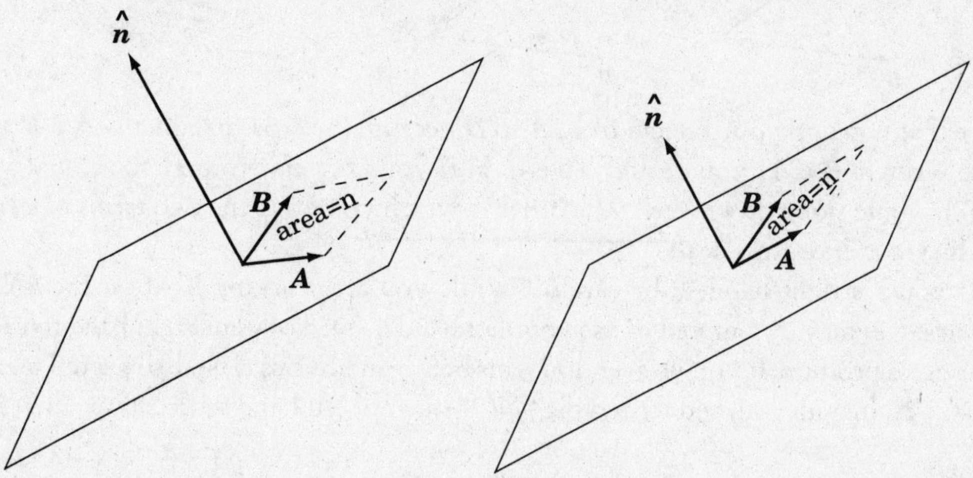

The Right-Hand Rule

You may have noticed an ambiguity here. The two vectors $\boldsymbol{A}$ and $\boldsymbol{B}$ always lie on a common plane and there are two directions perpendicular to this plane: "up" and "down."

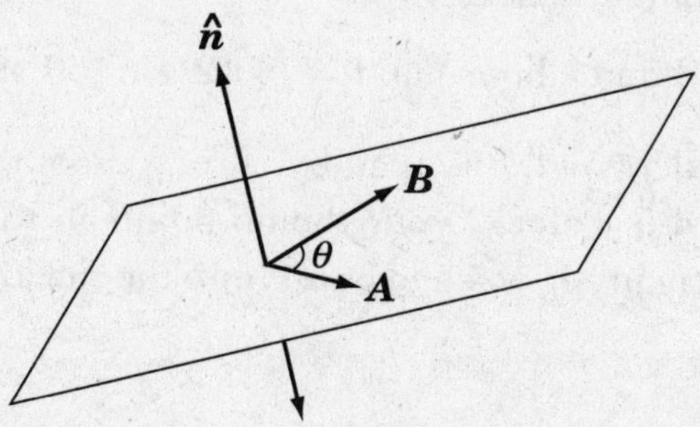

There is no real reason why we should choose the "up" or the "down" direction as the right one, but it's important that we remain consistent. To that end, everybody follows the convention known as the **right-hand rule**. In order to find the cross product, $\boldsymbol{A} \times \boldsymbol{B}$:

> Place the two vectors so their tails are at the same point. Align your right hand along the first vector, $\boldsymbol{A}$, such that the base of your palm is at the tail of the vector, and your fingertips are pointing towards the tip. Then curl your fingers via the small angle towards the second vector, $\boldsymbol{B}$. If $\boldsymbol{B}$ is in a clockwise direction from $\boldsymbol{A}$, you'll find you have to flip your hand over to make this work. The direction in which your thumb is pointing is the direction of $\boldsymbol{n}$, and the direction of $\boldsymbol{A} \times \boldsymbol{B}$.

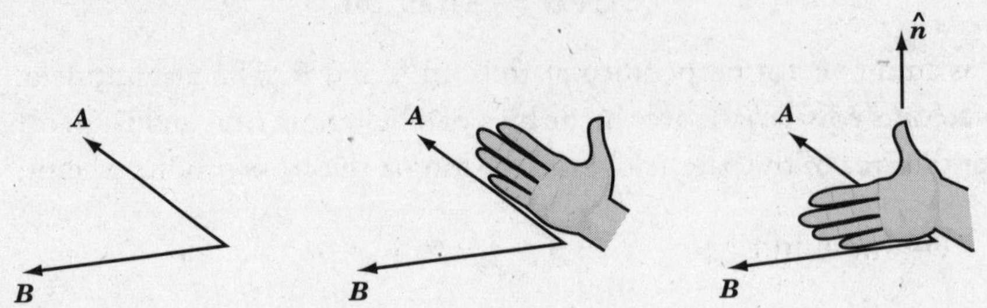

Note that you curl your fingers from $\boldsymbol{A}$ to $\boldsymbol{B}$ because the cross product is $\boldsymbol{A} \times \boldsymbol{B}$. If it were written $\boldsymbol{B} \times \boldsymbol{A}$, you would have to curl your fingers from $\boldsymbol{B}$ to $\boldsymbol{A}$, and your thumb would point downward. The order in which you write the two terms of a cross product matters a great deal!

If you are right-handed, be careful! While you are working hard on the SAT II Physics, you may be tempted to use your left hand instead of your right hand to calculate a cross product. If you've ever done this before in physics class, bring a red marker into the exam with you, and write a big "NO" on your left hand when the exam begins.

Example

> Suppose once again that the minute hand of a clock is a vector of magnitude 4 and the hour hand is a vector of magnitude 2. If, at 5 o'clock, one were to take the cross product of the minute hand × the hour hand, what would the resultant vector be?

First of all, let's calculate the magnitude of the cross product vector. The angle between the hour hand and the minute hand is 150°:

$$\text{minute hand} \times \text{hour hand} = (4)(2)\sin 150°\boldsymbol{n} = 4\boldsymbol{n}$$

Using the right hand rule, you'll find that, by curling the fingers of your right hand from 12 o'clock toward 5 o'clock, your thumb points in toward the clock. So the resultant vector has a magnitude of 4 and points into the clock.

Key Formulas

Dot Product	$\boldsymbol{A} \cdot \boldsymbol{B} = AB\cos\theta$
Cross Product	$\boldsymbol{A} \times \boldsymbol{B} = AB\sin\theta\hat{\boldsymbol{n}}$
Magnitude	$A = \sqrt{A_x^2 + A_y^2}$
Direction	$\theta = \arctan\dfrac{\boldsymbol{A}_y}{\boldsymbol{A}_x}$
X-, *Y*-Components	$\boldsymbol{A}_x = A\cos\theta\hat{\boldsymbol{x}}$ $\boldsymbol{A}_y = A\sin\theta\hat{\boldsymbol{y}}$
Vector Addition	$\boldsymbol{A} + \boldsymbol{B} = (A_x + B_x)\hat{\boldsymbol{x}} + (A_y + B_y)\hat{\boldsymbol{y}}$

Practice Questions

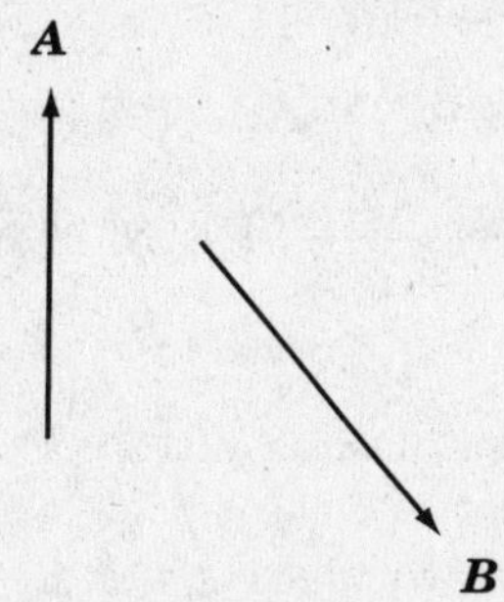

1. Which of the following vectors best represents the vector $\boldsymbol{A} + \boldsymbol{B}$?

(A) 
(B) 
(C) 
(D) 
(E) 

2. Vector $\boldsymbol{A}$ has a magnitude of 5 in the leftward direction and $\boldsymbol{B}$ has a magnitude of 2 in the rightward direction. What is the value of $2\boldsymbol{A} - \boldsymbol{B}$?

   (A) 12 in the leftward direction
   (B) 10 in the leftward direction
   (C) 8 in the leftward direction
   (D) 8 in the rightward direction
   (E) 12 in the rightward direction

3. When the tail of vector $\boldsymbol{A}$ is set at the origin of the xy-axis, the tip of $\boldsymbol{A}$ reaches (3,6). When the tail of vector $\boldsymbol{B}$ is set at the origin of the xy-axis, the tip of $\boldsymbol{B}$ reaches (–1,5). If the tail of vector $\boldsymbol{A} - \boldsymbol{B}$ were set at the origin of the xy-axis, what point would its tip touch?

   (A) (2,11)
   (B) (2,1)
   (C) (–2,7)
   (D) (4,1)
   (E) (4,11)

4. $\boldsymbol{A}$ and $\boldsymbol{B}$ are vectors, and θ is the angle between them. What can you do to maximize $\boldsymbol{A} \cdot \boldsymbol{B}$?

   I. Maximize the magnitude of $\boldsymbol{A}$
   II. Maximize the magnitude of $\boldsymbol{B}$
   III. Set θ to 90°

   (A) None of the above
   (B) I only
   (C) III only
   (D) I and II only
   (E) I, II, and III

5. Which of the following statements is NOT true about $\boldsymbol{A} \times \boldsymbol{B}$?

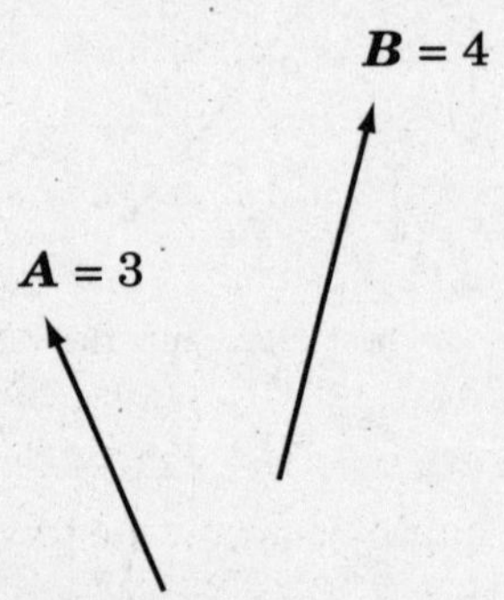

   (A) It is a vector that points into the page
   (B) It has a magnitude that is less than or equal to 12
   (C) It has no component in the plane of the page
   (D) The angle it makes with $\boldsymbol{B}$ is less than the angle it makes with $\boldsymbol{A}$
   (E) It is the same as $-\boldsymbol{B} \times \boldsymbol{A}$

Explanations

1. (A)

By adding $\boldsymbol{A}$ to $\boldsymbol{B}$ using the tip-to-tail method, we can see that (A) is the correct answer.

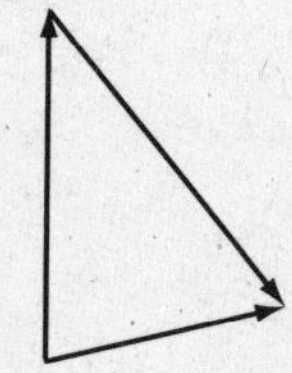

2. (A)

The vector 2$\boldsymbol{A}$ has a magnitude of 10 in the leftward direction. Subtracting $\boldsymbol{B}$, a vector of magnitude 2 in the rightward direction, is the same as adding a vector of magnitude 2 in the leftward direction. The resultant vector, then, has a magnitude of 10 + 2 =12 in the leftward direction.

3. (D)

To subtract one vector from another, we can subtract each component individually. Subtracting the x-components of the two vectors, we get $3 - -1 = 4$, and subtracting the y-components of the two vectors, we get $6 - 5 = 1$. The resultant vector therefore has an x-component of 4 and a y-component of 1, so that if its tail is at the origin of the xy-axis, its tip would be at (4,1).

4. (D)

The dot product of $\boldsymbol{A}$ and $\boldsymbol{B}$ is given by the formula $\boldsymbol{A} \cdot \boldsymbol{B} = AB \cos \theta$. This increases as either A or B increases. However, $\cos \theta = 0$ when $\theta = 90°$, so this is not a way to maximize the dot product. Rather, to maximize $\boldsymbol{A} \cdot \boldsymbol{B}$ one should set θ to 0° so $\cos \theta = 1$.

5. (D)

Let's take a look at each answer choice in turn. Using the right-hand rule, we find that $\boldsymbol{A} \times \boldsymbol{B}$ is indeed a vector that points into the page. We know that the magnitude of $\boldsymbol{A} \times \boldsymbol{B}$ is $\boldsymbol{AB} \sin \theta$, where θ is the angle between the two vectors. Since $\boldsymbol{AB} = 12$, and since $\sin \theta \leq 1$, we know that $\boldsymbol{A} \times \boldsymbol{B}$ cannot possibly be greater than 12. As a cross product vector, $\boldsymbol{A} \times \boldsymbol{B}$ is perpendicular to both $\boldsymbol{A}$ and $\boldsymbol{B}$. This means that it has no component in the plane of the page. It also means that both $\boldsymbol{A}$ and $\boldsymbol{B}$ are at right angles with the cross product vector, so neither angle is greater than or less than the other. Last, $\boldsymbol{B} \times \boldsymbol{A}$ is a vector of the same magnitude as $\boldsymbol{A} \times \boldsymbol{B}$, but it points in the opposite direction. By negating $\boldsymbol{B} \times \boldsymbol{A}$, we get a vector that is identical to $\boldsymbol{A} \times \boldsymbol{B}$.

Vectors

Kinematics

KINEMATICS DERIVES ITS NAME FROM THE Greek word for "motion," *kinema*. Before we can make any headway in physics, we have to be able to describe how bodies move. Kinematics provides us with the language and the mathematical tools to describe motion, whether the motion of a charging pachyderm or a charged particle. As such, it provides a foundation that will help us in all areas of physics. Kinematics is most intimately connected with dynamics: while kinematics describes motion, dynamics explains the causes for this motion.

Chapter Contents

Displacement *45*
Speed, Velocity, and Acceleration *47*
Kinematics with Graphs *50*
1-D Motion with Uniform Acceleration . . . *58*
2-D Motion with Uniform Acceleration . . . *61*
Key Formulas *62*
Practice Questions *62*
Explanations *64*

Displacement

Displacement is a vector quantity, commonly denoted by the vector $\boldsymbol{s}$, that reflects an object's change in spatial position. The displacement of an object that moves from point A to point B is a vector whose tail is at A and whose tip is at B. Displacement deals only with the separation between points A and B, and not with the path the object followed between points A and B. By contrast, the **distance** that the object travels is equal to the length of path AB.

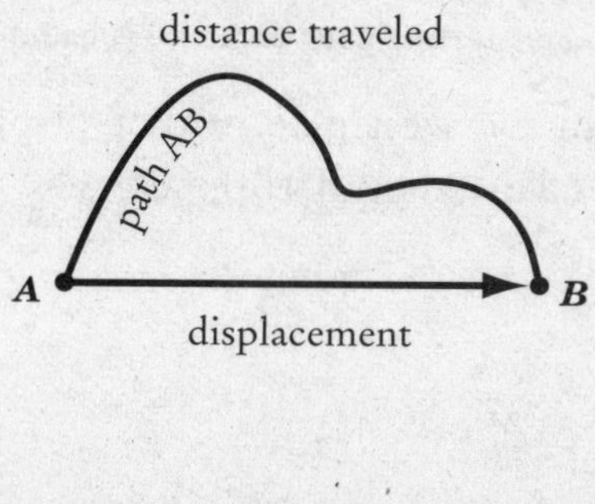

Students often mistake displacement for distance, and the SAT II may well call for you to distinguish between the two. A question favored by test-makers everywhere is to ask the displacement of an athlete who has run a lap on a 400-meter track. The answer, of course, is zero: after running a lap, the athlete is back where he or she started. The distance traveled by the athlete, and not the displacement, is 400 meters.

Example

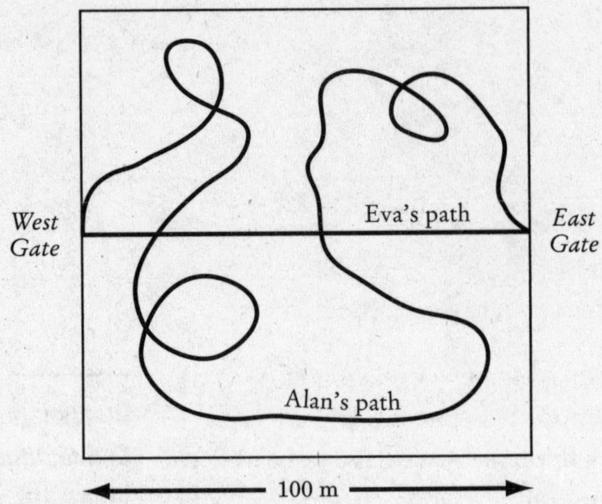

Alan and Eva are walking through a beautiful garden. Because Eva is very worried about the upcoming SAT II Physics test, she takes no time to smell the flowers and instead walks on a straight path from the west garden gate to the east gate, a distance of 100 meters. Alan, unconcerned about the SAT II Literature test, meanders off the straight path to smell all the flowers in sight. When Alan and Eva meet at the east gate, who has walked a greater distance? What are their displacements?

Since Eva took the direct path between the west and east garden gates and Alan took an indirect path, Alan has traveled a much greater distance than Eva. Yet, as we have discussed, displacement is a vector quantity that measures the distance separating the starting point from the ending point: the path taken between the two points is irrelevant. So Alan and Eva both have the same displacement: 100 meters east of the west gate. Note that, because displacement is a vector quantity, it is not enough to say that the displacement is 100 meters: you must also state the direction of that displacement. The distance that Eva has traveled is exactly equal to the magnitude of her displacement: 100 meters.

After reaching the east gate, Eva and Alan notice that the gate is locked, so they must turn around and exit the garden through the west gate. On the return trip, Alan again wanders off to smell the flowers, and Eva travels the path directly between the gates. At the center of the garden, Eva stops to throw a penny into a fountain. At this point, what is her displacement from her starting point at the west gate?

Eva is now 50 meters from the west gate, so her displacement is 50 meters, even though she has traveled a total distance of 150 meters.

Kinematics

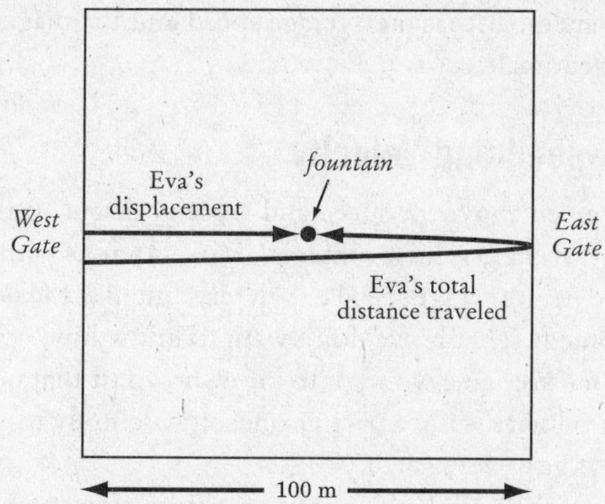

When Alan and Eva reconvene at the west gate, their displacements are both zero, as they both began and ended their garden journey at the west gate. The moral of the story? Always to take time to smell the flowers!

Speed, Velocity, and Acceleration

Along with displacement, **velocity** and **acceleration** round out the holy trinity of kinematics. As you'll see, all three are closely related to one another, and together they offer a pretty complete understanding of motion. **Speed**, like distance, is a scalar quantity that won't come up too often on the SAT II, but it might trip you up if you don't know how to distinguish it from velocity.

Speed and Velocity

As distance is to displacement, so speed is to velocity: the crucial difference between the two is that speed is a scalar and velocity is a vector quantity. In everyday conversation, we usually say speed when we talk about how fast something is moving. However, in physics, it is often important to determine the direction of this motion, so you'll find velocity come up in physics problems far more frequently than speed.

A common example of speed is the number given by the speedometer in a car. A speedometer tells us the car's speed, not its velocity, because it gives only a number and not a direction. Speed is a measure of the distance an object travels in a given length of time:

$$\text{average speed} = \frac{\text{distance traveled}}{\text{time elapsed}} = \frac{\Delta x}{\Delta t}$$

Velocity is a vector quantity defined as rate of change of the displacement vector over time:

$$\text{average velocity} = \frac{\text{change in displacement}}{\text{time elapsed}} = \frac{\Delta \mathbf{s}}{\Delta t}$$

It is important to remember that the average speed and the magnitude of the average velocity may not be equivalent.

Instantaneous Speed and Velocity

The two equations given above for speed and velocity discuss only the *average* speed and *average* velocity over a given time interval. Most often, as with a car's speedometer, we are not interested in an average speed or velocity, but in the **instantaneous velocity** or speed at a given moment. That is, we don't want to know how many meters an object covered in the past ten seconds; we want to know how fast that object is moving *right now.* Instantaneous velocity is not a tricky concept: we simply take the equation above and assume that Δt is very, very small.

Most problems on the SAT II ask about an object's instantaneous velocity rather than its average velocity or speed over a given time frame. Unless a question specifically asks you about the average velocity or speed over a given time interval, you can safely assume that it is asking about the instantaneous velocity at a given moment.

Example

Which of the follow sentences contains an example of instantaneous velocity?

(A) "The car covered 500 kilometers in the first 10 hours of its northward journey."
(B) "Five seconds into the launch, the rocket was shooting upward at 5000 meters per second."
(C) "The cheetah can run at 70 miles per hour."
(D) "Moving at five kilometers per hour, it will take us eight hours to get to the base camp."
(E) "Roger Bannister was the first person to run one mile in less than four minutes."

Instantaneous velocity has a magnitude and a direction, and deals with the velocity at a particular instant in time. All three of these requirements are met only in (B). (A) is an example of average velocity, (C) is an example of instantaneous speed, and both (D) and (E) are examples of average speed.

Acceleration

Speed and velocity only deal with movement at a constant rate. When we speed up, slow down, or change direction, we want to know our **acceleration**. Acceleration is a vector quantity that measures the rate of change of the velocity vector with time:

$$\text{average acceleration} = \frac{\text{change in velocity}}{\text{time elapsed}} = \frac{\Delta \mathbf{v}}{\Delta t}$$

Applying the Concepts of Speed, Velocity, and Acceleration

With these three definitions under our belt, let's apply them to a little story of a zealous high school student called Andrea. Andrea is due to take the SAT II Physics at the ETS building 10 miles due east from her home. Because she is particularly concerned with sleeping as much as possible before the test, she practices the drive the day before so she knows exactly how long it will take and how early she must get up.

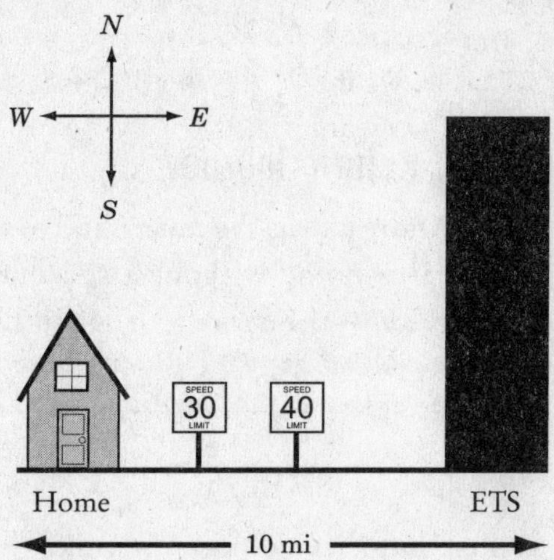

Instantaneous Velocity

After starting her car, she zeros her odometer so that she can record the exact distance to the test center. Throughout the drive, Andrea is cautious of her speed, which is measured by her speedometer. At first she is careful to drive at exactly 30 miles per hour, as advised by the signs along the road. Chuckling to herself, she notes that her instantaneous velocity—a vector quantity—is 30 miles per hour due east.

Average Acceleration

Along the way, Andrea sees a new speed limit sign of 40 miles per hour, so she accelerates. Noting with her trusty wristwatch that it takes her two seconds to change from 30 miles per hour due east to 40 miles per hour due east, Andrea calculates her average acceleration during this time frame:

$$\begin{aligned}\text{average acceleration} &= \frac{40 \text{ mi/hr} - 30 \text{ mi/hr}}{2 \text{ s}} \\ &= \frac{10 \text{ mi/hr}}{2 \text{ s}} \cdot \frac{3600 \text{ s}}{1 \text{ hr}} \\ &= 18,000 \text{ mi/hr}^2 \text{ east}\end{aligned}$$

This may seem like an outrageously large number, but in terms of meters per second squared, the standard units for measuring acceleration, it comes out to 0.22 m/s^2.

Average Velocity: One-Way

After reaching the tall, black ETS skyscraper, Andrea notes that the test center is exactly 10 miles from her home and that it took her precisely 16 minutes to travel between the two locations. She does a quick calculation to determine her average velocity during the trip:

$$\begin{aligned}\text{average velocity} &= \frac{10\text{ mi}}{16\text{ min}} \cdot \frac{60\text{ min}}{1\text{ hr}} \\ &= 37.5\text{ mi/hr due east}\end{aligned}$$

Average Speed and Velocity: Return Journey

Satisfied with her little exercise, Andrea turns the car around to see if she can beat her 16-minute time. Successful, she arrives home without a speeding ticket in 15 minutes. Andrea calculates her average speed for the entire journey to ETS and back home:

$$\begin{aligned}\text{average speed} &= \frac{20\text{ mi}}{16\text{ min} + 15\text{ min}} \cdot \frac{60\text{ min}}{1\text{ hr}} \\ &= 38.7\text{ mi/hr}\end{aligned}$$

Is this the same as her average velocity? Andrea reminds herself that, though her odometer reads 20 miles, her net displacement—and consequently her average velocity over the entire length of the trip—is zero. The SAT II is not going to get her with any trick questions like that!

Andrea goes to bed that night, rests well, wakes up the next morning, drives to the ETS building, and scores an 800 on the SAT II. She gets admitted to Harvard, and everyone lives happily ever after.

Kinematics with Graphs

Since you are not allowed to use calculators, the SAT II Physics places a heavy emphasis on qualitative problems. A common way of testing kinematics qualitatively is to present you with a graph plotting position vs. time, velocity vs. time, or acceleration vs. time and to ask you questions about the motion of the object represented by the graph. Because the SAT II Physics is entirely made up of multiple-choice questions, you won't need to know how to draw graphs; you'll just have to interpret the data presented in them.

Knowing how to read such graphs quickly and accurately will not only help you solve problems of this sort, it will also help you visualize the often-abstract realm of kinematic equations. In the examples that follow, we will examine the movement of an ant running back and forth along a line.

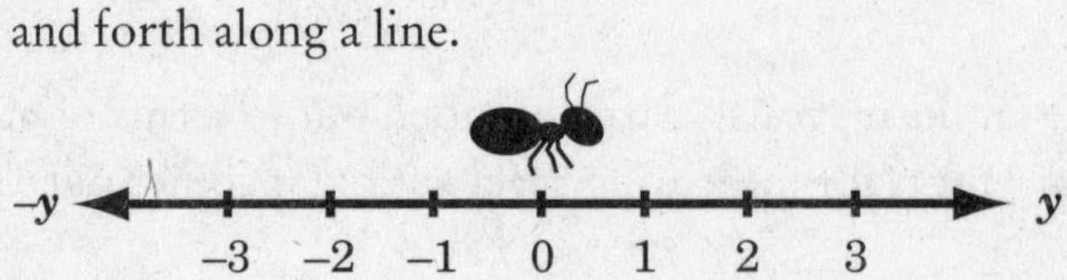

Position vs. Time Graphs

Position vs. time graphs give you an easy and obvious way of determining an object's displacement at any given time, and a subtler way of determining that object's velocity at any given time. Let's put these concepts into practice by looking at the following graph charting the movements of our friendly ant.

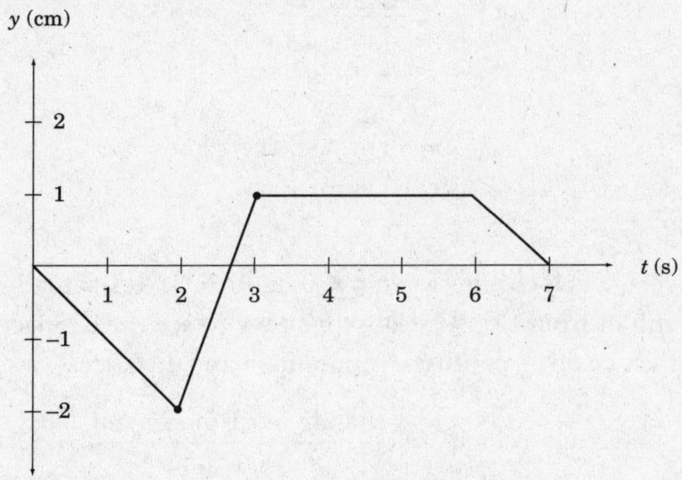

Any point on this graph gives us the position of the ant at a particular moment in time. For instance, the point at (2,–2) tells us that, two seconds after it started moving, the ant was two centimeters to the left of its starting position, and the point at (3,1) tells us that, three seconds after it started moving, the ant is one centimeter to the right of its starting position.

Let's read what the graph can tell us about the ant's movements. For the first two seconds, the ant is moving to the left. Then, in the next second, it reverses its direction and moves quickly to $y = 1$. The ant then stays still at $y = 1$ for three seconds before it turns left again and moves back to where it started. Note how concisely the graph displays all this information!

Calculating Velocity

We know the ant's displacement, and we know how long it takes to move from place to place. Armed with this information, we should also be able to determine the ant's velocity, since velocity measures the rate of change of displacement over time. If displacement is given here by the vector y, then the velocity of the ant is

$$\boldsymbol{v} = \frac{\Delta y}{\Delta t}$$

If you recall, the slope of a graph is a measure of rise over run; that is, the amount of change in the y-direction divided by the amount of change in the x-direction. In our graph, Δy is the change in the y-direction and Δt is the change in the x-direction, so $\boldsymbol{v}$ is a measure of the slope of the graph. *For any position vs. time graph, the velocity at time* t *is equal to the slope of the line at* t. In a graph made up of straight lines, like the one above, we can easily calculate the slope at each point on the graph, and hence know the instantaneous velocity at any given time.

We can tell that the ant has a velocity of zero from $t = 3$ to $t = 6$, because the slope of the line at these points is zero. We can also tell that the ant is cruising along at the fastest speed between $t = 2$ and $t = 3$, because the position vs. time graph is steepest between these points. Calculating the ant's average velocity during this time interval is a simple matter of dividing rise by run, as we've learned in math class.

$$\begin{aligned} \text{velocity} &= \frac{y_{\text{final}}\ y_{\text{initial}}}{t_{\text{final}}\ t_{\text{initial}}} \\ &= \frac{1(-2)\text{cm}}{32\ \text{s}} \\ &= 3\ \text{cm/s to the right} \end{aligned}$$

Average Velocity

How about the average velocity between $t = 0$ and $t = 3$? It's actually easier to sort this out with a graph in front of us, because it's easy to see the displacement at $t = 0$ and $t = 3$, and so that we don't confuse displacement and distance.

$$\begin{aligned} \text{average velocity} &= \frac{\text{change in displacement}}{\text{elapsed time}} \\ &= \frac{10\ \text{cm}}{30\ \text{s}} \\ &= 0.33\ \text{cm/s to the right} \end{aligned}$$

Average Speed

Although the total displacement in the first three seconds is one centimeter to the right, the total distance traveled is two centimeters to the left, and then three centimeters to the right, for a grand total of five centimeters. Thus, the average speed is not the same as the average velocity of the ant. Once we've calculated the total distance traveled by the ant, though, calculating its average speed is not difficult:

$$\frac{5\ \text{cm}}{3\ \text{s}} = 1.67\ \text{cm/s}$$

Curved Position vs. Time Graphs

This is all well and good, but how do you calculate the velocity of a curved position vs. time graph? Well, the bad news is that you'd need calculus. The good news is that the SAT II doesn't expect you to use calculus, so if you are given a curved position vs. time graph, you will only be asked qualitative questions and won't be expected to make any calculations. A few points on the graph will probably be labeled, and you will have to identify which point has the greatest or least velocity. Remember, the point with the greatest slope has the greatest velocity, and the point with the least slope has the least velocity. The turning points of the graph, the tops of the "hills" and the bottoms of the "valleys" where the slope is zero, have zero velocity.

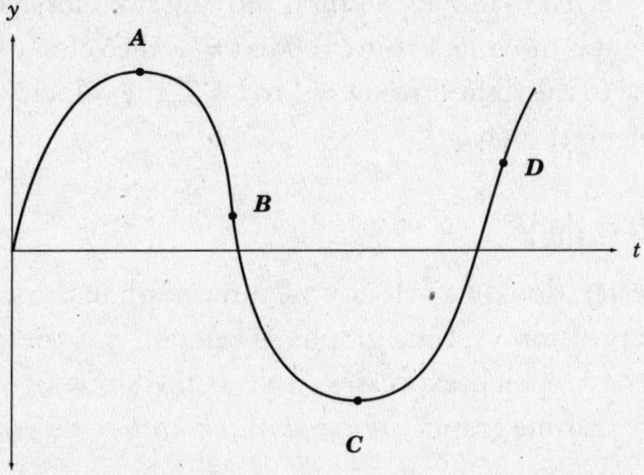

In this graph, for example, the velocity is zero at points *A* and *C*, greatest at point *D*, and smallest at point *B*. The velocity at point *B* is smallest because the slope at that point is negative. Because velocity is a vector quantity, the velocity at *B* would be a large negative number. However, the speed at *B* is greater even than the speed at *D*: speed is a scalar quantity, and so it is always positive. The slope at *B* is even steeper than at *D*, so the speed is greatest at *B*.

Velocity vs. Time Graphs

Velocity vs. time graphs are the most eloquent kind of graph we'll be looking at here. They tell us very directly what the velocity of an object is at any given time, and they provide subtle means for determining both the position and acceleration of the same object over time. The "object" whose velocity is graphed below is our ever-industrious ant, a little later in the day.

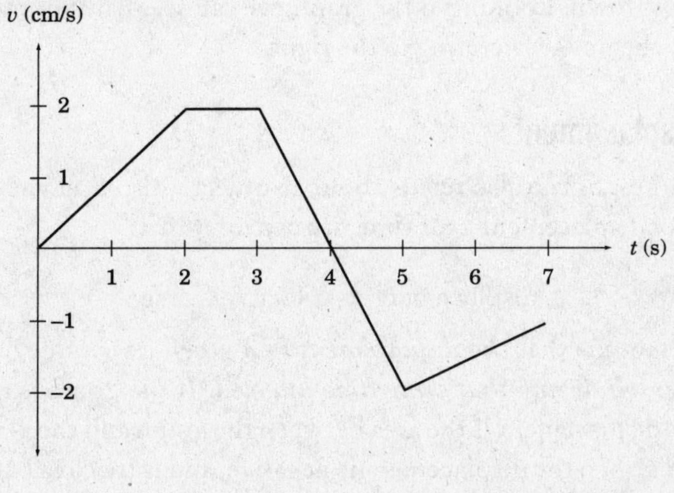

We can learn two things about the ant's velocity by a quick glance at the graph. First, we can tell exactly how fast it is going at any given time. For instance, we can see that, two seconds after it started to move, the ant is moving at 2 cm/s. Second, we can tell in which direction the ant is moving. From $t = 0$ to $t = 4$, the velocity is positive, meaning that the ant is moving to the right. From $t = 4$ to $t = 7$, the velocity is negative, meaning that the ant is moving to the left.

Calculating Acceleration

We can calculate acceleration on a velocity vs. time graph in the same way that we calculate velocity on a position vs. time graph. Acceleration is the rate of change of the velocity vector, $\Delta \mathbf{v}/\Delta \mathbf{t}$, which expresses itself as the slope of the velocity vs. time graph. For a velocity vs. time graph, *the acceleration at time* t *is equal to the slope of the line at* t.

What is the acceleration of our ant at $t = 2.5$ and $t = 4$? Looking quickly at the graph, we see that the slope of the line at $t = 2.5$ is zero and hence the acceleration is likewise zero. The slope of the graph between $t = 3$ and $t = 5$ is constant, so we can calculate the acceleration at $t = 4$ by calculating the average acceleration between $t = 3$ and $t = 5$:

$$\begin{aligned} \text{acceleration} &= \frac{v_{\text{final}}\ v_{\text{initial}}}{t_{\text{final}}\ t_{\text{initial}}} \\ &= \frac{-2(2)\ \text{cm/s}}{53\ \text{s}} \\ &= -2\ \text{cm/s}^2 \end{aligned}$$

The minus sign tells us that acceleration is in the leftward direction, since we've defined the y-coordinates in such a way that right is positive and left is negative. At $t = 3$, the ant is moving to the right at 2 cm/s, so a leftward acceleration means that the ant begins to slow down. Looking at the graph, we can see that the ant comes to a stop at $t = 4$, and then begins accelerating to the right.

Calculating Displacement

Velocity vs. time graphs can also tell us about an object's displacement. Because velocity is a measure of displacement over time, we can infer that:

$$\text{displacement} = \text{velocity} \times \text{time}$$

Graphically, this means that *the displacement in a given time interval is equal to the area under the graph during that same time interval.* If the graph is above the t-axis, then the positive displacement is the area between the graph and the t-axis. If the graph is below the t-axis, then the displacement is negative, and is the area between the graph and the t-axis. Let's look at two examples to make this rule clearer.

First, what is the ant's displacement between $t = 2$ and $t = 3$? Because the velocity is constant during this time interval, the area between the graph and the t-axis is a rectangle of width 1 and height 2.

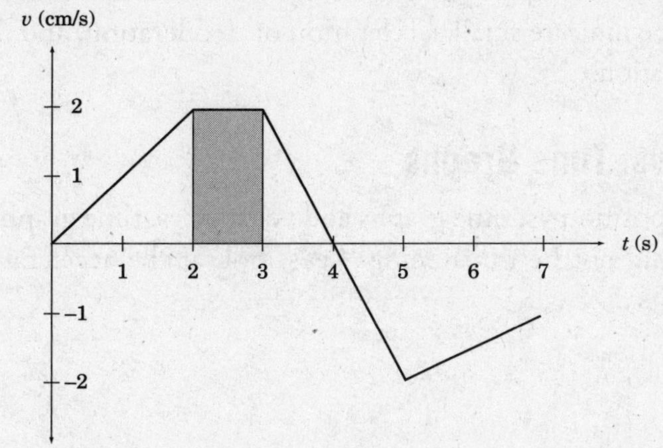

The displacement between $t = 2$ and $t = 3$ is the area of this rectangle, which is 1 cm/s × 2 s = 2 cm to the right.

Next, consider the ant's displacement between $t = 3$ and $t = 5$. This portion of the graph gives us two triangles, one above the t-axis and one below the t-axis.

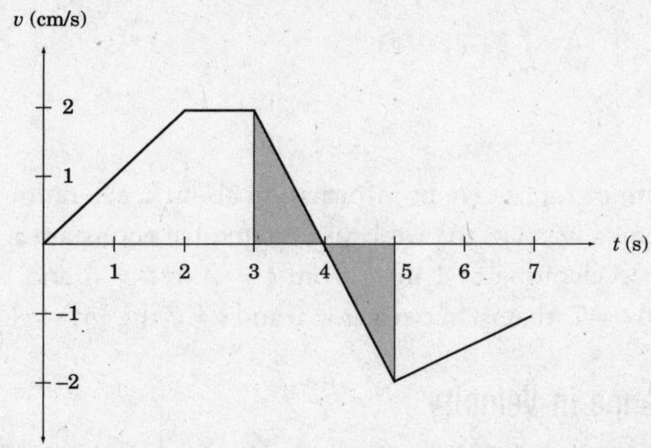

Both triangles have an area of ½(1 s)(2 cm/s) = 1 cm. However, the first triangle is above the t-axis, meaning that displacement is positive, and hence to the right, while the second triangle is below the t-axis, meaning that displacement is negative, and hence to the left. The total displacement between $t = 3$ and $t = 5$ is:

$$1\,\text{cm} - 1\,\text{cm} = 0$$

In other words, at $t = 5$, the ant is in the same place as it was at $t = 3$.

Curved Velocity vs. Time Graphs

As with position vs. time graphs, velocity vs. time graphs may also be curved. Remember that regions with a steep slope indicate rapid acceleration or deceleration, regions with a gentle slope indicate small acceleration or deceleration, and the turning points have zero acceleration.

Acceleration vs. Time Graphs

After looking at position vs. time graphs and velocity vs. time graphs, acceleration vs. time graphs should not be threatening. Let's look at the acceleration of our ant at another point in its dizzy day.

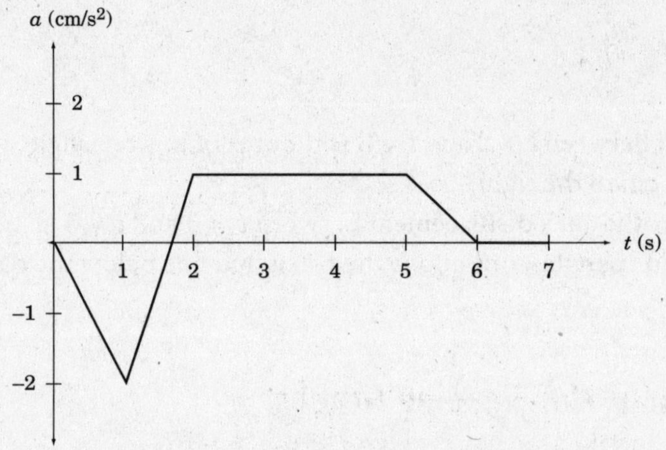

Acceleration vs. time graphs give us information about acceleration and about velocity. The SAT II generally sticks to problems that involve a constant acceleration. In this graph, the ant is accelerating at 1 m/s^2 from $t = 2$ to $t = 5$ and is not accelerating between $t = 6$ and $t = 7$; that is, between $t = 6$ and $t = 7$ the ant's velocity is constant.

Calculating Change in Velocity

Acceleration vs. time graphs tell us about an object's velocity in the same way that velocity vs. time graphs tell us about an object's displacement. *The change in velocity in a given time interval is equal to the area under the graph during that same time interval.* Be careful: the area between the graph and the t-axis gives the *change* in velocity, not the final velocity or average velocity over a given time period.

What is the ant's change in velocity between $t = 2$ and $t = 5$? Because the acceleration is constant during this time interval, the area between the graph and the t-axis is a rectangle of height 1 and length 3.

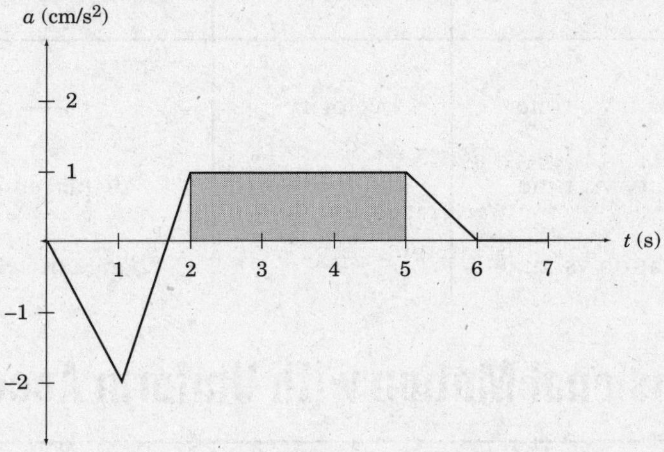

The area of the shaded region, and consequently the change in velocity during this time interval, is $1 \text{ cm/s}^2 \cdot 3 \text{ s} = 3 \text{ cm/s}$ to the right. This doesn't mean that the velocity at $t = 5$ is 3 cm/s; it simply means that the velocity is 3 cm/s greater than it was at $t = 2$. Since we have not been given the velocity at $t = 2$, we can't immediately say what the velocity is at $t = 5$.

Summary of Rules for Reading Graphs

You may have trouble recalling when to look for the slope and when to look for the area under the graph. Here are a couple handy rules of thumb:

1. **The slope on a given graph is equivalent to the quantity we get by dividing the y-axis by the x-axis.** For instance, the y-axis of a position vs. time graph gives us displacement, and the x-axis gives us time. Displacement divided by time gives us velocity, which is what the slope of a position vs. time graph represents.

2. **The area under a given graph is equivalent to the quantity we get by multiplying the x-axis and the y-axis.** For instance, the y-axis of an acceleration vs. time graph gives us acceleration, and the x-axis gives us time. Acceleration multiplied by time gives us the change in velocity, which is what the area between the graph and the x-axis represents.

We can summarize what we know about graphs in a table:

graph	slope	area under the graph
position vs. time	velocity	-----
velocity vs. time	acceleration	displacement
acceleration vs. time	-----	change in velocity

One-Dimensional Motion with Uniform Acceleration

Many introductory physics problems can be simplified to the special case of uniform motion in one dimension with constant acceleration. That is, most problems will involve objects moving in a straight line whose acceleration doesn't change over time. For such problems, there are five variables that are potentially relevant: the object's position, x; the object's initial velocity, v_0; the object's final velocity, v; the object's acceleration, a; and the elapsed time, t. If you know any three of these variables, you can solve for a fourth. Here are the five **kinematic equations** that you should memorize and hold dear to your heart:

$$x = x_0 + \frac{1}{2}(v + v_0)t$$
$$v = v_0 + at$$
$$x = x_0 + v_0 t + \frac{1}{2}at^2$$
$$x = x_0 + vt - \frac{1}{2}at^2$$
$$v^2 = v_0^2 + 2a(x - x_0)$$

The variable x_0 represents the object's position at $t = 0$. Usually, $x_0 = 0$.

You'll notice there are five equations, each of which contain four of the five variables we mentioned above. In the first equation, a is missing; in the second, x is missing; in the third, v is missing; in the fourth, v_0 is missing; and in the fifth, t is missing. You'll find that in any kinematics problem, you will know three of the five variables, you'll have to solve for a fourth, and the fifth will play no role in the problem. That means you'll have to choose the equation that doesn't contain the variable that is irrelavent to the problem.

Learning to Read Verbal Clues

Problems will often give you variables like t or x, and then give you verbal clues regarding velocity and acceleration. You have to learn to translate such phrases into kinematics-equation-speak:

When They Say . . .	They Mean . . .
". . . starts from rest . . ."	$v_0 = 0$
". . . moves at a constant velocity . . ."	$a = 0$
". . . comes to rest . . . "	$v = 0$

Very often, problems in kinematics on the SAT II will involve a body falling under the influence of gravity. You'll find people throwing balls over their heads, at targets, and even off the leaning tower of Pisa. Gravitational motion is uniformly accelerated motion: the only acceleration involved is the constant pull of gravity, –9.8 m/s^2 toward the center of the Earth. When dealing with this constant, called g, it is often convenient to round it off to –10 m/s^2.

Example

> A student throws a ball up in the air with an initial velocity of 12 m/s and then catches it as it comes back down to him. What is the ball's velocity when he catches it? How high does the ball travel? How long does it take the ball to reach its highest point?

Before we start writing down equations and plugging in numbers, we need to choose a coordinate system. This is usually not difficult, but it is vitally important. Let's make the origin of the system the point where the ball is released from the student's hand and begins its upward journey, and take the up direction to be positive and the down direction to be negative.

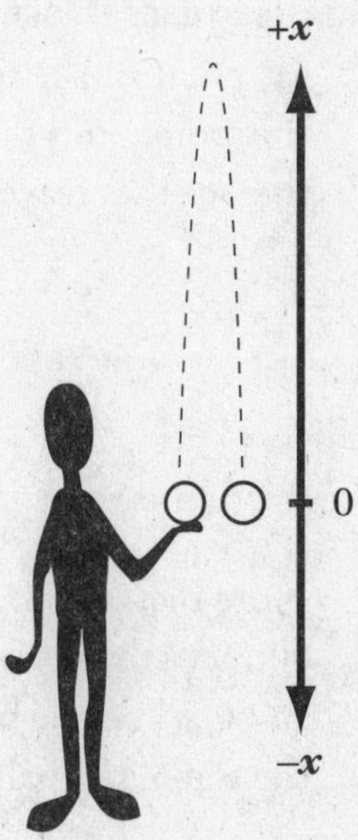

We could have chosen other coordinate systems—for instance, we could have made the origin the ground on which the student is standing—but our choice of coordinate system is convenient because in it, $x_0 = 0$, so we won't have to worry about plugging a value for x_0 into our equation. It's usually possible, and a good idea, to choose a coordinate system that eliminates x_0. Choosing the up direction as positive is simply more intuitive, and thus less likely to lead us astray. It's generally wise also to choose your coordinate system so that more variables will be positive numbers than negative ones, simply because positive numbers are easier to deal with.

Kinematics

What is the ball's velocity when he catches it? We can determine the answer to this question without any math at all. We know the initial velocity, $v_0 = 12$ m/s, and the acceleration due to gravity, $a \approx -10$ m/s^2, and we know that the displacement is $x = 0$ since the ball's final position is back in the student's hand where it started. We need to know the ball's final velocity, v, so we should look at the kinematic equation that leaves out time, t:

$$v^2 = v_0^2 + 2a(x - x_0)$$

Because both x and x_0 are zero, the equation comes out to $v^2 = v_0^2$. But don't be hasty and give the answer as 12 m/s: remember that we devised our coordinate system in such a way that the down direction is negative, so the ball's final velocity is –12 m/s.

How high does the ball travel? We know that at the top of the ball's trajectory its velocity is zero. That means that we know that $v_0 = 12$ m/s, $v = 0$, and $a \approx -10$ m/s^2, and we need to solve for x:

$$\begin{aligned} v^2 &= v_0^2 + 2a(x - x_0) \\ x &= \frac{v^2 - v_0^2}{2a} \\ &= \frac{(-12\text{m/s})}{(2)(-10\text{m/s}^2)} \\ &= 7.2\text{m} \end{aligned}$$

How long does it take the ball to reach its highest point? Having solved for x at the highest point in the trajectory, we now know all four of the other variables related to this point, and can choose any one of the five equations to solve for t. Let's choose the one that leaves out x:

$$\begin{aligned} v &= v_0 + at \\ t &= \frac{v - v_0}{a} \\ &= \frac{-12\text{m/s}}{-10\text{m/s}^2} \\ &= 1.2\text{s} \end{aligned}$$

Note that there are certain convenient points in the ball's trajectory where we can extract a third variable that isn't mentioned explicitly in the question: we know that $x = 0$ when the ball is at the level of the student's hand, and we know that $v = 0$ at the top of the ball's trajectory.

Two-Dimensional Motion With Uniform Acceleration

If you've got the hang of 1-D motion, you should have no trouble at all with 2-D motion. The motion of any object moving in two dimensions can be broken into x- and y-components. Then it's just a matter of solving two separate 1-D kinematic equations.

The most common problems of this kind on the SAT II involve projectile motion: the motion of an object that is shot, thrown, or in some other way launched into the air. Note that the motion or trajectory of a projectile is a parabola.

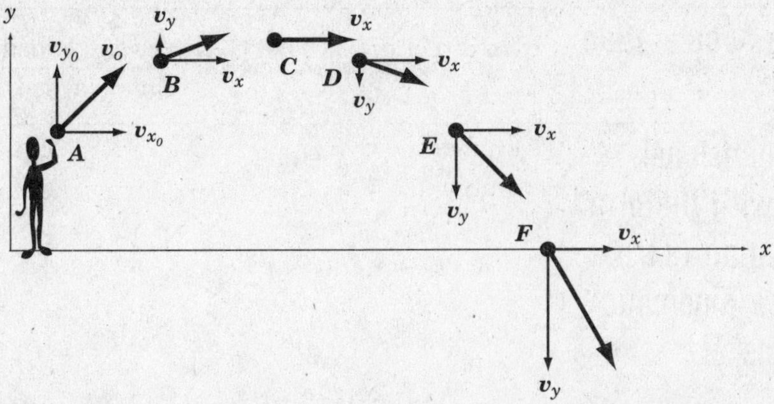

If we break this motion into x- and y-components, the motion becomes easy to understand. In the y direction, the ball is thrown upward with an initial velocity of v_{y_0} and experiences a constant downward acceleration of $g = -9.8$ m/s^2. This is exactly the kind of motion we examined in the previous section: if we ignore the x-component, the motion of a projectile is identical to the motion of an object thrown directly up in the air.

In the x direction, the ball is thrown forward with an initial velocity of v_{x_0} and there is no acceleration acting in the x direction to change this velocity. We have a very simple situation where $a_x = 0$ and v_x is constant.

The SAT II will probably not expect you to do much calculating in questions dealing with projectile motion. Most likely, it will ask about the relative velocity of the projectile at different points in its trajectory. We can calculate the x- and y-components separately and then combine them to find the velocity of the projectile at any given point:

$$v = \sqrt{v_x^2 + v_y^2}$$

Because v_x is constant, the speed will be greater or lesser depending on the magnitude of v_y. To determine where the speed is least or greatest, we follow the same method as we would with the one-dimensional example we had in the previous section. That means that the speed of the projectile in the figure above is at its greatest at position F, and at its least at position C. We also know that the speed is equal at position B and position D, and at position A and position E.

The key with two-dimensional motion is to remember that you are not dealing with one complex equation of motion, but rather with two simple equations.

Key Formulas

Average Speed	$\text{average speed} = \dfrac{\text{distance traveled}}{\text{time elapsed}} = \dfrac{\Delta x}{\Delta t}$
Average Velocity	$\text{average velocity} = \dfrac{\text{change in displacement}}{\text{time elapsed}} = \dfrac{\Delta \mathbf{s}}{\Delta t}$
Average Acceleration	$\text{average acceleration} = \dfrac{\text{change in velocity}}{\text{time elapsed}} = \dfrac{\Delta \mathbf{v}}{\Delta t}$
One-Dimensional Motion with Uniform Acceleration (a.k.a. "The Five Kinematic Equations")	$x = x_0 + \frac{1}{2}(v + v_0)t$ $v = v_0 + at$ $x = x_0 + v_0 t + \frac{1}{2}at^2$ $x = x_0 + vt - \frac{1}{2}at^2$ $v^2 = v_0^2 + 2a(x - x_0)$
Velocity of Two-Dimensional Projectiles	$v = \sqrt{v_x^2 + v_y^2}$

Practice Questions

1. An athlete runs four laps of a 400 m track. What is the athlete's total displacement?

   (A) –1600 m
   (B) –400 m
   (C) 0 m
   (D) 400 m
   (E) 1600 m

2. Which of the following statements contains a reference to displacement?

   I. "The town is a five mile drive along the winding country road."
   II. "The town sits at an altitude of 940 m."
   III. "The town is ten miles north, as the crow flies."

   (A) I only
   (B) III only
   (C) I and III only
   (D) II and III only
   (E) I, II, and III

Kinematics

Questions 3 and 4 refer to a car that travels from point A to point B in four hours, and then from point B back to point A in six hours. The road between point A and point B is perfectly straight, and the distance between the two points is 240 km.

3. What is the car's average velocity?

(A) 0 km/h
(B) 48 km/h
(C) 50 km/h
(D) 60 km/h
(E) 100 km/h

4. What is the car's average speed?

(A) 0 km/h
(B) 48 km/h
(C) 50 km/h
(D) 60 km/h
(E) 100 km/h

5. A ball is dropped from the top of a building. Taking air resistance into account, which best describes the speed of the ball while it is moving downward?

(A) It will increase until it reaches the speed of light
(B) It will increase at a steady rate
(C) It will remain constant
(D) It will decrease
(E) Its rate of acceleration will decrease until the ball moves at a constant speed

6. A car accelerates steadily so that it goes from a velocity of 20 m/s to a velocity of 40 m/s in 4 seconds. What is its acceleration?

(A) 0.2 m/s^2 (B) 4 m/s^2 (C) 5 m/s^2 (D) 10 m/s^2 (E) 80 m/s^2

Questions 7 and 8 relate to the graph of velocity vs. time of a moving particle plotted at right.

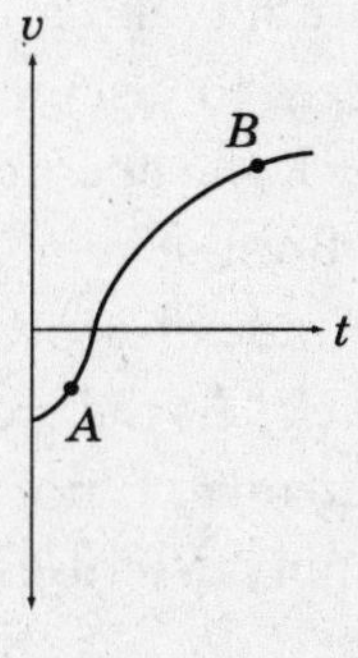

7. What is the acceleration and displacement of the particle at point A?

(A) Acceleration decreasing, displacement decreasing
(B) Acceleration constant, displacement decreasing
(C) Acceleration increasing, displacement decreasing
(D) Acceleration decreasing, displacement increasing
(E) Acceleration increasing, displacement increasing

8. How do the acceleration and displacement of the particle at point B compare to the acceleration and displacement of the particle at point A?

(A) Acceleration is less, displacement is less
(B) Acceleration is less, displacement is the same
(C) Acceleration is less, displacement is greater
(D) Acceleration is greater, displacement is less
(E) Acceleration is greater, displacement is greater

Kinematics

9. A sprinter starts from rest and accelerates at a steady rate for the first 50 m of a 100 m race, and then continues at a constant velocity for the second 50 m of the race. If the sprinter runs the 100 m in a time of 10 s, what is his instantaneous velocity when he crosses the finish line?

(A) 5 m/s
(B) 10 m/s
(C) 12 m/s
(D) 15 m/s
(E) 20 m/s

10. A woman runs 40 m to the north in 6.0 s, and then 30 m to the east in 4.0 s. What is the magnitude of her average velocity?

(A) 5.0 m/s
(B) 6.0 m/s
(C) 6.7 m/s
(D) 7.0 m/s
(E) 7.5 m/s

Explanations

1. (C)

Displacement is a vector quantity that measures the distance between the starting point and ending point, not taking the actual path traveled into account. At the end of four laps, the athlete will be back at the starting line for the track, so the athlete's total displacement will be zero.

2. (D)

Displacement is a vector quantity, measuring the separation between two points without taking into account the route taken between those two points. Statement I refers to distance, not displacement, since the five-mile distance is along a winding road, and does not describe a straight-line path.

Both statements II and III, however, contain a reference to displacement. The altitude of a town is a measure of the straight-line distance between the town and sea level. "As the crow flies" is a common way of saying "in a straight-line path." Neither statement II nor III describes a certain route between the two points in question: they simply describe how far apart those two points are.

3. (A)

Average velocity is a measure of total displacement divided by total time. Total displacement is the distance separating the starting point and the finishing point. Since the car both starts and finishes at point A, its total displacement is zero, so its average velocity is also zero.

4. (B)

Average speed is a measure of total distance traveled divided by the total time of the trip. Solving this problem calls for a single calculation:

$$\text{speed} = \frac{\text{total distance}}{\text{total time}} = \frac{480 \text{ km}}{10 \text{ hr}}$$
$$= 48 \text{ km/hr}$$

5. (E)

The force of air resistance against a ball increases as the ball accelerates. At a certain point, the force of air resistance will be equal to the force of gravity, and the net force acting on the ball will be zero. At this point, its velocity will remain constant. This velocity is known as an object's "terminal velocity," and it explains why, in real life, many falling objects don't continue accelerating all the way to the ground.

6. (C)

Acceleration is a measure of the change in velocity over time. The car's change in velocity is 40 – 20 = 20 m/s. Since this change in velocity takes place over 4 seconds, the car's acceleration is

$$\frac{20 \text{ m/s}}{4 \text{ s}} = 5 \text{ m/s}^2$$

7. (C)

Point *A* is below the *t*-axis, which means that the velocity is negative. Since velocity is the change in displacement over time, we can conclude that if the velocity is negative, then the displacement is decreasing.

Acceleration is given by the slope of the graph. Since the line at point *A* has a positive slope, we know that the acceleration is increasing.

8. (C)

Acceleration is given by the slope of the line. As we can see, the slope is greater at point *A* than at point *B*, so the acceleration is less at point *B*.

The change in displacement is given by the area between the graph and the *t*-axis:

As we can see, between points A and B, a great deal more of the graph is above the t-axis than below it. This means that, overall, displacement is positive between these two points.

9. (D)

We know the total distance the sprinter covers, and we know the total time. However, since the acceleration isn't uniform, we can't calculate the velocity quite so simply. Rather, we need two equations, one for the first 50 meters of the race, and another for the second 50 meters. In the first fifty meters, the sprinter accelerates from an initial velocity of $v_0 = 0$ to a final velocity of v in an amount of time, t_1. We can express this relationship using the kinematic equation that leaves out velocity, and then solve for t:

$$x = x_0 + \frac{1}{2}(v + v_0)t_1$$
$$50 \text{ m} = \frac{1}{2}vt_1$$
$$t_1 = \frac{100\text{m}}{v}$$

In the last 50 meters of the race, the sprinter runs with a constant velocity of v, covering a distance of $x = 50$ m in a time t_2. Solving for t_2, we find:

$$t_2 = \frac{50 \text{ m}}{v}$$

We know that the total time of the race, $t_1 + t_2 = 10$ s. With this in mind, we can add the two sprint times together and solve for v:

$$10 \text{ s} = \frac{100 \text{ m}}{v} + \frac{50 \text{ m}}{v} = \frac{150 \text{ m}}{v}$$
$$v = \frac{150 \text{ m}}{10 \text{ s}} = 15 \text{ m/s}$$

10. (A)

Average velocity is given by the total displacement divided by the total time elapsed. The displacement is not simply 30 + 40 = 70 m, however, since the woman doesn't run in a straight line path. The 40 m north and the 30 m east are at right angles to one another, so we can use the Pythagorean theorem to determine that the total displacement is in fact 50 m. Her displacement is 50 m over a total time of 10 s, so her average velocity is 5.0 m/s.

Dynamics

Chapter Contents

What Are Forces? **67**
Newton's Laws **68**
Problem Solving with Newton's Laws **72**
Types of Forces **74**
Key Formulas **80**
Practice Questions **80**
Explanations **83**

WHEREAS KINEMATICS IS THE STUDY OF objects in motion, **dynamics** is the study of the *causes* of motion. In other words, kinematics covers the "what" of motion, while dynamics covers the "how" and "why." **Forces** are the lifeblood of dynamics: objects move and change their motion under the influence of different forces. Our main emphasis will be on Newton's three laws, which succinctly summarize everything you need to know about dynamics.

Dynamics questions on the SAT II often call upon your knowledge of kinematics and vectors, but these questions will probably be simpler than the problems you've encountered in your physics class. Because you won't be asked to do any math that would require a calculator, you should focus on mastering the concepts that lie behind the math.

What Are Forces?

Whenever we lift something, push something, or otherwise manipulate an object, we are exerting a force. A force is defined very practically as a push or a pull—essentially it's what makes things move. A force is a vector quantity, as it has both a magnitude and a direction.

In this chapter, we will use the example of pushing a box along the floor to illustrate many concepts about forces, with the assumption that it's a pretty intuitive model that you will have little trouble imagining. If not, rest assured you will acquire *loads* of experience when you move into college!

Physicists use simple pictures called **free-body diagrams** to illustrate the forces acting on an object. In these diagrams, the forces acting *on* a body are drawn as vectors originating from the center of the object. Following is a free-body diagram of you pushing a box into your new college dorm with force $\boldsymbol{F}$.

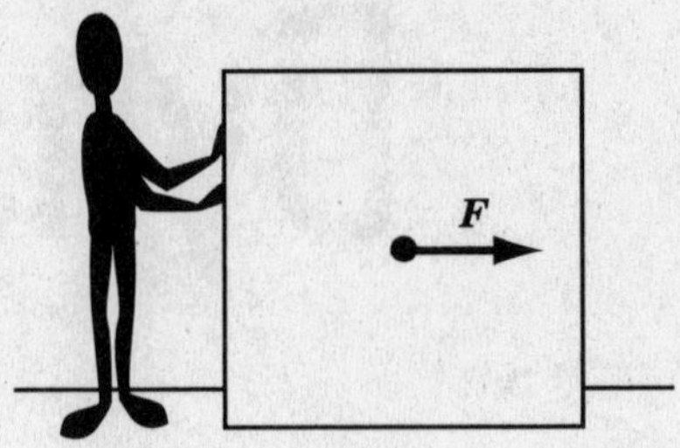

Because force is a vector quantity, it follows the rules of vector addition. If your evil roommate comes and pushes the box in the opposite direction with exactly the same magnitude of force (force $-F$), the net force on the box is zero.

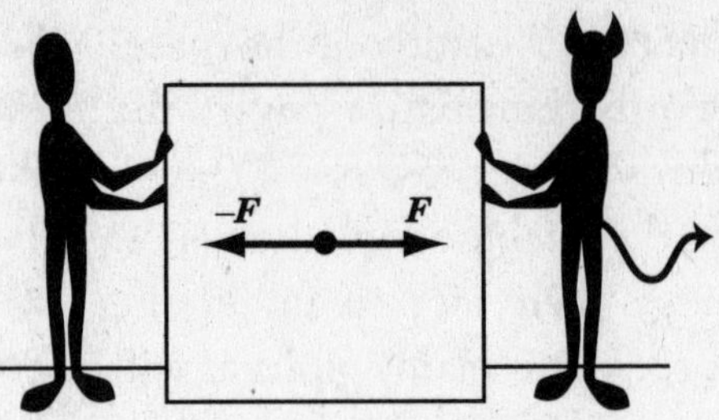

Newton's Laws

Isaac Newton first published his three laws of motion in 1687 in his monumental *Mathematical Principles of Natural Philosophy.* In these three simple laws, Newton sums up everything there is to know about dynamics. This achievement is just one of the many reasons why he is considered one of the greatest physicists in history.

While a multiple-choice exam can't ask you to write down each law in turn, there is a good chance you will encounter a problem where you are asked to choose which of Newton's laws best explains a given physical process. You will also be expected to make simple calculations based on your knowledge of these laws. But by far the most important reason for mastering Newton's laws is that, without them, thinking about dynamics is impossible. For that reason, we will dwell at some length on describing how these laws work qualitatively.

Newton's First Law

Newton's First Law describes how forces relate to motion:

> *An object at rest remains at rest, unless acted upon by a net force. An object in motion remains in motion, unless acted upon by a net force.*

A soccer ball standing still on the grass does not move until someone kicks it. An ice hockey puck will continue to move with the same velocity until it hits the boards, or someone else hits it. Any change in the velocity of an object is evidence of a net force acting on that object. A world without forces would be much like the images we see of the insides of spaceships, where astronauts, pens, and food float eerily about.

Remember, since velocity is a vector quantity, a change in velocity can be a change either in the magnitude or the direction of the velocity vector. A moving object upon which no net force is acting doesn't just maintain a constant speed—it also moves in a straight line.

But what does Newton mean by a *net* force? The net force is the sum of the forces acting on a body. Newton is careful to use the phrase "net force," because an object at rest will stay at rest if acted upon by forces with a sum of zero. Likewise, an object in motion will retain a constant velocity if acted upon by forces with a sum of zero.

Consider our previous example of you and your evil roommate pushing with equal but opposite forces on a box. Clearly, force is being applied to the box, but the two forces on the box cancel each other out exactly: $\boldsymbol{F} + -\boldsymbol{F} = 0$. Thus the net force on the box is zero, and the box does not move.

Yet if your other, good roommate comes along and pushes alongside you with a force $\boldsymbol{R}$, then the tie will be broken and the box will move. The net force is equal to:

$$\boldsymbol{F} + -\boldsymbol{F} + \boldsymbol{R} = \boldsymbol{R}$$

Note that the velocity of the box, $\boldsymbol{v}$, is in the same direction as the net force.

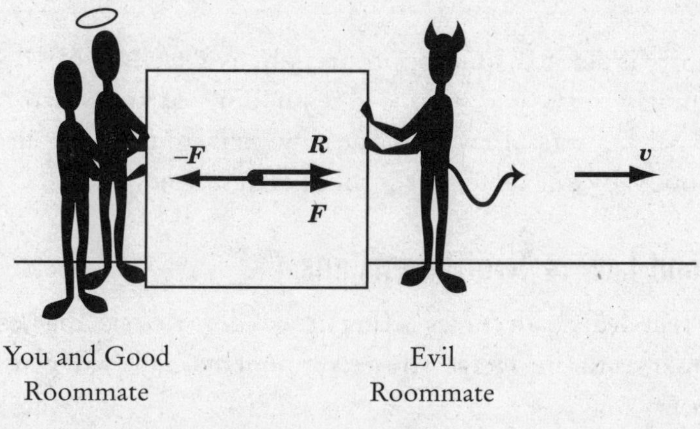

Inertia

The First Law is sometimes called the law of **inertia**. We define inertia as the tendency of an object to remain at a constant velocity, or its resistance to being accelerated. Inertia is a fundamental property of all matter and is important to the definition of **mass**.

Newton's Second Law

To understand **Newton's Second Law**, you must understand the concept of mass. Mass is an intrinsic scalar quantity: it has no direction and is a property of an object, not of the object's location. Mass is a measurement of a body's inertia, or its resistance to being accelerated. The words "mass" and "matter" are related: a handy way of thinking about mass is as a measure of how much matter there is in an object, how much "stuff" it's made out of. Although in everyday language we use the words "mass" and "**weight**" interchangeably, they refer to two different, but related, quantities in physics. We will expand upon the relation between mass and weight later in this chapter, after we have finished our discussion of Newton's laws.

We already have some intuition from everyday experience as to how mass, force, and acceleration relate. For example, we know that the more force we exert on a bowling ball, the faster it will roll. We also know that if the same force were exerted on a basketball, the basketball would move faster than the bowling ball because the basketball has less mass. This intuition is quantified in Newton's Second Law:

$$\boldsymbol{F} = m\boldsymbol{a}$$

Stated verbally, Newton's Second Law says that the net force, $\boldsymbol{F}$, acting on an object causes the object to accelerate, $\boldsymbol{a}$. Since $\boldsymbol{F} = m\boldsymbol{a}$ can be rewritten as $\boldsymbol{a} = \boldsymbol{F}/m$, you can see that the magnitude of the acceleration is directly proportional to the net force and inversely proportional to the mass, m. Both force and acceleration are vector quantities, and the acceleration of an object will always be in the *same* direction as the net force.

The unit of force is defined, quite appropriately, as a **newton** (N). Because acceleration is given in units of m/s^2 and mass is given in units of kg, Newton's Second Law implies that $1\ N = 1\ kg \cdot m/s^2$. In words, one newton is the force required to accelerate a one-kilogram body, by one meter per second, each second.

Newton's Second Law in Two Dimensions

With a problem that deals with forces acting in two dimensions, the best thing to do is to break each force vector into its x- and y-components. This will give you two equations instead of one:

$$\boldsymbol{F}_x = m\boldsymbol{a}_x$$
$$\boldsymbol{F}_y = m\boldsymbol{a}_y$$

The component form of Newton's Second Law tells us that the component of the net force in the $\boldsymbol{x}$ direction is directly proportional to the resulting component of the acceleration in the $\boldsymbol{x}$ direction, and likewise for the y-component.

Dynamics

Newton's Third Law

Newton's Third Law has become a cliché. The Third Law tells us that:

To every action, there is an equal and opposite reaction.

What this tells us in physics is that every push or pull produces not one, but two forces. In any exertion of force, there will always be two objects: the object exerting the force and the object on which the force is exerted. Newton's Third Law tells us that when object A exerts a force $\boldsymbol{F}$ on object B, object B will exert a force $-\boldsymbol{F}$ on object A. When you push a box forward, you also feel the box pushing back on your hand. If Newton's Third Law did not exist, your hand would feel nothing as it pushed on the box, because there would be no reaction force acting on it.

Anyone who has ever played around on skates knows that when you push forward on the wall of a skating rink, you recoil backwards.

skater pushes on wall — skater recoils off wall

Newton's Third Law tells us that the force that the skater exerts on the wall, $\boldsymbol{F}_{skater}$, is exactly equal in magnitude and opposite in direction to the force that the wall exerts on the skater, $\boldsymbol{F}_{wall}$. The harder the skater pushes on the wall, the harder the wall will push back, sending the skater sliding backward.

Newton's Third Law at Work

Here are three other examples of Newton's Third Law at work, variations of which often pop up on the SAT II:

> You push down with your hand on a desk, and the desk pushes upwards with a force equal in magnitude to your push.
>
> A brick is in free fall. The brick pulls the Earth upward with the same force that the Earth pulls the brick downward.
>
> When you walk, your feet push the Earth backward. In response, the Earth pushes your feet forward, which is the force that moves you on your way.

Dynamics

The second example may seem odd: the Earth doesn't move upward when you drop a brick. But recall Newton's Second Law: the acceleration of an object is inversely proportional to its mass ($\boldsymbol{a} = \boldsymbol{F}/\text{m}$). The Earth is about 10^{24} times as massive as a brick, so the brick's downward acceleration of -9.8 m/s^2 is about 10^{24} times as great as the Earth's upward acceleration. The brick exerts a force on the Earth, but the effect of that force isn't noticeable. For the same reason, we don't notice the earth moving backward when we take a walk.

Problem Solving with Newton's Laws

Dynamics problem solving in physics class usually involves difficult calculations that take into account a number of vectors on a free-body diagram. The SAT II won't expect you to make any difficult calculations, and the test will usually include the free-body diagrams that you need. Your task will usually be to interpret free-body diagrams rather than to draw them.

Example 1

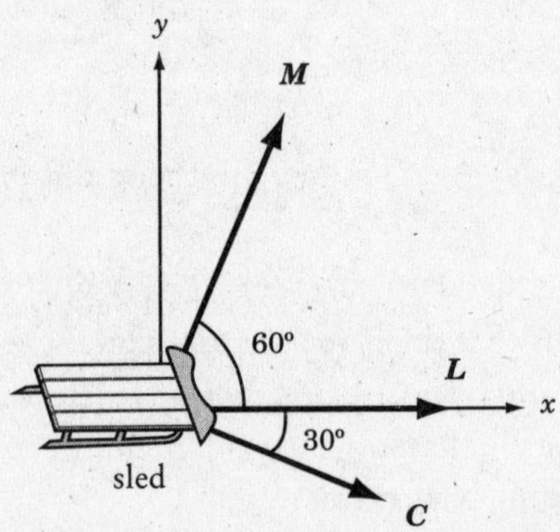

The three stooges are dragging a 10 kg sled across a frozen lake. Moe pulls with force $\boldsymbol{M}$, Larry pulls with force $\boldsymbol{L}$, and Curly pulls with force $\boldsymbol{C}$. If the sled is moving in the $\boldsymbol{x}$ direction, and both Moe and Larry are exerting a force of 10 N, what is the magnitude of the force Curly is exerting? Assuming that friction is negligible, what is the acceleration of the sled? (Note: sin 30 = cos 60 = 0.500 and sin 60 = cos 30 = 0.866.)

The figure above gives us a free-body diagram that shows us the direction in which all forces are acting, but we should be careful to note that vectors in the diagram are not drawn to scale: we cannot estimate the magnitude of $\boldsymbol{C}$ simply by comparing it to $\boldsymbol{M}$ and $\boldsymbol{L}$.

What is the magnitude of the force Curly is exerting? Since we know that the motion of the sled is in the $\boldsymbol{x}$ direction, the net force, $\boldsymbol{M} + \boldsymbol{L} + \boldsymbol{C}$, must also be in the $\boldsymbol{x}$ direction. And since the sled is not moving in the $\boldsymbol{y}$ direction, the y-component of the net force must be zero. Because the y-component of Larry's force is zero, this implies:

$$\boldsymbol{M}_y + \boldsymbol{C}_y = 0$$

where $\boldsymbol{M}_y$ is the y-component of $\boldsymbol{M}$ and $\boldsymbol{C}_y$ is the y-component of $\boldsymbol{C}$. We also know:

$$M_y = M \sin\theta = (10 \text{ N}) \sin 60 = 8.660 \text{ N}$$
$$C_y = C \sin\theta = C \sin -30 = -0.500C$$

If we substitute these two equations for $\boldsymbol{M}_y$ and $\boldsymbol{C}_y$ into the equation $\boldsymbol{M}_y + \boldsymbol{C}_y = 0$, we have:

$$8.660 \text{ N} - 0.500C = 0$$
$$C = 4.33 \text{ N}$$

What is the acceleration of the sled? According to Newton's Second Law, the acceleration of the sled is $\boldsymbol{a} = \boldsymbol{F}/\text{m}$. We know the sled has a mass of 10 kg, so we just need to calculate the magnitude of the net force in the $\boldsymbol{x}$-direction.

$$\begin{aligned} M_x + L_x + C_x &= (10 \text{ N}) \cos 60 + 10 \text{ N} + (4.33 \text{ N}) \cos -30 \\ &= 0.500(10 \text{ N}) + 10 \text{ N} + 0.866(4.33 \text{ N}) \\ &= 18.750 \text{ N} \end{aligned}$$

Now that we have calculated the magnitude of the net force acting on the sled, a simple calculation can give us the sled's acceleration:

$$\begin{aligned} \boldsymbol{a} &= \frac{\boldsymbol{F}}{m} = \frac{18.750 \text{ N}}{10 \text{ kg}} \\ &= 1.875 \text{ m/s}^2 \end{aligned}$$

We have been told that the sled is moving in the $\boldsymbol{x}$ direction, so the acceleration is also in the $\boldsymbol{x}$ direction.

This example problem illustrates the importance of vector components. For the SAT II, you will need to break vectors into components on any problem that deals with vectors that are not all parallel or perpendicular. As with this example, however, the SAT II will always provide you with the necessary trigonometric values.

Example 2

Each of the following free-body diagrams shows the forces, *F*, acting on a particle and the particle's velocity, *v*. All forces represented in the diagrams are of the same magnitude.

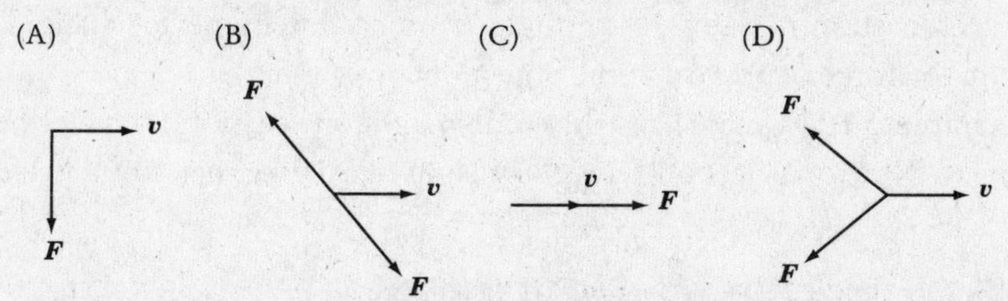

1. **In which diagram is neither the speed nor the direction of the particle being changed?**
2. **In which diagram is the speed but not the direction of the particle being changed?**
3. **In which diagram is the direction but not the speed of the particle being changed?**
4. **In which diagram are both the speed and direction of the particle being changed?**

Dynamics

The answer to question 1 is (B). The two forces in that diagram cancel each other out, so the net force on the particle is zero. The velocity of a particle only changes under the influence of a net force. The answer to question 2 is (C). The net force is in the same direction as the particle's motion, so the particle continues to accelerate in the same direction. The answer to question 3 is (A). Because the force is acting perpendicular to the particle's velocity, it does not affect the particle's speed, but rather acts to pull the particle in a circular orbit. Note, however, that the speed of the particle only remains constant if the force acting on the particle remains perpendicular to it. As the direction of the particle changes, the direction of the force must also change to remain perpendicular to the velocity. This rule is the essence of circular motion, which we will examine in more detail later in this book. The answer to question 4 is (D). The net force on the particle is in the opposite direction of the particle's motion, so the particle slows down, stops, and then starts accelerating in the opposite direction.

Types of Forces

There are a number of forces that act in a wide variety of cases and have been given specific names. Some of these, like friction and the normal force, are so common that we're hardly aware of them as distinctive forces. It's important that you understand how and when these forces function, because questions on the SAT II often make no mention of them explicitly, but expect you to factor them into your calculations. Some of these forces will also play an important role in the chapter on special problems in mechanics.

Weight

Although the words "weight" and "mass" are often interchangeable in everyday language, these words refer to two different quantities in physics. The mass of an object is a property of the object itself, which reflects its resistance to being accelerated. The weight of an object is a measure of the gravitational force being exerted upon it, and so it varies depending on the gravitational force acting on the object. Mass is a scalar quantity measured in kilograms, while weight is a vector quantity measuring force, and is represented in newtons. Although an object's mass never changes, its weight depends on the force of gravity in the object's environment.

For example, a 10 kg mass has a different weight on the moon than it does on Earth. According to Newton's Second Law, the weight of a 10 kg mass on Earth is

$$\begin{aligned} \boldsymbol{F} &= m\boldsymbol{g}_{\text{earth}} \\ &= (10 \text{ kg})(9.8 \text{ m/s}^2) \\ &= 98 \text{ N} \end{aligned}$$

This force is directed toward the center of the Earth. On the moon, the acceleration due to gravity is roughly one-sixth that on Earth. Therefore, the weight of a 10 kg mass on the moon is only about 16.3 N toward the center of the moon.

The Normal Force

The **normal force** always acts perpendicular (or "normal") to the surface of contact between two objects. The normal force is a direct consequence of Newton's Third Law. Consider the example of a 10 kg box resting on the floor. The force of gravity causes the box to push down upon the ground with a force, ***W***, equal to the box's weight. Newton's Third Law dictates that the floor must apply an equal and opposite force, $\mathbf{N} = -\mathbf{W}$, to the box. As a result, the net force on the box is zero and, as we would expect, the box remains at rest. If there were no normal force pushing the box upward, there would be a net force acting downward on the box, and the box would accelerate downward.

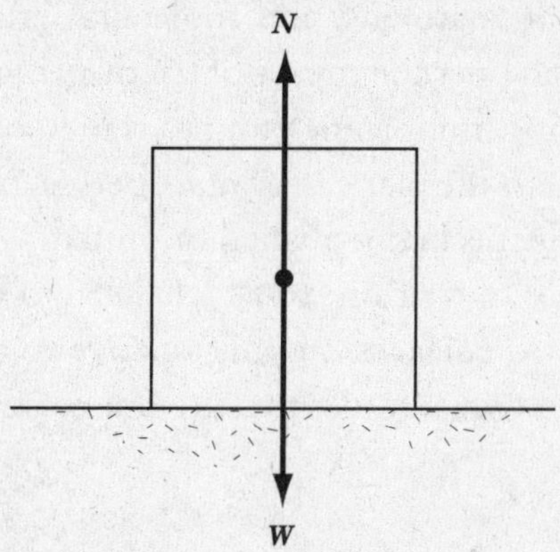

Be careful not to confuse the normal force vector ***N*** with the abbreviation for newtons, N. It can be a bit confusing that both are denoted by the same letter of the alphabet, but they are two totally different entities.

Example

> A person pushes downward on a box of weight ***W*** with a force ***F***. What is the normal force, ***N***, acting on the box?

The total force pushing the box toward the ground is $\mathbf{W} + \mathbf{F}$. From Newton's Third Law, the normal force exerted on the box by the floor has the same magnitude as $\mathbf{W} + \mathbf{F}$ but is directed upwards. Therefore, the net force on the box is zero and the box remains at rest.

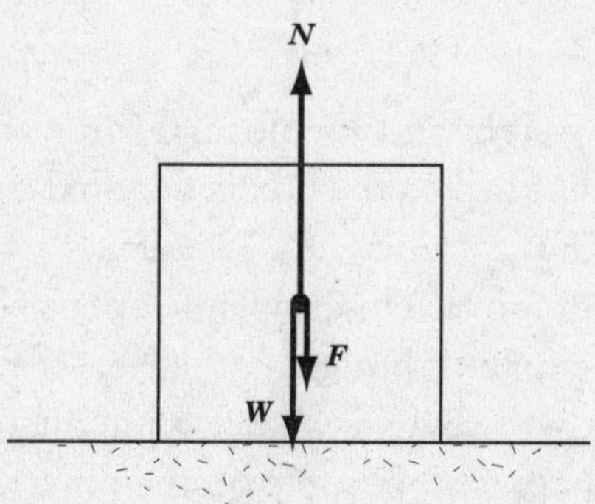

Dynamics

Friction

Newton's First Law tells us that objects in motion stay in motion unless a force is acting upon them, but experience tells us that when we slide coins across a table, or push boxes along the floor, they slow down and come to a stop. This is not evidence that Newton was wrong; rather, it shows that there is a force acting upon the coin or the box to slow its motion. This is the force of **friction**, which is at work in every medium but a vacuum, and is the bugbear of students pushing boxes across the sticky floors of dorm rooms everywhere.

Roughly speaking, frictional forces are caused by the roughness of the materials in contact, deformations in the materials, and molecular attraction between materials. You needn't worry too much over the causes of friction, though: the SAT II isn't going to test you on them. The most important thing to remember about frictional forces is that they are always parallel to the plane of contact between two surfaces, and opposite to the direction that the object is being pushed or pulled.

There are two main types of friction: **static friction** and **kinetic friction**. Kinetic friction is the force between two surfaces moving relative to one another, whereas static friction is the force between two surfaces that are not moving relative to one another.

Static Friction

Imagine, once more, that you are pushing a box along a floor. When the box is at rest, it takes some effort to get it to start moving at all. That's because the force of static friction is resisting your push and holding the box in place.

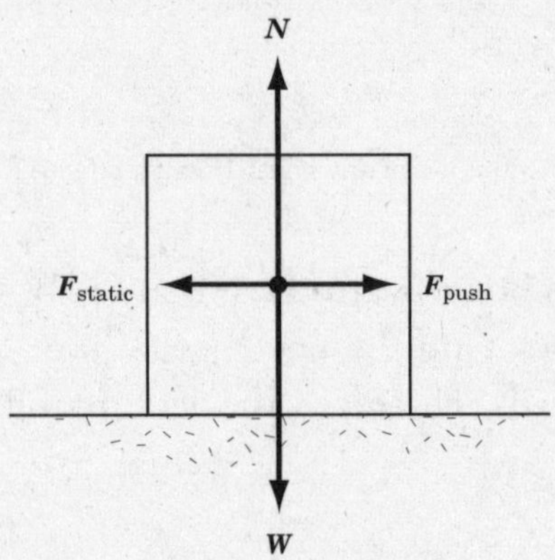

In the diagram above, the weight and the normal force are represented as $\boldsymbol{W}$ and $\boldsymbol{N}$ respectively, and the force applied to the box is denoted by $\boldsymbol{F}_{push}$. The force of static friction is represented by $\boldsymbol{F}_{static}$, where $\boldsymbol{F}_{static} = \boldsymbol{F}_{push}$. The net force on the box is zero, and so the box does not move. This is what happens when you are pushing on the box, but not hard enough to make it budge.

Static friction is only at work when the net force on an object is zero, and hence when $\boldsymbol{F}_{static} = -\boldsymbol{F}_{push}$. If there is a net force on the object, then that object will be in motion, and kinetic rather than static friction will oppose its motion.

Kinetic Friction

The force of static friction will only oppose a push up to a point. Once you exert a strong enough force, the box will begin to move. However, you still have to keep pushing with a strong, steady force to keep it moving along, and the box will quickly slide to a stop if you quit pushing. That's because the force of kinetic friction is pushing in the opposite direction of the motion of the box, trying to bring it to a rest.

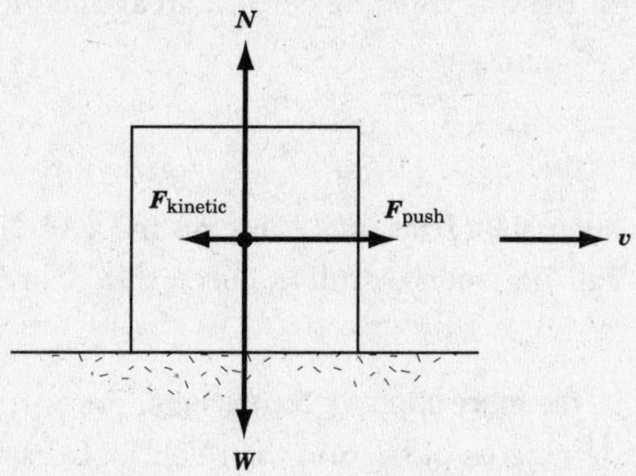

Though the force of kinetic friction will always act in the opposite direction of the force of the push, it need not be equal in magnitude to the force of the push. In the diagram above, the magnitude of $\boldsymbol{F}_{\text{kinetic}}$ is less than the magnitude of $\boldsymbol{F}_{\text{push}}$. That means that the box has a net force in the direction of the push, and the box accelerates forward. The box is moving at velocity $\boldsymbol{v}$ in the diagram, and will speed up if the same force is steadily applied to it. If $\boldsymbol{F}_{\text{push}}$ were equal to $-\boldsymbol{F}_{\text{kinetic}}$, the net force acting on the box would be zero, and the box would move at a steady velocity of $\boldsymbol{v}$, since Newton's First Law tells us that an object in motion will remain in motion if there is no net force acting on it. If the magnitude of $\boldsymbol{F}_{\text{push}}$ were less than the magnitude of $\boldsymbol{F}_{\text{kinetic}}$, the net force would be acting against the motion, and the box would slow down until it came to a rest.

The Coefficients of Friction

The amount of force needed to overcome the force of static friction on an object, and the magnitude of the force of kinetic friction on an object, are both proportional to the normal force acting on the object in question. We can express this proportionality mathematically as follows:

$$F_{\text{kinetic}} = \mu_k N = \mu_k mg$$
$$F_{\text{static, max}} = \mu_s N = \mu_s mg$$

where μ_k is the **coefficient of kinetic friction**, μ_s is the **coefficient of static friction**, and N is the magnitude of the normal force. The coefficients of kinetic and static friction are constants of proportionality that vary from object to object.

Dynamics

Note that the equation for static friction is for the *maximum* value of the static friction. This is because the force of static friction is never greater than the force pushing on an object. If a box has a mass of 10 kg and $\mu_s = 0.5$, then:

$$F_{\text{static, max}} = 0.5(10\text{ kg})(9.8\text{ m/s}^2) = 49\text{ N}$$

If you push this box with a force less than 49 newtons, the box will not move, and consequently the net force on the box must be zero. If an applied force $\boldsymbol{F}_{\text{push}}$ is less than $\boldsymbol{F}_{\text{static, max}}$, then $\boldsymbol{F}_{\text{static}} = -\boldsymbol{F}_{\text{push}}$.

Three Reminders

Whenever you need to calculate a frictional force on the SAT II, you will be told the value of μ, which will fall between 0 and 1. Three things are worth noting about frictional forces:

1. **The smaller μ is, the more slippery the surface.** For instance, ice will have much lower coefficients of friction than Velcro. In cases where $\mu = 0$, the force of friction is zero, which is the case on ideal frictionless surfaces.
2. **The coefficient of kinetic friction is smaller than the coefficient of static friction.** That means it takes more force to start a stationary object moving than to keep it in motion. The reverse would be illogical: imagine if you could push on an object with a force greater than the maximum force of static friction but less than the force of kinetic friction. That would mean you could push it hard enough to get it to start moving, but as soon as it starts moving, the force of kinetic friction would push it backward.
3. **Frictional forces are directly proportional to the normal force.** That's why it's harder to slide a heavy object along the floor than a light one. A light coin can slide several meters across a table because the kinetic friction, proportional to the normal force, is quite small.

Dynamics

Example

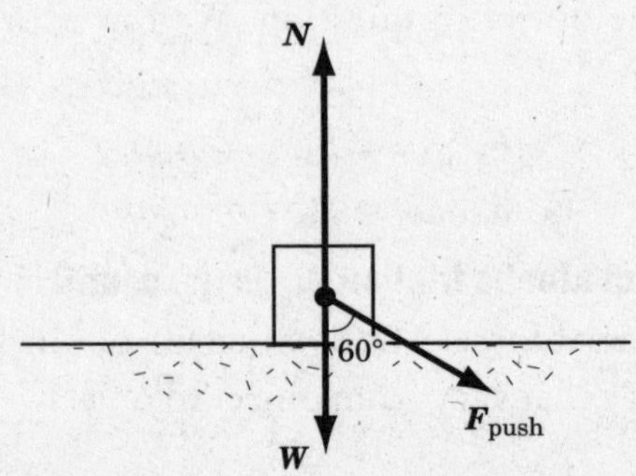

A student pushes a box that weighs 15 N with a force of 10 N at a 60° angle to the perpendicular. The maximum coefficient of static friction between the box and the floor is 0.4. Does the box move? Note that sin 60° = 0.866 and cos 60° = 0.500.

In order to solve this problem, we have to determine whether the horizontal component of $\boldsymbol{F}_{\text{push}}$ is of greater magnitude than the maximum force of static friction.

We can break the $\boldsymbol{F}_{\text{push}}$ vector into horizontal and vertical components. The vertical component will push the box harder into the floor, increasing the normal force, while the horizontal component will push against the force of static friction. First, let's calculate the vertical component of the force so that we can determine the normal force, $\boldsymbol{N}$, of the box:

$$\boldsymbol{F}_{\text{push, y}} = \boldsymbol{F}_{\text{push}} \cos 60° = 0.500(10 \text{ N}) = 5.0 \text{ N}$$

If we add this force to the weight of the box, we find that the normal force is 15 + 5.0 = 20 N. Thus, the maximum force of static friction is:

$$\boldsymbol{F}_{\text{static, max}} = \mu_s N = 0.4(20 \text{ N}) = 8.0 \text{ N}$$

The force pushing the box forward is the horizontal component of $\boldsymbol{F}_{\text{push}}$, which is:

$$\boldsymbol{F}_{\text{push, x}} = F_{\text{push}} \sin 60° = 0.866(10 \text{ N}) = 8.66 \text{ N}$$

As we can see, this force is just slightly greater than the maximum force of static friction opposing the push, so the box will slide forward.

Tension

Consider a box being pulled by a rope. The person pulling one end of the rope is not in contact with the box, yet we know from experience that the box will move in the direction that the rope is pulled. This occurs because the force the person exerts on the rope is transmitted to the box.

The force exerted on the box from the rope is called the **tension** force, and comes into play whenever a force is transmitted across a rope or a cable. The free-body diagram below shows us a box being pulled by a rope, where $\boldsymbol{W}$ is the weight of the box, $\boldsymbol{N}$ is the normal force, $\boldsymbol{T}$ is the tension force, and $\boldsymbol{F}_f$ is the frictional force.

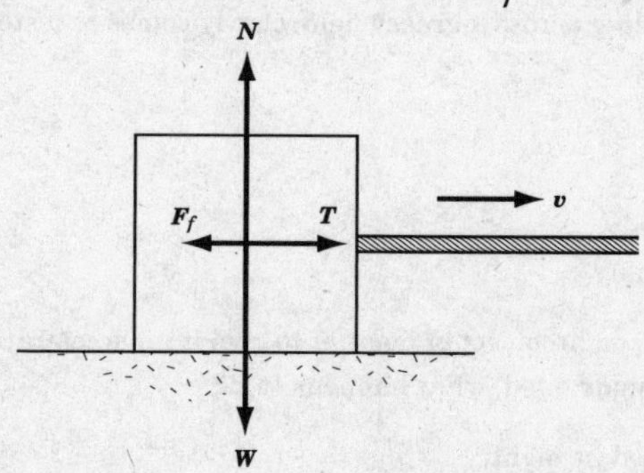

In cases like the diagram above, it's very easy to deal with the force of tension by treating the situation just as if there were somebody behind the box pushing on it. We'll find the force of tension coming up quite a bit in the chapter on special problems in mechanics, particularly when we deal with pulleys.

Key Formulas

Newton's Second Law	$\boldsymbol{F} = m\boldsymbol{a}$
Formula for Force of Kinetic Friction	$F_{\text{kinetic}} = \mu_k N$
Formula for Force of Maximum Static Friction	$F_{\text{static, max}} = \mu_s N$

Practice Questions

1. Each of the figures below shows a particle moving with velocity $\boldsymbol{v}$, and with one or two forces of magnitude F acting upon it. In which of the figures will $\boldsymbol{v}$ remain constant?

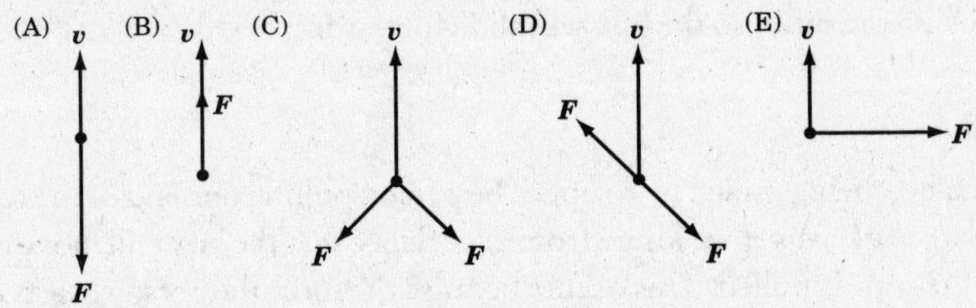

2. In which of the following examples is a net force of zero acting on the object in question?

I. A car drives around a circular racetrack at a constant speed
II. A person pushes on a door to hold it shut
III. A ball, rolling across a grassy field, slowly comes to a stop

(A) I only
(B) II only
(C) III only
(D) I and II only
(E) I and III only

3. A force $\boldsymbol{F}$ is acting on an object of mass m to give it an acceleration of $\boldsymbol{a}$. If m is halved and $\boldsymbol{F}$ is quadrupled, what happens to $\boldsymbol{a}$?

(A) It is divided by eight
(B) It is divided by two
(C) It remains unchanged
(D) It is multiplied by two
(E) It is multiplied by eight

4. A force F_1 pushes on an object of mass 10 kg with a force of 5 N to the right. A force F_2 pushes on the same object with a force of 15 N to the left. What is the acceleration of the object?

(A) 0.3 m/s² to the left
(B) 0.5 m/s² to the left
(C) 1 m/s² to the left
(D) 1.5 m/s² to the left
(E) 10 m/s² to the left

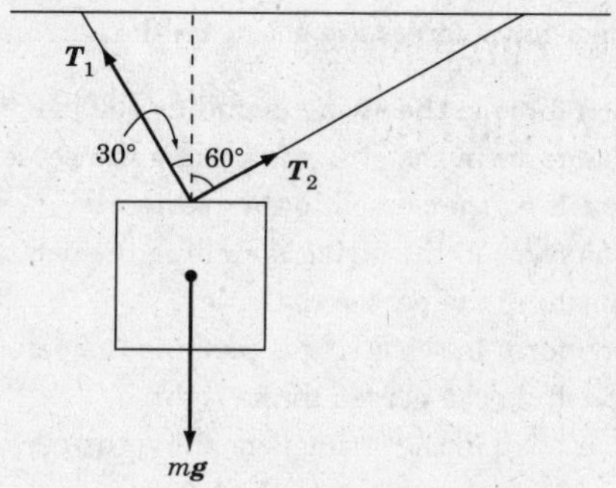

5. In the figure above, a block is suspended from two ropes, so that it hangs motionless in the air. If the magnitude of T_2 is 10.0 N, what is the magnitude of T_1? Note that sin 30 = cos 60 = 0.500, and sin 60 = cos 30 = 0.866.

(A) 0.433 N
(B) 0.500 N
(C) 0.866 N
(D) 10.0 N
(E) 17.3 N

6. In scenario 1, a person pulls with a force F on a string of length $2d$ that is connected to a spring scale. The other end of the spring scale is connected to a post by a string of length d. In scenario 2, the person pulls on the string of length $2d$ with a force of F, and a second person stands where the post was in scenario 1, and also pulls with a force of F. If the spring scale reads 50 N in scenario 1, what does the spring scale read in scenario 2?

(A) 50 N
(B) 67 N
(C) 100 N
(D) 133 N
(E) 150 N

Dynamics

7. In the figure above, a person is dragging a box attached to a string along the ground. Both the person and the box are moving to the right with a constant velocity, v. What horizontal forces are acting on the person?

(A) The tension force in the string is pulling the person to the left.
(B) The tension force in the string is pulling the person to the left, and the Earth is pushing the person to the right.
(C) The tension force in the string is pulling the person to the left, and the Earth is pushing the person to the left.
(D) The tension force in the string is pushing the person to the right, and the Earth is pushing the person to the right.
(E) The tension force in the string is pushing the person to the right, and the Earth is pushing the person to the left.

8. What is the weight of a man whose mass is 80 kg?

(A) 8.1 N
(B) 70.2 N
(C) 80 N
(D) 89.8 N
(E) 784 N

9. A 50 kg crate rests on the floor. The coefficient of static friction is 0.5. The force parallel to the floor needed to move the crate is most nearly:

(A) 25 N
(B) 50 N
(C) 125 N
(D) 250 N
(E) 500 N

10. A person is pushing an object of mass m along the ground with a force $\boldsymbol{F}$. The coefficient of kinetic friction between the object and the ground is μ_k. The object is accelerating, but then the person stops pushing and the object slides to a halt. The person then starts pushing on the object again with a force $\boldsymbol{F}$, but the object doesn't budge. The maximum coefficient of static friction between the object and the ground is μ_s. Which of the following statements is true?

(A) $F > \frac{\mu_k}{\mu_s}$
(B) $\mu_k > \mu_s$
(C) $\mu_k mg < F \le \mu_s mg$
(D) $\mu_s mg = F$
(E) The scenario described is physically impossible

Dynamics

Explanations

1. (D)

According to Newton's First Law, an object maintains a constant velocity if the net force acting on it is zero. Since the two forces in (D) cancel each other out, the net force on the particle is zero.

2. (B)

Newton's First Law tells us that a net force of zero is acting on an object if that object maintains a constant velocity. The car going around the racetrack in statement I has a constant speed, but since its direction is constantly changing (as it's going in a circle), its velocity is also changing, and so the net force acting on it isn't zero.

The person in statement II exerts a force on the door, but neither she nor the door actually moves: the force is exerted so as to hold the door in place. If the door isn't moving, its velocity is constant at zero, and so the net force acting on the door must also be zero.

Though no one is pushing on the soccer ball in statement III, some force must be acting on it if it slows down and comes to a stop. This is a result of the force of friction between the ball and the grass: if there were no friction, the ball would keep rolling. Since the net force is zero only in statement II, (B) is the correct answer.

3. (E)

Newton's Second Law tells us that $\boldsymbol{F} = m\boldsymbol{a}$. From this we can infer that $\boldsymbol{a} = \boldsymbol{F} / m$. Since $\boldsymbol{F}$ is directly proportional to $\boldsymbol{a}$, quadrupling $\boldsymbol{F}$ will also quadruple $\boldsymbol{a}$. And since m is inversely proportional to $\boldsymbol{a}$, halving m will double $\boldsymbol{a}$. We're quadrupling $\boldsymbol{a}$ and then doubling $\boldsymbol{a}$, which means that, ultimately, we're multiplying $\boldsymbol{a}$ by eight.

4. (C)

Newton's Second Law tells us that $\boldsymbol{F}_{\text{net}} = m\boldsymbol{a}$. The net force acting on the object is: 15 N left – 5 N right = 10 N left. With that in mind, we can simply solve for $\boldsymbol{A}$:

$$\begin{aligned}\boldsymbol{a} &= \frac{\boldsymbol{F}_{\text{net}}}{m} = \frac{10 \text{ N to the left}}{10 \text{ kg}} \\ &= 1 \text{ m/s}^2 \text{ to the left}\end{aligned}$$

5. (E)

Since the block is motionless, the net force acting on it must be zero. That means that the component of $\boldsymbol{T}_1$ that pulls the block to the left must be equal and opposite to the component of $\boldsymbol{T}_2$ that pulls the block to the right. The component pulling the block to the right is $\boldsymbol{T}_2 \sin 60 = (0.866)(10.0 \text{ N})$. The component pulling the block to the left is $\boldsymbol{T}_1 \sin 30 = 0.500\ \boldsymbol{T}_1$. With these components, we can solve for $\boldsymbol{T}_1$:

$$\begin{aligned}T_1 &= \frac{(0.866)(10.0 \text{ N})}{0.500} = (2.00 \text{ N})(0.866) \\ &= 17.3 \text{ N}\end{aligned}$$

6. (A)

In both cases, the spring scale isn't moving, which means that the net force acting on it is zero. If the person in scenario 1 is pulling the spring scale to the right with force $\boldsymbol{F}$, then there must be a tension force of $\boldsymbol{F}$ in the string attaching the spring scale to the post in order for the spring scale to remain motionless. That means that the same forces are acting on the spring scale in both scenarios, so if the spring scale reads 50 N in scenario 1, then it must also read 50 N in scenario 2. Don't be fooled by the lengths of the pieces of string. Length has no effect on the tension force in a string.

7. (B)

Solving this problem demands an understanding of Newton's Third Law. Since the person exerts a force to pull the string to the right, the string must exert an equal and opposite force to pull the person to the left. Further, we know that the person moves at a constant velocity, so the net force acting on the person is zero. That means there must be a force pushing the person to the right to balance the string's reaction force pulling to the left. That other force is the reaction force of the Earth: the person moves forward by pushing the Earth to the left, and the Earth in turn pushes the person to the right. This may sound strange, but it's just a fancy way of saying "the person is walking to the right."

8. (E)

The weight of any object is the magnitude of the force of gravity acting upon it. In the case of the man, this force has a magnitude of:

$$\begin{aligned} W &= mg = (80 \text{ kg})(9.8 \text{ m/s}^2) \\ &= 784 \text{ N} \end{aligned}$$

9. (D)

The force needed to move the crate is equal and opposite to the maximum force of static friction, $F_s = \mu_s mg$, where μ_s is the coefficient of static friction. Therefore, the magnitude of the force parallel to the floor is

$$\begin{aligned} F &= \mu_s mg = (0.5)(50 \text{ kg})(10 \text{ m/s}^2) \\ &= 250 \text{ N} \end{aligned}$$

10. (C)

When the person is pushing on the moving box, the box accelerates, meaning that $\boldsymbol{F}$ is greater than the force of kinetic friction, $\mu_k mg$. When the box is at rest, the person is unable to make the box move, which means that the maximum force of static friction, μ_s, is greater than or equal to $\boldsymbol{F}$.

You may be tempted by (D): the box isn't moving, so the force of static friction perfectly balances out the pushing force exerted by the person. However, μ_s is the *maximum* coefficient of static friction. The force of static friction is always only enough to resist the pushing force, so it's possible that the person could apply a greater force and still not make the object budge. Also, note that (B) states a physical impossibility. The coefficient of static friction is always greater than the coefficient of kinetic friction.

Work, Energy, and Power

Chapter Contents

Work 85
Energy 91
Forms of Energy 92
Power 97
Key Formulas 98
Practice Questions 99
Explanations 101

THERE ARE A NUMBER OF TECHNICAL TERMS in physics that have a nontechnical equivalent in ordinary usage. An example we saw in the previous chapter is force. We can talk about force in conversation without meaning a push or a pull that changes the velocity of an object, but it's easy to see that that technical definition has something in common with the ordinary use of the word force. The same is true with "**work**," "**energy**," and "**power**." All three of these words have familiar connotations in ordinary speech, but in physics they take on a technical meaning. As with force, the ordinary meaning of these words provides us with some hint as to their meaning in physics. However, we shouldn't rely too heavily on our intuition, since, as we shall see, there are some significant divergences from what common sense tells us.

The related phenomena of work, energy, and power find their way into a good number of questions on the SAT II. And energy, like force, finds its way into almost every aspect of physics, so a mastery of this subject matter is very important. The **conservation of energy** is one of the most important laws of physics, and conveniently serves as a tool to sort out many a head-splitting physics problem.

Work

When we are told that a person pushes on an object with a certain force, we only know how hard the person pushes: we don't know what the pushing accomplishes. Work, W, a scalar quantity that measures the product of the force exerted on an object and the resulting displacement of that object, is a measure of what an applied force accomplishes. The harder you push an object, and the farther that object travels, the more work you have done. In general, we say that work is done *by* a force, or *by* the object or person exerting the force, *on* the object on which the force is acting. Most simply, work is the product of force times displacement. However, as you may have remarked, both force and displacement are vector quantities, and so the direction of these vectors comes into play when calculating the work done by a given force. Work is measured in units of **joules** (J), where $1\text{ J} = 1\text{ N}\cdot\text{m} = 1\text{ kg}\cdot\text{m}^2/\text{s}^2$.

Work When Force and Displacement Are Parallel

When the force exerted on an object is in the same direction as the displacement of the object, calculating work is a simple matter of multiplication. Suppose you exert a force of 10 N on a box in the northward direction, and the box moves 5 m to the north. The work you have done on the box is $10\text{ N} \times 5\text{ m} = 50\text{ N}\cdot\text{m} = 50\text{ J}$. If force and displacement are parallel to one another, then the work done by a force is simply the product of the magnitude of the force and the magnitude of the displacement.

Work When Force and Displacement Are Not Parallel

Unfortunately, matters aren't quite as simple as scalar multiplication when the force and displacement vectors aren't parallel. In such a case, we define work as the product of the displacement of a body and the component of the force in the direction of that displacement. For instance, suppose you push a box with a force $\boldsymbol{F}$ along the floor for a distance **s**, but rather than pushing it directly forward, you push on it at a downward angle of 45º. The work you do on the box is not equal to $\boldsymbol{F} \times \boldsymbol{s}$, the magnitude of the force times the magnitude of the displacement. Rather, it is equal to $\boldsymbol{F}_s \times \boldsymbol{s}$, the magnitude of the force exerted in the direction of the displacement times the magnitude of the displacement.

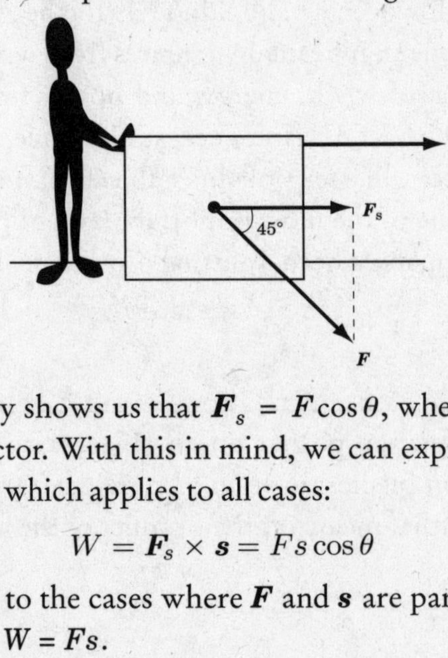

Some simple trigonometry shows us that $\boldsymbol{F}_s = F\cos\theta$, where θ is the angle between the $\boldsymbol{F}$ vector and the **s** vector. With this in mind, we can express a general formula for the work done by a force, which applies to all cases:

$$W = \boldsymbol{F}_s \times \boldsymbol{s} = Fs\cos\theta$$

This formula also applies to the cases where $\boldsymbol{F}$ and **s** are parallel, since in those cases, $\theta = 0$, and $\cos\theta = 1$, so $W = Fs$.

Dot Product

What the formula above amounts to is that work is the dot product of the force vector and the displacement vector. As we recall, the dot product of two vectors is the product of the magnitudes of the two vectors multiplied by the cosine of the angle between the two vectors. So the most general vector definition of work is:

$$W = \boldsymbol{F} \cdot \boldsymbol{s} = Fs\cos\theta$$

Review

The concept of work is actually quite straightforward, as you'll see with a little practice. You just need to bear a few simple points in mind:

- If force and displacement are both in the same direction, the work done is the product of the magnitudes of force and displacement.
- If force and displacement are at an angle to one another, you need to calculate the component of the force that points in the direction of the displacement, or the component of the displacement that points in the direction of the force. The work done is the product of the one vector and the component of the other vector.
- If force and displacement are perpendicular, no work is done.

Because of the way work is defined in physics, there are a number of cases that go against our everyday intuition. Work is not done whenever a force is exerted, and there are certain cases in which we might think that a great deal of work is being done, but in fact no work is done at all! Let's look at some examples that might be tested on the SAT II:

- A force that instantaneously causes an object to move, but does not act on the object while it is in motion, does no work. This is because the object's displacement when it is in contact with the force is effectively zero. For example, when you kick a ball, you do no work on the ball.
- You do work on a 10 kg mass when you lift it off the ground, but you do no work to hold the same mass stationary in the air. As you strain to hold the mass in the air, you are actually making sure that it is not displaced. Consequently, the work you do to hold it is zero.
- Displacement is a vector quantity that is not the same thing as distance traveled. When a jogger runs a lap of a circular track, her displacement is zero. Consequently, she has done no work.

- When a force is perpendicular to the direction of an object's motion, this force does no work on the object. For example, say you swing a tethered ball in a circle overhead, as in the diagram below. The tension force, $\boldsymbol{T}$, is always perpendicular to the velocity, $\boldsymbol{v}$, of the ball, and so the rope does no work on the ball.

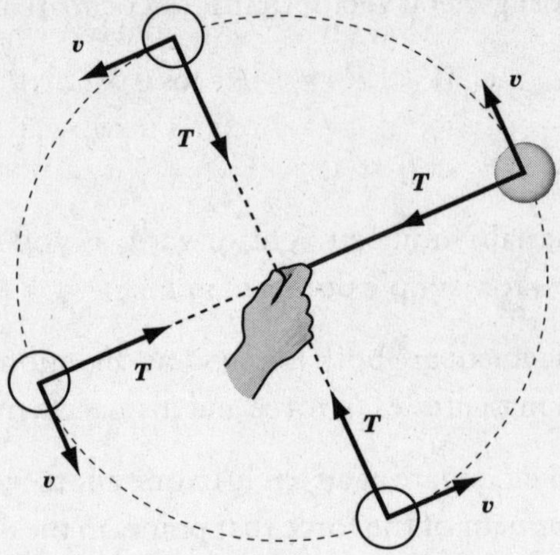

Example

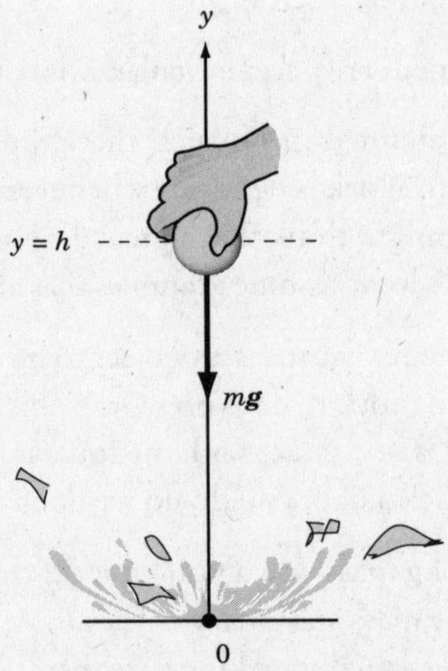

A water balloon of mass m is dropped from a height h. What is the work done on the balloon by gravity? How much work is done by gravity if the balloon is thrown horizontally from a height h with an initial velocity of v_0?

What is the work done on the balloon by gravity? Since the gravitational force of $-m\boldsymbol{g}$ is in the same direction as the water balloon's displacement, $-\boldsymbol{h}$, the work done by the gravitational force on the ball is the force times the displacement, or $W = mgh$, where $\boldsymbol{g} = -9.8 \text{ m/s}^2$.

How much work is done by gravity if the balloon is thrown horizontally from a height *h* with an initial velocity of v_0? The gravitational force exerted on the balloon is still $-mg$, but the displacement is different. The balloon has a displacement of $-h$ in the y direction and d (see the figure below) in the x direction. But, as we recall, the work done on the balloon by gravity is not simply the product of the magnitudes of the force and the displacement. We have to multiply the force by the component of the displacement that is parallel to the force. The force is directed downward, and the component of the displacement that is directed downward is $-h$. As a result, we find that the work done by gravity is mgh, just as before!

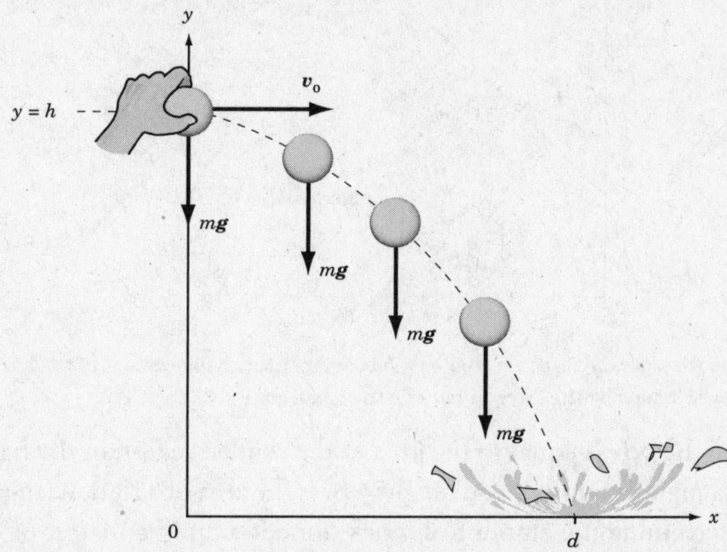

The work done by the force of gravity is the same if the object falls straight down or if it makes a wide parabola and lands 100 m to the east. This is because the force of gravity does no work when an object is transported horizontally, because the force of gravity is perpendicular to the horizontal component of displacement.

Work Problems with Graphs

There's a good chance the SAT II may test your understanding of work by asking you to interpret a graph. This graph will most likely be a force vs. position graph, though there's a chance it may be a graph of $F\cos\theta$ vs. position. Don't let the appearance of trigonometry scare you: the principle of reading graphs is the same in both cases. In the latter case, you'll be dealing with a graphic representation of a force that isn't acting parallel to the displacement, but the graph will have already taken this into account. Bottom line: all graphs dealing with work will operate according to the same easy principles. The most important thing that you need to remember about these graphs is:

The work done in a force vs. displacement graph is equal to the area between the graph and the x-axis during the same interval.

If you recall your kinematics graphs, this is exactly what you would do to read velocity on an acceleration vs. time graph, or displacement on a velocity vs. time graph. In fact, whenever you want a quantity that is the product of the quantity measured by the y-axis and the quantity measured by the x-axis, you can simply calculate the area between the graph and the x-axis.

Example

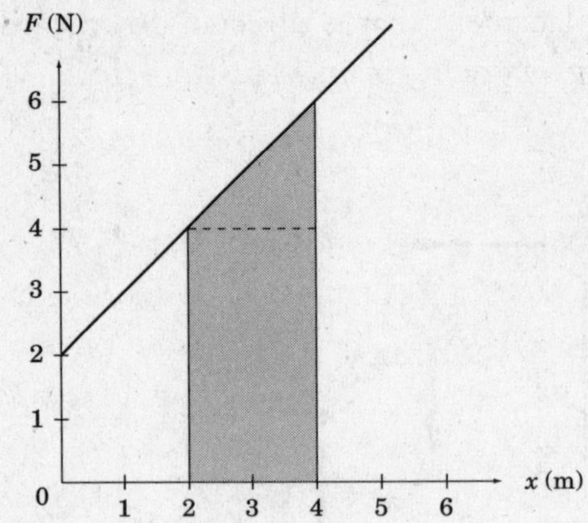

The graph above plots the force exerted on a box against the displacement of the box. What is the work done by the force in moving the box from $x = 2$ to $x = 4$?

The work done on the box is equal to the area of the shaded region in the figure above, or the area of a rectangle of width 2 and height 4 plus the area of a right triangle of base 2 and height 2. Determining the amount of work done is simply a matter of calculating the area of the rectangle and the area of the triangle, and adding these two areas together:

$$2 \cdot 4 + \frac{1}{2} \cdot 2 \cdot 2 = 10 \text{ J}$$

Curved Force vs. Position Graphs

If the SAT II throws you a curved force vs. position graph, don't panic. You won't be asked to calculate the work done, because you can't do that without using calculus. Most likely, you'll be asked to estimate the area beneath the curve for two intervals, and to select the interval in which the most, or least, work was done. In the figure below, more work was done between $x = 6$ and $x = 8$ than between $x = 2$ and $x = 4$, because the area between the graph and the x-axis is larger for the interval between $x = 6$ and $x = 8$.

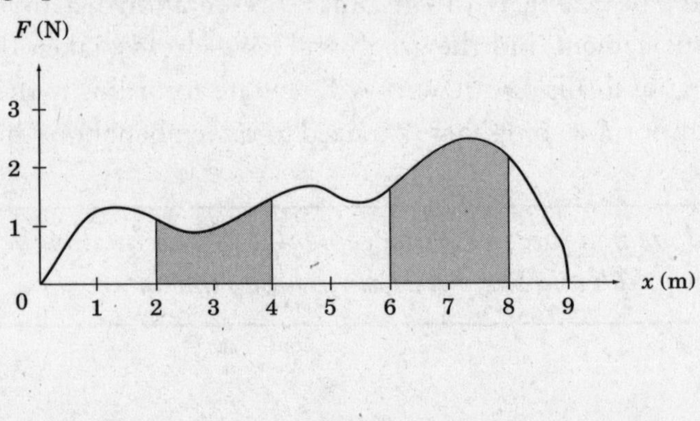

Energy

Energy is one of the central concepts of physics, and one of the most difficult to define. One of the reasons we have such a hard time defining it is because it appears in so many different forms. There is the **kinetic** and **potential energy** of kinematic motion, the **thermal energy** of heat reactions, the chemical energy of your discman batteries, the **mechanical energy** of a machine, the elastic energy that helps you launch rubber bands, the electrical energy that keeps most appliances on this planet running, and even mass energy, the strange phenomenon that Einstein discovered and that has been put to such devastating effect in the atomic bomb. This is only a cursory list: energy takes on an even wider variety of forms.

How is it that an electric jolt, a loud noise, and a brick falling to the ground can all be treated using the same concept? Well, one way of defining energy is as a capacity to do work: any object or phenomenon that is capable of doing work contains and expends a certain amount of energy. Because anything that can exert a force or have a force exerted on it can do work, we find energy popping up wherever there are forces.

Energy, like work, is measured in joules (J). In fact, work is a measure of the transfer of energy. However, there are forms of energy that do not involve work. For instance, a box suspended from a string is doing no work, but it has **gravitational potential energy** that will turn into work as soon as the string is cut. We will look at some of the many forms of energy shortly. First, let's examine the important law of conservation of energy.

Conservation of Energy

As the name suggests, the law of conservation of energy tells us that the energy in the universe is constant. Energy cannot be made or destroyed, only changed from one form to another form. Energy can also be transferred via a force, or as heat. For instance, let's return to the example mentioned earlier of the box hanging by a string. As it hangs motionless, it has gravitational potential energy, a kind of latent energy. When we cut the string, that energy is converted into **kinetic energy**, or work, as the force of gravity acts to pull the box downward. When the box hits the ground, that kinetic energy does not simply disappear. Rather, it is converted into sound and heat energy: the box makes a loud thud and the impact between the ground and the box generates a bit of heat.

This law applies to any closed system. A closed system is a system where no energy leaves the system and goes into the outside world, and no energy from the outside world enters the system. It is virtually impossible to create a truly closed system on Earth, since energy is almost always dissipated through friction, heat, or sound, but we can create close approximations. Objects sliding over ice or air hockey tables move with a minimal amount of friction, so the energy in these systems remains nearly constant. Problems on the SAT II that quiz you on the conservation of energy will almost always deal with frictionless surfaces, since the law of conservation of energy applies only to closed systems.

The law of conservation of energy is important for a number of reasons, one of the most fundamental being that it is so general: it applies to the whole universe and extends across all time. For the purposes of the SAT II, it helps you solve a number of problems that would be very difficult otherwise. For example, you can often determine an object's velocity quite easily by using this law, while it might have been very difficult or even impossible using only kinematic equations. We will see this law at work later in this chapter, and again when we discuss elastic and inelastic collisions in the chapter on linear momentum.

Forms of Energy

Though energy is always measured in joules, and though it can always be defined as a capacity to do work, energy manifests itself in a variety of different forms. These various forms pop up all over the SAT II, and we will look at some additional forms of energy when we discuss electromagnetism, relativity, and a number of other specialized topics. For now, we will focus on the kinds of energy you'll find in mechanics problems.

Kinetic Energy

Kinetic energy is the energy a body in motion has by virtue of its motion. We define energy as the capacity to do work, and a body in motion is able to use its motion to do work. For instance, a cue ball on a pool table can use its motion to do work on the eight ball. When the cue ball strikes the eight ball, the cue ball comes to a stop and the eight ball starts moving. This occurs because the cue ball's kinetic energy has been transferred to the eight ball.

There are two main types of kinetic energy: translational and rotational. **Translational kinetic energy** is the energy of a particle moving in space and is defined in terms of the particle's mass, m, and velocity, v:

$$KE = \frac{1}{2}mv^2$$

For instance, a cue ball of mass 0.5 kg moving at a velocity of 2 m/s has a kinetic energy of ½(0.5 kg)(2 m/s)^2 = 1 J.

The Work-Energy Theorem

If you recall, work is a measure of the transfer of energy. An object that has a certain amount of work done on it has that amount of energy transferred to it. This energy moves the object over a certain distance with a certain force; in other words, it is kinetic energy. This handy little fact is expressed in the **work-energy theorem**, which states that the net work done on an object is equal to the object's change in kinetic energy:

$$W = \Delta KE$$

For example, say you apply a force to a particle, causing it to accelerate. This force does positive work on the particle and increases its kinetic energy. Conversely, say you apply a force to decelerate a particle. This force does negative work on the particle and decreases its kinetic energy. If you know the forces acting on an object, the work-energy theorem provides a convenient way to calculate the velocity of a particle.

Example

> **A hockey puck of mass 1 kg slides across the ice with an initial velocity of 10 m/s. There is a 1 N force of friction acting against the puck. What is the puck's velocity after it has glided 32 m along the ice?**

If we know the puck's kinetic energy after it has glided 32 m, we can calculate its velocity. To determine its kinetic energy at that point, we need to know its initial kinetic energy, and how much that kinetic energy changes as the puck glides across the ice.

First, let's determine the initial kinetic energy of the puck. We know the puck's initial mass and initial velocity, so we just need to plug these numbers into the equation for kinetic energy:

$$KE = \frac{1}{2}mv^2 = \frac{1}{2}(1\text{ kg})(10\text{ m/s})^2 = 50\text{ J}$$

The friction between the puck and the ice decelerates the puck. The amount of work the ice does on the puck, which is the product of the force of friction and the puck's displacement, is negative.

$$W = \boldsymbol{F} \cdot \boldsymbol{s} = (-1\text{ N})(32\text{ m}) = 32\text{ J}$$

The work done on the puck decreases its kinetic energy, so after it has glided 32 m, the kinetic energy of the puck is 50 – 32 = 18 J. Now that we know the final kinetic energy of the puck, we can calculate its final velocity by once more plugging numbers into the formula for kinetic energy:

$$\begin{aligned} KE &= \frac{1}{2}mv^2 \\ 18\text{ J} &= \frac{1}{2}(1\text{ kg})v^2 \\ v^2 &= (36\text{ m/s})^2 \\ v &= 6\text{ m/s} \end{aligned}$$

We could also have solved this problem using Newton's Second Law and some kinematics, but the work-energy theorem gives us a quicker route to the same answer.

Potential Energy

As we said before, work is the process of energy transfer. In the example above, the kinetic energy of the puck was transferred into the heat and sound caused by friction.

There are a great number of objects, though, that spend most of their time neither doing work nor having work done on them. This book in your hand, for instance, is not doing any work right now, but the second you drop it—whoops!—the force of gravity does some work on it, generating kinetic energy. Now pick up the book and let's continue.

Potential energy, U, is a measure of an object's unrealized potential to have work done on it, and is associated with that object's position in space, or its configuration in relation to other objects. Any work done on an object converts its potential energy into kinetic energy, so the net work done on a given object is equal to the negative change in its potential energy:

$$W = -\Delta U$$

Be very respectful of the minus sign in this equation! It may be tempting to think that the work done on an object increases its potential energy, but the opposite is true. Work converts potential energy into other forms of energy, usually kinetic energy. Remove the minus sign from the equation above, and you are in direct violation of the law of conservation of energy!

There are many forms of potential energy, each of which is associated with a different type of force. The SAT II usually confines itself to gravitational potential energy and the potential energy of a compressed spring. We will review gravitational potential energy in this section, and the potential energy of a spring in the next chapter.

Gravitational Potential Energy

Gravitational potential energy registers the potential for work done on an object by the force of gravity. For example, say that you lift a water balloon to height h above the ground. The work done by the force of gravity as you lift the water balloon is the force of gravity, $-mg$, times the water balloon's displacement, h. So the work done by the force of gravity is $W = -mgh$. Note that there is a negative amount of work done, since the water balloon is being lifted upward, in the opposite direction of the force of gravity.

By doing $-mgh$ joules of work on the water balloon, you have increased its gravitational potential energy by mgh joules (recall the equation $W = -\Delta U$). In other words, you have increased its potential to accelerate downwards and cause a huge splash. Because the force of gravity has the potential to do mgh joules of work on the water balloon at height h, we say that the water balloon has mgh joules of gravitational potential energy.

$$U_g = mgh$$

For instance, a 50 kg mass held at a height of 4 m from the ground has a gravitational potential energy of:

$$U_g = mgh = (50\text{ kg})(9.8\text{ m/s}^2)(4\text{ m}) = 1960\text{ J}$$

The most important thing to remember is that *the higher an object is off the ground, the greater its gravitational potential energy.*

Work, Energy, and Power

Mechanical Energy

We now have equations relating work to both kinetic and potential energy:

$$W = \Delta KE$$
$$W = -\Delta U$$

Combining these two equations gives us this important result:

$$\Delta KE + \Delta U = 0$$

Or, alternatively:

$$\Delta KE = -\Delta U$$

As the kinetic energy of a system increases, its potential energy decreases by the same amount, and vice versa. As a result, the sum of the kinetic energy and the potential energy in a system is constant. We define this constant as E, the mechanical energy of the system:

$$KE + U = E$$

This law, the conservation of mechanical energy, is one form of the more general law of conservation of energy, and it's a handy tool for solving problems regarding projectiles, pulleys, springs, and inclined planes. However, mechanical energy is *not* conserved in problems involving frictional forces. When friction is involved, a good deal of the energy in the system is dissipated as heat and sound. The conservation of mechanical energy only applies to closed systems.

Example 1

A student drops an object of mass 10 kg from a height of 5 m. What is the velocity of the object when it hits the ground? Assume, for the purpose of this question, that $g = -10$ m/s^2.

Before the object is released, it has a certain amount of gravitational potential energy, but no kinetic energy. When it hits the ground, it has no gravitational potential energy, since $h = 0$, but it has a certain amount of kinetic energy. The mechanical energy, E, of the object remains constant, however. That means that the potential energy of the object before it is released is equal to the kinetic energy of the object when it hits the ground.

When the object is dropped, it has a gravitational potential energy of:

$$mgh = (10\text{ kg})(-10\text{ m/s}^2)(-5\text{ m}) = 500\text{ J}$$

By the time it hits the ground, all this potential energy will have been converted to kinetic energy. Now we just need to solve for v:

$$\frac{1}{2}mv^2 = 500\text{ J}$$
$$v^2 = \frac{2(500\text{ J})}{10\text{ kg}}$$
$$= (100\text{ m/s})^2$$
$$v = 10\text{ m/s}$$

Example 2

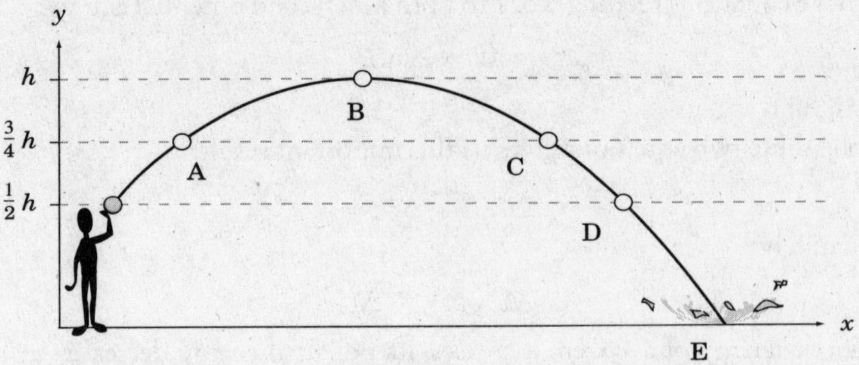

Consider the above diagram of the trajectory of a thrown tomato:

1. At what point is the potential energy greatest?
2. At what point is the kinetic energy the least?
3. At what point is the kinetic energy greatest?
4. At what point is the kinetic energy decreasing and the potential energy increasing?
5. At what point are the kinetic energy and the potential energy equal to the values at position A?

The answer to question 1 is point B. At the top of the tomato's trajectory, the tomato is the greatest distance above the ground and hence has the greatest potential energy.

The answer to question 2 is point B. At the top of the tomato's trajectory, the tomato has the smallest velocity, since the *y*-component of the velocity is zero, and hence the least kinetic energy. Additionally, since mechanical energy is conserved in projectile motion, we know that the point where the potential energy is the greatest corresponds to the point where the kinetic energy is smallest.

The answer to question 3 is point E. At the bottom of its trajectory, the tomato has the greatest velocity and thus the greatest kinetic energy.

The answer to question 4 is point A. At this point, the velocity is decreasing in magnitude and the tomato is getting higher in the air. Thus, the kinetic energy is decreasing and the potential energy is increasing.

The answer to question 5 is point C. From our study of kinematics, we know that the speed of a projectile is equal at the same height in the projectile's ascent and descent. Therefore, the tomato has the same kinetic energy at points A and C. Additionally, since the tomato has the same height at these points, its potential energy is the same at points A and C.

Keep this example in mind when you take the SAT II, because it is likely that a similar question will appear on the test.

Thermal Energy

There are many cases where the energy in a system seems simply to have disappeared. Usually, this is because that energy has been turned into sound and heat. For instance, a coin sliding across a table slows down and comes to a halt, but in doing so, it produces the sound energy of the coin scraping along the table and the heat energy of friction. Rub your hands together briskly and you will feel that friction causes heat.

We will discuss thermal energy, or heat, in greater detail in Chapter 9, but it's worth noting here that it is the most common form of energy produced in energy transformations. It's hard to think of an energy transformation where no heat is produced. Take these examples:

- Friction is at work everywhere, and friction produces heat.
- Electric energy produces heat: a light bulb produces far more heat than it does light.
- When people talk about burning calories, they mean it quite literally: exercise is a way of converting food energy into heat.
- Sounds fade to silence because the sound energy is gradually converted into the heat of the vibrating air molecules. In other words, if you shout very loudly, you make the air around you warmer!

Power

Power is an important physical quantity that frequently, though not always, appears on the SAT II. Mechanical systems, such as engines, are not limited by the amount of work they can do, but rather by the rate at which they can perform the work. Power, *P*, is defined as the rate at which work is done, or the rate at which energy is transformed. The formula for average power is:

$$\frac{\Delta W}{\Delta t} \text{ or } \frac{\Delta E}{\Delta t}$$

Power is measured in units of watts (W), where 1 W = 1 J/s.

Example

A piano mover pushes on a piano with a force of 100 N, moving it 9 m in 12 s. With how much power does the piano mover push?

Power is a measure of the amount of work done in a given time period. First we need to calculate how much work the piano mover does, and then we divide that quantity by the amount of time the work takes.

$$W = Fs = (100\ \text{N})(9\ \text{m}) = 900\ \text{J}$$
$$P = \frac{\Delta W}{\Delta t} = \frac{900\ \text{J}}{12\ \text{s}} = 75\ \text{W}$$

Be careful not to confuse the symbol for watts, W, with the symbol for work, *W*.

Instantaneous Power

Sometimes we may want to know the instantaneous power of an engine or person, the amount of power output by that person at any given instant. In such cases, there is no value for Δt to draw upon. However, when a steady force is applied to an object, the change in the amount of work done on the object is the product of the force and the change in that object's displacement. Bearing this in mind, we can express power in terms of force and velocity:

$$\begin{aligned} P &= \frac{\Delta W}{\Delta t} \\ &= \boldsymbol{F}\frac{\Delta \boldsymbol{s}}{\Delta t} \\ &= \boldsymbol{F} \cdot \boldsymbol{v} \end{aligned}$$

Key Formulas

Work	$W = \boldsymbol{F} \cdot \boldsymbol{s} = Fs\cos\theta$
Work Done by Gravity	$W = mgh$
Kinetic Energy	$KE = \frac{1}{2}mv^2$
Work-Energy Theorem	$W = \Delta KE$
Potential Energy	$W = -\Delta U$
Gravitational Potential Energy	$U_g = mgh$
Mechanical Energy	$E = KE + U$
Average Power	$\frac{\Delta W}{\Delta t}$ or $\frac{\Delta E}{\Delta t}$
Instantaneous Power	$P = \boldsymbol{F} \cdot \boldsymbol{v}$

Practice Questions

1. How much work does a person do in pushing a box with a force of 10 N over a distance of 4.0 m in the direction of the force?

   (A) 0.4 J (B) 4.0 J (C) 40 J (D) 400 J (E) 4000 J

2. A person pushes a 10 kg box at a constant velocity over a distance of 4 m. The coefficient of kinetic friction between the box and the floor is 0.3. How much work does the person do in pushing the box?

   (A) 12 J
   (B) 40 J
   (C) 75 J
   (D) 120 J
   (E) 400 J

3. How much work does the force of gravity do in pulling a 10 kg box down a 30º inclined plane of length 8.0 m? Note that sin 30 = cos 60 = 0.500 and cos 30 = sin 60 = 0.866.

   (A) 40 J
   (B) 69 J
   (C) 400 J
   (D) 690 J
   (E) 800 J

4. How much work does a person do in pushing a box with a force of 20 N over a distance of 8.0 m in the direction of the force?

   (A) 1.6 J
   (B) 16 J
   (C) 160 J
   (D) 1600 J
   (E) 16000 J

5. The figure at right is a force vs. displacement graph, showing the amount of force applied to an object by three different people. Al applies force to the object for the first 4 m of its displacement, Betty applies force from the 4 m point to the 6 m point, and Chuck applies force from the 6 m point to the 8 m point. Which of the three does the most work on the object?

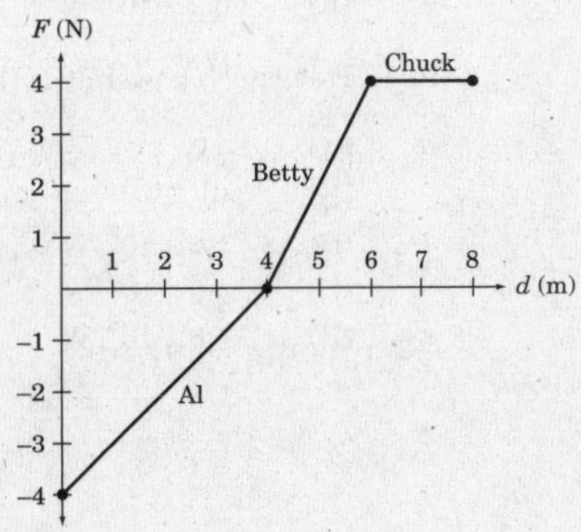

   (A) Al
   (B) Betty
   (C) Chuck
   (D) Al and Chuck do the same amount of work
   (E) Betty and Chuck do the same amount of work

6. When a car's speed doubles, what happens to its kinetic energy?

(A) It is quartered
(B) It is halved
(C) It is unchanged
(D) It is doubled
(E) It is quadrupled

7. A worker does 500 J of work on a 10 kg box. If the box transfers 375 J of heat to the floor through the friction between the box and the floor, what is the velocity of the box after the work has been done on it?

(A) 5 m/s
(B) 10 m/s
(C) 12.5 m/s
(D) 50 m/s
(E) 100 m/s

8. A person on the street wants to throw an 8 kg book up to a person leaning out of a window 5 m above street level. With what velocity must the person thrown the book so that it reaches the person in the window?

(A) 5 m/s
(B) 8 m/s
(C) 10 m/s
(D) 40 m/s
(E) 50 m/s

Questions 9 and 10 refer to a forklift lifting a crate of mass 100 kg at a constant velocity to a height of 8 m over a time of 4 s. The forklift then holds the crate in place for 20 s.

9. How much power does the forklift exert in lifting the crate?

(A) 0 W
(B) 2.0×10^3 W
(C) 3.2×10^3 W
(D) 2.0×10^4 W
(E) 3.2×10^4 W

10. How much power does the forklift exert in holding the crate in place?

(A) 0 W
(B) 400 W
(C) 1.6×10^3 W
(D) 4.0×10^3 W
(E) 1.6×10^4 W

Explanations

1. (C)

When the force is exerted in the direction of motion, work is simply the product of force and displacement. The work done is (10 N)(4.0 m) = 40 J.

2. (D)

The work done on the box is the force exerted multiplied by the box's displacement. Since the box travels at a constant velocity, we know that the net force acting on the box is zero. That means that the force of the person's push is equal and opposite to the force of friction. The force of friction is given by μN, where μ is the coefficient of kinetic friction and N is the normal force. The normal force is equal to the weight of the box, which is mg = (10 kg)(10 m/s^2) = 100 N. With all this in mind, we can solve for the work done on the box:

$$W = \mu N(4\text{ m}) = (0.3)(100\text{ N})(4\text{ m})$$
$$= 120\text{ J}$$

3. (C)

The work done by the force of gravity is the dot product of the displacement of the box and the force of gravity acting on the box. That means that we need to calculate the component of the force of gravity that is parallel to the incline. This is mg sin 30 = (10 kg)(10 m/s^2) sin 30. Thus, the work done is

$$W = mg\sin 30(8.0\text{ m}) = (10\text{ kg})(10\text{ m/s}^2)(0.5)(8.0\text{ m})$$
$$= 400\text{ J}$$

4. (C)

This is the same question as question 1. We were hoping that with different numbers and line spacing you wouldn't notice. The ETS does that too sometimes.

5. (C)

On a force vs. displacement graph, the amount of work done is the area between the graph and the x-axis. The work Al does is the area of the 4 × –4 right triangle:

$$\frac{1}{2}(-4\text{ N})(4\text{ m}) = -8\text{ J}$$

The amount of work Betty does is equal to the area of a triangle of length 2 and height 4:

$$\frac{1}{2}(4\text{ N})(2\text{ m}) = 4\text{ J}$$

The amount of work done by Chuck is equal to the area of a rectangle of length 2 and height 4: 4 × 2 = 8 J. We can conclude that Chuck did the most work.

Don't be fooled by (D): the force exerted by Al is in the opposite direction of the object's displacement, so he does negative work on the object.

6. (E)
The formula for kinetic energy is $KE = \frac{1}{2}mv^2$. Since the car's kinetic energy is directly proportional to the square of its velocity, doubling the velocity would mean quadrupling its kinetic energy.

7. (A)
The work-energy theorem tells us that the amount of work done on an object is equal to the amount of kinetic energy it gains, and the amount of work done by an object is equal to the amount of kinetic energy it loses. The box gains 500 J of kinetic energy from the worker's push, and loses 375 J of kinetic energy to friction, for a net gain of 125 J. Kinetic energy is related to velocity by the formula $KE = \frac{1}{2}mv^2$, so we can get the answer by plugging numbers into this formula and solving for v:

$$KE = \frac{1}{2}mv^2 \qquad v^2 = \frac{2KE}{m}$$

$$v = \sqrt{\frac{2KE}{m}} = \sqrt{\frac{2(125\text{ J})}{10\text{ kg}}} = \sqrt{25\text{ J/kg}} = 5\text{ m/s}$$

8. (C)
When the book reaches the person in the window, it will have a gravitational potential energy of $U = mgh$. In order for the book to reach the window, then, it must leave the hands of the person at street level with at least that much kinetic energy. Kinetic energy is given by the formula $KE = \frac{1}{2}mv^2$, so we can solve for v by making $KE = U$:

$$\frac{1}{2}mv^2 = mgh \qquad v^2 = 2gh$$

$$v = \sqrt{2(10\text{ m/s}^2)(5\text{ m})} = \sqrt{100\text{ m}^2/\text{s}^2} = 10\text{ m/s}$$

9. (B)
Power is a measure of work divided by time. In turn, work is a measure of force multiplied by displacement. Since the crate is lifted with a constant velocity, we know that the net force acting on it is zero, and so the force exerted by the forklift must be equal and opposite to the weight of the crate, which is $(100\text{ kg})(10\text{ m/s}^2) = 1.0 \times 10^3$ N. From this, we can calculate the power exerted by the forklift:

$$P = \frac{W}{t} = \frac{(1.0 \times 10^3)(8.0\text{ m})}{4.0\text{ s}} = 2.0 \times 10^3\text{ W}$$

10. (A)
Power is measured as work divided by time, and work is the dot product of force and displacement. While the crate is being held in the air, it is not displaced, so the displacement is zero. That means the forklift does no work, and thus exerts no power.

Special Problems in Mechanics

Chapter Contents

3-Step Approach to Problem Solving.... 103
Pulleys 104
Inclined Planes 110
Springs 116
Pendulums 125
Key Formulas...................... 129
Practice Questions................ 129
Explanations 132

THE "SPECIAL PROBLEMS" WE WILL ADDRESS in this chapter deal with four common mechanical systems: pulleys, inclined planes, springs, and pendulums. These systems pop up on many mechanics problems on the SAT II, and it will save you time and points if you familiarize yourself with their quirks. These systems obey the same mechanical rules as the rest of the world, and we will only introduce one principle (Hooke's Law) that hasn't been covered in the previous three chapters. However, there are a number of problem-solving techniques that are particular to these sorts of problems, and mastering them will help you get through these problems quickly and easily.

The Three-Step Approach to Problem Solving

The systems we will look at in this chapter won't test your knowledge of obscure formulas so much as your problem solving abilities. The actual physics at work on these systems is generally quite simple—it rarely extends beyond Newton's three laws and a basic understanding of work and energy—but you'll need to apply this simple physics in imaginative ways.

There are three general steps you can take when approaching any problem in mechanics. Often the problems are simple enough that these steps are unnecessary. However, with the special problems we will tackle in this chapter, following these steps carefully may save you many times over on the SAT II. The three steps are:

1. **Ask yourself how the system will move:** Before you start writing down equations and looking at answer choices, you should develop an intuitive

sense of what you're looking at. In what direction will the objects in the system move? Will they move at all? Once you know what you're dealing with, you'll have an easier time figuring out how to approach the problem.

2. **Choose a coordinate system:** Most systems will only move in one dimension: up and down, left and right, or on an angle in the case of inclined planes. Choose a coordinate system where one direction is negative, the other direction is positive, and, if necessary, choose an origin point that you label 0. Remember: no coordinate system is right or wrong in itself, some are just more convenient than others. The important thing is to be strictly consistent once you've chosen a coordinate system, and to be mindful of those subtle but crucial minus signs!
3. **Draw free-body diagrams:** Most students find mechanics easier than electromagnetism for the simple reason that mechanics problems are easy to visualize. Free-body diagrams allow you to make the most of this advantage. Make sure you've accounted for all the forces acting on all the bodies in the system. Make ample use of Newton's Third Law, and remember that for systems at rest or at a constant velocity, the net force acting on every body in the system must be zero.

Students too often think that physics problem solving is just a matter of plugging the right numbers into the right equations. The truth is, physics problem solving is more a matter of determining what those right numbers and right equations are. These three steps should help you do just that. Let's look at some mechanical systems.

Pulleys

Pulleys are simple machines that consist of a rope that slides around a disk, called a block. Their main function is to change the direction of the tension force in a rope. The pulley systems that appear on the SAT II almost always consist of idealized, massless and frictionless pulleys, and idealized ropes that are massless and that don't stretch. These somewhat unrealistic parameters mean that:

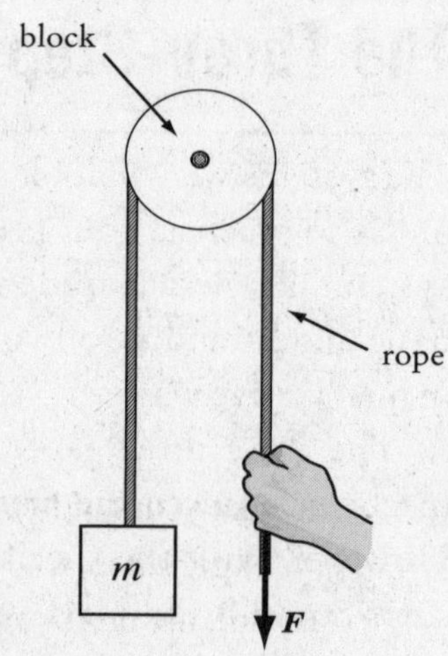

1. The rope slides without any resistance over the pulley, so that the pulley changes the direction of the tension force without changing its magnitude.
2. You can apply the law of conservation of energy to the system without worrying about the energy of the rope and pulley.
3. You don't have to factor in the mass of the pulley or rope when calculating the effect of a force exerted on an object attached to a pulley system.

The one exception to this rule is the occasional problem you might find regarding the torque applied to a pulley block. In such a problem, you will have to take the pulley's mass into account. We'll deal with this special case in Chapter 7, when we look at torque.

The Purpose of Pulleys

We use pulleys to lift objects because they reduce the amount of force we need to exert. For example, say that you are applying force $\boldsymbol{F}$ to the mass in the figure below. How does $\boldsymbol{F}$ compare to the force you would have to exert in the absence of a pulley?

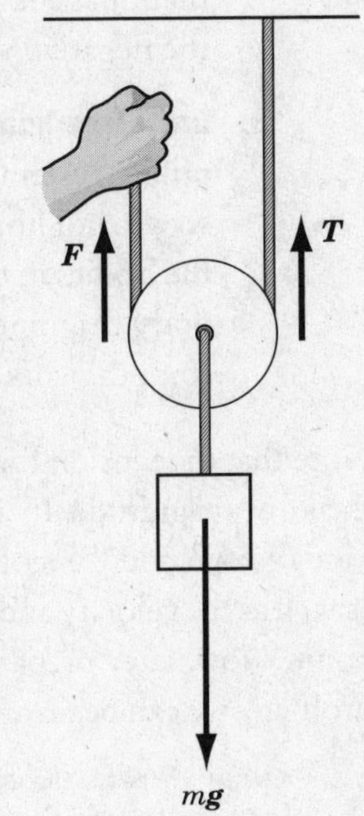

To lift mass m at a constant velocity without a pulley, you would have to apply a force equal to the mass's weight, or a force of mg upward. Using a pulley, the mass must still be lifted with a force of mg upward, but this force is distributed between the tension of the rope attached to the ceiling, $\boldsymbol{T}$, and the tension of the rope gripped in your hand, $\boldsymbol{F}$:, as at right.

Because there are two ropes pulling the block, and hence the mass, upward, there are two equal upward forces, $\boldsymbol{F}$ and $\boldsymbol{T}$. We know that the sum of these forces is equal to the gravitational force pulling the mass down, so $\boldsymbol{F} + \boldsymbol{T} = 2\boldsymbol{F} = mg$ or, $\boldsymbol{F} = mg/2$. Therefore, you need to pull with only one half the force you would have to use to lift mass m if there were no pulley.

Standard Pulley Problem

The figure at right represents a pulley system where masses m and M are connected by a rope over a massless and frictionless pulley. Note that $M > m$ and both masses are at the same height above the ground. The system is initially held at rest, and is then released. We will learn to calculate the acceleration of the masses, the velocity of mass m when it moves a distance h, and the work done by the tension force on mass m as it moves a distance h.

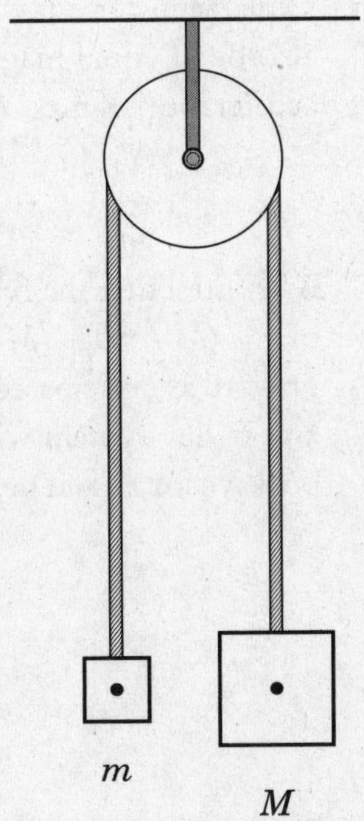

Before we start calculating values for acceleration, velocity, and work, let's go through the three steps for problem solving:

1. **Ask yourself how the system will move:** From experience, we know that the heavy mass, M, will fall, lifting the smaller mass, m. Because the masses are connected, we know that the velocity of mass m is equal in magnitude to the velocity of mass M, but opposite in direction. Likewise, the acceleration of mass m is equal in magnitude to the acceleration of mass M, but opposite in direction.

2. **Choose a coordinate system:** Some diagrams on the SAT II will provide a coordinate system for you. If they don't, choose one that will simplify your calculations. In this case, let's follow the standard convention of saying that up is the positive *y* direction and down is the negative *y* direction.

3. **Draw free-body diagrams:** We know that this pulley system will accelerate when released, so we shouldn't expect the net forces acting on the bodies in the system to be zero. Your free-body diagram should end up looking something like the figure at right.

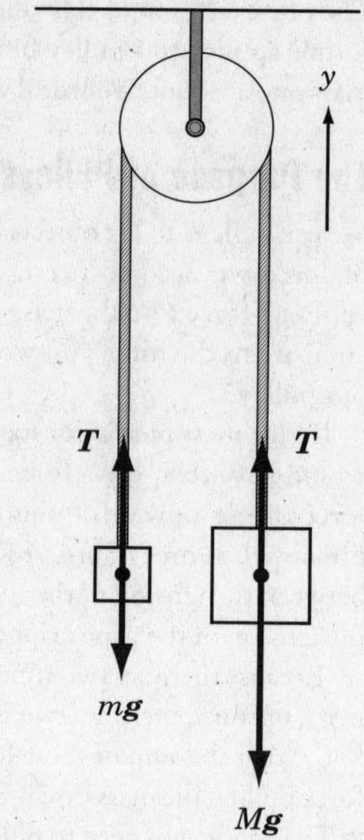

Note that the tension force, ***T***, on each of the blocks is of the same magnitude. In any nonstretching rope (the only kind of rope you'll encounter on the SAT II), the tension, as well as the velocity and acceleration, is the same at every point. Now, after preparing ourselves to understand the problem, we can begin answering some questions.

1. What is the acceleration of mass *M*?
2. What is the velocity of mass *m* after it travels a distance *h*?
3. What is the work done by the force of tension in lifting mass *m* a distance *h*?

1. What is the acceleration of mass *M*? Because the acceleration of the rope is of the same magnitude at every point in the rope, the acceleration of the two masses will also be of equal magnitude. If we label the acceleration of mass *m* as ***a***, then the acceleration of mass ***M*** is $-\boldsymbol{a}$. Using Newton's Second Law we find:

$$\text{for mass } M\text{: } T - Mg = -Ma$$
$$\text{for mass } m\text{: } T - mg = ma$$

By subtracting the first equation from the second, we find $(M - m)g = (M + m)a$ or $a = (M - m)g/(M + m)$. Because $M - m > 0$, a is positive and mass *m* accelerates upward as anticipated. This result gives us a general formula for the acceleration of any pulley system with unequal masses, ***M*** and *m*. Remember, the acceleration is positive for *m* and negative for ***M***, since *m* is moving up and ***M*** is going down.

$$a = \frac{g(M - m)}{M + m}$$

2. **What is the velocity of mass *m* after it travels a distance *h*?** We could solve this problem by plugging numbers into the kinematics equations, but as you can see, the formula for the acceleration of the pulleys is a bit unwieldy, so the kinematics equations may not be the best approach. Instead, we can tackle this problem in terms of energy. Because the masses in the pulley system are moving up and down, their movement corresponds with a change in gravitational potential energy. Because mechanical energy, $\boldsymbol{E}$, is conserved, we know that any change in the potential energy, U, of the system will be accompanied by an equal but opposite change in the kinetic energy, $\boldsymbol{KE}$, of the system.

$$\Delta KE = -\Delta U$$

Remember that since the system begins at rest, $KE_{\text{initial}} = 0$. As the masses move, mass M loses Mgh joules of potential energy, whereas mass m gains mgh joules of potential energy. Applying the law of conservation of mechanical energy, we find:

$$\frac{1}{2}Mv^2 + \frac{1}{2}mv^2 = -(mgh - Mgh)$$

$$\frac{1}{2}(M+m)v^2 = (M-m)gh$$

$$v = \sqrt{2gh\frac{(M-m)}{M+m}}$$

Mass m is moving in the positive y direction.

We admit it: the above formula is pretty scary to look at. But since the SAT II doesn't allow calculators, you almost certainly will not have to calculate precise numbers for a mass's velocity. It's less important that you have this exact formula memorized, and more important that you understand the principle by which it was derived. You may find a question that involves a derivation of this or some related formula, so it's good to have at least a rough understanding of the relationship between mass, displacement, and velocity in a pulley system.

3. **What is the work done by the force of tension in lifting mass *m* a distance *h*?** Since the tension force, $\boldsymbol{T}$, is in the same direction as the displacement, $\boldsymbol{h}$, we know that the work done is equal to hT. But what is the magnitude of the tension force? We know that the sum of forces acting on m is $T - mg$ which is equal to ma. Therefore, $T = m(g - a)$. From the solution to question 1, we know that $a = g(M - m)/(M + m)$, so substituting in for a, we get:

$$W = hT = m(g-a)h = m(g - \frac{g(M-m)}{M+m}h)$$

$$= mgh(1 - \frac{M-m}{M+m})$$

Special Problems

A Pulley on a Table

Now imagine that masses m and M are in the following arrangement:

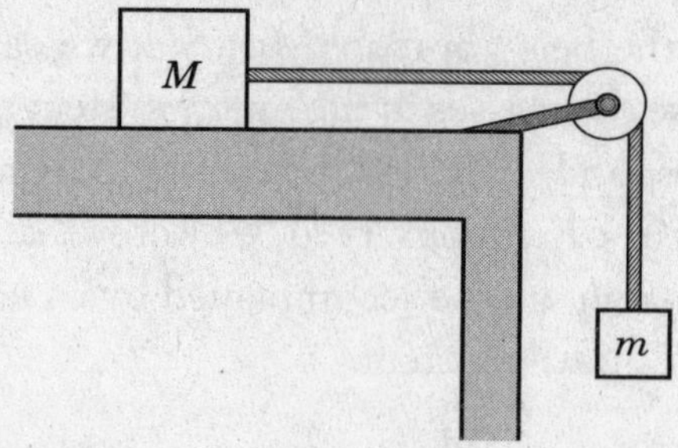

Let's assume that mass M has already begun to slide along the table, and its movement is opposed by the force of kinetic friction, $\mathbf{F}_{fr} = \mu N$, where μ is the coefficient of kinetic friction, and N is the normal force acting between the mass and the table. If the mention of friction and normal forces frightens you, you might want to flip back to Chapter 3 and do a little reviewing.

So let's approach this problem with our handy three-step problem solving method:

1. **Ask yourself how the system will move:** First, we know that mass m is falling and dragging mass M off the table. The force of kinetic friction opposes the motion of mass M. We also know, since both masses are connected by a non-stretching rope, that the two masses must have the same velocity and the same acceleration.
2. **Choose a coordinate system:** For the purposes of this problem, it will be easier if we set our coordinate system relative to the rope rather than to the table. If we say that the x-axis runs parallel to the rope, this means the x-axis will be the up-down axis for mass m and the left-right axis for mass M. Further, we can say that gravity pulls in the negative x direction. The y-axis, then, is perpendicular to the rope, and the positive y direction is away from the table.
3. **Draw free-body diagrams:** The above description of the coordinate system may be a bit confusing. That's why a diagram can often be a lifesaver.

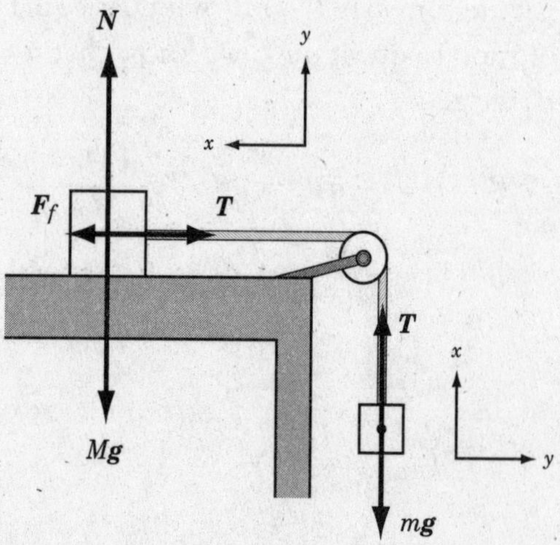

Special Problems

Given this information, can you calculate the acceleration of the masses? If you think analytically and don't panic, you can. Since they are attached by a rope, we know that both masses have the same velocity, and hence the same acceleration, a. We also know the net force acting on both masses: the net force acting on mass M is $\mu MG - T$, and the net force acting on mass m is $T - mg$. We can then apply Newton's Second Law to both of the masses, giving us two equations involving a:

$$\text{for mass } M\text{: } \mu Mg - T = Ma$$
$$\text{for mass } m\text{: } T - mg = ma$$

Adding the two equations, we find $\mu Mg - mg = (M + m)a$. Solving for a, we get:

$$a = \frac{g(\mu M - m)}{m + M}$$

Since m is moving downward, a must be negative. Therefore, $\mu M < m$.

How Complex Formulas Will Be Tested on the SAT II Physics

It is highly unlikely that the SAT II will ask a question that involves remembering and then plugging numbers into an equation like this one. Remember: the SAT II places far less emphasis on math than your high school physics class. The test writers don't want to test your ability to recall a formula or do some simple math. Rather, they want to determine whether you understand the formulas you've memorized. Here are some examples of the kinds of questions you might be asked regarding the pulley system in the free-body diagram above:

1. **Which of the following five formulas represents the acceleration of the pulley system?** You would then be given five different mathematical formulas, one of which is the correct formula. The test writers would not expect you to have memorized the correct formula, but they would expect you to be able to derive it.

2. **Which of the following is a way of maximizing the system's acceleration?** You would then be given options like "maximize M and m and minimize μ," or "maximize μ and m and minimize M." With such a question, you don't even need to know the correct formula, but you do need to understand how the pulley system works. The downward motion is due to the gravitational force on m and is opposed by the force of friction on M, so we would maximize the downward acceleration by maximizing m and minimizing M and μ.

3. **If the system does not move, which of the following must be true?** You would then be given a number of formulas relating M, m, and μ. The idea behind such a question is that the system does not move if the downward force on m is less than or equal to the force of friction on M, so $mg \le \mu Mg$.

These examples are perhaps less demanding than a question that expects you to derive or recall a complex formula and then plug numbers into it, but they are still difficult questions. In fact, they are about as difficult as mechanics questions on the SAT II will get.

Inclined Planes

What we call wedges or slides in everyday language are called **inclined planes** in physics-speak. From our experience on slides during recess in elementary school, sledding down hills in the winter, and skiing, we know that when people are placed on slippery inclines, they slide down the slope. We also know that slides can sometimes be sticky, so that when you are at the top of the incline, you need to give yourself a push to overcome the force of static friction. As you descend a sticky slide, the force of kinetic friction opposes your motion. In this section, we will consider problems involving inclined planes both with and without friction. Since they're simpler, we'll begin with frictionless planes.

Frictionless Inclined Planes

Suppose you place a 10 kg box on a frictionless 30° inclined plane and release your hold, allowing the box to slide to the ground, a horizontal distance of d meters and a vertical distance of h meters.

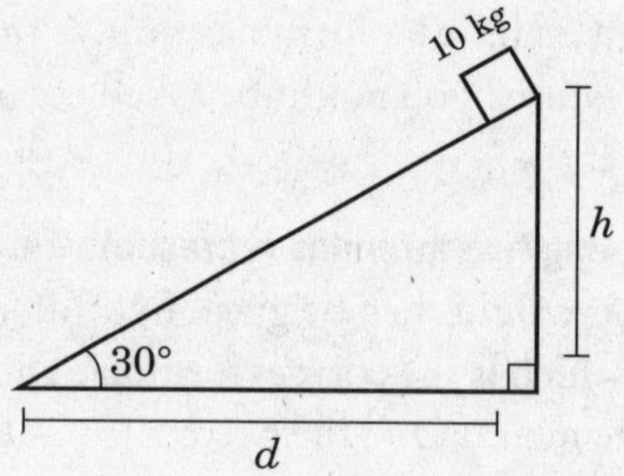

Before we continue, let's follow those three important preliminary steps for solving problems in mechanics:

1. **Ask yourself how the system will move:** Because this is a frictionless plane, there is nothing to stop the box from sliding down to the bottom. Experience suggests that the steeper the incline, the faster an object will slide, so we can expect the acceleration and velocity of the box to be affected by the angle of the plane.
2. **Choose a coordinate system:** Because we're interested in how the box slides along the inclined plane, we would do better to orient our coordinate system to the slope of the plane. The x-axis runs parallel to the plane, where downhill is the positive x direction, and the y-axis runs perpendicular to the plane, where up is the positive y direction.

3. **Draw free-body diagrams:** The two forces acting on the box are the force of gravity, acting straight downward, and the normal force, acting perpendicular to the inclined plane, along the y-axis. Because we've oriented our coordinate system to the slope of the plane, we'll have to resolve the vector for the gravitational force, $\mathbf{mg}$, into its x- and y-components. If you recall what we learned about vector decomposition in Chapter 1, you'll know you can break $\mathbf{mg}$ down into a vector of magnitude cos 30° in the negative y direction and a vector of magnitude sin 30° in the positive x direction. The result is a free-body diagram that looks something like this:

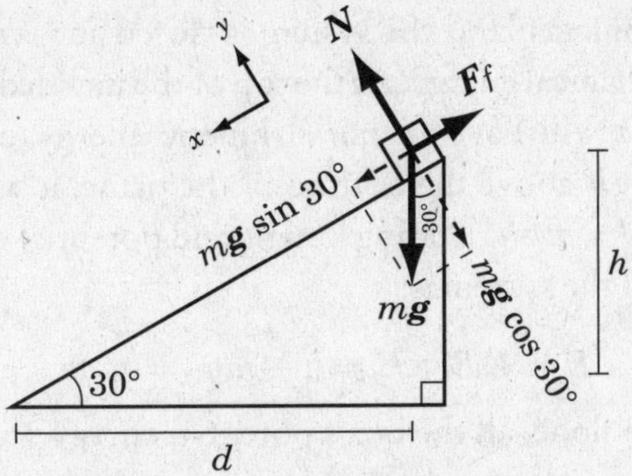

Decomposing the $\mathbf{mg}$ vector gives a total of three force vectors at work in this diagram: the y-component of the gravitational force and the normal force, which cancel out; and the x-component of the gravitational force, which pulls the box down the slope. Note that the steeper the slope, the greater the force pulling the box down the slope.

Now let's solve some problems. For the purposes of these problems, take the acceleration due to gravity to be $g = 10$ m/s^2. Like the SAT II, we will give you the values of the relevant trigonometric functions: cos 30 = sin 60 = 0.866, cos 60 = sin 30 = 0.500.

1. What is the magnitude of the normal force?
2. What is the acceleration of the box?
3. What is the velocity of the box when it reaches the bottom of the slope?
4. What is the work done on the box by the force of gravity in bringing it to the bottom of the plane?

1. **What is the magnitude of the normal force?** The box is not moving in the y direction, so the normal force must be equal to the y-component of the gravitational force. Calculating the normal force is then just a matter of plugging a few numbers in for variables in order to find the y-component of the gravitational force:

$$\begin{aligned} N &= mg\cos 30 \\ &= (10\text{ kg})(10\text{ m/s}^2)(0.866) \\ &= 86.6\text{ N} \end{aligned}$$

2. **What is the acceleration of the box?** We know that the force pulling the box in the positive x direction has a magnitude of mg sin 30. Using Newton's Second Law, $F = ma$, we just need to solve for a:

$$\begin{aligned} ma &= mg \sin 30 \\ a &= g \sin 30 \\ &= (10 \text{ m/s}^2)(0.500) \\ &= 5 \text{ m/s}^2 \end{aligned}$$

3. **What is the velocity of the box when it reaches the bottom of the slope?** Because we're dealing with a frictionless plane, the system is closed and we can invoke the law of conservation of mechanical energy. At the top of the inclined plane, the box will not be moving and so it will have an initial kinetic energy of zero ($KE_{\text{initial}} = 0$). Because it is a height h above the bottom of the plane, it will have a gravitational potential energy of $U = mgh$. Adding kinetic and potential energy, we find that the mechanical energy of the system is:

$$E = KE + U = 0 + mgh = mgh$$

At the bottom of the slope, all the box's potential energy will have been converted into kinetic energy. In other words, the kinetic energy, $\frac{1}{2}mv^2$, of the box at the bottom of the slope is equal to the potential energy, mgh, of the box at the top of the slope. Solving for v, we get:

$$\begin{aligned} v &= \sqrt{2gh} \\ &= \sqrt{2(10 \text{ m/s}^2)h} \\ &= 4.47\sqrt{h} \end{aligned}$$

3. **What is the work done on the box by the force of gravity in bringing it to the bottom of the inclined plane?** The fastest way to solve this problem is to appeal to the work-energy theorem, which tells us that the work done on an object is equal to its change in kinetic energy. At the top of the slope the box has no kinetic energy, and at the bottom of the slope its kinetic energy is equal to its potential energy at the top of the slope, mgh. So the work done on the box is:

$$\begin{aligned} W = mgh &= (10 \text{ kg})(10 \text{ m/s}^2)h \\ &= 100h \text{ J} \end{aligned}$$

Note that the work done is independent of how steep the inclined plane is, and is only dependent on the object's change in height when it slides down the plane.

Frictionless Inclined Planes with Pulleys

Let's bring together what we've learned about frictionless inclined planes and pulleys on tables into one exciting über-problem:

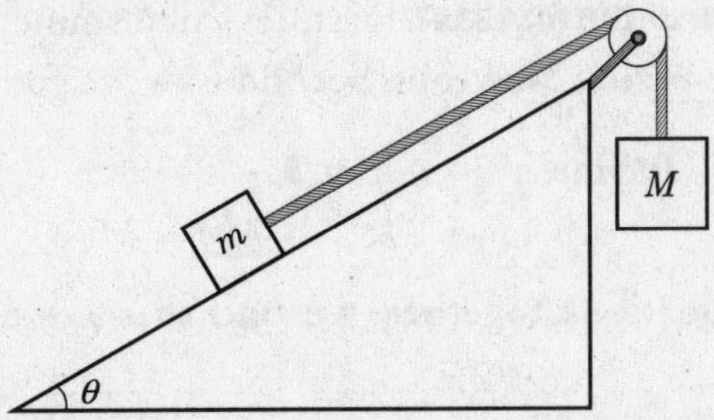

Assume for this problem that $M > m \sin\theta$—that is, mass M will pull mass m up the slope. Now let's ask those three all-important preliminary questions:

1. **Ask yourself how the system will move:** Because the two masses are connected by a rope, we know that they will have the same velocity and acceleration. We also know that the tension in the rope is constant throughout its length. Because $M > m \sin\theta$, we know that when the system is released from rest, mass M will move downward and mass m will slide up the inclined plane.
2. **Choose a coordinate system:** Do the same thing here that we did with the previous pulley-on-a-table problem. Make the x-axis parallel to the rope, with the positive x direction being up for mass M and downhill for mass m, and the negative x direction being down for mass M and uphill for mass m. Make the y-axis perpendicular to the rope, with the positive y-axis being away from the inclined plane, and the negative y-axis being toward the inclined plane.
3. **Draw free-body diagrams:** We've seen how to draw free-body diagrams for masses suspended from pulleys, and we've seen how to draw free-body diagrams for masses on inclined planes. All we need to do now is synthesize what we already know:

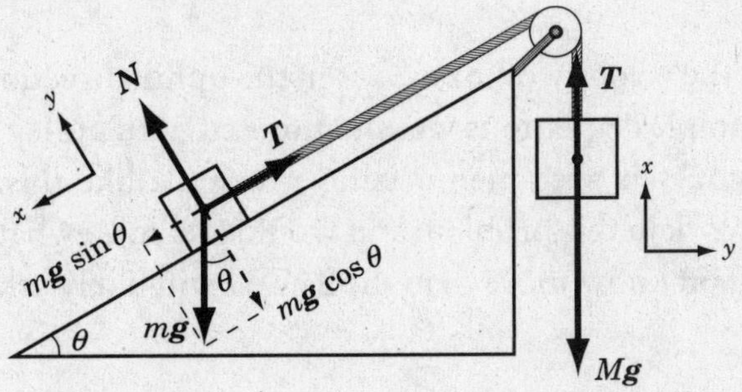

Now let's tackle a couple of questions:

1. What is the acceleration of the masses?
2. What is the velocity of mass m after mass M has fallen a distance h?

1. What is the acceleration of the masses? First, let's determine the net force acting on each of the masses. Applying Newton's Second Law we get:

$$\text{for mass } m\text{: } mg\sin\theta - T = ma$$
$$\text{for mass } M\text{: } T - Mg = Ma$$

Adding these two equations together, we find that $mg\sin\theta - Mg = (m+M)a$. Solving for a, we get:

$$a = \frac{g(m\sin\theta - M)}{(m+M)}$$

Because $M > m\sin\theta$, the acceleration is negative, which, as we defined it, is down for mass M and uphill for mass m.

2. What is the velocity of mass *m* after mass *M* has fallen a distance *h*? Once again, the inclined plane is frictionless, so we are dealing with a closed system and we can apply the law of conservation of mechanical energy. Since the masses are initially at rest, $KE_{\text{initial}} = 0$. Since mass M falls a distance h, its potential energy changes by $-Mgh$. If mass M falls a distance h, then mass m must slide the same distance up the slope of the inclined plane, or a vertical distance of $h\sin\theta$. Therefore, mass m's potential energy increases by $mgh\sin\theta$. Because the sum of potential energy and kinetic energy cannot change, we know that the kinetic energy of the two masses increases precisely to the extent that their potential energy decreases. We have all we need to scribble out some equations and solve for v:

$$\frac{1}{2}Mv^2 + \frac{1}{2}mv^2 - Mgh + mgh\sin\theta = 0$$
$$\frac{1}{2}(m+M)v^2 = gh(M - m\sin\theta)$$
$$v = \sqrt{\frac{2gh(M - m\sin\theta)}{m+M}}$$

Finally, note that the velocity of mass m is in the uphill direction.

As with the complex equations we encountered with pulley systems above, you needn't trouble yourself with memorizing a formula like this. If you understand the principles at work in this problem and would feel somewhat comfortable deriving this formula, you know more than the SAT II will likely ask of you.

Inclined Planes With Friction

There are two significant differences between frictionless inclined plane problems and inclined plane problems where friction is a factor.

1. **There's an extra force to deal with.** The force of friction will oppose the downhill component of the gravitational force.
2. **We can no longer rely on the law of conservation of mechanical energy.** Because energy is being lost through the friction between the mass and the inclined plane, we are no longer dealing with a closed system. Mechanical energy is not conserved.

Consider the 10 kg box we encountered in our example of a frictionless inclined plane. This time, though, the inclined plane has a coefficient of kinetic friction of $\mu = 0.5$. How will this additional factor affect us? Let's follow three familiar steps:

1. **Ask yourself how the system will move:** If the force of gravity is strong enough to overcome the force of friction, the box will accelerate down the plane. However, because there is a force acting against the box's descent, we should expect it to slide with a lesser velocity than it did in the example of the frictionless plane.
2. **Choose a co-ordinate system:** There's no reason not to hold onto the co-ordinate system we used before: the positive *x* direction is down the slope, and the positive *y* direction is upward, perpendicular to the slope.
3. **Draw free-body diagrams:** The free-body diagram will be identical to the one we drew in the example of the frictionless plane, except we will have a vector for the force of friction in the negative *x* direction.

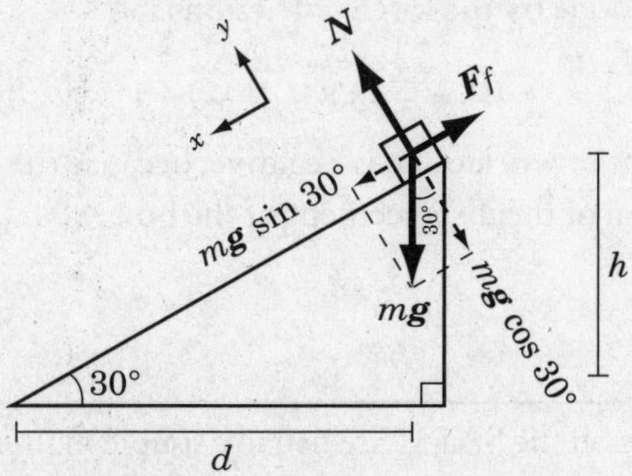

Now let's ask some questions about the motion of the box.

1. What is the force of kinetic friction acting on the box?
2. What is the acceleration of the box?
3. What is the work done on the box by the force of kinetic friction?

What is the force of kinetic friction acting on the box? The normal force acting on the box is 86.6 N, exactly the same as for the frictionless inclined plane. The force of kinetic friction is defined as $F_f = \mu N$, so plugging in the appropriate values for μ and N:

$$\begin{aligned} F_f = \mu N &= 0.500 \cdot 86.6 \text{ N} \\ &= 43.3 \text{ N} \end{aligned}$$

Remember, though, that the force of friction is exerted in the negative x direction, so the correct answer is –43.3 N.

What is the acceleration of the box? The net force acting on the box is the difference between the downhill gravitational force and the force of friction: $F = mg \sin 30 - F_f$. Using Newton's Second Law, we can determine the net force acting on the box, and then solve for a:

$$\begin{aligned} ma &= mg \sin 30 - F_f \\ (10 \text{ kg})a &= (10 \text{ kg})(10 \text{ m/s}^2)(0.500) - 43.3 \text{ N} \\ a &= \frac{50 \text{ N} - 43.3 \text{ N}}{10 \text{ kg}} \\ &= 0.67 \text{ m/s}^2 \end{aligned}$$

Because $mg \sin 30 > F_f$, the direction of the acceleration is in the downhill direction.

What is the work done on the box by the force of kinetic friction? Since $W = \mathbf{F} \cdot \mathbf{d}$, the work done by the force of friction is the product of the force of friction and the displacement of the box in the direction that the force is exerted. Because the force of friction is exerted in the negative x direction, we need to find the displacement of the box in the x direction. We know that it has traveled a horizontal distance of d and a vertical distance of h. The Pythagorean theorem then tells us that the displacement of the box is $\sqrt{d^2 + h^2}$. Recalling that the force of friction is –43.3 N, we know that the work done by the force of friction is

$$W = -43.3\sqrt{d^2 + h^2} \text{ J}$$

Note that the amount of work done is negative, because the force of friction acts in the opposite direction of the displacement of the box.

Springs

Questions about springs on the SAT II are usually simple matters of a mass on a spring oscillating back and forth. However, spring motion is the most interesting of the four topics we will cover here because of its generality. The **harmonic motion** that springs exhibit applies equally to objects moving in a circular path and to the various wave phenomena that we'll study later in this book. So before we dig in to the nitty-gritty of your typical SAT II spring questions, let's look at some general features of harmonic motion.

Oscillation and Harmonic Motion

Consider the following physical phenomena:

- When you drop a rock into a still pond, the rock makes a big splash, which causes ripples to spread out to the edges of the pond.
- When you pluck a guitar string, the string vibrates back and forth.
- When you rock a small boat, it wobbles to and fro in the water before coming to rest again.
- When you stretch out a spring and release it, the spring goes back and forth between being compressed and being stretched out.

There are just a few examples of the widespread phenomenon of **oscillation**. Oscillation is the natural world's way of returning a system to its **equilibrium position**, the stable position of the system where the net force acting on it is zero. If you throw a system off-balance, it doesn't simply return to the way it was; it oscillates back and forth about the equilibrium position.

A system oscillates as a way of giving off energy. A system that is thrown off-kilter has more energy than a system in its equilibrium position. To take the simple example of a spring, a stretched-out spring will start to move as soon as you let go of it: that motion is evidence of kinetic energy that the spring lacks in its equilibrium position. Because of the law of conservation of energy, a stretched-out spring cannot simply return to its equilibrium position; it must release some energy in order to do so. Usually, this energy is released as thermal energy caused by friction, but there are plenty of interesting exceptions. For instance, a plucked guitar string releases sound energy: the music we hear is the result of the string returning to its equilibrium position.

The movement of an oscillating body is called harmonic motion. If you were to graph the position, velocity, or acceleration of an oscillating body against time, the result would be a sinusoidal wave; that is, some variation of a $y = a \sin bx$ or a $y = a \cos bx$ graph. This generalized form of harmonic motion applies not only to springs and guitar strings, but to anything that moves in a **cycle**. Imagine placing a pebble on the edge of a turntable, and watching the turntable rotate while looking at it from the side. You will see the pebble moving back and forth in one dimension. The pebble will appear to oscillate just like a spring: it will appear to move fastest at the middle of its trajectory and slow to a halt and reverse direction as it reaches the edge of its trajectory.

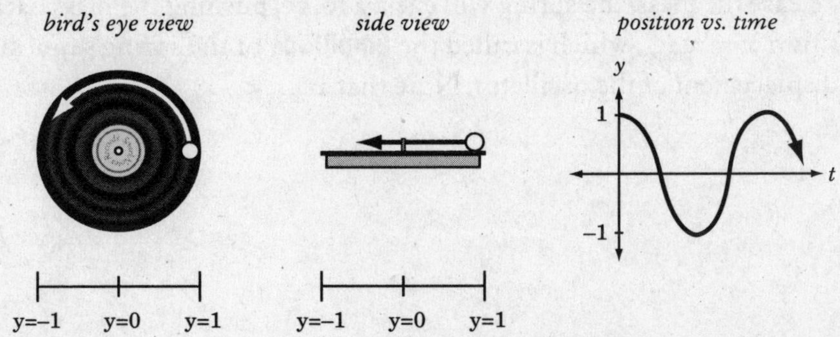

Special Problems

This example serves two purposes. First, it shows you that the oscillation of springs is just one of a wide range of phenomena exhibiting harmonic motion. Anything that moves in a cyclic pattern exhibits harmonic motion. This includes the light and sound waves without which we would have a lot of trouble moving about in the world. Second, we bring it up because the SAT II has been known to test students on the nature of the horizontal or vertical component of the motion of an object in circular motion. As you can see, circular motion viewed in one dimension is harmonic motion.

Though harmonic motion is one of the most widespread and important of physical phenomena, your understanding of it will not be taxed to any great extent on the SAT II. In fact, beyond the motion of springs and pendulums, everything you will need to know will be covered in this book in the chapter on Waves (see page 301). The above discussion is mostly meant to fit your understanding of the oscillation of springs into a wider context.

The Oscillation of a Spring

Now let's focus on the harmonic motion exhibited by a spring. To start with, we'll imagine a mass, m, placed on a frictionless surface, and attached to a wall by a spring. In its equilibrium position, where no forces act upon it, the mass is at rest. Let's label this equilibrium position $x = 0$. Intuitively, you know that if you compress or stretch out the spring it will begin to oscillate.

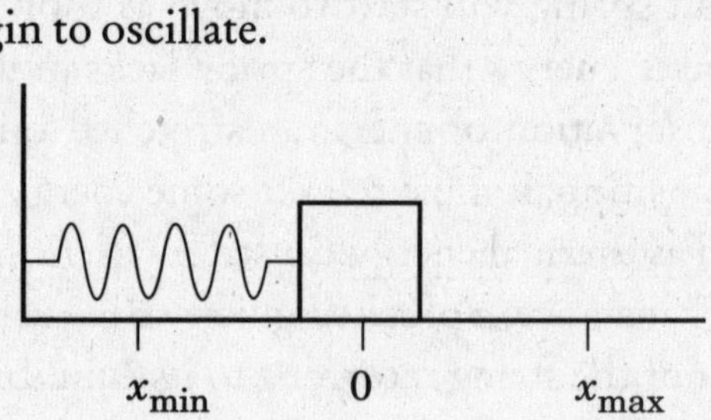

Suppose you push the mass toward the wall, compressing the spring, until the mass is in position $x = x_{min}$.

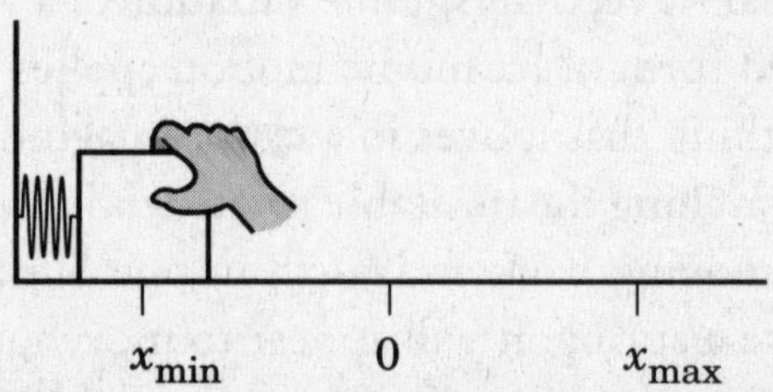

When you release the mass, the spring will exert a force, pushing the mass back until it reaches position $x = x_{max}$, which is called the **amplitude** of the spring's motion, or the maximum displacement of the oscillator. Note that $x_{min} = -x_{max}$.

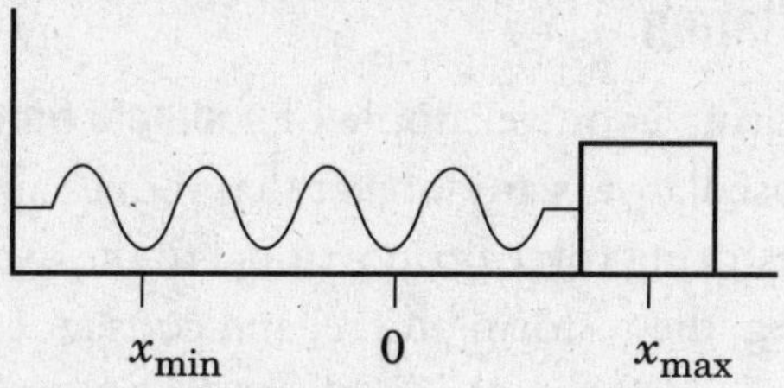

By that point, the spring will be stretched out, and will be exerting a force to pull the mass back in toward the wall. Because we are dealing with an idealized frictionless surface, the mass will not be slowed by the force of friction, and will oscillate back and forth repeatedly between x_{min} and x_{max}.

Hooke's Law

This is all well and good, but we can't get very far in sorting out the amplitude, the velocity, the energy, or anything else about the mass's motion if we don't understand the manner in which the spring exerts a force on the mass attached to it. The force, $\mathbf{F}$, that the spring exerts on the mass is defined by **Hooke's Law**:

$$\mathbf{F} = -k\mathbf{x}$$

where $\mathbf{x}$ is the spring's displacement from its equilibrium position and k is a constant of proportionality called the **spring constant**. The spring constant is a measure of "springiness": a greater value for k signifies a "tighter" spring, one that is more resistant to being stretched.

The greater k is, the tighter the spring is.

Hooke's Law tells us that the further the spring is displaced from its equilibrium position ($\mathbf{x}$) the greater the force the spring will exert in the direction of its equilibrium position ($\mathbf{F}$). We call $\mathbf{F}$ a **restoring force**: it is always directed towards equilibrium. Because F and x are directly proportional, a graph of F vs. x is a line with slope $-k$.

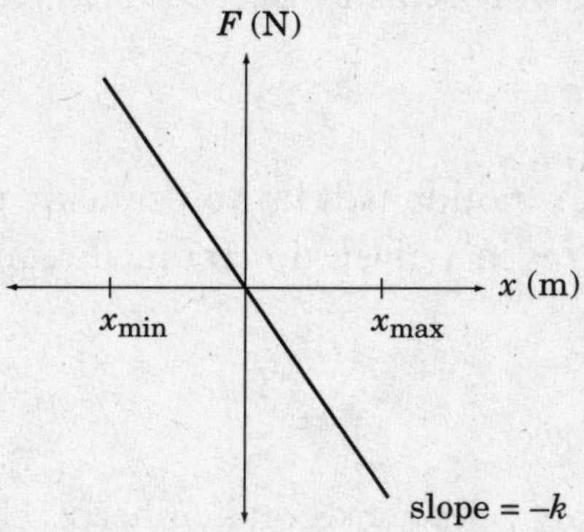

Simple Harmonic Oscillation

A mass oscillating on a spring is one example of a **simple harmonic oscillator**. Specifically, a simple harmonic oscillator is any object that moves about a stable equilibrium point and experiences a restoring force proportional to the oscillator's displacement.

For an oscillating spring, the restoring force, and consequently the acceleration, are greatest and positive at x_{min}. These quantities decrease as x approaches the equilibrium position and are zero at $x = 0$. The restoring force and acceleration—which are now negative—increase in magnitude as x approaches x_{max} and are maximally negative at x_{max}.

Important Properties of a Mass on a Spring

There are a number of important properties related to the motion of a mass on a spring, all of which are fair game for the SAT II. Remember, though: the test makers have no interest in testing your ability to recall complex formulas and perform difficult mathematical operations. You may be called upon to know the simpler of these formulas, but not the complex ones. As we mentioned at the end of the section on pulleys, it's less important that you memorize the formulas and more important that you understand what they mean. If you understand the principle, there probably won't be any questions that will stump you.

Period of Oscillation

The period of oscillation, T, of a spring is the amount of time it takes for a spring to complete a round-trip or cycle. Mathematically, the period of oscillation of a simple harmonic oscillator described by Hooke's law is:

$$T = 2\pi\sqrt{\frac{m}{k}}$$

This equation tells us that as the mass of the block, m, increases and the spring constant, k, decreases, the period increases. In other words, a heavy mass attached to an easily stretched spring will oscillate back and forth very slowly, while a light mass attached to a resistant spring will oscillate back and forth very quickly.

Frequency

The frequency of the spring's motion tells us how quickly the object is oscillating, or how many cycles it completes in a given timeframe. Frequency is inversely proportional to period:

$$f = \frac{1}{T}$$

Frequency is given in units of cycles per second, or hertz (Hz).

Potential Energy

The potential energy of a spring (U_s) is sometimes called elastic energy, because it results from the spring being stretched or compressed. Mathematically, U_s is defined by:

$$U_s = \frac{1}{2}kx^2$$

The potential energy of a spring is greatest when the coil is maximally compressed or stretched, and is zero at the equilibrium position.

Kinetic Energy

The SAT II will not test you on the motion of springs involving friction, so for the purposes of the test, the mechanical energy of a spring is a conserved quantity. As we recall, mechanical energy is the sum of the kinetic energy and potential energy.

At the points of maximum compression and extension, the velocity, and hence the kinetic energy, is zero and the mechanical energy is equal to the potential energy, $U_s = \frac{1}{2} kx^2$.

At the equilibrium position, the potential energy is zero, and the velocity and kinetic energy are maximized. The kinetic energy at the equilibrium position is equal to the mechanical energy:

$$KE_{\text{max}} = \frac{1}{2}mv^2 = \frac{1}{2}kx^2_{\text{min}}$$

From this equation, we can derive the maximum velocity:

$$v_{\text{max}} = x\sqrt{\frac{k}{m}}$$

You won't need to know this equation, but it might be valuable to note that the velocity increases with a large displacement, a resistant spring, and a small mass.

Summary

It is highly unlikely that the formulas discussed above will appear on the SAT II. More likely, you will be asked conceptual questions such as: at what point in a spring's oscillation is the kinetic or potential energy maximized or minimized, for instance. The figure below summarizes and clarifies some qualitative aspects of simple harmonic oscillation. Your qualitative understanding of the relationship between force, velocity, and kinetic and potential energy in a spring system is far more likely to be tested than your knowledge of the formulas discussed above.

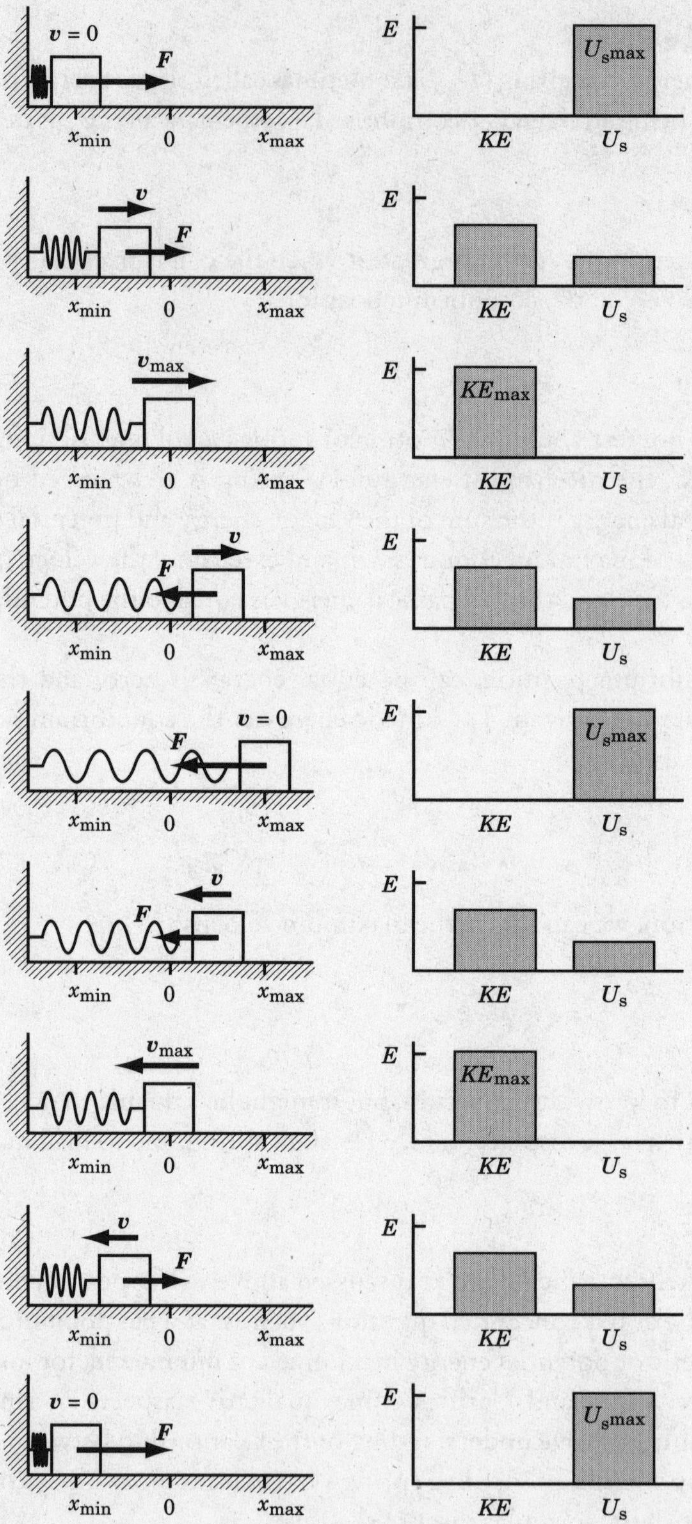

In this figure, ***v*** represents velocity, ***F*** represents force, *KE* represents kinetic energy, and U_s represents potential energy.

Vertical Oscillation of Springs

Now let's consider a mass attached to a spring that is suspended from the ceiling. Questions of this sort have a nasty habit of coming up on the SAT II. The oscillation of the spring when compressed or extended won't be any different, but we now have to take gravity into account.

Equilibrium Position

Because the mass will exert a gravitational force to stretch the spring downward a bit, the equilibrium position will no longer be at $x = 0$, but at $x = -h$, where h is the vertical displacement of the spring due to the gravitational pull exerted on the mass. The equilibrium position is the point where the net force acting on the mass is zero; in other words, the point where the upward restoring force of the spring is equal to the downward gravitational force of the mass.

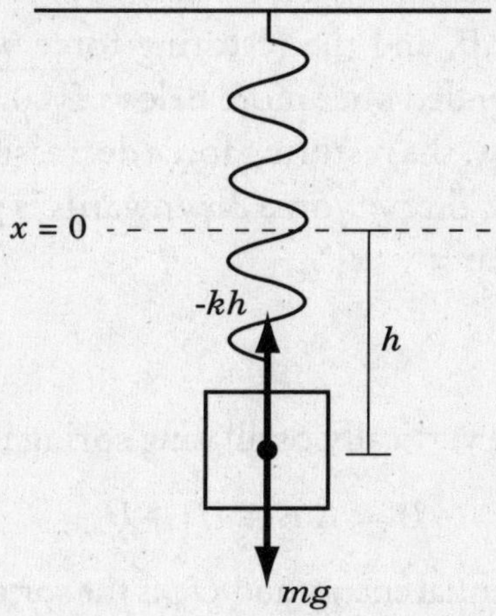

Combining the restoring force, $F = -kh$, and the gravitational force, $F = mg$, we can solve for h:

$$-kh = mg$$
$$-h = \frac{mg}{k}$$

Since m is in the numerator and k in the denominator of the fraction, the mass displaces itself more if it has a large weight and is suspended from a lax spring, as intuition suggests.

A Vertical Spring In Motion

If the spring is then stretched a distance d, where $d < h$, it will oscillate between $x_{max} = -h - d$ and $x_{min} = -h + d$.

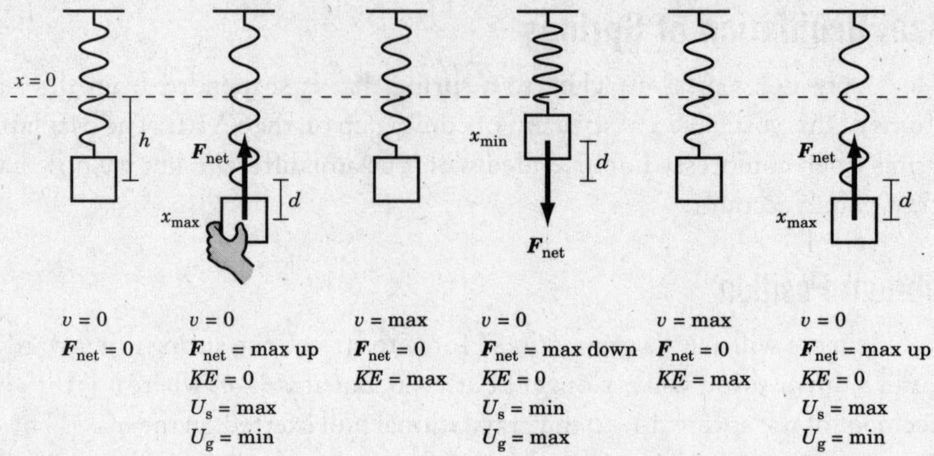

Throughout the motion of the mass, the force of gravity is constant and downward. The restoring force of the spring is always upwards, because even at x_{min} the mass is below the spring's initial equilibrium position of $x = 0$. Note that if d were greater than h, x_{min} would be above $x = 0$, and the restoring force would act in the downward direction until the mass descended once more below $x = 0$.

According to Hooke's Law, the restoring force decreases in magnitude as the spring is compressed. Consequently, the net force downwards is greatest at $x = x_{min}$ and the net force upwards is greatest at $x = x_{max}$.

Energy

The mechanical energy of the vertically oscillating spring is:

$$E = KE + U_g + U_s$$

where U_g is gravitational potential energy and U_s is the spring's (elastic) potential energy.

Note that the velocity of the block is zero at $x = x_{min}$ and $x = x_{max}$, and maximized at the equilibrium position, $x = -h$. Consequently, the kinetic energy of the spring is zero for $x = x_{max}$ and $x = x_{min}$ and is greatest at $x = -h$. The gravitational potential energy of the system increases with the height of the mass. The elastic potential energy of the spring is greatest when the spring is maximally extended at x_{max} and decreases with the extension of the spring.

How This Knowledge Will Be Tested

Most of the questions on the SAT II that deal with spring motion will ask qualitatively about the energy or velocity of a vertically oscillating spring. For instance, you may be shown a diagram capturing one moment in a spring's trajectory and asked about the relative magnitudes of the gravitational and elastic potential energies and kinetic energy. Or you may be asked at what point in a spring's trajectory the velocity is maximized. The answer, of course, is that it is maximized at the equilibrium position. It is far less likely that you will be asked a question that involves any sort of calculation.

Pendulums

A **pendulum** is defined as a mass, or bob, connected to a rod or rope, that experiences simple harmonic motion as it swings back and forth without friction. The equilibrium position of the pendulum is the position when the mass is hanging directly downward.

Consider a pendulum bob connected to a massless rope or rod, that is held at an angle θ_{max} from the horizontal. If you release the mass, then the system will swing to position $-\theta_{max}$ and back again.

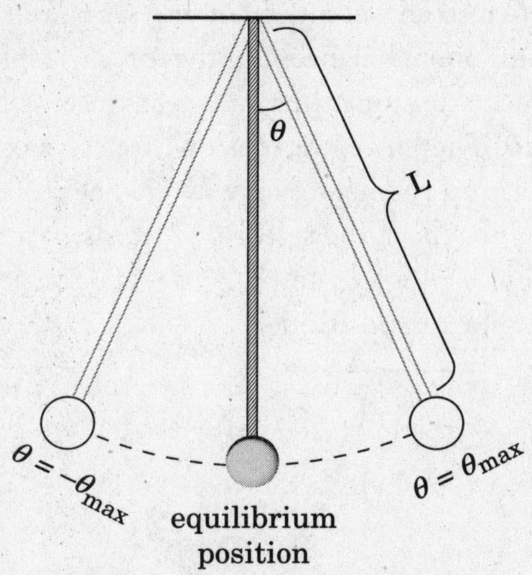

The oscillation of a pendulum is much like that of a mass on a spring. However, there are significant differences, and many a student has been tripped up by trying to apply the principles of a spring's motion to pendulum motion.

Properties of Pendulum Motion

As with springs, there are a number of properties of pendulum motion that you might be tested on, from frequency and period to kinetic and potential energy. Let's apply our three-step method of approaching special problems in mechanics and then look at the formulas for some of those properties.

1. **Ask yourself how the system will move:** It doesn't take a rocket scientist to surmise that when you release the pendulum bob it will accelerate toward the equilibrium position. As it passes through the equilibrium position, it will slow down until it reaches position $-\theta$, and then accelerate back. At any given moment, the velocity of the pendulum bob will be perpendicular to the rope. The pendulum's trajectory describes an arc of a circle, where the rope is a radius of the circle and the bob's velocity is a line tangent to the circle.

2. **Choose a co-ordinate system:** We want to calculate the forces acting on the pendulum at any given point in its trajectory. It will be most convenient to choose a y-axis that runs parallel to the rope. The x-axis then runs parallel to the instantaneous velocity of the bob so that, at any given moment, the bob is moving along the x-axis.
3. **Draw free-body diagrams:** Two forces act on the bob: the force of gravity, $\boldsymbol{F} = m\boldsymbol{g}$, pulling the bob straight downward and the tension of the rope, $\boldsymbol{F}_T$, pulling the bob upward along the y-axis. The gravitational force can be broken down into an x component, $mg \sin\theta$, and a y component, $mg \cos\theta$. The y component balances out the force of tension—the pendulum bob doesn't accelerate along the y-axis—so the tension in the rope must also be $mg \cos\theta$. Therefore, the tension force is maximum for the equilibrium position and decreases with θ. The restoring force is $mg \sin\theta$, so, as we might expect, the restoring force is greatest at the endpoints of the oscillation, $\theta = \pm\theta_{max}$ and is zero when the pendulum passes through its equilibrium position.

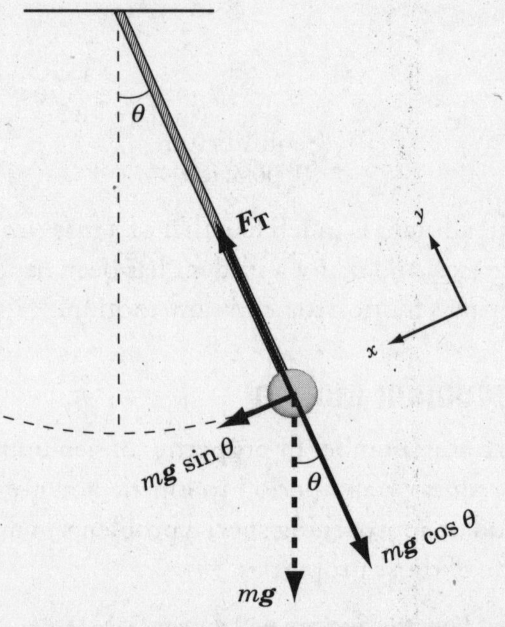

You'll notice that the restoring force for the pendulum, $mg \sin\theta$, is not directly proportional to the displacement of the pendulum bob, θ, which makes calculating the various properties of the pendulum very difficult. Fortunately, pendulums usually only oscillate at small angles, where $\sin\theta \approx \theta$. In such cases, we can derive more straightforward formulas, which are admittedly only approximations. However, they're good enough for the purposes of the SAT II.

Period

The period of oscillation of the pendulum, T, is defined in terms of the acceleration due to gravity, g, and the length of the pendulum, L:

$$T = 2\pi\sqrt{\frac{L}{g}}$$

This is a pretty scary-looking equation, but there's really only one thing you need to gather from it: the longer the pendulum rope, the longer it will take for the pendulum to oscillate back and forth. You should also note that the mass of the pendulum bob and the angle of displacemet play no role in determining the period of oscillation.

Energy

The mechanical energy of the pendulum is a conserved quantity. The potential energy of the pendulum, mgh, increases with the height of the bob; therefore the potential energy is minimized at the equilibrium point and is maximized at $\theta = \pm\theta_{max}$. Conversely, the kinetic energy and velocity of the pendulum are maximized at the equilibrium point and minimized when $\theta = \pm\theta_{max}$.

The figure below summarizes this information in a qualitative manner, which is the manner in which you are most likely to find it on the SAT II. In this figure, v signifies velocity, F_r signifies the restoring force, F_T signifies the tension in the pendulum string, U signifies potential energy, and KE signifies kinetic energy.

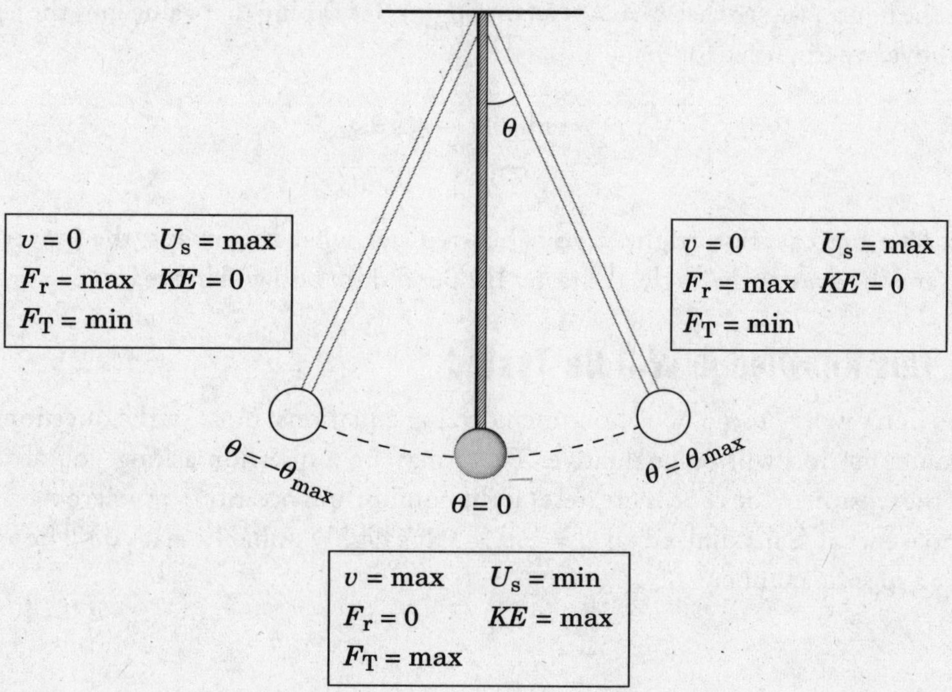

Velocity

Calculating the velocity of the pendulum bob at the equilibrium position requires that we arrange our coordinate system so that the height of the bob at the equilibrium position is zero. Then the total mechanical energy is equal to the kinetic energy at the equilibrium point where $U = 0$. The total mechanical energy is also equal to the total potential energy at $\pm\theta_{max}$ where $KE = 0$. Putting these equalities together, we get

$$E = \frac{1}{2}mv^2 = mgh$$

But what is h?

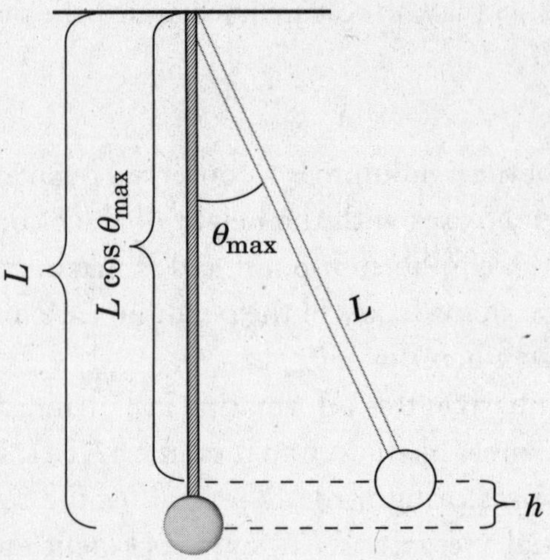

From the figure, we see that $h = L - L\cos(-\theta_{max})$. If we plug that value into the equation above, we can solve for v:

$$\frac{1}{2}mv^2 = mgL(1 - \cos\theta_{max})$$

$$v = \sqrt{2gL(1 - \cos\theta_{max})}$$

Don't let a big equation frighten you. Just register what it conveys: the longer the string and the greater the angle, the faster the pendulum bob will move.

How This Knowledge Will Be Tested

Again, don't worry too much about memorizing equations: most of the questions on pendulum motion will be qualitative. There may be a question asking you at what point the tension in the rope is greatest (at the equilibrium position) or where the bob's potential energy is maximized (at $\theta = \pm\theta_{max}$). It's highly unlikely that you'll be asked to give a specific number.

Key Formulas

Hooke's Law	$F = -kx$
Period of Oscillation of a Spring	$T = 2\pi\sqrt{\frac{m}{k}}$
Frequency	$f = \frac{1}{T}$
Potential Energy of a Spring	$U_s = \frac{1}{2}kx^2$
Velocity of a Spring at the Equilibrium Position	$v_{\text{max}} = x\sqrt{\frac{k}{m}}$
Period of Oscillation of a Pendulum	$T = 2\pi\sqrt{\frac{L}{g}}$
Velocity of a Pendulum Bob at the Equilibrium Position	$v = \sqrt{2gL(1 - \cos\theta_{\text{max}})}$

Practice Questions

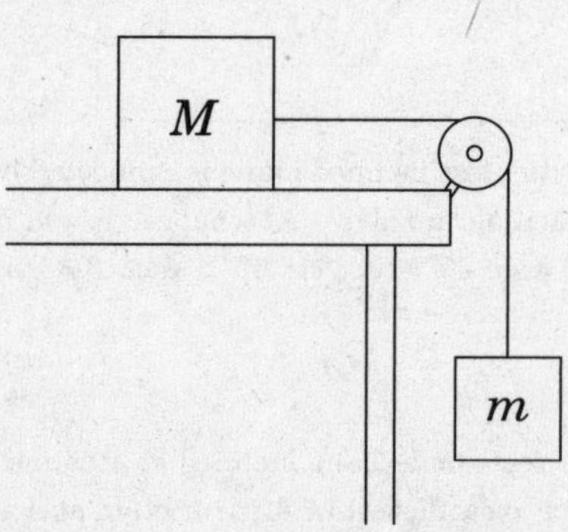

1. Two masses, m and M, are connected to a pulley system attached to a table, as in the diagram above. What is the minimum value for the coefficient of static friction between mass M and the table if the pulley system does not move?

   (A) m/M (B) M/m (C) $g\,(m/M)$ (D) $g\,(M/m\}$ (E) $g(M - m)$

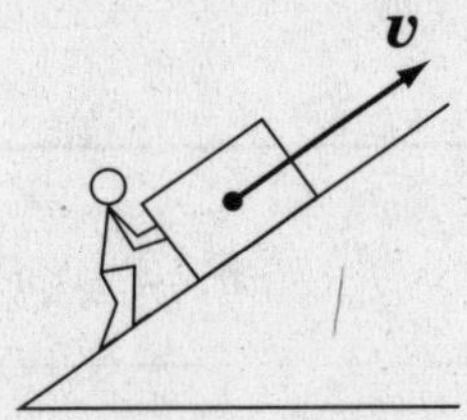

2. A mover pushes a box up an inclined plane, as shown in the figure above. Which of the following shows the direction of the normal force exerted by the plane on the box?

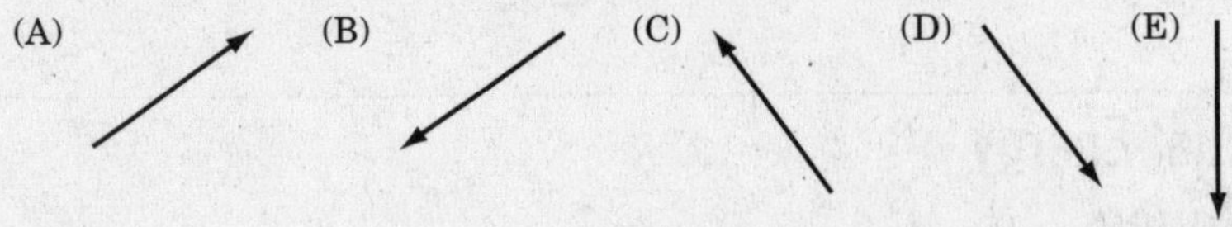

3. Consider a block sliding down a frictionless inclined plane with acceleration a. If we double the mass of the block, what is its acceleration?

(A) $a/4$ (B) $a/2$ (C) a (D) $2a$ (E) $4a$

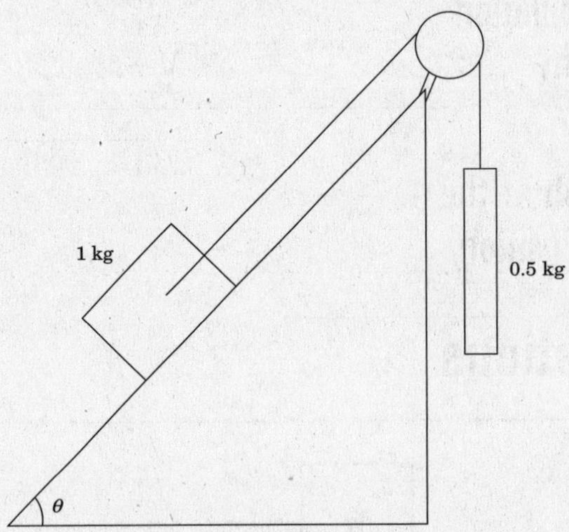

4. A 1 kg mass on a frictionless inclined plane is connected by a pulley to a hanging 0.5 kg mass, as in the diagram above. At what angle will the system be in equilibrium? $\cos 30° = \sin 60° = \sqrt{3}$, $\cos 60° = \sin 30° = 1/2$, $\cos 45° = \sin 45° = 1/\sqrt{2}$.

(A) 0° (B) –30° (C) 30° (D) 45° (E) 60°

5. An object of mass m rests on a plane inclined at an angle of θ. What is the maximum value for the coefficient of static friction at which the object will slide down the incline?

(A) $\sin\theta - \cos\theta$
(B) $\cos\theta - \sin\theta$
(C) $mg \sin\theta$
(D) $(\sin\theta)/(\cos\theta)$
(E) $\sin\theta + \cos\theta$

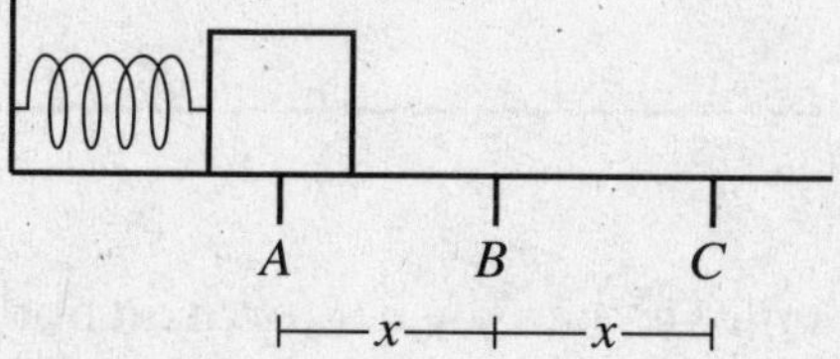

6. A mass on a frictionless surface is attached to a spring. The spring is compressed from its equilibrium position, *B*, to point *A*, a distance *x* from *B*. Point *C* is also a distance *x* from *B*, but in the opposite direction. When the mass is released and allowed to oscillated freely, at what point or points is its velocity maximized?

(A) *A*
(B) *B*
(C) *C*
(D) Both *A* and *C*
(E) Both *A* and *B*

7. An object of mass 3 kg is attached to a spring of spring constant 50 N/m. How far is the equilibrium position of this spring system from the point where the spring exerts no force on the object?

(A) 0.15 m
(B) 0.3 m
(C) 0.5 m
(D) 0.6 m
(E) 1.5 m

Questions 8–10 refer to a pendulum in its upward swing. That is, the velocity vector for the pendulum is pointing in the direction of *E*.

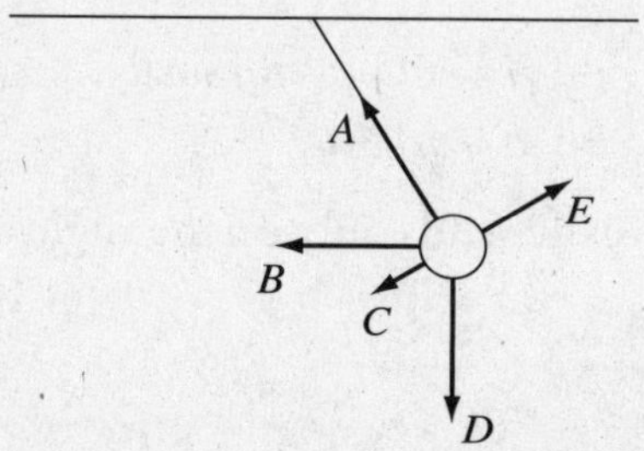

8. What is the direction of the force of gravity on the pendulum bob?

(A) *A* (B) *B* (C) *C* (D) *D* (E) *E*

9. What is the direction of the net force acting on the pendulum?

(A) *A* (B) *B* (C) *C* (D) *D* (E) *E*

10. If the pendulum string is suddenly cut, what is the direction of the velocity vector of the pendulum bob the moment it is released?

(A) *A* (B) *B* (C) *C* (D) *D* (E) *E*

Explanations

1. (A)

If the pulley system doesn't move, then the net force on both masses is zero. For mass m, that means that the force of gravity, mg, pulling it downward, is equal to the force of tension in the rope, pulling it upward. If the force of tension pulling mass m upward is mg, then the force of tension pulling mass M toward the edge of the table is also mg. That means that the force of static friction resisting the pull of the rope must also equal mg. The force of static friction for mass M is μMg, where μ is the coefficient of static friction. Since this force must be equal to mg, we can readily solve for μ:

$$\mu M g = mg$$
$$\mu = \frac{m}{M}$$

2. (C)

The normal force is always normal, i.e. perpendicular, to the surface that exerts it, and in a direction such that one of its components opposes gravity. In this case, the inclined plane's surface exerts the force, so the normal force vector must be perpendicular to the slope of the incline, and in the upward direction.

3. (A)

The acceleration of any particle due to the force of gravity alone doesn't depend on the mass, so the answer is (A). Whether or not the mass is on an inclined plane doesn't matter in the least bit! We can prove this by calculating the acceleration mathematically:

$$F = ma = mg\sin\theta$$
$$a = g\sin\theta$$

As you can see, the acceleration depends only on the angle of the incline, and not on the mass of the block.

4. (C)

The system will be in equilibrium when the net force acting on the 1 kg mass is equal to zero. A free-body diagram of the forces acting on the 1 kg mass shows that it is in equilibrium when the force of tension in the pulley rope is equal to $mg\sin\theta$, where m = 1 kg and θ is the angle of the inclined plane.

Since the system is in equilibrium, the force of tension in the rope must be equal and opposite to the force of gravity acting on the 0.5 kg mass. The force of gravity on the 0.5 kg mass,

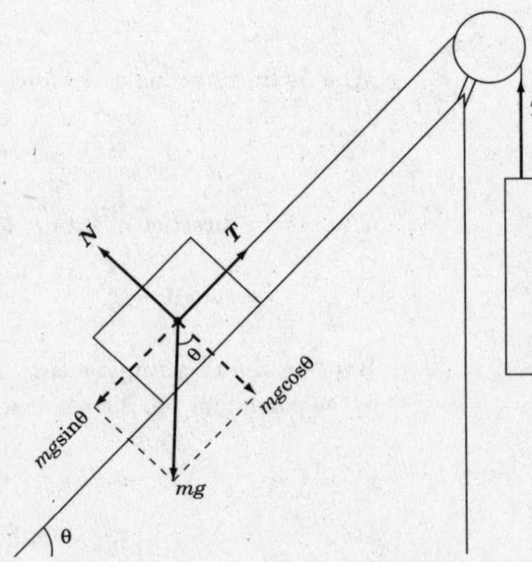

and hence the force of tension in the rope, has a magnitude of 0.5 g. Knowing that the force of tension is equal to $mg \sin \theta$, we can now solve for θ:

$$mg \sin\theta = 0.5g$$
$$(1 \text{ kg}) \sin\theta = 0.5$$
$$\sin\theta = 0.5$$
$$\theta = 30°$$

5. (D)

The best way to approach this problem is to draw a free-body diagram:

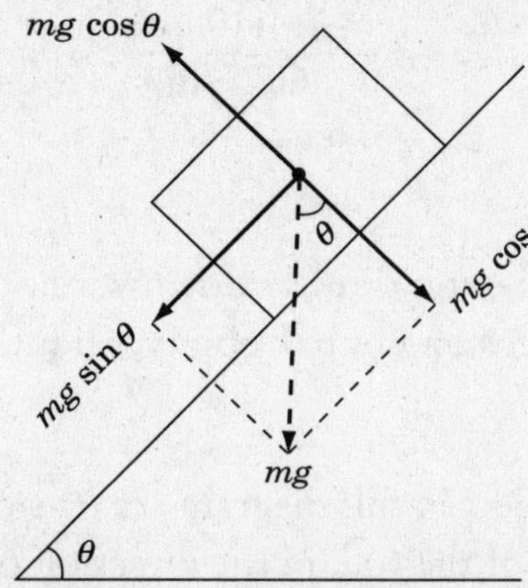

From the diagram, we can see that there is a force of $mg \sin \theta$ pulling the object down the incline. The force of static friction is given by μN, where μ is the coefficient of static friction and N is the normal force. If the object is going to move, then $mg \sin \theta > \mu N$. From the diagram, we can also see that $N = mg \cos \theta$, and with this information we can solve for μ:

$$\mu mg \cos\theta < mg \sin\theta$$
$$\mu < \frac{\sin\theta}{\cos\theta}$$

This inequality tells us that the maximum value of μ is $\sin \theta / \cos \theta$.

6. (B)

The velocity of a spring undergoing simple harmonic motion is a maximum at the equilibrium position, where the net force acting on the spring is zero.

7. (D)

The equilibrium position is the position where the net force acting on the object is zero. That would be the point where the downward force of gravity, mg, is perfectly balanced out by the upward spring force, kx, where k is the spring constant and x is the object's displacement. To solve this problem, we need to equate the two formulas for force and solve for x:

$$\begin{aligned} kx &= mg \\ x &= \frac{mg}{k} \\ &= \frac{(3\text{ kg})(10\text{ m/s}^2)}{50\text{ N/m}} \\ &= 0.6\text{ m} \end{aligned}$$

8. (D)

The force of gravity always operates directly downward on the surface of the earth. It doesn't matter what other forces act upon the body. Thus the answer is (D).

9. (C)

The forces acting upon the object in this diagram are tension and gravity. The force of tension is along the direction of the rod, in the direction of A. The force of gravity is directly downward, in the direction of D. The net force acting on the pendulum bob is the vector sum of these two forces, namely (C).

10. (E)

Since the instantaneous velocity of the pendulum bob is in the direction of E, that is the path that the object will travel along. Eventually, the force of gravity will cause the pendulum bob to fall downward, but the question only asks you for the instantaneous velocity of the bob the moment it is released.

Linear Momentum

THE CONCEPT OF **linear momentum** is closely tied to the concept of force—in fact, Newton first defined his Second Law not in terms of mass and acceleration, but in terms of momentum. Like energy, linear momentum is a conserved quantity in closed systems, making it a very handy tool for solving problems in mechanics. On the whole, it is useful to analyze systems in terms of energy when there is an exchange of potential energy and kinetic energy. Linear momentum, however, is useful in those cases where there is no clear measure for potential energy. In particular, we will use the law of **conservation of momentum** to determine the outcome of collisions between two bodies.

Chapter Contents

What is Linear Momentum? *135*
Impulse *136*
Conservation of Momentum *138*
Collisions *141*
Center of Mass *145*
Key Formulas *149*
Practice Questions *150*
Explanations *153*

What is Linear Momentum?

Linear momentum is a vector quantity defined as the product of an object's mass, m, and its velocity, $\boldsymbol{v}$. Linear momentum is denoted by the letter $\boldsymbol{p}$ and is called "momentum" for short:

$$\boldsymbol{p} = m\boldsymbol{v}$$

Note that a body's momentum is always in the same direction as its velocity vector. The units of momentum are kg · m/s.

Fortunately, the way that we use the word "momentum" in everyday life is consistent with the definition of momentum in physics. For example, we say that a BMW driving 20 miles per hour has less momentum than the same car speeding on the highway at 80 miles per hour. Additionally, we know that if a large truck and a BMW travel at the same speed on a highway, the truck has a greater momentum than the BMW, because the truck has greater mass. Our everyday usage reflects the definition given above, that momentum is proportional to mass and velocity.

Linear Momentum and Newton's Second Law

In Chapter 3, we introduced Newton's Second Law as $\boldsymbol{F} = m\boldsymbol{a}$. However, since acceleration can be expressed as $\Delta\boldsymbol{v}/\Delta t$, we could equally well express Newton's Second Law as $\boldsymbol{F} = m\Delta\boldsymbol{v}/\Delta t$. Substituting $\boldsymbol{p}$ for $m\boldsymbol{v}$, we find an expression of Newton's Second Law in terms of momentum:

$$\bar{\boldsymbol{F}} = \frac{\Delta \boldsymbol{p}}{\Delta t}$$

In fact, this is the form in which Newton first expressed his Second Law. It is more flexible than $\boldsymbol{F} = m\boldsymbol{a}$ because it can be used to analyze systems where not just the velocity, but also the mass of a body changes, as in the case of a rocket burning fuel.

Impulse

The above version of Newton's Second Law can be rearranged to define the **impulse,** $\boldsymbol{J}$, delivered by a constant force, $\boldsymbol{F}$. Impulse is a vector quantity defined as the product of the force acting on a body and the time interval during which the force is exerted. If the force changes during the time interval, $\boldsymbol{F}$ is the average net force over that time interval. The impulse caused by a force during a specific time interval is equal to the body's change of momentum during that time interval: impulse, effectively, is a measure of change in momentum.

$$\boldsymbol{J} = \boldsymbol{F}\Delta t = \Delta \boldsymbol{p}$$

The unit of impulse is the same as the unit of momentum, kg · m/s.

Example

> A soccer player kicks a 0.1 kg ball that is initially at rest so that it moves with a velocity of 20 m/s. What is the impulse the player imparts to the ball? If the player's foot was in contact with the ball for 0.01 s, what was the force exerted by the player's foot on the ball?

What is the impulse the player imparts to the ball? Since impulse is simply the change in momentum, we need to calculate the difference between the ball's initial momentum and its final momentum. Since the ball begins at rest, its initial velocity, and hence its initial momentum, is zero. Its final momentum is:

$$\begin{aligned}\boldsymbol{p} = m\boldsymbol{v} &= (0.1\text{ kg})(20\text{ m/s})\\ &= 2\text{ kg}\cdot\text{m/s}\end{aligned}$$

Because the initial momentum is zero, the ball's change in momentum, and hence its impulse, is 2 kg · m/s.

What was the force exerted by the player's foot on the ball? Impulse is the product of the force exerted and the time interval over which it was exerted. It follows, then, that $\boldsymbol{F} = \boldsymbol{J}/\Delta t$. Since we have already calculated the impulse and have been given the time interval, this is an easy calculation:

$$\begin{aligned} \boldsymbol{F} &= \frac{\boldsymbol{J}}{\Delta t} = \frac{2 \text{ kg} \cdot \text{m/s}}{0.01 \text{ s}} \\ &= 200 \text{ N} \end{aligned}$$

Impulse and Graphs

The SAT II may also present you with a force vs. time graph, and ask you to calculate the impulse. There is a single, simple rule to bear in mind for calculating the impulse in force vs. time graphs:

The impulse caused by a force during a specific time interval is equal to the area underneath the force vs. time graph during the same interval.

If you recall, whenever you are asked to calculate the quantity that comes from multiplying the units measured by the y-axis with the units measured by the x-axis, you do so by calculating the area under the graph for the relevant interval.

Example

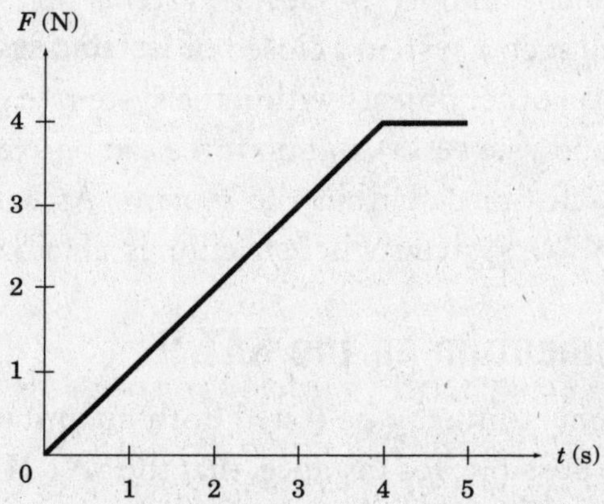

What is the impulse delivered by the force graphed in the figure above between $t = 0$ and $t = 5$?

The impulse over this time period equals the area of a triangle of height 4 and base 4 plus the area of a rectangle of height 4 and width 1. A quick calculation shows us that the impulse is:

$$\boldsymbol{J} = \frac{1}{2} \cdot 4 \cdot 4 + 4 \cdot 1 = 12 \text{ kg} \cdot \text{m/s}$$

Conservation of Momentum

If we combine Newton's Third Law with what we know about impulse, we can derive the important and extremely useful law of conservation of momentum.

Newton's Third Law tells us that, to every action, there is an equal and opposite reaction. If object *A* exerts a force $\boldsymbol{F}$ on object *B*, then object *B* exerts a force $-\boldsymbol{F}$ on object *A*. The net force exerted between objects *A* and *B* is zero.

The impulse equation, $\boldsymbol{J} = \boldsymbol{F}\Delta t = \Delta\boldsymbol{p}$, tells us that if the net force acting on a system is zero, then the impulse, and hence the change in momentum, is zero. Because the net force between the objects *A* and *B* that we discussed above is zero, the momentum of the system consisting of objects *A* and *B* does not change.

Suppose object *A* is a cue ball and object *B* is an eight ball on a pool table. If the cue ball strikes the eight ball, the cue ball exerts a force on the eight ball that sends it rolling toward the pocket. At the same time, the eight ball exerts an equal and opposite force on the cue ball that brings it to a stop. Note that both the cue ball and the eight ball each experience a change in momentum. However, the sum of the momentum of the cue ball and the momentum of the eight ball remains constant throughout. While the initial momentum of the cue ball, $\boldsymbol{p}_A$, is not the same as its final momentum, $\boldsymbol{p}_A{}'$, and the initial momentum of the eight ball, $\boldsymbol{p}_B$, is not the same as its final momentum, $\boldsymbol{p}_B{}'$, the initial momentum of the two balls combined is equal to the final momentum of the two balls combined:

$$\boldsymbol{p}_A + \boldsymbol{p}_B = \boldsymbol{p}_A{}' + \boldsymbol{p}_B{}'$$

The conservation of momentum only applies to systems that have no external forces acting upon them. We call such a system a closed or **isolated system**: objects within the system may exert forces on other objects within the system (e.g., the cue ball can exert a force on the eight ball and vice versa), but no force can be exerted between an object outside the system and an object within the system. As a result, conservation of momentum does not apply to systems where friction is a factor.

Conservation of Momentum on the SAT II

The conservation of momentum may be tested both quantitatively and qualitatively on the SAT II. It is quite possible, for instance, that the SAT II will contain a question or two that involves a calculation based on the law of conservation of momentum. In such a question, "conservation of momentum" will not be mentioned explicitly, and even "momentum" might not be mentioned. Most likely, you will be asked to calculate the velocity of a moving object after a collision of some sort, a calculation that demands that you apply the law of conservation of momentum.

Alternately, you may be asked a question that simply demands that you identify the law of conservation of momentum and know how it is applied. The first example we will look at is of this qualitative type, and the second example is of a quantitative conservation of momentum question.

Example 1

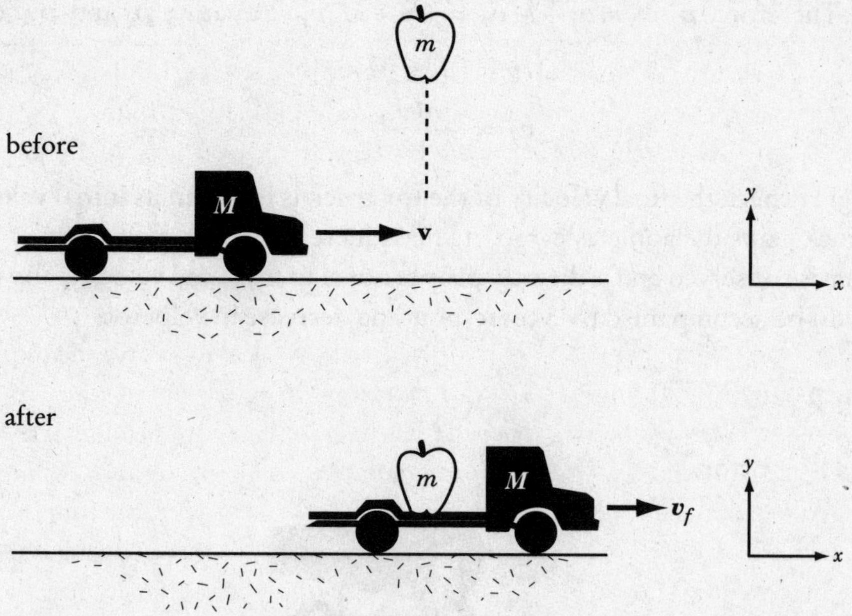

An apple of mass m falls into the bed of a moving toy truck of mass M. Before the apple lands in the car, the car is moving at constant velocity v on a frictionless track. Which of the following laws would you use to find the speed of the toy truck after the apple has landed?

(A) Newton's First Law
(B) Newton's Second Law
(C) Kinematic equations for constant acceleration
(D) Conservation of mechanical energy
(E) Conservation of linear momentum

Although the title of the section probably gave the solution away, we phrase the problem in this way because you'll find questions of this sort quite a lot on the SAT II. You can tell a question will rely on the law of conservation of momentum for its solution if you are given the initial velocity of an object and are asked to determine its final velocity after a change in mass or a collision with another object.

Some Supplemental Calculations

But how would we use conservation of momentum to find the speed of the toy truck after the apple has landed?

First, note that the net force acting in the x direction upon the apple and the toy truck is zero. Consequently, linear momentum in the x direction is conserved. The initial momentum of the system in the x direction is the momentum of the toy truck, $\boldsymbol{p}_i = M\boldsymbol{v}$.

Once the apple is in the truck, both the apple and the truck are traveling at the same speed, $\boldsymbol{v}_f$. Therefore, $\boldsymbol{p}_f = m\boldsymbol{v}_f + M\boldsymbol{v}_f = (m + M)\boldsymbol{v}_f$. Equating $\boldsymbol{p}_i$ and $\boldsymbol{p}_f$, we find:

$$M\boldsymbol{v} = (m + M)\boldsymbol{v}_f$$

$$\boldsymbol{v}_f = \frac{M\boldsymbol{v}}{m + M}$$

As we might expect, the final velocity of the toy truck is less than its initial velocity. As the toy truck gains the apple as cargo, its mass increases and it slows down. Because momentum is conserved and is directly proportional to mass and velocity, any increase in mass must be accompanied by a corresponding decrease in velocity.

Example 2

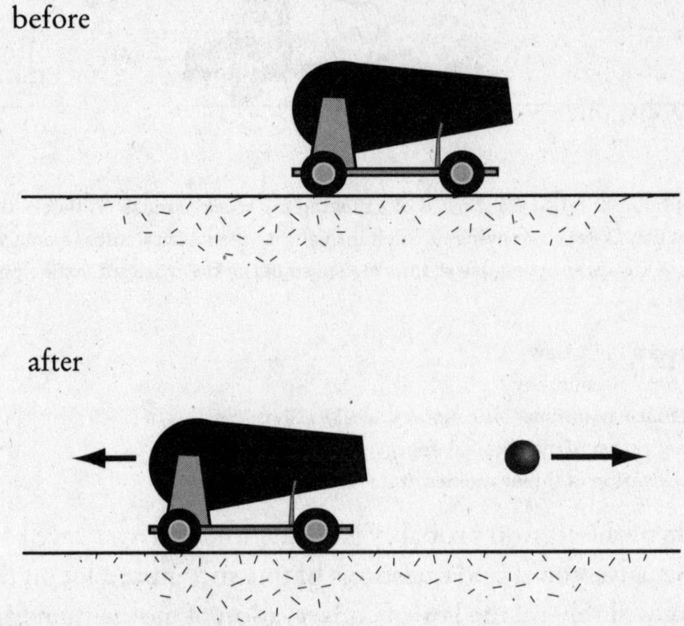

A cannon of mass 1000 kg launches a cannonball of mass 10 kg at a velocity of 100 m/s. At what speed does the cannon recoil?

Questions involving firearms recoil are a common way in which the SAT II may test your knowledge of conservation of momentum. Before we dive into the math, let's get a clear picture of what's going on here. Initially the cannon and cannonball are at rest, so the total momentum of the system is zero. No external forces act on the system in the horizontal direction, so the system's linear momentum in this direction is constant. Therefore the momentum of the system both before and after the cannon fires must be zero.

Now let's make some calculations. When the cannon is fired, the cannonball shoots forward with momentum (10 kg)(100 m/s) = 1000 kg · m/s. To keep the total momentum of the system at zero, the cannon must then recoil with an equal momentum:

$$p_{\text{cannon}} = mv_{\text{cannon}}$$
$$1000 \text{ kg} \cdot \text{m/s} = (1000 \text{ kg})v_{\text{cannon}}$$
$$v_{\text{cannon}} = 1 \text{ m/s}$$

Any time a gun, cannon, or an artillery piece releases a projectile, it experiences a "kick" and moves in the opposite direction of the projectile. The more massive the firearm, the slower it moves.

Collisions

A **collision** occurs when two or more objects hit each other. When objects collide, each object feels a force for a short amount of time. This force imparts an impulse, or changes the momentum of each of the colliding objects. But if the system of particles is isolated, we know that momentum is conserved. Therefore, while the momentum of each individual particle involved in the collision changes, the total momentum of the system remains constant.

The procedure for analyzing a collision depends on whether the process is **elastic** or **inelastic**. Kinetic energy is conserved in elastic collisions, whereas kinetic energy is converted into other forms of energy during an inelastic collision. In both types of collisions, momentum is conserved.

Elastic Collisions

Anyone who plays pool has observed elastic collisions. In fact, perhaps you'd better head over to the pool hall right now and start studying! Some kinetic energy is converted into sound energy when pool balls collide—otherwise, the collision would be silent—and a very small amount of kinetic energy is lost to friction. However, the dissipated energy is such a small fraction of the ball's kinetic energy that we can treat the collision as elastic.

Equations for Kinetic Energy and Linear Momentum

Let's examine an elastic collision between two particles of mass m_1 and m_2, respectively. Assume that the collision is head-on, so we are dealing with only one dimension—you are unlikely to find two-dimensional collisions of any complexity on the SAT II. The velocities of the particles before the elastic collision are v_1 and v_2, respectively. The velocities of the particles after the elastic collision are v_1' and v_2'. Applying the law of conservation of kinetic energy, we find:

$$\frac{1}{2}m_1 v_1^2 + \frac{1}{2}m_2 v_2^2 = \frac{1}{2}m_1 v_1'^2 + \frac{1}{2}m_2 v_2'^2$$

Applying the law of conservation of linear momentum:

$$m_1v_1 + m_2v_2 = m_1v_1' + m_2v_2'$$

These two equations put together will help you solve any problem involving elastic collisions. Usually, you will be given quantities for m_1, m_2, v_1 and v_2, and can then manipulate the two equations to solve for v_1' and v_2'.

Example

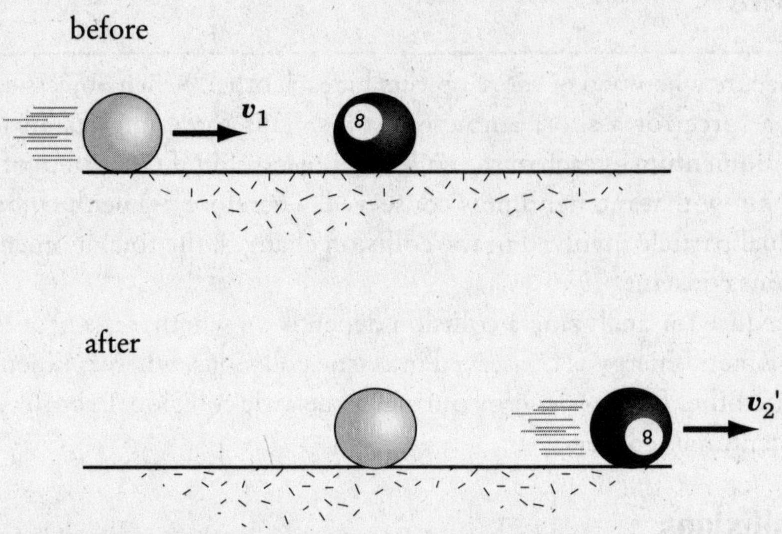

A pool player hits the eight ball, which is initially at rest, head-on with the cue ball. Both of these balls have the same mass, and the velocity of the cue ball is initially v_1. What are the velocities of the two balls after the collision? Assume the collision is perfectly elastic.

Substituting $m_1 = m_2 = m$ and $v_2 = 0$ into the equation for conservation of kinetic energy we find:

$$\frac{1}{2}mv_1^2 = \frac{1}{2}m(v_1'^2 + v_2'^2)$$
$$v_1^2 = v_1'^2 + v_2'^2$$

Applying the same substitutions to the equation for conservation of momentum, we find:

$$mv_1 = mv_1' + mv_2'$$
$$v_1 = v_1' + v_2'$$

If we square this second equation, we get:

$$v_1^2 = v_1'^2 + v_2'^2 + 2v_1'v_2'$$

By subtracting the equation for kinetic energy from this equation, we get:

$$2v_1'v_2' = 0$$

The only way to account for this result is to conclude that $v_1' = 0$ and consequently $v_1 = v_2'$. In plain English, the cue ball and the eight ball swap velocities: after the balls collide, the cue ball stops and the eight ball shoots forward with the initial velocity of the cue ball. This is the simplest form of an elastic collision, and also the most likely to be tested on the SAT II.

Inelastic Collisions

Most collisions are inelastic because kinetic energy is transferred to other forms of energy—such as thermal energy, potential energy, and sound—during the collision process. If you are asked to determine if a collision is elastic or inelastic, calculate the kinetic energy of the bodies before and after the collision. If kinetic energy is not conserved, then the collision is inelastic. Momentum is conserved in all inelastic collisions.

On the whole, inelastic collisions will only appear on the SAT II qualitatively. You may be asked to identify a collision as inelastic, but you won't be expected to calculate the resulting velocities of the objects involved in the collision. The one exception to this rule is in the case of **completely inelastic collisions.**

Completely Inelastic Collisions

A completely inelastic collision, also called a "perfectly" or "totally" inelastic collision, is one in which the colliding objects stick together upon impact. As a result, the velocity of the two colliding objects is the same after they collide. Because $v_1' = v_2' = v'$, it is possible to solve problems asking about the resulting velocities of objects in a completely inelastic collision using only the law of conservation of momentum.

Example

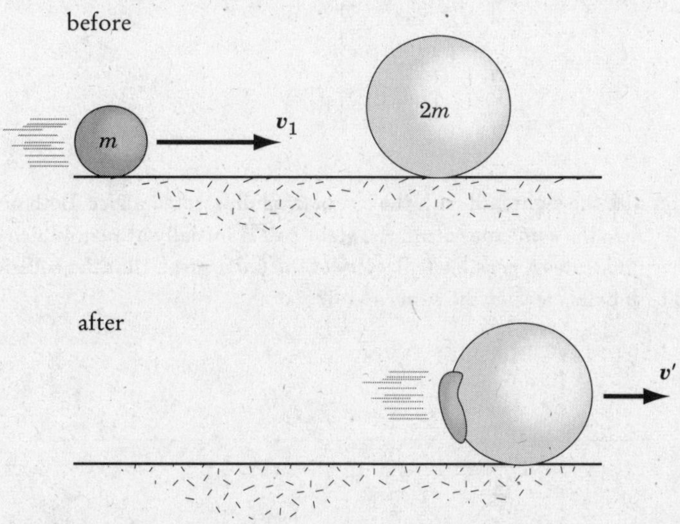

Two gumballs, of mass m and mass $2m$ respectively, collide head-on. Before impact, the gumball of mass m is moving with a velocity v_1, and the gumball of mass $2m$ is stationary. What is the final velocity, v', of the gumball wad?

First, note that the gumball wad has a mass of $m + 2m = 3m$. The law of conservation of momentum tells us that $mv_1 = 3mv'$, and so $v' = v_1/3$. Therefore, the final gumball wad moves in the same direction as the first gumball, but with one third of its velocity.

Collisions in Two Dimensions

Two-dimensional collisions, while a little more involved than the one-dimensional examples we've looked at so far, can be treated in exactly the same way as their one-dimensional counterparts. Momentum is still conserved, as is kinetic energy in the case of elastic collisions. The significant difference is that you will have to break the trajectories of objects down into x- and y-components. You will then be able to deal with the two components separately: momentum is conserved in the x direction, and momentum is conserved in the y direction. Solving a problem of two-dimensional collision is effectively the same thing as solving two problems of one-dimensional collision.

Because the SAT II generally steers clear of making you do too much math, it's unlikely that you'll be faced with a problem where you need to calculate the final velocities of two objects that collide two-dimensionally. However, questions that test your understanding of two-dimensional collisions qualitatively are perfectly fair game.

Example

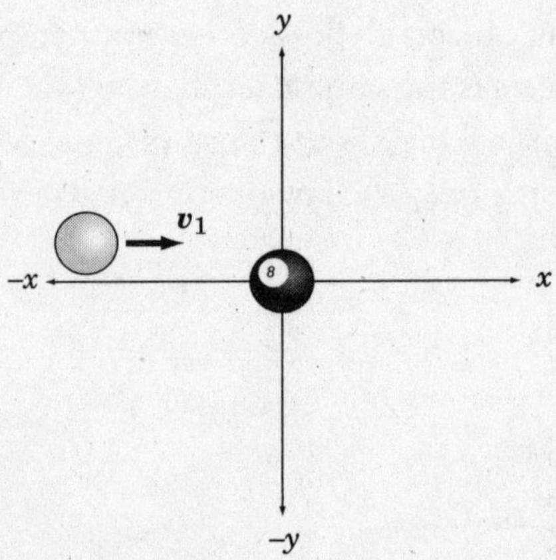

A pool player hits the eight ball with the cue ball, as illustrated above. Both of the billiard balls have the same mass, and the eight ball is initially at rest. Which of the figures below illustrates a possible trajectory of the balls, given that the collision is elastic and both balls move at the same speed?

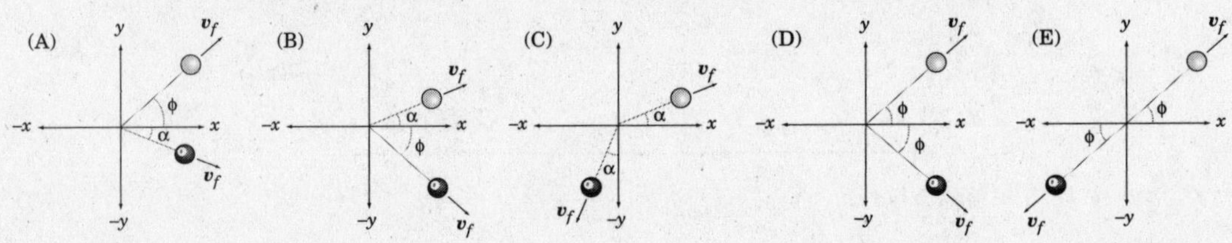

The correct answer choice is (D), because momentum is not conserved in any of the other figures. Note that the initial momentum in the *y* direction is zero, so the momentum of the balls in the *y* direction after the collision must also be zero. This is only true for choices (D) and (E). We also know that the initial momentum in the *x* direction is positive, so the final momentum in the *x* direction must also be positive, which is not true for (E).

Center of Mass

When calculating trajectories and collisions, it's convenient to treat extended bodies, such as boxes and balls, as point masses. That way, we don't need to worry about the shape of an object, but can still take into account its mass and trajectory. This is basically what we do with free-body diagrams. We can treat objects, and even systems, as point masses, even if they have very strange shapes or are rotating in complex ways. We can make this simplification because there is always a point in the object or system that has the same trajectory as the object or system as a whole would have if all its mass were concentrated in that point. That point is called the object's or system's **center of mass**.

Consider the trajectory of a diver jumping into the water. The diver's trajectory can be broken down into the translational movement of his center of mass, and the rotation of the rest of his body about that center of mass.

A human being's center of mass is located somewhere around the pelvic area. We see here that, though the diver's head and feet and arms can rotate and move gracefully in space, the center of mass in his pelvic area follows the inevitable parabolic trajectory of a body moving under the influence of gravity. If we wanted to represent the diver as a point mass, this is the point we would choose.

Our example suggests that Newton's Second Law can be rewritten in terms of the motion of the center of mass:

$$\boldsymbol{F}_{\text{net}} = M\boldsymbol{a}_{\text{cm}}$$

Put in this form, the Second Law states that the net force acting on a system, $\boldsymbol{F}_{\text{net}}$, is equal to the product of the total mass of the system, M, and the acceleration of the center of mass, $\boldsymbol{a}_{\text{cm}}$. Note that if the net force acting on a system is zero, then the center of mass does not accelerate.

Similarly, the equation for linear momentum can be written in terms of the velocity of the center of mass:

$$\boldsymbol{p} = M\boldsymbol{v}_{\text{cm}}$$

You will probably never need to plug numbers into these formulas for the SAT II, but it's important to understand the principle: the rules of dynamics and momentum apply to systems as a whole just as they do to bodies.

Calculating the Center of Mass

The center of mass of an object of uniform density is the body's geometric center. Note that the center of mass does not need to be located within the object itself. For example, the center of mass of a donut is in the center of its hole.

For a System of Two Particles

For a collection of particles, the center of mass can be found as follows. Consider two particles of mass m_1 and m_2 separated by a distance d:

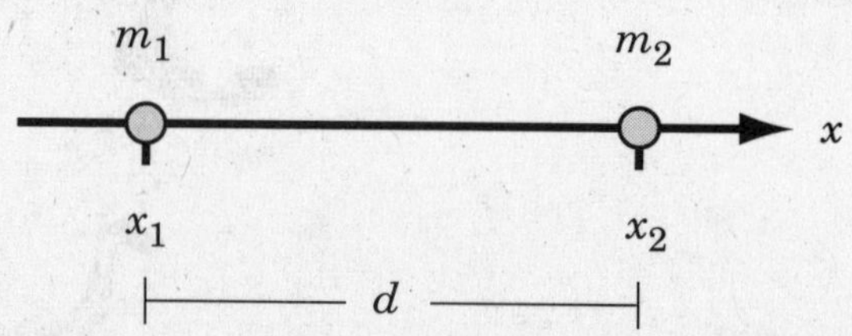

If you choose a coordinate system such that both particles fall on the x-axis, the center of mass of this system, x_{cm}, is defined by:

$$x_{\text{cm}} = \frac{m_1 x_1 + m_2 x_2}{m_1 + m_2}$$

For a System in One Dimension

We can generalize this definition of the center of mass for a system of n particles on a line. Let the positions of these particles be $x_1, x_2, \ldots, x_n$. To simplify our notation, let M be the total mass of all n particles in the system, meaning $M = m_1 + m_2 + \ldots + m_n$. Then, the center of mass is defined by:

$$x_{\text{cm}} = \frac{m_1x_1 + m_2x_2 + \cdots + m_nx_n}{M}$$

For a System in Two Dimensions

Defining the center of mass for a two-dimensional system is just a matter of reducing each particle in the system to its x- and y-components. Consider a system of n particles in a random arrangement of x-coordinates $x_1, x_2, \ldots, x_n$ and y-coordinates $y_1, y_2, \ldots, y_n$. The x-coordinate of the center of mass is given in the equation above, while the y-coordinate of the center of mass is:

$$y_{\text{cm}} = \frac{m_1y_1 + m_2y_2 + \cdots + m_ny_n}{M}$$

How Systems Will Be Tested on the SAT II Physics

The formulas we give here for systems in one and two dimensions are general formulas to help you understand the principle by which the center of mass is determined. Rest assured that for the SAT II, you'll never have to plug in numbers for mass and position for a system of several particles. However, your understanding of center of mass may be tested in less mathematically rigorous ways.

For instance, you may be shown a system of two or three particles and asked explicitly to determine the center of mass for the system, either mathematically or graphically. Another example, which we treat below, is that of a system consisting of two parts, where one part moves relative to the other. In this cases, it is important to remember that the center of mass of the system as a whole doesn't move.

Example

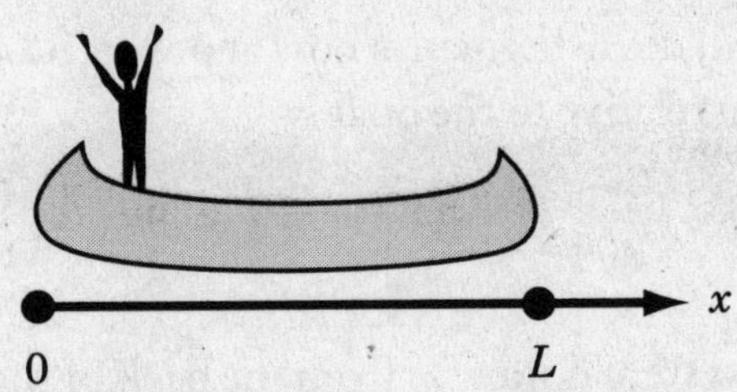

A fisherman stands at the back of a perfectly symmetrical boat of length L. The boat is at rest in the middle of a perfectly still and peaceful lake, and the fisherman has a mass ¼ that of the boat. If the fisherman walks to the front of the boat, by how much is the boat displaced?

If you've ever tried to walk from one end of a small boat to the other, you may have noticed that the boat moves backward as you move forward. That's because there are no external forces acting on the system, so the system as a whole experiences no net force. If we recall the equation $\boldsymbol{F}_{\text{net}} = M\boldsymbol{a}_{\text{cm}}$, the center of mass of the system cannot move if there is no net force acting on the system. The fisherman can move, the boat can move, but the system as a whole must maintain the same center of mass. Thus, as the fisherman moves forward, the boat must move backward to compensate for his movement.

Because the boat is symmetrical, we know that the center of mass of the boat is at its geometrical center, at $x = {}^{L}\!/_{2}$. Bearing this in mind, we can calculate the center of mass of the system containing the fisherman and the boat:

$$x_{\text{cm}} = \frac{(\frac{m}{4})(0) + m\frac{L}{2}}{m + \frac{1}{4}m} = \frac{2}{5}L$$

Now let's calculate where the center of mass of the fisherman-boat system is relative to the boat after the fisherman has moved to the front. We know that the center of mass of the fisherman-boat system hasn't moved relative to the water, so its displacement with respect to the boat represents how much the boat has been displaced with respect to the water.

In the figure below, the center of mass of the boat is marked by a dot, while the center of mass of the fisherman-boat system is marked by an x.

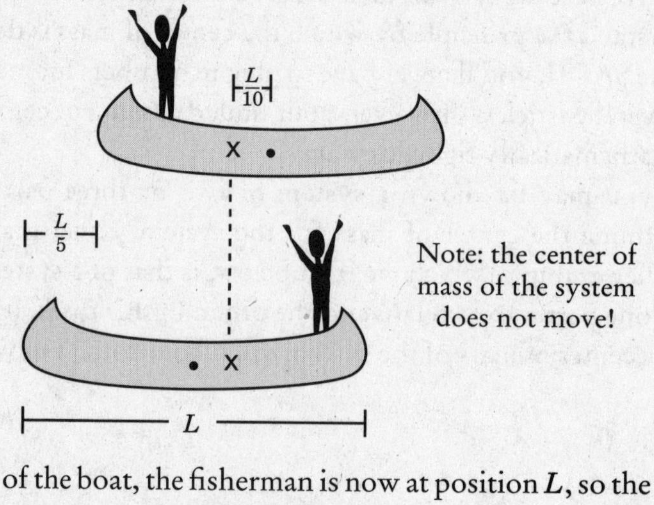

At the front end of the boat, the fisherman is now at position L, so the center of mass of the fisherman-boat system relative to the boat is

$$x_{\text{cm}} = \frac{(\frac{m}{4})(L) + m\frac{L}{2}}{m + \frac{1}{4}m} = \frac{3}{5}L$$

The center of mass of the system is now $^3\!/_5$ from the back of the boat. But we know the center of mass hasn't moved, which means the boat has moved backward a distance of $^1\!/_5\,L$, so that the point $^3\!/_5\,L$ is now located where the point $^2\!/_5\,L$ was before the fisherman began to move.

Key Formulas

Linear Momentum	$\boldsymbol{p} = m\boldsymbol{v}$
Impulse of a Constant Force	$\boldsymbol{J} = \boldsymbol{F}\Delta t = \Delta \boldsymbol{p}$
Conservation of Energy for an Elastic Collision of Two Particles	$\frac{1}{2}m_1 v_1^2 + \frac{1}{2}m_2 v_2^2 = \frac{1}{2}m_1 v_1{\prime}^2 + \frac{1}{2}m_2 v_2{\prime}^2$
Conservation of Momentum for a Collision of Two Particles	$m_1 v_1 + m_2 v_2 = m_1 v_1{\prime} + m_2 v_2{\prime}$
Center of Mass for a System of *n* Particles	$x_{\text{cm}} = \frac{m_1 x_1 + m_2 x_2 + \cdots + m_n x_n}{M}$ $y_{\text{cm}} = \frac{m_1 y_1 + m_2 y_2 + \cdots + m_n y_n}{M}$
Acceleration of the Center of Mass	$\boldsymbol{F}_{\text{net}} = M\boldsymbol{a}_{\text{cm}}$
Velocity of the Center of Mass	$\boldsymbol{p} = M\boldsymbol{v}_{\text{cm}}$

Practice Questions

1. An athlete of mass 70.0 kg applies a force of 500 N to a 30.0 kg luge, which is initially at rest, over a period of 5.00 s before jumping onto the luge. Assuming there is no friction between the luge and the track on which it runs, what is its velocity when the athlete jumps on?

   (A) 12.5 m/s
   (B) 25.0 m/s
   (C) 35.7 m/s
   (D) 83.3 m/s
   (E) 100 m/s

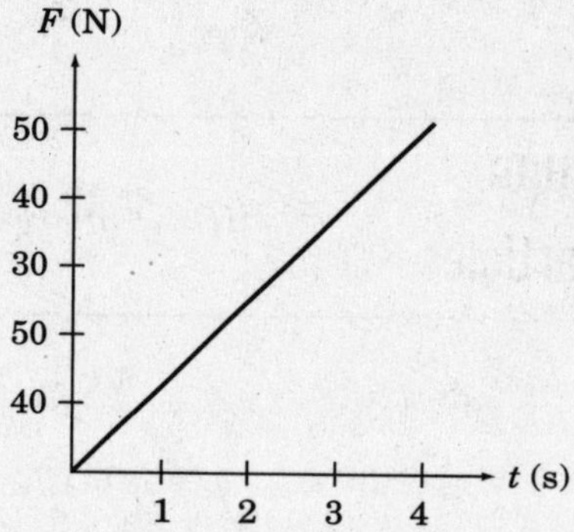

2. The graph above shows the amount of force applied to an initially stationary 20 kg curling rock over time. What is the velocity of the rock after the force has been applied to it?

   (A) 1.25 m/s
   (B) 5 m/s
   (C) 10 m/s
   (D) 25 m/s
   (E) 50 m/s

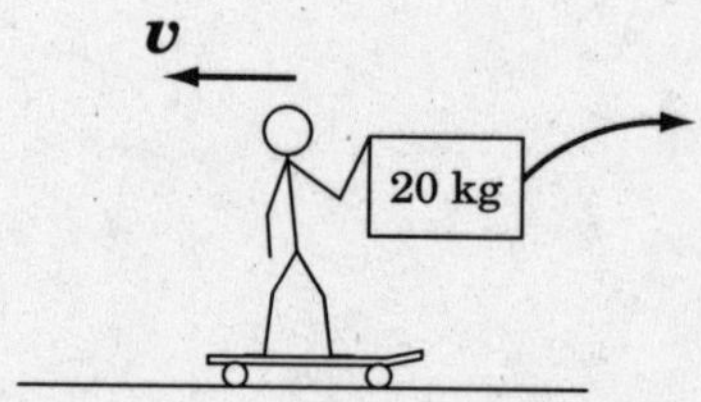

3. A 60 kg man holding a 20 kg box rides on a skateboard at a speed of 7 m/s. He throws the box behind him, giving it a velocity of 5 m/s. What is his velocity after throwing the object?

   (A) 8 m/s
   (B) 9 m/s
   (C) 10 m/s
   (D) 11 m/s
   (E) 12 m/s

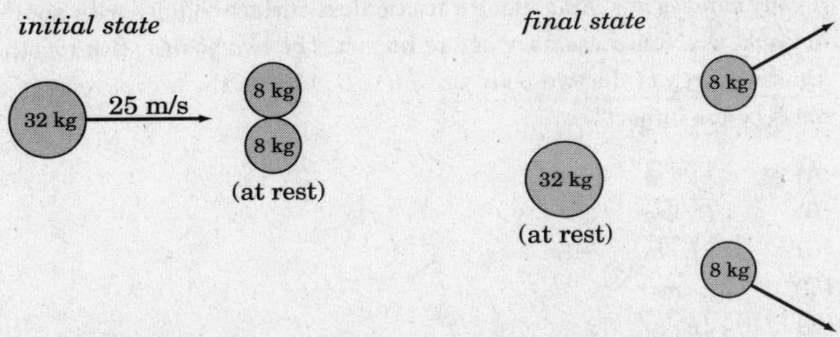

4. A scattering experiment is done with a 32 kg disc and two 8 kg discs on a frictionless surface. In the initial state of the experiment, the heavier disc moves in the x direction with velocity $v = 25$ m/s toward the lighter discs, which are at rest. The discs collide elastically. In the final state, the heavy disc is at rest and the two smaller discs scatter outward with the same speed. What is the x-component of the velocity of each of the 8 kg discs in the final state?

   (A) 12.5 m/s
   (B) 16 m/s
   (C) 25 m/s
   (D) 50 m/s
   (E) 100 m/s

5. An moving object has kinetic energy $KE = 100$ J and momentum $p = 50$ kg · m/s. What is its mass?

   (A) 2 kg
   (B) 4 kg
   (C) 6.25 kg
   (D) 12.5 kg
   (E) 25 kg

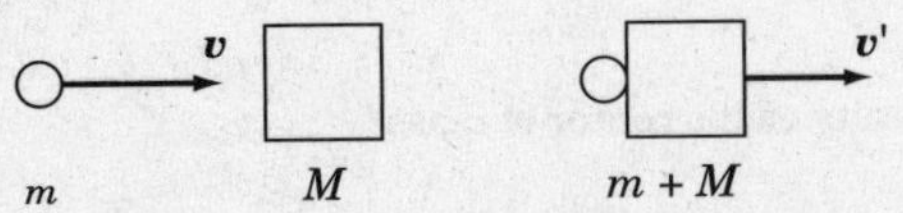

6. An object of mass m moving with a velocity v collides with another object of mass M. If the two objects stick together, what is their velocity?

   (A) $\frac{M}{m+M}v$
   (B) $\frac{m}{m+M}v$
   (C) $\frac{m+M}{m}v$
   (D) $\frac{m+M}{M}v$
   (E) Zero

7. A body of mass m sliding along a frictionless surface collides with another body of mass m, which is stationary before impact. The two bodies stick together. If the kinetic energy of the two-body system is E, what is the initial velocity of the first mass before impact?

(A) $\sqrt{E/2m}$
(B) $\sqrt{E/2m}$
(C) $\sqrt{2E/m}$
(D) $\sqrt{E/m}$
(E) $\sqrt{2E/m}$

8. A hockey puck of mass m is initially at rest on a frictionless ice rink. A player comes and hits the puck, imparting an impulse of J. If the puck then collides with another object of mass M at rest and sticks to it, what is the final velocity of the two-body system?

(A) $\frac{J}{m}$

(B) $\frac{J}{M}$

(C) $\frac{J}{m+M}$

(D) $\frac{m+M}{J}$

(E) $\frac{M}{J}$

Questions 9 and 10 refer to two 1 kg masses moving toward each other, one with velocity v_1 = 10 m/s, the other with velocity v_2 = 20 m/s.

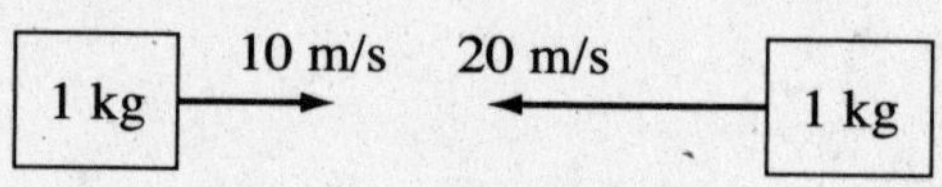

9. What is the velocity of the center of mass?

(A) 0 m/s
(B) 5 m/s to the left
(C) 10 m/s to the left
(D) 15 m/s to the left
(E) 20 m/s to the left

10. What is the total energy of the system?

(A) 50 J
(B) 150 J
(C) 200 J
(D) 250 J
(E) 400 J

Explanations

1. (B)
The athlete imparts a certain impulse to the luge over the 5 s period that is equal to $F\Delta t$. This impulse tells us the change in momentum for the luge. Since the luge starts from rest, this change in momentum gives us the total momentum of the luge:

$$J = F\Delta t = (500\text{ N})(5\text{ s})$$
$$= 2500\text{ kg}\cdot\text{m/s}$$

The total momentum of the luge when the athlete jumps on is 2500 kg · m/s. Momentum is the product of mass and velocity, so we can solve for velocity by dividing momentum by the combined mass of the athlete and the luge:

$$v = \frac{p}{m} = \frac{2500\text{ kg}\cdot\text{m/s}}{100\text{ kg}}$$
$$= 25\text{ m/s}$$

2. (B)
The area under a force vs. time graph tells us the impulse given to the rock. Since the rock is motionless at $t = 0$, the impulse given to the rock is equal to the rock's total momentum. The area under the graph is a triangle of height 50N and length 4 s:

$$p = \frac{1}{2}(50\text{ N})(4\text{ s}) = 100\text{ kg}\cdot\text{m/s}$$

Calculating the rock's velocity, then, is simply a matter of dividing its momentum by its mass:

$$v = \frac{p}{m} = \frac{100\text{ kg}\cdot\text{m/s}}{20\text{ kg}}$$
$$= 5\text{ m/s}$$

3. (D)
This is a conservation of momentum problem. The initial momentum of the system must be equal to the final momentum. The initial momentum of the system is:

$$p = mv = (60\text{ kg} + 20\text{ kg})(7\text{ m/s})$$
$$= 560\text{ kg}\cdot\text{m/s}$$

The final momentum of the system is the sum of the momentum of the box and of the skateboarder. Since the box is thrown in the opposite direction of the skateboard's initial momentum, it will have a negative momentum. Because the final momentum and the initial momentum are equal, we know that the final momentum of the skateboarder minus the momentum of the box will equal 560 kg · m/s. With this information, we can solve for v, the skateboarder's final velocity:

$$\begin{aligned} p_{\text{skateboarder}} + p_{\text{box}} &= 560 \text{ kg} \cdot \text{m/s} \\ (60 \text{ kg})v - (20 \text{ kg})(5 \text{ m/s}) &= 560 \text{ kg} \cdot \text{m/s} \\ (60 \text{ kg})v &= 660 \text{ kg} \cdot \text{m/s} \\ v &= 11 \text{ m/s} \end{aligned}$$

4. (D)

The law of conservation of linear momentum tells us that the x-component of the system's momentum must be equal before and after the collision. The x-component of the system's momentum before the collision is the momentum of the large disc. The x-component of the system's momentum after the collision is the x-component of the momentum of both of the smaller discs put together. Since momentum is $\mathbf{p} = m\mathbf{v}$, and since the larger disc has twice the mass of the two smaller discs put together, that means that the velocity of the two smaller discs must be twice the velocity of the larger disc; that is, 50 m/s.

5. (D)

We have equations for kinetic energy, $KE = \frac{1}{2}mv^2$, and momentum, $p = mv$, both of which include variables for mass and velocity. If we first solve for velocity, we can then plug that value into the equation and solve for mass:

$$\begin{aligned} \frac{1}{2}mv^2 &= 2mv = 100 \\ v^2 &= 4v \\ v(v-4) &= 0 \\ v &= 4 \text{ m/s} \end{aligned}$$

If $v = 4$ m/s, then we can plug this value into the equation for momentum to find that $p = 4m = 50$ kg · m/s, and conclude that $m = 12.5$ kg.

6. (B)

The law of conservation of momentum tells us that the initial momentum of the system is equal to the final momentum of the system. The initial momentum is $p = mv$, and the final momentum is $p' = (m + M)v'$, where v' is the final velocity of the two objects. Knowing that $p = p'$, we can solve for v':

$$(m+M)v' = mv$$
$$v' = \frac{m}{m+M}v$$

7. (E)

Momentum is conserved in this collision. If the mass is moving with velocity v before impact and the two-mass system is moving with velocity v' after impact, we know that $mv = 2mv'$. We also know that the kinetic energy of the two-body system is $E = \frac{1}{2}mv'$. If we solve for v', we find:

$$\frac{1}{2}mv'^2 = E$$
$$v'^2 = \frac{2E}{m}$$
$$v' = \sqrt{\frac{2E}{m}}$$

From the equation $mv = 2mv'$, we can conclude that the initial velocity of the first body, v, is double v'. If the value for v' is given in terms of KE in the equation above, then the value of v is simply twice that.

8. (C)

Impulse is defined as the change in momentum. Since the hockey puck is initially at rest, its change in momentum is simply its momentum after it has been set in motion. In other words, the momentum of the puck in motion is equal to J.

When the puck collides with the other object, momentum is conserved, so the system of the puck and the other object also has a momentum of J. This momentum is equal to the mass, $m + M$, of the system, multiplied by the velocity of the two-body system, v'. Solving for v' is now quite easy:

$$J = (m+M)v'$$
$$v' = \frac{J}{m+M}$$

9. (B)

The velocity of the center of mass of the system is the same as the total velocity of the system. To find the total velocity of the system, we need to find the total momentum of the system and divide it by the total mass of the system.

The momentum of the first mass is $p_1 = 10 \text{ kg} \cdot \text{m/s}$ to the right, and the momentum of the second mass is $p_2 = 20 \text{ kg} \cdot \text{m/s}$ to the left. Therefore, the total momentum of the system is $p_1 + p_2 = 10 \text{ kg} \cdot \text{m/s}$ to the left. Since the total mass of the system is 2 kg, we can find the total velocity of the system by dividing its momentum by its mass:

$$\begin{aligned} \boldsymbol{v} &= \frac{\boldsymbol{p}}{m} = \frac{10 \text{ kg} \cdot \text{m/s to the left}}{2 \text{ kg}} \\ &= 5 \text{ m/s to the left} \end{aligned}$$

10. (D)

The only energy in the system is the kinetic energy of the two masses. These can be determined through two easy calculations:

$$KE_1 = \frac{1}{2}mv_1^2 = \frac{1}{2}(1 \text{ kg})(10 \text{ m/s})^2 = 50 \text{ J}$$

$$KE_2 = \frac{1}{2}mv_2^2 = \frac{1}{2}(1 \text{ kg})(20 \text{ m/s})^2 = 200 \text{ J}$$

Adding these two energies together, we find that the total energy of the system is 50 J + 200 J = 250 J.

Rotational Motion

Chapter Contents

Important Definitions **158**
Rotational Kinematics **160**
Frequency and Period **162**
Rotational Dynamics **166**
Kinetic Energy **173**
Angular Momentum **174**
Key Formulas **177**
Practice Questions **178**
Explanations **181**

UNTIL THIS CHAPTER, we have focused almost entirely on **translational motion**, the motion of bodies moving through space. But there is a second kind of motion, called **rotational motion**, which deals with the rotation of a body about its center of mass. The movement of any object can be described through the combination of translational motion of the object's center of mass and its rotational motion about that center of mass. For example, look at the diver jumping into the water that we saw in the previous chapter.

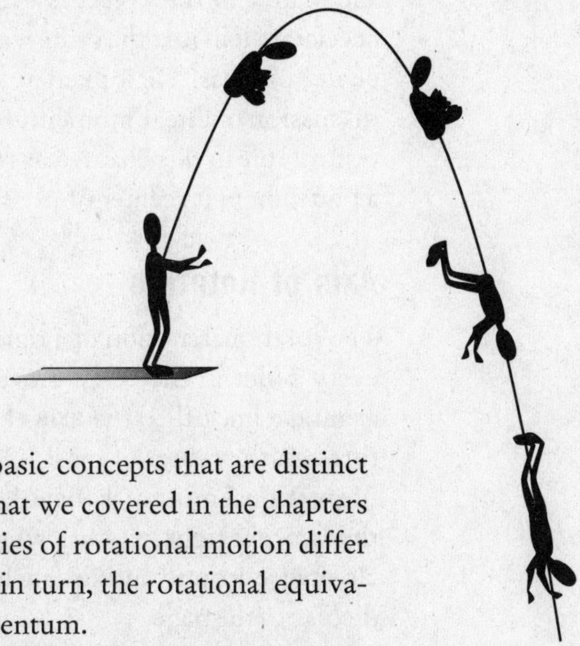

The diver's translational motion is the parabolic trajectory of her center of mass. However, if that were the only motion of the diver's body, diving competitions would be considerably more boring. What astonishes fans and impresses judges is the grace and fluidity of the rotational motion of the diver's arms, legs, feet, etc., about that center of mass.

You will find that rotational motion and translational motion have a lot in common. In fact, aside from a few basic differences, the mechanics of rotational motion are identical to those of translational motion. We'll begin this chapter by introducing some basic concepts that are distinct to rotational motion. After that, we will recapitulate what we covered in the chapters on translational motion, explaining how the particularities of rotational motion differ from their translational counterparts. We will examine, in turn, the rotational equivalents for kinematic motion, dynamics, energy, and momentum.

There will be at most one or two questions on rotational motion on any given SAT II test. On the whole, they tend to center around the concepts of torque and equilibrium.

Rotational Motion

Important Definitions

There are a few basic physical concepts that are fundamental to a proper understanding of rotational motion. With a steady grasp of these concepts, you should encounter no major difficulties in making the transition between the mechanics of translational motion and of rotational motion.

Rigid Bodies

The questions on rotational motion on the SAT II deal only with **rigid bodies**. A rigid body is an object that retains its overall shape, meaning that the particles that make up the rigid body stay in the same position relative to one another. A pool ball is one example of a rigid body since the shape of the ball is constant as it rolls and spins. A wheel, a record, and a top are other examples of rigid bodies that commonly appear in questions involving rotational motion. By contrast, a slinky is not a rigid body, because its coils expand, contract, and bend, so that its motion would be considerably more difficult to predict if you were to spin it about.

Center of Mass

The **center of mass** of an object, in case you have forgotten, is the point about which all the matter in the object is evenly distributed. A net force acting on the object will accelerate it in just the same way as if all the mass of the object were concentrated in its center of mass. We looked at the concept of center of mass in the previous chapter's discussion of linear momentum. The concept of center of mass will play an even more central role in this chapter, as rotational motion is essentially defined as the rotation of a body about its center of mass.

Axis of Rotation

The rotational motion of a rigid body occurs when every point in the body moves in a circular path around a line called the **axis of rotation**, which cuts through the center of mass. One familiar example of rotational motion is that of a spinning wheel. In the figure at right, we see a wheel rotating counterclockwise around an axis labeled O that is perpendicular to the page.

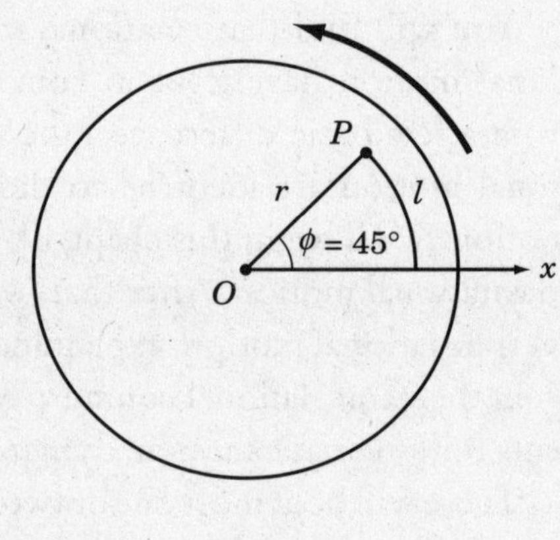

As the wheel rotates, every point in the rigid body makes a circle around the axis of rotation, O.

Radians

We're all very used to measuring angles in degrees, and know perfectly well that there are 360° in a circle, 90° in a right angle, and so on. You've probably noticed that 360 is also a convenient number because so many other numbers divide into it. However, this is a totally arbitrary system that has its origins in the Ancient Egyptian calendar which was based on a 360-day year.

It makes far more mathematical sense to measure angles in **radians** (rad). If we were to measure the arc of a circle that has the same length as the radius of that circle, then one radian would be the angle made by two radii drawn to either end of the arc.

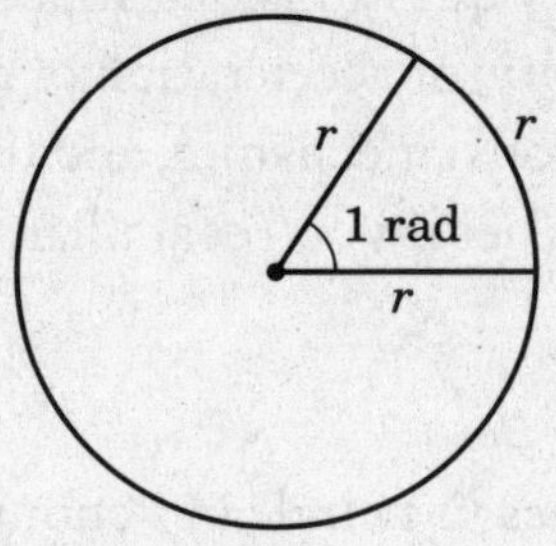

Converting between Degrees and Radians

It is unlikely that the SAT II will specifically ask you to convert between degrees and radians, but it will save you time and headaches if you can make this conversion quickly and easily. Just remember this formula:

$$y^\circ = \frac{180x}{\pi}\text{rad}$$

You'll quickly get used to working in radians, but at right is a conversion table for the more commonly occurring angles.

Value in degrees	Value in radians
30	π/6
45	π/4
60	π/3
90	π/2
180	π
360	2π

Calculating the Length of an Arc

The advantage of using radians instead of degrees, as will quickly become apparent, is that the radian is based on the nature of angles and circles themselves, rather than on the arbitrary fact of how long it takes our Earth to circle the sun.

For example, calculating the length of any arc in a circle is much easier with radians than with degrees. We know that the circumference of a circle is given by $P = 2\pi r$, and we know that there are 2π radians in a circle. If we wanted to know the length, l, of the arc described by any angle θ, we would know that this arc is a fraction of the perimeter, $(\theta/2\pi)P$. Because $P = 2\pi r$, the length of the arc would be:

$$l = \frac{\theta}{2\pi}P = \frac{\theta}{2\pi}2\pi r = \theta r$$

Rotational Kinematics

You are now going to fall in love with the word "angular." You'll find that for every term in kinematics that you're familiar with, there's an "angular" counterpart: **angular displacement, angular velocity, angular acceleration**, etc. And you'll find that, "angular" aside, very little changes when dealing with rotational kinematics.

Angular Position, Displacement, Velocity, and Acceleration

The SAT II is unlikely to have any questions that simply ask you to calculate the angular position, displacement, velocity, or acceleration of a rotating body. However, these concepts form the basis of rotational mechanics, and the questions you *will* encounter on the SAT II will certainly be easier if you're familiar with these fundamentals.

Angular Position

By convention, we measure angles in a circle in a counterclockwise direction from the positive x-axis. The **angular position** of a particle is the angle, ϕ, made between the line connecting that particle to the origin, O, and the positive x-axis, measured counterclockwise. Let's take the example of a point P on a rotating wheel:

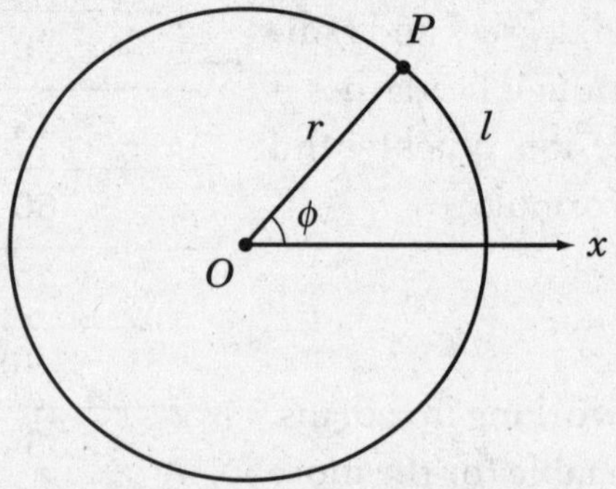

In this figure, point P has an angular position of ϕ. Note that every point on the line $\overline{OP}$ has the same angular position: the angular position of a point does not depend on how far that point is from the origin, O.

We can relate the angular position of P to the length of the arc of the circle between P and the x-axis by means of an easy equation:

$$\phi = \frac{l}{r}$$

In this equation, l is the length of the arc, and r is the radius of the circle.

Angular Displacement

Now imagine that the wheel is rotated so that every point on line $\overline{OP}$ moves from an initial angular position of ϕ_i to a final angular position of ϕ_f. The **angular displacement**, θ, of line $\overline{OP}$ is:

$$\theta = \phi_f - \phi_i$$

For example, if you rotate a wheel counterclockwise such that the angular position of line $\overline{OP}$ changes from $\phi_i = 45° = \pi/4$ to $\phi_f = 135° = 3\pi/4$, as illustrated below, then the angular displacement of line $\overline{OP}$ is 90° or $\pi/2$ radians.

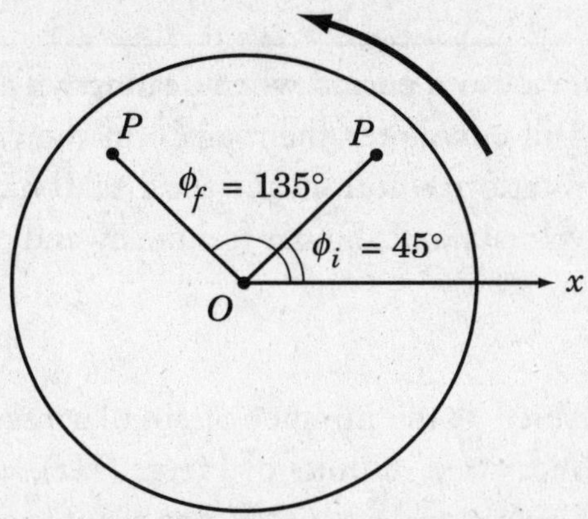

For line $\overline{OP}$ to move in the way described above, every point along the line must rotate 90° counterclockwise. By definition, the particles that make up a rigid body must stay in the same relative position to one another. As a result, the angular displacement is the same for every point in a rotating rigid body.

Also note that the angular distance a point has rotated may or may not equal that point's angular displacement. For example, if you rotate a record 45° clockwise and then 20° counterclockwise, the angular displacement of the record is 25°, although the particles have traveled a total angular distance of 65°. Hopefully, you've already had it hammered into your head that distance and displacement are not the same thing: well, the same distinction applies with angular distance and angular displacement.

Angular Velocity

Angular velocity, ω, is defined as the change in the angular displacement over time. Average angular velocity, $\bar{\omega}$, is defined by:

$$\bar{\omega} = \frac{\Delta\theta}{\Delta t}$$

Angular velocity is typically given in units of rad/s. As with angular displacement, the angular velocity of every point on a rotating object is identical.

Angular Acceleration

Angular acceleration, α, is defined as the rate of change of angular velocity over time. Average angular acceleration, $\bar{\alpha}$, is defined by:

$$\bar{\alpha} = \frac{\Delta\omega}{\Delta t}$$

Angular acceleration is typically given in units of rad/s^2.

Frequency and Period

You've encountered frequency and period when dealing with springs and simple harmonic motion, and you will encounter them again in the chapter on waves. These terms are also relevant to rotational motion, and the SAT II has been known to test the relation between angular velocity and angular frequency and period.

Angular frequency

Angular frequency, f, is defined as the number of circular revolutions in a given time interval. It is commonly measured in units of Hertz (Hz), where 1 Hz = 1 s^{-1}. For example, the second hand on a clock completes one revolution every 60 seconds and therefore has an angular frequency of $\frac{1}{60}$ Hz.

The relationship between frequency and angular velocity is:

$$f = \frac{\omega}{2\pi}$$

For example, the second hand of a clock has an angular velocity of $\omega = \Delta\theta/\Delta t = 2\pi/60$ s. Plugging that value into the equation above, we get

$$f = \frac{2\pi/(60 \text{ s})}{2\pi} = \frac{1}{60 \text{ s}} = 1/60 \text{ Hz}$$

which we already determined to be the frequency of the second hand of a clock.

Angular period

Angular period, T, is defined as the time required to complete one revolution and is related to frequency by the equation:

$$T = \frac{1}{f}$$

Since we know that the frequency of the second hand is $\frac{1}{60}$ Hz, we can quickly see that the period of the second hand is 60 s. It takes 60 seconds for the second hand to complete a revolution, so the period of the second hand is 60 seconds. Period and angular velocity are related by the equation

$$T = \frac{2\pi}{\omega}$$

Example

> The Earth makes a complete rotation around the sun once every 365.25 days. What is the Earth's angular velocity?

The question tells us that the Earth has a period of $T = 365.25$ days. If we plug this value into the equation relating period and angular velocity, we find:

$$\omega = \frac{2\pi}{T} = \frac{2\pi}{365.25 \text{ days}}$$
$$= 1.7202 \times 10^{-2} \text{ rad/day}$$

Note, however, that this equation only gives us the Earth's angular velocity in terms of radians per day. In terms of radians per second, the correct answer is:

$$1.7202 \times 10^{-2} \text{ rad/day}\left(\frac{1 \text{ day}}{8.64 \times 10^{4}}\right) = 1.9910 \times 10^{-7} \text{ rad/s}$$

Relation of Angular Variables to Linear Variables

At any given moment, a rotating particle has an instantaneous linear velocity and an instantaneous linear acceleration. For instance, a particle P that is rotating counter-clockwise will have an instantaneous velocity in the positive y-direction at the moment it is at the positive x-axis. In general, a rotating particle has an instantaneous velocity that is tangent to the circle described by its rotation and an instantaneous acceleration that points toward the center of the circle.

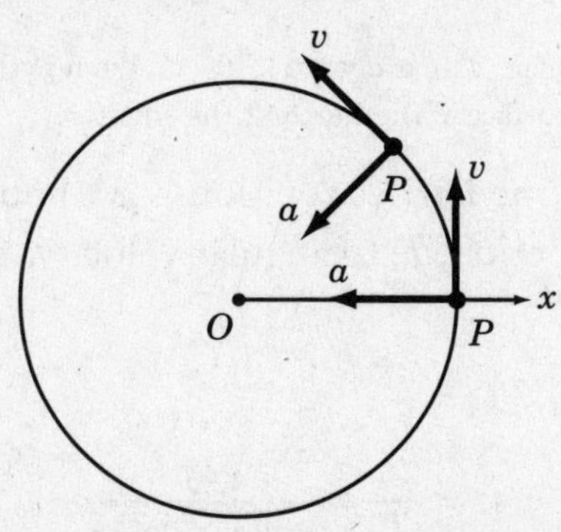

On the SAT II, you may be called upon to determine a particle's linear velocity or acceleration given its angular velocity or acceleration, or vice versa. Let's take a look at how this is done.

Distance

We saw earlier that the angular position, ϕ, of a rotating particle is related to the length of the arc, l, between the particle's present position and the positive x-axis by the equation $\phi = l / r$, or $l = \phi r$. Similarly, for any angular displacement, θ, we can say that the length, l, of the arc made by a particle undergoing that displacement is

$$l = \theta r$$

Note that the length of the arc gives us a particle's distance traveled rather than its displacement, since displacement is a vector quantity measuring only the straight-line distance between two points, and not the length of the route traveled between those two points.

Velocity and Acceleration

Given the relationship we have determined between arc distance traveled, l, and angular displacement, θ, we can now find expressions to relate linear and angular velocity and acceleration.

We can express the instantaneous linear velocity of a rotating particle as $v = l/t$, where l is the distance traveled along the arc. From this formula, we can derive a formula relating linear and angular velocity:

$$\begin{aligned} v &= \frac{l}{t} = \frac{\theta r}{t} = \frac{\theta}{t} r \\ &= \omega r \end{aligned}$$

In turn, we can express linear acceleration as $a = v/t$, giving us this formula relating linear and angular acceleration:

$$\begin{aligned} a &= \frac{v}{t} = \frac{\omega r}{t} = \frac{\omega}{t} r \\ &= \alpha r \end{aligned}$$

Example

> The radius of the Earth is approximately 6.4×10^6 m. What is the instantaneous velocity of a point on the surface of the Earth at the equator?

We know that the period of the Earth's rotation is 24 hours, or 8.64×10^4 seconds. From the equation relating period, T, to angular velocity, ω, we can find the angular velocity of the Earth:

$$\begin{aligned} T &= \frac{2\pi}{\omega} \\ \omega &= \frac{2\pi}{T} = \frac{2\pi}{8.64 \times 10^4 \text{ s}} \\ &= 7.27 \times 10^{-5} \text{ rad/s} \end{aligned}$$

Now that we know the Earth's angular velocity, we simply plug that value into the equation for linear velocity:

$$\begin{aligned} v = \omega r &= (7.27 \times 10^{-5} \text{rad/s})(6.4 \times 10^6 \text{m}) \\ &= 4.7 \times 10^2 \text{m/s} \end{aligned}$$

They may not notice it, but people living at the equator are moving faster than the speed of sound!

Equations of Rotational Kinematics

In Chapter 2 we defined the kinematic equations for bodies moving at constant acceleration. As we have seen, there are very clear rotational counterparts for linear displacement, velocity, and acceleration, so we are able to develop an analogous set of five equations for solving problems in rotational kinematics:

$$\phi = \phi_0 + \frac{1}{2}(\omega + \omega_0)t$$
$$\omega = \omega_0 + \alpha t$$
$$\phi = \phi_0 + \omega_0 t + \frac{1}{2}\alpha t^2$$
$$\phi = \phi_0 + \omega t - \frac{1}{2}\alpha t^2$$
$$\omega^2 = \omega_0^2 + 2\alpha(\phi - \phi_0)$$

In these equations, ω_0 is the object's initial angular velocity at its initial position, ϕ_0.

Any questions on the SAT II that call upon your knowledge of the kinematic equations will almost certainly be of the translational variety. However, it's worth noting just how deep the parallels between translational and rotational kinematics run.

Vector Notation of Rotational Variables

Angular velocity and angular acceleration are vector quantities; the equations above define their magnitudes but not their directions. Given that objects with angular velocity or acceleration are moving in a circle, how do we determine the direction of the vector? It may seem strange, but the direction of the vector for angular velocity or acceleration is actually perpendicular to the plane in which the object is rotating.

We determine the direction of the angular velocity vector using the **right-hand rule**. Take your right hand and curl your fingers along the path of the rotating particle or body. Your thumb then points in the direction of the angular velocity of the body. Note that the angular velocity is along the body's axis of rotation.

The figure below illustrates a top spinning counterclockwise on a table. The right-hand rule shows that its angular velocity is in the upward direction. Note that if the top were rotating clockwise, then its angular velocity would be in the downward direction.

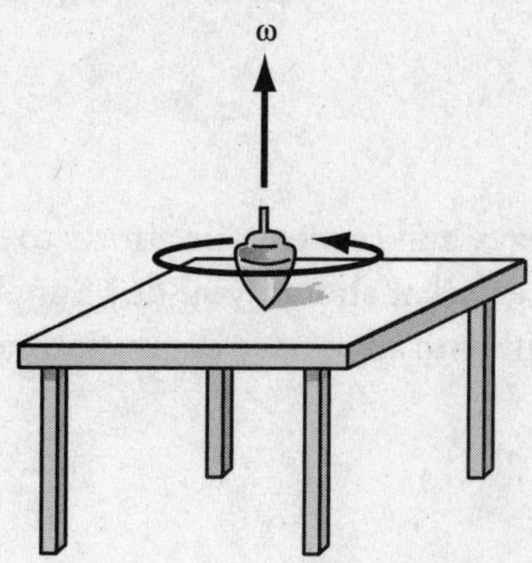

To find the direction of a rigid body's angular acceleration, you must first find the direction of the body's angular velocity. Then, if the magnitude of the angular velocity is increasing, the angular acceleration is in the same direction as the angular velocity vector. On the other hand, if the magnitude of the angular velocity is decreasing, then the angular acceleration points in the direction opposite the angular velocity vector.

Rotational Dynamics

Just as we have rotational counterparts for displacement, velocity, and acceleration, so do we have rotational counterparts for force, mass, and Newton's Laws. As with angular kinematics, the key here is to recognize the striking similarity between rotational and linear dynamics, and to learn to move between the two quickly and easily.

Torque

If a net force is applied to an object's center of mass, it will not cause the object to rotate. However, if a net force is applied to a point other than the center of mass, it will affect the object's rotation. Physicists call the effect of force on rotational motion **torque**.

Torque Defined

Consider a lever mounted on a wall so that the lever is free to move around an axis of rotation O. In order to lift the lever, you apply a force $\boldsymbol{F}$ to point P, which is a distance r away from the axis of rotation, as illustrated below.

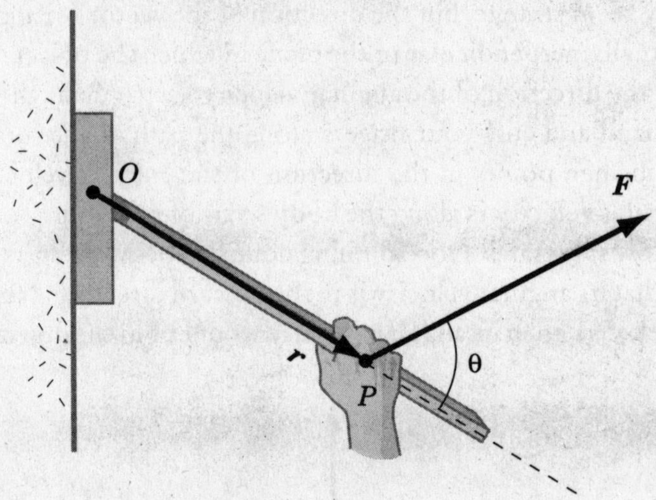

Suppose the lever is very heavy and resists your efforts to lift it. If you want to put all you can into lifting this lever, what should you do? Simple intuition would suggest, first of all, that you should lift with all your strength. Second, you should grab onto the end of the lever, and not a point near its axis of rotation. Third, you should lift in a direction that is perpendicular to the lever: if you pull very hard away from the wall, or push very hard toward the wall, the lever won't rotate at all.

Let's summarize. In order to maximize torque, you need to:

1. Maximize the magnitude of the force, $\boldsymbol{F}$, that you apply to the lever.
2. Maximize the distance, r, from the axis of rotation of the point on the lever to which you apply the force.
3. Apply the force in a direction perpendicular to the lever.

We can apply these three requirements to an equation for torque, τ:

$$\tau = Fr \sin \theta$$

In this equation, θ is the angle made between the vector for the applied force and the lever.

Torque Defined in Terms of Perpendicular Components

There's another way of thinking about torque that may be a bit more intuitive than the definition provided above. Torque is the product of the distance of the applied force from the axis of rotation and the component of the applied force that is perpendicular to the lever arm. Or, alternatively, torque is the product of the applied force and the component of the length of the lever arm that runs perpendicular to the applied force.

We can express these relations mathematically as follows:

$$\tau = \boldsymbol{F}_{\perp}\boldsymbol{r} = \boldsymbol{F}\boldsymbol{r}_{\perp}$$

where $\boldsymbol{F}_{\perp}$ and $\boldsymbol{r}_{\perp}$ are defined below.

Torque Defined as a Vector Quantity

Torque, like angular velocity and angular acceleration, is a vector quantity. Most precisely, it is the cross product of the displacement vector, $\boldsymbol{r}$, from the axis of rotation to the point where the force is applied, and the vector for the applied force, $\boldsymbol{F}$.

$$\boldsymbol{\tau} = \boldsymbol{r} \times \boldsymbol{F}$$

To determine the direction of the torque vector, use the right-hand rule, curling your fingers around from the $\boldsymbol{r}$ vector over to the $\boldsymbol{F}$ vector. In the example of lifting the lever, the torque would be represented by a vector at O pointing out of the page.

Example

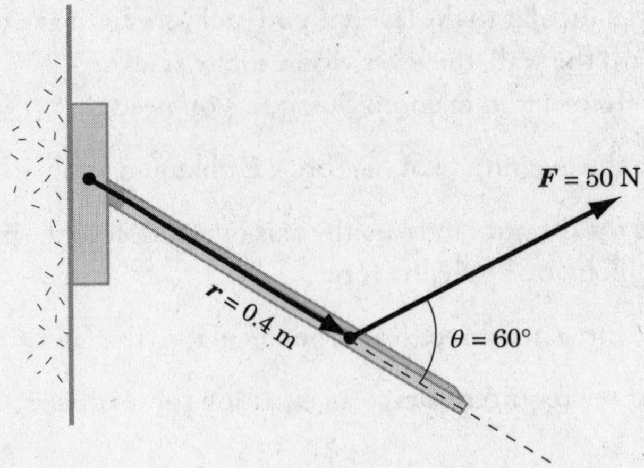

A student exerts a force of 50 N on a lever at a distance 0.4 m from its axis of rotation. The student pulls at an angle that is 60° above the lever arm. What is the torque experienced by the lever arm?

Let's plug these values into the first equation we saw for torque:

$$\tau = Fr \sin\theta = (50\ \text{N})(0.4\ \text{m}) \sin 60^\circ$$
$$= 17.3\ \text{N} \cdot \text{m}$$

This vector has its tail at the axis of rotation, and, according to the right-hand rule, points out of the page.

Newton's First Law and Equilibrium

Newton's Laws apply to torque just as they apply to force. You will find that solving problems involving torque is made a great deal easier if you're familiar with how to apply Newton's Laws to them. The First Law states:

If the net torque acting on a rigid object is zero, it will rotate with a constant angular velocity.

The most significant application of Newton's First Law in this context is with regard to the concept of **equilibrium**. When the net torque acting on a rigid object is zero, and that object is not already rotating, it will not begin to rotate.

When the SAT II tests you on equilibrium, it will usually present you with a system where more than one torque is acting upon an object, and will tell you that the object is not rotating. That means that the net torque acting on the object is zero, so that the sum of all torques acting in the clockwise direction is equal to the sum of all torques acting in the counterclockwise direction. A typical SAT II question will ask you to determine the magnitude of one or more forces acting on a given object that is in equilibrium.

Example

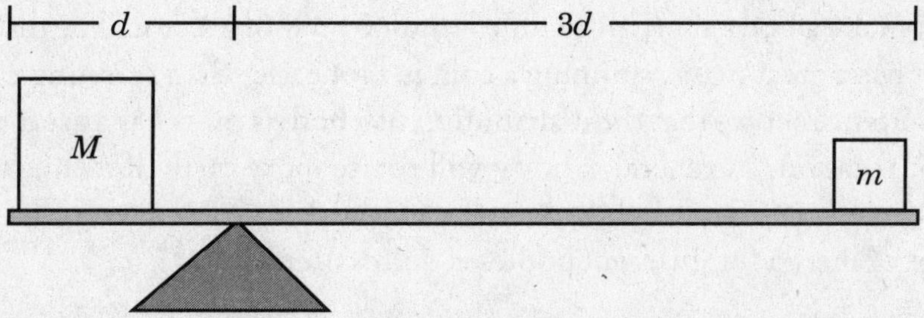

Two masses are balanced on the scale pictured above. If the bar connecting the two masses is horizontal and massless, what is the weight of mass m in terms of M?

Since the scale is not rotating, it is in equilibrium, and the net torque acting upon it must be zero. In other words, the torque exerted by mass M must be equal and opposite to the torque exerted by mass m. Mathematically,

$$\begin{aligned} \tau_M + \tau_m &= 0 \\ Mgd + (-mg(3d)) &= 0 \\ 3mgd &= Mgd \\ m &= \frac{M}{3} \end{aligned}$$

Because m is three times as far from the axis of rotation as M, it applies three times as much torque per mass. If the two masses are to balance one another out, then M must be three times as heavy as m.

Newton's Second Law

We have seen that acceleration has a rotational equivalent in angular acceleration, α, and that force has a rotational equivalent in torque, τ. Just as the familiar version of Newton's Second Law tells us that the acceleration of a body is proportional to the force applied to it, the rotational version of Newton's Second Law tells us that the angular acceleration of a body is proportional to the torque applied to it.

Of course, force is also proportional to mass, and there is also a rotational equivalent for mass: the **moment of inertia**, I, which represents an object's resistance to being rotated. Using the three variables, τ, I, and α, we can arrive at a rotational equivalent for Newton's Second Law:

$$\tau_{\text{net}} = I\boldsymbol{\alpha}$$

As you might have guessed, the real challenge involved in the rotational version of Newton's Second Law is sorting out the correct value for the moment of inertia.

Moment of Inertia

What might make a body more difficult to rotate? First of all, it will be difficult to set in a spin if it has a great mass : spinning a coin is a lot easier than spinning a lead block. Second, experience shows that the distribution of a body's mass has a great effect on its potential for rotation. In general, a body will rotate more easily if its mass is concentrated near the axis of rotation, but the calculations that go into determining the precise moment of inertia for different bodies is quite complex.

Moment of inertia for a single particle

Consider a particle of mass m that is tethered by a massless string of length r to point O, as pictured below:

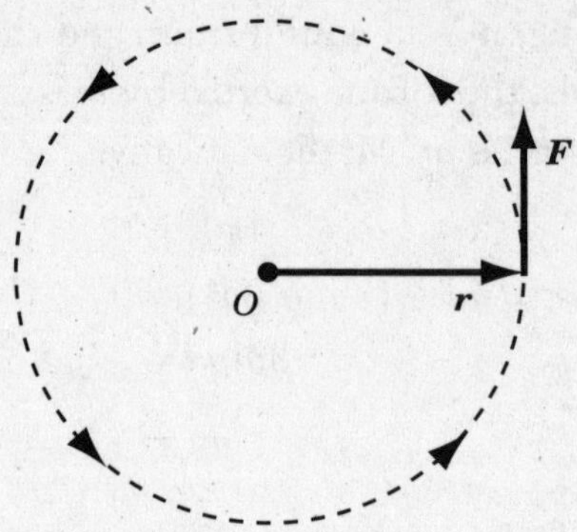

The torque that produces the angular acceleration of the particle is $\tau = rF$, and is directed out of the page. From the linear version of Newton's Second Law, we know that $\boldsymbol{F} = ma$ or $\boldsymbol{F} = m\,\alpha r$. If we multiply both sides of this equation by r, we find:

$$\tau = mr^2\alpha$$

If we compare this equation to the rotational version of Newton's Second Law, we see that the moment of inertia of our particle must be mr^2.

Moment of inertia for rigid bodies

Consider a wheel, where every particle in the wheel moves around the axis of rotation. The net torque on the wheel is the sum of the torques exerted on each particle in the wheel. In its most general form, the rotational version of Newton's Second Law takes into account the moment of inertia of each individual particle in a rotating system:

$$\boldsymbol{\tau}_{\text{net}} = (\Sigma mr^2)\boldsymbol{\alpha}$$

Of course, adding up the radius and mass of every particle in a system is very tiresome unless the system consists of only two or three particles. The moment of inertia for more complex systems can only be determined using calculus. The SAT II doesn't expect you to know calculus, so it will give you the moment of inertia for a complex body whenever the need arises. For your own reference, however, here is the moment of inertia for a few common shapes.

Moments of Inertia

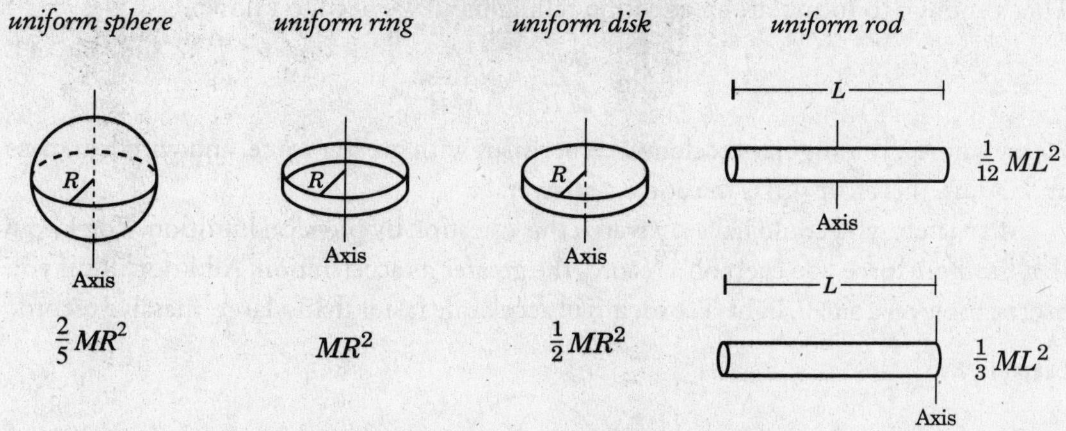

In these figures, M is the mass of the rigid body, R is the radius of round bodies, and L is the distance on a rod between the axis of rotation and the end of the rod. Note that the moment of inertia depends on the shape and mass of the rigid body, as well as on its axis of rotation, and that for most objects, the moment of inertia is a multiple of MR^2.

Example 1

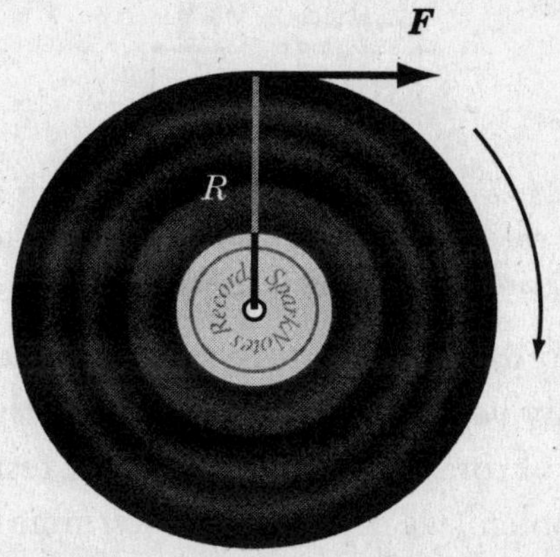

A record of mass M and radius R is free to rotate around an axis through its center, O. A tangential force $\boldsymbol{F}$ is applied to the record. What must one do to maximize the angular acceleration?

(A) Make $\boldsymbol{F}$ and M as large as possible and R as small as possible
(B) Make M as large as possible and $\boldsymbol{F}$ and R as small as possible.
(C) Make $\boldsymbol{F}$ as large as possible and M and R as small as possible.
(D) Make R as large as possible and $\boldsymbol{F}$ and M as small as possible.
(E) Make $\boldsymbol{F}$, M, and R as large as possible.

To answer this question, you don't need to know exactly what a disc's moment of inertia is—you just need to be familiar with the general principle that it will be some multiple of MR^2.

The rotational version of Newton's Second Law tells us that $\tau = I\alpha$, and so $\alpha = FR / I$. Suppose we don't know what I is, but we know that it is some multiple of MR^2. That's enough to formulate an equation telling us all we need to know:

$$\alpha \propto \frac{FR}{MR^2} = \frac{F}{MR}$$

As we can see, the angular acceleration increases with greater force, and with less mass and radius, therefore (C) is the correct answer.

Alternately, you could have answered this question by physical intuition. You know that the more force you exert on a record, the greater its acceleration. Additionally, if you exert a force on a small, light record, it will accelerate faster than a large, massive record.

Example 2

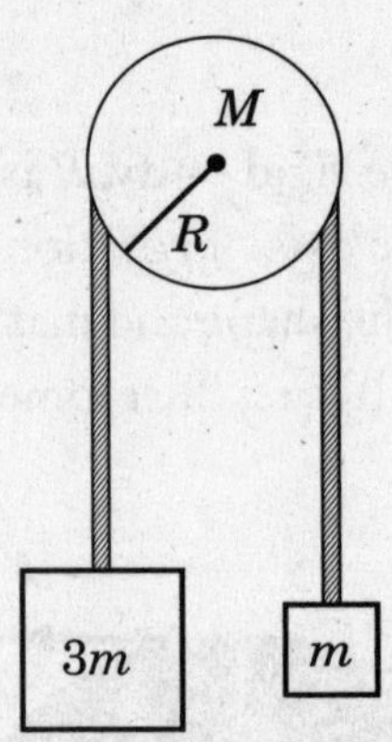

The masses in the figure above are initially held at rest and are then released. If the mass of the pulley is M, what is the angular acceleration of the pulley? The moment of inertia of a disk spinning around its center is $\frac{1}{2}MR^2$.

This is the only situation on the SAT II where you may encounter a pulley that is not considered massless. Usually you can ignore the mass of the pulley block, but it matters when your knowledge of rotational motion is being tested.

In order to solve this problem, we first need to determine the net torque acting on the pulley, and then use Newton's Second Law to determine the pulley's angular acceleration. The weight of each mass is transferred to the tension in the rope, and the two forces of tension on the pulley block exert torques in opposite directions as illustrated below:

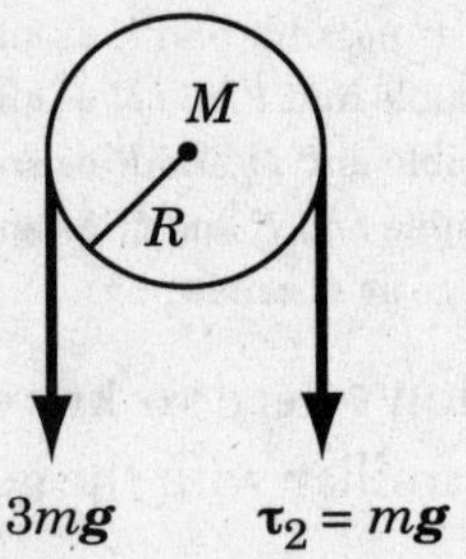

To calculate the net torque, we simply need to get the sum of all torques. Let's take the counterclockwise direction to be positive and the clockwise direction to be negative:

$$\tau_{\text{net}} = 3mg - mg = 2mg$$

Because $\tau = I\alpha$, we can conclude that $\alpha = \tau/I$. Since we know the net torque acting on the pulley block and we know its moment of inertia, we can calculate its angular acceleration:

$$\begin{aligned}\alpha &= \frac{\tau}{I} = \frac{2mg}{\frac{1}{2}MR^2} \\ &= \frac{4mg}{MR^2}\end{aligned}$$

Because α is positive, we know that the pulley will spin in the counterclockwise direction and the $3m$ block will drop.

Kinetic Energy

There is a certain amount of energy associated with the rotational motion of a body, so that a ball rolling down a hill does not accelerate in quite the same way as a block sliding down a frictionless slope. Fortunately, the formula for rotational kinetic energy, much like the formula for translational kinetic energy, can be a valuable problem-solving tool.

The kinetic energy of a rotating rigid body is:

$$KE = \frac{1}{2}I\omega^2$$

Considering that I is the rotational equivalent for mass and ω is the rotational equivalent for velocity, this equation should come as no surprise.

An object, such as a pool ball, that is spinning as it travels through space, will have both rotational and translational kinetic energy:

$$KE = \frac{1}{2}Mv_{\text{cm}} + \frac{1}{2}I\omega^2$$

In this formula, M is the total mass of the rigid body and v_{cm} is the velocity of its center of mass.

This equation comes up most frequently in problems involving a rigid body that is rolling along a surface without sliding. Unlike a body sliding along a surface, there is no kinetic friction to slow the body's motion. Rather, there is static friction as each point of the rolling body makes contact with the surface, but this static friction does no work on the rolling object and dissipates no energy.

Example

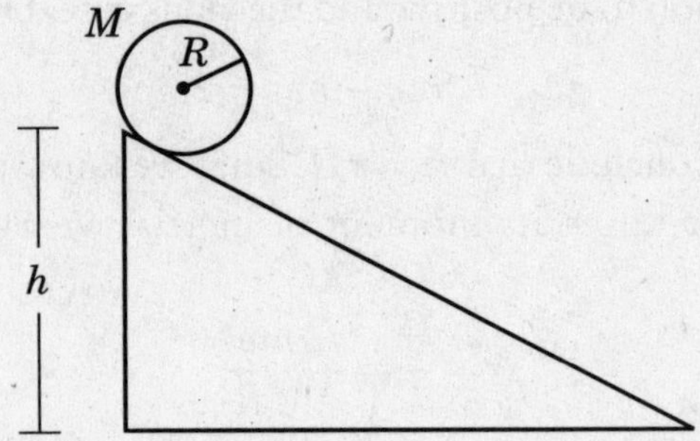

A wheel of mass M and radius R is released from rest and rolls to the bottom of an inclined plane of height h without slipping. What is its velocity at the bottom of the incline? The moment of inertia of a wheel of mass M and radius R rotating about an axis through its center of mass is $\frac{1}{2}MR^2$.

Because the wheel loses no energy to friction, we can apply the law of conservation of mechanical energy. The change in the wheel's potential energy is $-mgh$. The change in the wheel's kinetic energy is $\frac{1}{2}Mv_{cm}^2 + \frac{1}{2}I\omega^2$. Applying conservation of mechanical energy:

$$-Mgh + \frac{1}{2}Mv^2 + \frac{1}{2}(\frac{1}{2}MR^2)(\frac{v^2}{R^2}) = 0$$

$$Mgh = (\frac{1}{2} + \frac{1}{4})Mv^2$$

$$v = \sqrt{\frac{4}{3}gh}$$

It's worth remembering that an object rolling down an incline will pick up speed more slowly than an object sliding down a frictionless incline. Rolling objects pick up speed more slowly because only some of the kinetic energy they gain is converted into translational motion, while the rest is converted into rotational motion.

Angular Momentum

The rotational analogue of linear momentum is **angular momentum**, $\boldsymbol{L}$. After torque and equilibrium, angular momentum is the aspect of rotational motion most likely to be tested on the SAT II. For the test, you will probably have to deal only with the angular momentum of a particle or body moving in a circular trajectory. In such a case, we can define angular momentum in terms of moment of inertia and angular velocity, just as we can define linear momentum in terms of mass and velocity:

$$\boldsymbol{L} = I\boldsymbol{\omega}$$

The angular momentum vector always points in the same direction as the angular velocity vector.

Angular Momentum of a Single Particle

Let's take the example of a tetherball of mass m swinging about on a rope of length r:

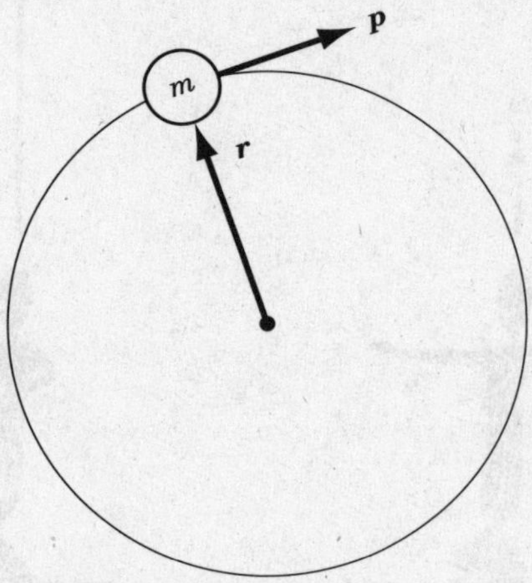

The tetherball has a moment of inertia of $I = mr^2$ and an angular velocity of $\omega = v/r$. Substituting these values into the formula for linear momentum we get:

$$L = I\omega = (mr^2)(v/r) \\ = mvr$$

This is the value we would expect from the cross product definition we saw earlier of angular momentum. The momentum, $\boldsymbol{p}=m\boldsymbol{v}$ of a particle moving in a circle is always tangent to the circle and perpendicular to the radius. Therefore, when a particle is moving in a circle,

$$L = pr\sin 90^\circ = pr = mvr$$

Newton's Second Law and Conservation of Angular Momentum

In the previous chapter, we saw that the net force acting on an object is equal to the rate of change of the object's momentum with time. Similarly, the net torque acting on an object is equal to the rate of change of the object's angular momentum with time:

$$\tau_{\text{net}} = \frac{\Delta L}{\Delta t}$$

If the net torque action on a rigid body is zero, then the angular momentum of the body is constant or conserved. The **law of conservation of angular momentum** is another one of nature's beautiful properties, as well as a very useful means of solving problems. It is likely that angular momentum will be tested in a conceptual manner on the SAT II.

Example

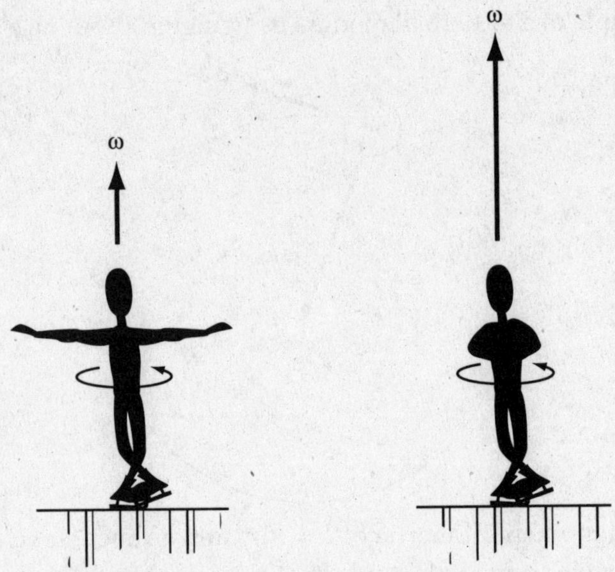

One of Brian Boitano's crowd-pleasing skating moves involves initiating a spin with his arms extended and then moving his arms closer to his body. As he does so, he spins at a faster and faster rate. Which of the following laws best explains this phenomenon?

(A) Conservation of Mechanical Energy
(B) Conservation of Angular Momentum
(C) Conservation of Linear Momentum
(D) Newton's First Law
(E) Newton's Second Law

Given the context, the answer to this question is no secret: it's (B), the conservation of angular momentum. Explaining why is the interesting part.

As Brian spins on the ice, the net torque acting on him is zero, so angular momentum is conserved. That means that $I\omega$ is a conserved quantity. I is proportional to R^2, the distance of the parts of Brian's body from his axis of rotation. As he draws his arms in toward his body, his mass is more closely concentrated about his axis of rotation, so I decreases. Because $I\omega$ must remain constant, ω must increase as I decreases. As a result, Brian's angular velocity increases as he draws his arms in toward his body.

Key Formulas

Angular Position	$\phi = \frac{l}{r}$
Definition of a Radian	1 revolution $= 2\pi\text{rad} = 360°$
Average Angular Velocity	$\bar{\omega} = \frac{\Delta\theta}{\Delta t}$
Average Angular Acceleration	$\bar{\alpha} = \frac{\Delta\alpha}{\Delta t}$
Angular Frequency	$f = \frac{\omega}{2\pi}$
Angular Period	$T = \frac{1}{f}$
Relations between Linear and Angular Variables	$s = \theta r$ $v = \omega r$ $a = \alpha r$
Equations for Rotational and Angular Kinematics with Constant Acceleration	$\phi = \phi_0 + \frac{1}{2}(\omega + \omega_0)t$ $\omega = \omega_0 + \alpha t$ $\phi = \phi_0 + \omega_0 t + \frac{1}{2}\alpha t^2$ $\phi = \phi_0 + \omega t - \frac{1}{2}\alpha t^2$ $\omega^2 = \omega_0^2 + 2\alpha(\phi - \phi_0)$
Torque As Trigonometric Function	$\tau = Fr\sin\theta$

Component Form of the Torque Equation	$\tau = \boldsymbol{F}_{\perp}\boldsymbol{r} = \boldsymbol{F}\boldsymbol{r}_{\perp}$
Torque As Cross Product	$\boldsymbol{\tau} = \boldsymbol{r} \times \boldsymbol{F}$
Newton's Second Law in Terms of Rotational Motion	$\boldsymbol{\tau}_{\text{net}} = I\boldsymbol{\alpha}$
Moment of Inertia	$I = \Sigma mr^2$
Kinetic Energy of Rotation	$KE = \frac{1}{2}I\omega^2$
Angular Momentum of a Particle	$\boldsymbol{L} = \boldsymbol{r} \times \boldsymbol{p}$
Component Form of the Angular Momentum of a Particle	$L = mrv_{\perp}$ $L = mr_{\perp}v$
Angular Momentum of a Rotating Rigid Body	$\boldsymbol{L} = I\boldsymbol{\omega}$

Rotational Motion

Practice Questions

1. The instantaneous velocity of a point on the outer edge of a disk with a diameter of 4 m that is rotating at 120 revolutions per minute is most nearly:

    (A) 4 m/s
    (B) 6 m/s
    (C) 12 m/s
    (D) 25 m/s
    (E) 50 m/s

2. A washing machine, starting from rest, accelerates within 3.14 s to a point where it is revolving at a frequency of 2.00 Hz. Its angular acceleration is most nearly:

    (A) 0.100 rad/s²
    (B) 0.637 rad/s²
    (C) 2.00 rad/s²
    (D) 4.00 rad/s²
    (E) 6.28 rad/s²

3. What is the direction of the angular velocity vector for the second hand of a clock going from 0 to 30 seconds?

(A) Outward from the clock face
(B) Inward toward the clock face
(C) Upward
(D) Downward
(E) To the right

4. Which of the following are means of maximizing the torque of a force applied to a rotating object?

I. Maximize the magnitude of the applied force
II. Apply the force as close as possible to the axis of rotation
III. Apply the force perpendicular to the displacement vector between the axis of rotation and the point of applied force

(A) I only
(B) II only
(C) I and II only
(D) I and III only
(E) I, II, and III

5. What is the torque on the pivot of a pendulum of length R and mass m, when the mass is at an angle θ?

(A) $m\frac{g}{R}\sin\theta$

(B) $m\frac{g}{R}\cos\theta$

(C) $mgR \sin\theta$

(D) $mgR \cos\theta$

(E) $mgR \tan\theta$

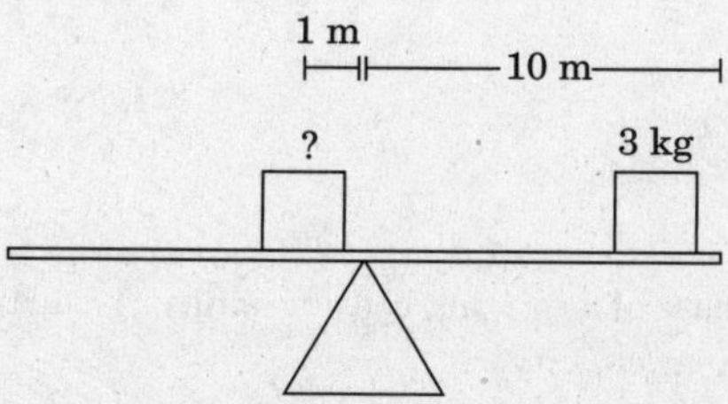

6. Two objects rest on a seesaw. The first object has a mass of 3 kg and rests 10 m from the pivot. The other rests 1 m from the pivot. What is the mass of the second object if the seesaw is in equilibrium?

(A) 0.3 kg (B) 3 kg (C) 10 kg (D) 30 kg (E) 50 kg

7. What is the angular acceleration of a 0.1 kg record with a radius of 0.1 m to which a torque of 0.05 N · m is applied? The moment of inertia of a disk spinning about its center is ½ MR^2.

(A) 0.1 rad/s^2
(B) 0.5 rad/s^2
(C) 1 rad/s^2
(D) 5 rad/s^2
(E) 10 rad/s^2

8. A disk of mass m and radius R rolls down an inclined plane of height h without slipping. What is the velocity of the disk at the bottom of the incline? The moment of inertia for a disk is ½ mR^2.

(A) $\sqrt{gh}$

(B) $\sqrt{\frac{4}{3}gh}$

(C) $\sqrt{2gh}$

(D) $2\sqrt{gh}$

(E) $2\sqrt{2gh}$

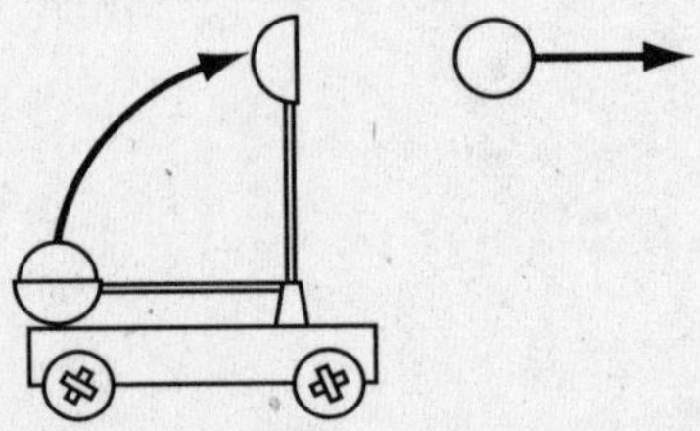

9. A catapult with a basket of mass 50 kg launches a 200 kg rock by swinging around from a horizontal to a vertical position with an angular velocity of 2.0 rad/s. Assuming the rest of the catapult is massless and the catapult arm is 10 m long, what is the velocity of the rock as it leaves the catapult?

(A) 10 m/s
(B) 20 m/s
(C) 25 m/s
(D) 50 m/s
(E) 100 m/s

10. How should the mass of a rotating body of radius r be distributed so as to maximize its angular velocity?

(A) The mass should be concentrated at the outer edge of the body
(B) The mass should be evenly distributed throughout the body
(C) The mass should be concentrated at the axis of rotation
(D) The mass should be concentrated at a point midway between the axis of rotation and the outer edge of the body
(E) Mass distribution has no impact on angular velocity

Explanations

1. (D)

An object that experiences 120 revolutions per minute experiences 2 revolutions per second; in other words, it rotates with a frequency of 2 Hz. We have formulas relating frequency to angular velocity and angular velocity to linear velocity, so solving this problem is simply a matter of finding an expression for linear velocity in terms of frequency. Angular and linear velocity are related by the formula $v = \omega r$, so we need to plug this formula into the formula relating frequency and angular velocity:

$$\begin{aligned} f &= \frac{\omega}{2\pi} = \frac{v}{2\pi r} \\ v &= 2\pi r f = 2\pi(2\text{ m})(2\text{ Hz}) \\ &= 8\pi\text{ m/s} \approx 25.1\text{ m/s} \end{aligned}$$

2. (D)

Frequency and angular velocity are related by the formula $\omega = 2\pi f$, and angular velocity and angular acceleration are related by the formula $\alpha = \Delta\omega\Delta t$. In order to calculate the washing machine's acceleration, then, we must calculate its angular velocity, and divide that number by the amount of time it takes to reach that velocity:

$$\begin{aligned} \alpha &= \frac{\Delta\omega}{\Delta t} = \frac{2\pi f}{\Delta t} = \frac{2(3.14)(2.00\text{ Hz})}{3.14\text{ s}} \\ &= 4.00\text{ rad/s}^2 \end{aligned}$$

3. (B)

You need to apply the right-hand rule in order to solve this problem. Extend the fingers of your right hand upward so that they point to the 0 second point on the clock face, and then curl them around so that they point downward to the 30 second point on the clock face. In order to do this, you'll find that your thumb must be pointing inward toward the clock face. This is the direction of the angular velocity vector.

4. (D)

The torque on an object is given by the formula $\boldsymbol{\tau} = \boldsymbol{r} \times \boldsymbol{F}$, where $\boldsymbol{F}$ is the applied force and $\boldsymbol{r}$ is the distance of the applied force from the axis of rotation. In order to maximize this cross product, we need to maximize the two quantities and insure that they are perpendicular to one another. Statement I maximizes $\boldsymbol{F}$ and statement III demands that $\boldsymbol{F}$ and $\boldsymbol{r}$ be perpendicular, but statement II minimizes $\boldsymbol{r}$ rather than maximizes it, so statement II is false.

5. (C)
The torque acting on the pendulum is the product of the force acting perpendicular to the radius of the pendulum and the radius, $\tau = F_{\perp}R$. A free-body diagram of the pendulum shows us that the force acting perpendicular to the radius is $F_{\perp} = mg\sin\theta$.

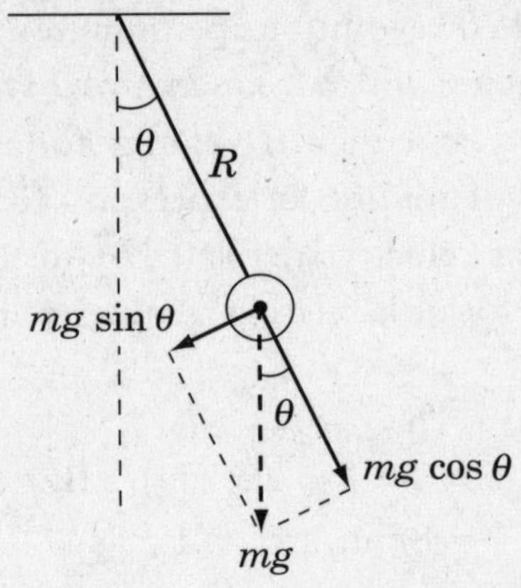

Since torque is the product of $F_{\perp}$ and R, the torque is $mgR\sin\theta$.

6. (D)
The seesaw is in equilibrium when the net torque acting on it is zero. Since both objects are exerting a force perpendicular to the seesaw, the torque is equal to $\tau = rF = rmg$. The 3 kg mass exerts a torque of $\tau = 30g$ N · m in the clockwise direction. The second mass exerts a torque in the counterclockwise direction. If we know this torque also has a magnitude of 30g N · m, we can solve for m:

$$\tau = rmg = 30g$$
$$mg = 30g$$
$$m = 30 \text{ kg}$$

7. (E)
The rotational equivalent of Newton's Second Law states that $\boldsymbol{\tau} = I\boldsymbol{\alpha}$. We are told that $\tau = 5.0 \times 10^{-2}$ N · m and $I = \frac{1}{2}MR^2$, so now we can solve for α:

$$\alpha = \frac{\tau}{I} = \frac{5.0 \times 10^{-2}\text{ N}\cdot\text{m}}{\frac{1}{2}(0.1\text{ kg})(0.1\text{ m})^2}$$
$$= 10 \text{ rad/s}^2$$

8. (B)
At the top of the incline, the disk has no kinetic energy, and a gravitational potential energy of mgh. At the bottom of the incline, all this gravitational potential energy has been converted into kinetic energy. However, in rolling down the hill, only some of this potential energy becomes translational kinetic energy, and the rest becomes rotational

kinetic energy. Translational kinetic energy is given by ½ mv^2 and rotational kinetic energy is given by ½ $I\omega^2$. We can express ω in terms of v and R with the equation $\omega = v/R$, and in the question we were told that $I = ½\, mR^2$. We now have all the information we need to solve for v:

$$\frac{1}{2}mv^2 + \frac{1}{2}I\omega^2 = mgh$$
$$\frac{1}{2}mv^2 + \frac{1}{2}(\frac{1}{2}mR^2)(\frac{v}{R})^2 = mgh$$
$$\frac{1}{2}mv^2 + \frac{1}{4}mv^2 = mgh$$
$$\frac{3}{4}v^2 = gh$$
$$v = \sqrt{\frac{4}{3}gh}$$

9. (C)

This is a conservation of momentum question. The angular momentum of the rock as it is launched is equal to its momentum after it's been launched. The momentum of the rock-basket system as it swings around is:

$$L = I\omega = (MR^2)\omega = ((250\text{ kg})(10\text{ m})^2)(2\text{ rad/s})$$
$$= 5.0 \times 10^4\text{ kg}\cdot\text{m}^2/\text{s}^2$$

The rock will have the same momentum as it leaves the basket. The angular momentum of a single particle is given by the formula $L = mvr$. Since L is conserved, we can manipulate this formula and solve for v:

$$v = \frac{L}{mr} = \frac{5.0 \times 10^4\text{kg}\cdot\text{m}^2/\text{s}^2}{(200\text{kg})(10\text{m})}$$
$$= 25\text{m/s}$$

Be sure to remember that the initial mass of the basket-rock system is 250 kg, while the final mass of the rock is only 200 kg.

10. (C)

Angular momentum, $L = I\omega$, is a conserved quantity, meaning that the greater I is, the less ω will be, and vice versa. In order to maximize angular velocity, then, it is necessary to minimize the moment of inertia. Since the moment of inertia is greater the farther the mass of a body is from its axis of rotation, we can maximize angular velocity by concentrating all the mass near the axis of rotation.

Circular Motion and Gravitation

NEWTON'S FIRST LAW TELLS US THAT objects will move in a straight line at a constant speed unless a net force is acting upon them. That rule would suggest that objects moving in a circle—whether they're tetherballs or planets—are under the constant influence of a changing force, since their trajectory is not in a straight line. We will begin by looking at the general features of circular motion and then move on to examine gravity, one of the principal sources of circular motion.

Chapter Contents

Uniform Circular Motion 185
Newton's Law of Universal Gravitation . . 189
Gravitational Potential Energy 193
Weightlessness 194
Kepler's Laws 196
Key Formulas 197
Practice Questions 198
Explanations 200

Uniform Circular Motion

Uniform circular motion occurs when a body moves in a circular path with constant speed. For example, say you swing a tethered ball overhead in a circle:

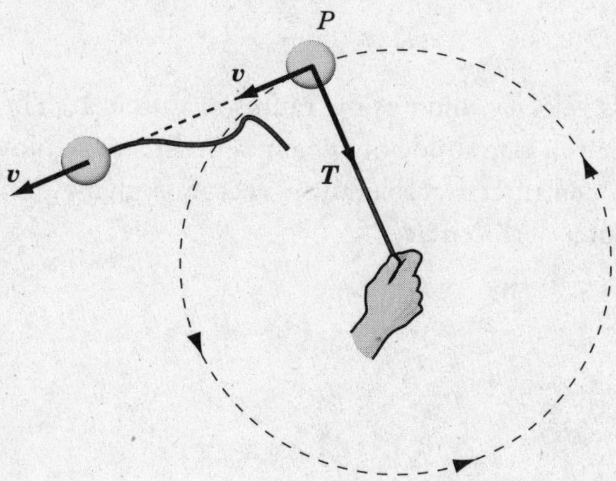

If we leave aside gravity for the moment, the only force acting on the ball is the force of tension, $\boldsymbol{T}$, of the string. This force is always directed radially inward along the string, toward your hand. In other words, the force acting on a tetherball traveling in a circular path is always directed toward the center of that circle.

Note that although the direction of the ball's velocity changes, the ball's velocity is constant in magnitude and is always tangent to the circle.

Centripetal Acceleration

From kinematics, we know that acceleration is the rate of change of the velocity vector with time. If we consider two points very close together on the ball's trajectory and calculate $\Delta \boldsymbol{v}$, we find that the ball's acceleration points inward along the radius of the circle.

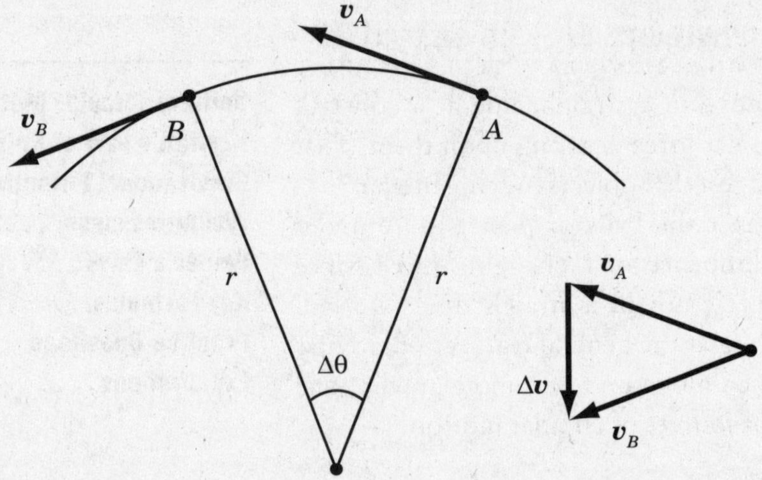

The acceleration of a body experiencing uniform circular motion is always directed toward the center of the circle, so we call that acceleration **centripetal acceleration**, $\boldsymbol{a}_c$. "Centripetal" comes from a Latin word meaning "center-seeking." We define the centripetal acceleration of a body moving in a circle as:

$$\boldsymbol{a}_c = \frac{\boldsymbol{v}^2}{r}$$

where $\boldsymbol{v}$ is the body's velocity, and r is the radius of the circle. The body's centripetal acceleration is constant in magnitude but changes in direction. Note that even though the direction of the centripetal acceleration vector is changing, the vector always points toward the center of the circle.

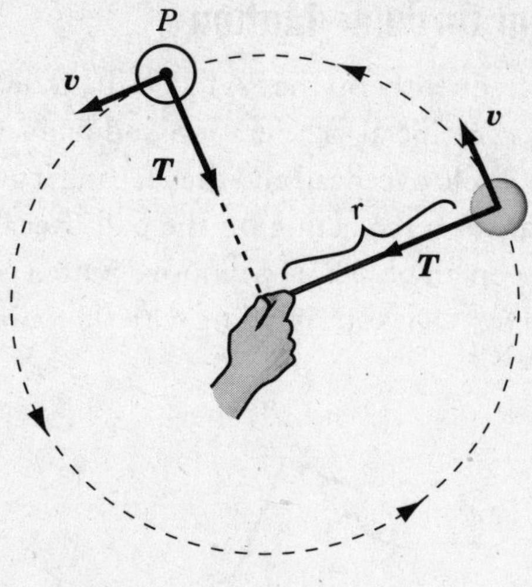

How This Knowledge Will Be Tested

Most of us are accustomed to think of "change" as a change in magnitude, so it may be counterintuitive to think of the acceleration vector as "changing" when its magnitude remains constant. You'll frequently find questions on the SAT II that will try to catch you sleeping on the nature of centripetal acceleration. These questions are generally qualitative, so if you bear in mind that the acceleration vector is constant in magnitude, has a direction that always points toward the center of the circle, and is always perpendicular to the velocity vector, you should have no problem at all.

Centripetal Force

Wherever you find acceleration, you will also find force. For a body to experience centripetal acceleration, a **centripetal force** must be applied to it. The vector for this force is similar to the acceleration vector: it is of constant magnitude, and always points radially inward to the center of the circle, perpendicular to the velocity vector.

We can use Newton's Second Law and the equation for centripetal acceleration to write an equation for the centripetal force that maintains an object's circular motion:

$$\boldsymbol{F} = m\boldsymbol{a}_c = \frac{m\boldsymbol{v}^2}{r}$$

Example

> A ball with a mass of 2 kg is swung in a circular path on a massless rope of length 0.5 m. If the ball's speed is 1 m/s, what is the tension in the rope?

The tension in the rope is what provides the centripetal force, so we just need to calculate the centripetal force using the equation above:

$$\boldsymbol{T} = \frac{m\boldsymbol{v}^2}{r} = \frac{(2\text{ kg})(1\text{ m/s})^2}{0.5\text{ m}}$$
$$= 4\text{ N}$$

Objects Released from Circular Motion

One concept that is tested frequently on the SAT II is the trajectory of a circling body when the force providing centripetal acceleration suddenly vanishes. For example, imagine swinging a ball in a circle overhead and then letting it go. As soon as you let go, there is no longer a centripetal force acting on the ball. Recall Newton's First Law: when no net force is acting on an object, it will move with a constant velocity. When you let go of the ball, it will travel in a straight line with the velocity it had when you let go of it.

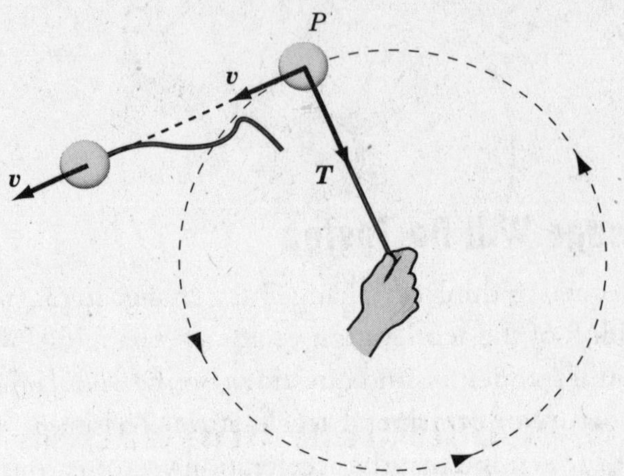

Example

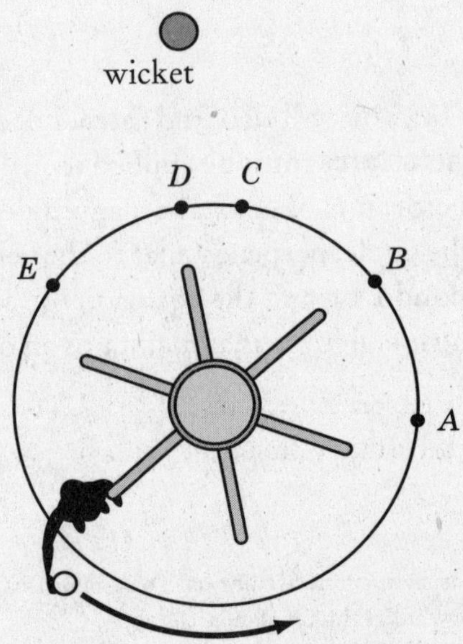

A student is standing on a merry-go-round that is rotating counterclockwise, as illustrated above. The student is given a ball and told to release it in such a way that it knocks over the wicket at the top of the diagram. At what point should the student release the ball?

When the student releases the ball, it will travel in a straight line, tangent to the circle. In order to hit the wicket, then, the student should release the ball at point B.

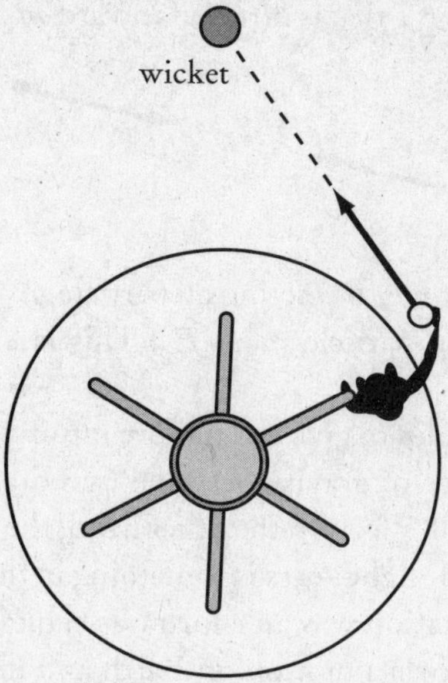

Newton's Law of Universal Gravitation

Newton's Law of Universal Gravitation is a fundamental physical law. We experience its effects everywhere on this planet, and it is the prime mover in the vast world of astronomy. It can also be expressed in a relatively simple mathematical formula on which the SAT II is almost certain to test you.

Gravitational Force

In 1687, Isaac Newton published his Law of Gravitation in *Philosophiae Naturalis Principia Mathematica*. Newton proposed that every body in the universe is attracted to every other body with a force that is directly proportional to the product of the bodies' masses and inversely proportional to the square of the bodies' separation. In terms of mathematical relationships, Newton's Law of Gravitation states that the force of gravity, $\boldsymbol{F}_g$, between two particles of mass m_1 and m_2 has a magnitude of:

$$F_g = G\frac{m_1 m_2}{r^2}$$

where r is the distance between the center of the two masses and G is the **gravitational constant**. The value of G was determined experimentally by Henry Cavendish in 1798:

$$G = 6.67 \times 10^{-11}\ \text{N} \cdot \text{m}^2/\text{kg}^2$$

The force of gravity is a vector quantity. Particle m_1 attracts particle m_2 with a force that is directed toward m_1, as illustrated in the figure below. Similarly, particle m_2 attracts particle m_1 with a force that is directed toward m_2.

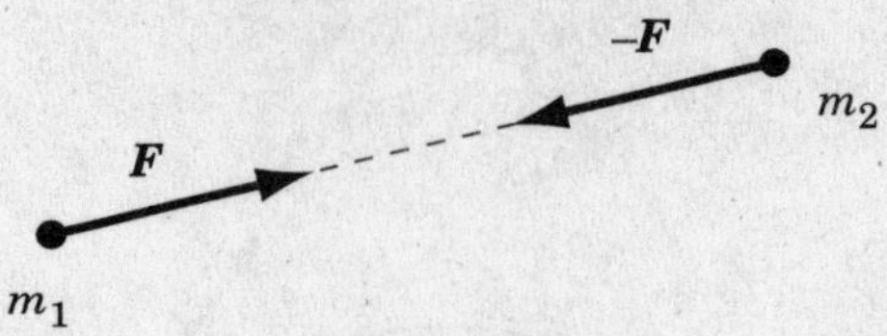

Note that the gravitational force, $\boldsymbol{F}_g$, acting on particle m_1 is equal and opposite to the gravitational force acting on particle m_2, $-\boldsymbol{F}_g$. This is a consequence of Newton's Third Law.

Let's consider two examples to give you a more intuitive feel for the strength of the gravitational force. The force of gravity between two oranges on opposite sides of a table is quite tiny, roughly 10^{-13} N. On the other hand, the gravitational force between two galaxies separated by 10^6 light years is something in the neighborhood of 10^{27} N!

Newton's Law of Gravitation was an enormous achievement, precisely because it synthesized the laws that govern motion on Earth and in the heavens. Additionally, Newton's work had a profound effect on philosophical thought. His research implied that the universe was a rational place that could be described by universal, scientific laws. But this is knowledge for another course. If you are interested in learning more about it, make sure to take a class on the history of science in college.

Gravity on the Surface of Planets

Previously, we noted that the acceleration due to gravity on Earth is 9.8 m/s² toward the center of the Earth. We can derive this result using Newton's Law of Gravitation.

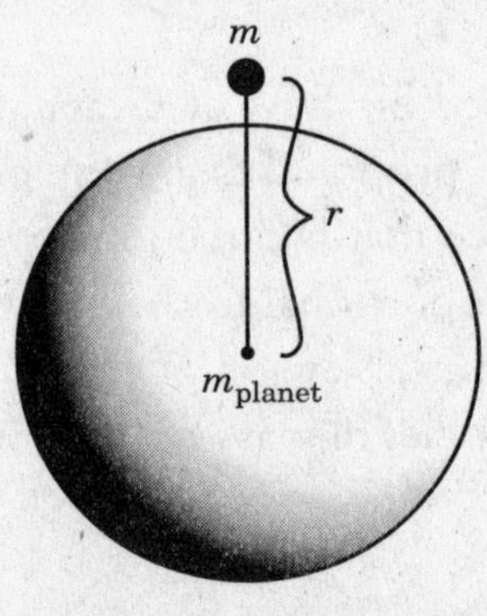

Consider the general case of a mass accelerating toward the center of a planet. Applying Newton's Second Law, we find:

$$F = m_{\text{object}}a = G\frac{m_{\text{object}}m_{\text{planet}}}{r^2}$$
$$a = G\frac{m_{\text{planet}}}{r^2}$$

Note that this equation tells us that acceleration is directly proportional to the mass of the planet and inversely proportional to the square of the radius. The mass of the object under the influence of the planet's gravitational pull doesn't factor into the equation. This is now pretty common knowledge, but it still trips up students on the SAT II: all objects under the influence of gravity, regardless of mass, fall with the same acceleration.

Acceleration on the Surface of the Earth

To find the acceleration due to gravity on the surface of the Earth, we must substitute values for the gravitational constant, the mass of the Earth, and the radius of the Earth into the equation above:

$$a = (6.67 \times 10^{-11}\ \mathrm{N \cdot m^2/kg^2})\frac{5.98 \times 10^{24}\ \mathrm{kg}}{(6.37 \times 10^{6}\ \mathrm{m})^2}$$
$$= 9.8\ \mathrm{m/s^2}$$

Not coincidentally, this is the same number we've been using in all those kinematic equations.

Acceleration Beneath the Surface of the Earth

If you were to burrow deep into the bowels of the Earth, the acceleration due to gravity would be different. This difference would be due not only to the fact that the value of r would have decreased. It would also be due to the fact that not all of the Earth's mass would be under you. The mass above your head wouldn't draw you toward the center of the Earth—quite the opposite—and so the value of m_{planet} would also decrease as you burrowed. It turns out that there is a linear relationship between the acceleration due to gravity and one's distance from the Earth's center when you are beneath the surface of the Earth. Burrow halfway to the center of the Earth and the acceleration due to gravity will be ½ g. Burrow three quarters of the way to the center of the Earth and the acceleration due to gravity will be ¼ g.

Orbits

The **orbit** of satellites—whether of artificial satellites or natural ones like moons and planets—is a common way in which the SAT II will test your knowledge of both uniform circular motion and gravitation in a single question.

How Do Orbits Work?

Imagine a baseball pitcher with a very strong arm. If he just tosses the ball lightly, it will fall to the ground right in front of him. If he pitches the ball at 100 miles per hour in a line horizontal with the Earth, it will fly something in the neighborhood of 80 feet before it hits the ground. By the same token, if he were to pitch the ball at 100,000 miles per hour in a line horizontal with the Earth, it will fly something in the neighborhood of 16 miles before it hits the ground. Now remember: the Earth is round, so if the ball flies far enough, the ball's downward trajectory will simply follow the curvature of the Earth until it makes a full circle of the Earth and hits the pitcher in the back of the head. A satellite in orbit is an object in free fall moving at a high enough velocity that it falls around the Earth rather than back down to the Earth.

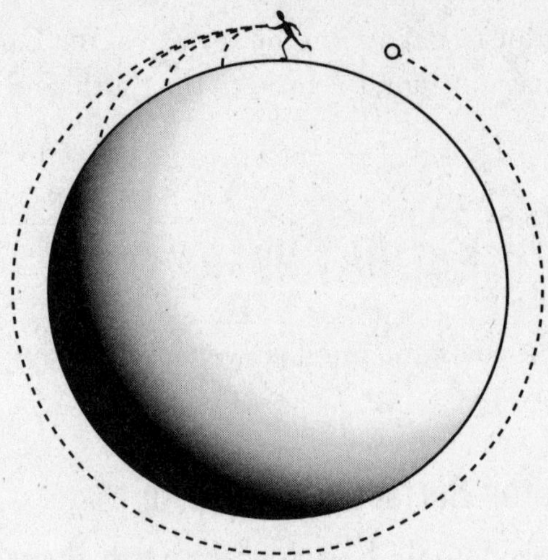

Gravitational Force and Velocity of an Orbiting Satellite

Let's take the example of a satellite of mass m_s orbiting the Earth with a velocity v. The satellite is a distance R from the center of the Earth, and the Earth has a mass of m_e.

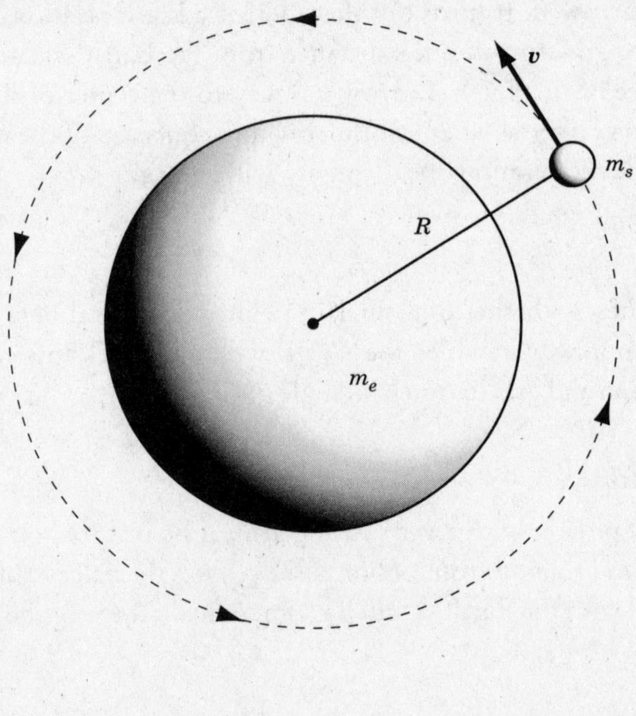

The centripetal force acting on the satellite is the gravitational force of the Earth. Equating the formulas for gravitational force and centripetal force we can solve for v:

$$G\frac{m_s m_e}{R^2} = \frac{m_s v^2}{R}$$

$$v = \sqrt{\frac{Gm_e}{R}}$$

As you can see, for a planet of a given mass, each radius of orbit corresponds with a certain velocity. That is, any object orbiting at radius R must be orbiting with a velocity of $\sqrt{Gm_e/R}$. If the satellite's speed is too small, then the satellite will fall back down to Earth. If the satellite's speed is too great, then the satellite will fly out into space.

Gravitational Potential Energy

In Chapter 4, we learned that the potential energy of a system is equal to the amount of work that must be done to arrange the system in that particular configuration. We also saw that **gravitational potential energy** depends on how high an object is off the ground: the higher an object is, the more work needs to be done to get it there.

Gravitational potential energy is not an absolute measure. It tells us the amount of work needed to move an object from some arbitrarily chosen reference point to the position it is presently in. For instance, when dealing with bodies near the surface of the Earth, we choose the ground as our reference point, because it makes our calculations easier. If the ground is $h = 0$, then for a height h above the ground an object has a potential energy of mgh.

Gravitational Potential in Outer Space

Off the surface of the Earth, there's no obvious reference point from which to measure gravitational potential energy. Conventionally, we say that an object that is an infinite distance away from the Earth has zero gravitational potential energy with respect to the Earth. Because a negative amount of work is done to bring an object closer to the Earth, gravitational potential energy is always a negative number when using this reference point.

The gravitational potential energy of two masses, m_1 and m_2, separated by a distance r is:

$$U = -G\frac{m_1 m_2}{r}$$

Example

A satellite of mass m_s is launched from the surface of the Earth into an orbit of radius $2r_e$, where r_e is the radius of the Earth. How much work is done to get it into orbit?

The work done getting the satellite from one place to another is equal to the change in the satellite's potential energy. If its potential energy on the surface of the Earth is U_1 and its potential energy when it is in orbit is U_2, then the amount of work done is:

$$\begin{aligned} W = U_2 - U_1 &= -G\frac{m_s m_e}{2r_e} - -G\frac{m_s m_e}{r_e} \\ &= G\frac{m_s m_e}{r_e} - G\frac{m_s m_e}{2r_e} \\ &= G\frac{m_s m_e}{2r_e} \end{aligned}$$

Energy of an Orbiting Satellite

Suppose a satellite of mass m_s is in orbit around the Earth at a radius R. We know the kinetic energy of the satellite is $KE = ½\, mv^2$. We also know that we can express centripetal force, F_c, as $F_c = mv^2/R$. Accordingly, we can substitute this equation into the equation for kinetic energy and get:

$$KE = \frac{1}{2}F_c R$$

Because F_c is equal to the gravitational force, we can substitute Newton's Law of Universal Gravitation in for F_c:

$$KE = G\frac{m_s m_e}{2R}$$

We know that the potential energy of the satellite is $U = -Gm_s m_e/R$, so the total energy of the satellite is the sum, $E = KE + U$:

$$E = -G\frac{m_s m_e}{R} + G\frac{m_s m_e}{2R} = -G\frac{m_s m_e}{2R}$$

Weightlessness

People rarely get to experience firsthand the phenomenon of **weightlessness**, but that doesn't keep the SAT II from testing you on it. There is a popular misconception that astronauts in satellites experience weightlessness because they are beyond the reach of the Earth's gravitational pull. If you already know this isn't the case, you're in a good position to answer correctly anything the SAT II may ask about weightlessness.

In order to understand how weightlessness works, let's look at the familiar experience of gaining and losing weight in an elevator. Suppose you bring a bathroom scale into the elevator with you to measure your weight.

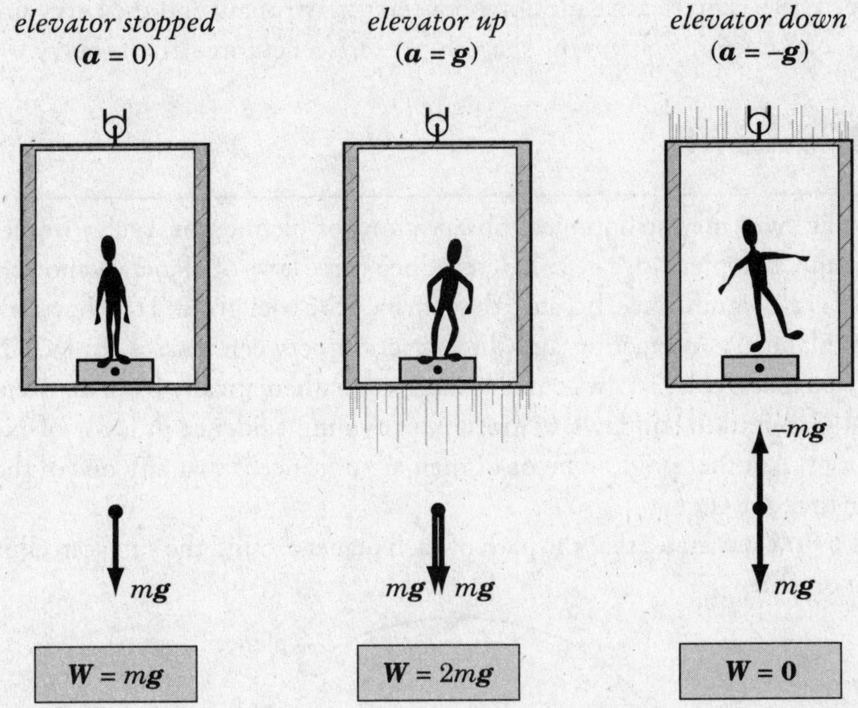

When the elevator is at rest, the scale will read your usual weight, $W = mg$, where m is your mass. When the elevator rises with an acceleration of g, you will be distressed to read that your weight is now $m(g + g) = 2mg$. If the elevator cable is cut so that the elevator falls freely with an acceleration of $-g$, then your weight will be $m(g - g) = 0$.

While in free fall in the elevator, if you were to take a pen out of your pocket and "drop" it, it would just hover in the air next to you. You, the pen, and the elevator are all falling at the same rate, so you are all motionless relative to one another. When objects are in free fall, we say that they experience weightlessness. You've probably seen images of astronauts floating about in space shuttles. This is not because they are free from the Earth's gravitational pull. Rather, their space shuttle is in orbit about the Earth, meaning that it is in a perpetual free fall. Because they are in free fall, the astronauts, like you in your falling elevator, experience weightlessness.

Weightless environments provide an interesting context for testing Newton's Laws. Newton's First Law tells us that objects maintain a constant velocity in the absence of a net force, but we're so used to being in an environment with gravity and friction that we never really see this law working to its full effect. Astronauts, on the other hand, have ample opportunity to play around with the First Law. For example, say that a weightless astronaut is eating lunch as he orbits the Earth in the space station.

If the astronaut releases his grasp on a tasty dehydrated strawberry, then the berry, like your pen, floats in midair exactly where it was "dropped." The force of gravity exerted by the Earth on the strawberry causes the strawberry to move in the same path as the spaceship. There is no relative motion between the astronaut and the berry unless the astronaut, or something else in the spaceship, exerts a net force on the berry.

Kepler's Laws

After poring over the astronomical observations of his mentor Tycho Brahe (1546–1601), Johannes Kepler (1571–1630) determined three laws of planetary motion. These laws are of great significance, because they formed the background to Newton's thinking about planetary interaction and the attraction between masses. In fact, Newton later showed that Kepler's Laws could be derived mathematically from his own Law of Universal Gravitation and laws of motion, providing evidence in favor of Newton's new theories. Another point in favor of their significance is that any one of them may appear on the SAT II.

Kepler's First Law states that the path of each planet around the sun is an ellipse with the sun at one focus.

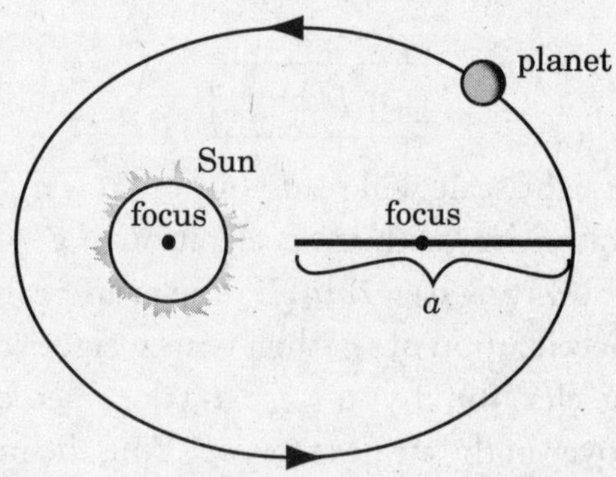

a = semimajor axis

Kepler's Second Law relates a planet's speed to its distance from the sun. Because the planets' orbits are elliptical, the distance from the sun varies. The Second Law states that if a line is drawn from the sun to the orbiting planet, then the area swept out by this line in a given time interval is constant. This means that when the planet is farthest from the sun it moves much more slowly than when it is closest to the sun.

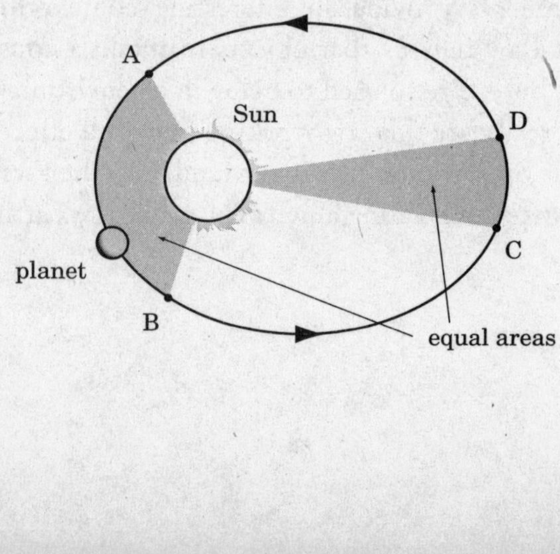

It is important to remember that although Kepler formulated this law in reference to planets moving around the sun, it also holds true for astronomical objects, like comets, that also travel in elliptical orbits around the sun.

Kepler's Third Law states that given the period, T, and semimajor axis, a, of a planet's elliptical orbit, the ratio T^2/a^3 is the same for every planet. The semimajor axis is the longer one, along which the two foci are located.

Example

Every 76 years, Halley's comet passes quite close by the Earth. At the most distant point in its orbit, it is much farther from the sun even than Pluto. Is the comet moving faster when it is closer to Earth or closer to Pluto?

According to Kepler's Second Law, objects that are closer to the sun orbit faster than objects that are far away. Therefore, Halley's comet must be traveling much faster when it is near the Earth than when it is off near Pluto.

Key Formulas

Centripetal Acceleration	$\mathbf{a}_c = \frac{\mathbf{v}^2}{r}$
Centripetal Force	$\mathbf{F} = \frac{m\mathbf{v}^2}{r}$
Newton's Law of Universal Gravitation	$F_g = G\frac{m_1 m_2}{r^2}$
Acceleration Due to Gravity at the Surface of a Planet	$a = G\frac{m_{\text{planet}}}{r^2}$
Velocity of a Satellite in Orbit	$v = \sqrt{\frac{Gm_e}{R}}$
Gravitational Potential Energy	$U = -G\frac{m_1 m_2}{r}$
Kinetic Energy of a Satellite in Orbit	$KE = G\frac{m_s m_e}{2R}$
Total Energy of a Satellite in Orbit	$E = -G\frac{m_s m_e}{2R}$
Kepler's Third Law	$\frac{T^2}{a^3} = \text{constant}$

Practice Questions

Questions 1–3 refer to a ball of mass m on a string of length R, swinging around in circular motion, with instantaneous velocity v and centripetal acceleration a.

1. What is the centripetal acceleration of the ball if the length of the string is doubled?

   (A) $a/4$ (B) $a/2$ (C) a (D) $2a$ (E) $4a$

2. What is the centripetal acceleration of the ball if the instantaneous velocity of the ball is doubled?

   (A) $a/4$ (B) $a/2$ (C) a (D) $2a$ (E) $4a$

3. What is the centripetal acceleration of the ball if its mass is doubled?

   (A) $a/4$ (B) $a/2$ (C) a (D) $2a$ (E) $4a$

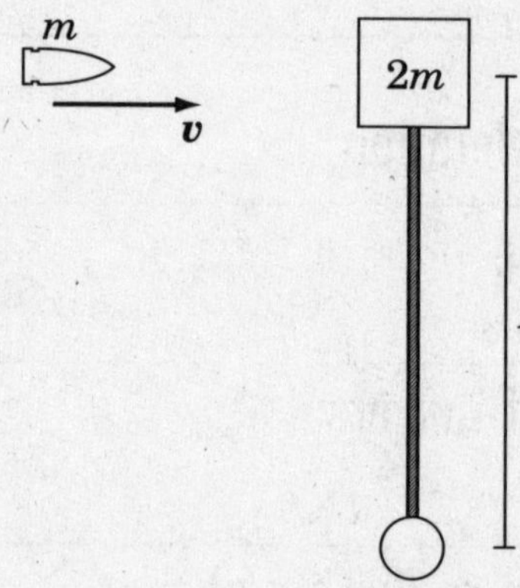

4. A bullet of mass m traveling at velocity v strikes a block of mass $2m$ that is attached to a rod of length R. The bullet collides with the block at a right angle and gets stuck in the block. The rod is free to rotate. What is the centripetal acceleration of the block after the collision?

   (A) v^2/R
   (B) $(1/2)v^2/R$
   (C) $(1/3)v^2/R$
   (D) $(1/4)v^2/R$
   (E) $(1/9)v^2/R$

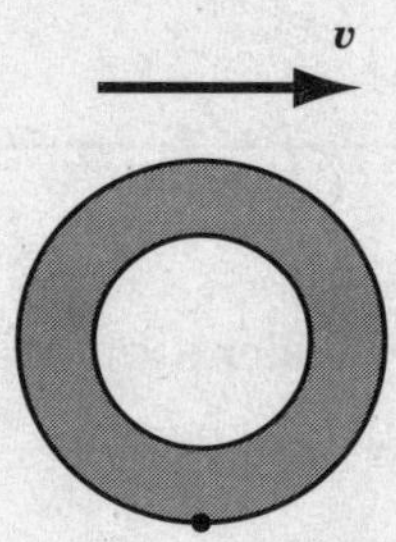

5. A car wheel drives over a pebble, which then sticks to the wheel momentarily as the wheel displaces it. What is the direction of the initial velocity of the pebble?

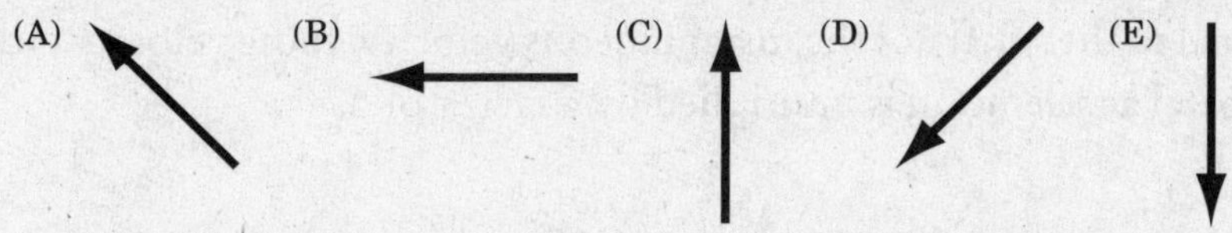

6. If we consider the gravitational force F between two objects of masses m_1 and m_2 respectively, separated by a distance R, and we double the distance between them, what is the new magnitude of the gravitational force between them?

(A) $F/4$ (B) $F/2$ (C) F (D) $2F$ (E) $4F$

7. If the Earth were compressed in such a way that its mass remained the same, but the distance around the equator were just one-half what it is now, what would be the acceleration due to gravity at the surface of the Earth?

(A) $g/4$ (B) $g/2$ (C) g (D) $2g$ (E) $4g$

8. A satellite orbits the Earth at a radius r and a velocity v. If the radius of its orbit is doubled, what is its velocity?

(A) $v/2$ (B) $v/\sqrt{2}$ (C) v (D) $\sqrt{2}\,v$ (E) $2v$

9. An object is released from rest at a distance of $2r_e$ from the center of the Earth, where r_e is the radius of the Earth. In terms of the gravitational constant (G), the mass of the Earth (M), and r_e, what is the velocity of the object when it hits the Earth?

(A) $\sqrt{GM/r_e}$
(B) GM / r_e
(C) $\sqrt{GM/2r_e}$
(D) $GM / 2r_e$
(E) $2GM / r_e$

10. Two planets, A and B, orbit a star. Planet A moves in an elliptical orbit whose semimajor axis has length a. Planet B moves in an elliptical orbit whose semimajor axis has a length of $9a$. If planet A orbits with a period T, what is the period of planet B's orbit?

(A) $729T$ (B) $27T$ (C) $3T$ (D) $T/3$ (E) $T/27$

Explanations

1. (B)

The equation for the centripetal acceleration is $a = v^2/r$. That is, acceleration is inversely proportional to the radius of the circle. If the radius is doubled, then the acceleration is halved.

2. (E)

From the formula $a = v^2/r$, we can see that centripetal acceleration is directly proportional to the square of the instantaneous velocity. If the velocity is doubled, then the centripetal acceleration is multiplied by a factor of 4.

3. (C)

The formula for centripetal acceleration is $a_c = v^2/r$. As you can see, mass has no influence on centripetal acceleration. If you got this question wrong, you were probably thinking of the formula for centripetal force: $F = mv^2/r$. Much like the acceleration due to gravity, centripetal acceleration is independent of the mass of the accelerating object.

4. (E)

The centripetal acceleration of the block is given by the equation $a = v'^2/R$, where v' is the velocity of the bullet-block system after the collision. We can calculate the value for v' by applying the law of conservation of linear momentum. The momentum of the bullet before it strikes the block is $p = mv$. After it strikes the block, the bullet-block system has a momentum of $p' = 3mv'$. Setting these two equations equal to one another, we find:

$$\begin{aligned} 3mv' &= mv \\ v' &= \frac{1}{3}v \end{aligned}$$

If we substitute $v' = (1/3)v$ into the equation $a = v'^2/R$, we find:

$$\begin{aligned} a &= \frac{v'^2}{R} \\ &= \frac{(\frac{1}{3}v)^2}{R} \\ &= \frac{1}{9}\frac{v^2}{R} \end{aligned}$$

5. (C)

The rotating wheel exerts a centripetal force on the pebble. That means that, initially, the pebble is drawn directly upward toward the center of the wheel. This might seem counter-intuitive: shouldn't the pebble shoot out behind the wheel? Relative to the wheel, yes. By the time the pebble has shot up into the air, the wheel will have moved on, so the pebble will seem to have been thrown behind the car wheel. However, relative to the ground, the pebble will simply shoot straight up in the air and back down again.

6. (A)

Newton's Law of Universal Gravitation tells us that the gravitational force between two objects is directly proportional to the masses of those two objects, and inversely proportional to the square of the distance between them. If that distance is doubled, then the gravitational force is divided by four.

7. (E)

Circumference and radius are related by the formula $C = 2\pi r$, so if the circumference of the earth were halved, so would the radius. The acceleration due to gravity at the surface of the earth is given by the formula:

$$a = \frac{GM}{r^2}$$

where M is the mass of the earth. This is just a different version Newton's Law of Universal Gravitation, where both sides of the equation are divided by m, the mass of the falling object. From this formula, we can see that a is inversely proportional to r^2. If the value of a is normally g, the value of a when r is halved must be $4g$.

8. (B)

To get a formula that relates orbital velocity and orbital radius, we need to equate the formulas for gravitational force and centripetal force, and then solve for v:

$$\frac{mv^2}{r} = \frac{GMm}{r^2}$$

$$v^2 = \frac{GM}{r}$$

$$v = \sqrt{\frac{GM}{r}}$$

From this formula, we can see that velocity is inversely proportional to the square root of r. If r is doubled, v is multiplied by $1/\sqrt{2}$.

Circular Motion

9. (A)

We can apply the law of conservation of energy to calculate that the object's change in potential energy is equal to its change in kinetic energy. The potential energy of an object of mass m at a distance r_e from a planet of mass M is $U = -GMm / r$. The change in potential energy for the object is:

$$\Delta U = -G\frac{Mm}{2r_e} - -G\frac{Mm}{r_e} = -G\frac{Mm}{2r_e} + 2G\frac{Mm}{2r_e}$$
$$= G\frac{Mm}{2r_e}$$

This change in potential energy represents the object's total kinetic energy, $KE = \frac{1}{2} mv^2$, when it hits the Earth. Equating change in potential energy and total kinetic energy, we can solve for v:

$$\frac{1}{2}mv^2 = G\frac{Mm}{2r_e}$$
$$v^2 = \frac{GM}{r_e}$$
$$v = \sqrt{\frac{GM}{r_e}}$$

10. (B)

Kepler's Third Law tells us that T^2/a^3 is a constant for every planet in a system. If we let xT be the value for the period of planet B's orbit, then we can solve for x using a bit of algebra:

$$\frac{(xT)^2}{(9a)^3} = \frac{T^2}{a^3}$$
$$x^2T^2 = 9^3T^2$$
$$x^2 = (3^2)^3$$
$$= (3^3)^2$$
$$x = 3^3$$
$$= 27$$

Thermal Physics

Chapter Contents

Heat and Temperature 204
The Kinetic Theory of Gases & the Ideal Gas Law 211
The Laws of Thermodynamics........ 216
Heat Engines 219
Key Formulas 221
Practice Questions 222
Explanations 224

Thermal physics is essentially the study of **heat**, **temperature**, and **heat transfer**. As we shall see—particularly when we look at the **Second Law of Thermodynamics**—these concepts have a far broader range of application than you may at first imagine. All of these concepts are closely related to **thermal energy**, which is one of the most important forms of energy. In almost every energy transformation, some thermal energy is produced in the form of heat. To take an example that by now should be familiar, friction produces heat. Rub your hands briskly together and you'll feel heat produced by friction.

When you slide a book along a table, the book will not remain in motion, as Newton's First Law would lead us to expect, because friction between the book and the table causes the book to slow down and stop. As the velocity of the book decreases, so does its kinetic energy, but this decrease is not a startling violation of the law of conservation of energy. Rather, the kinetic energy of the book is slowly transformed into thermal energy. Because friction acts over a relatively large distance, neither the table nor the book will be noticeably warmer. However, if you were somehow able to measure the heat produced through friction, you would find that the total heat produced in bringing the book to a stop is equal to the book's initial kinetic energy.

Technically speaking, thermal energy is the energy associated with the random vibration and movement of molecules. All matter consists of millions of tiny molecules, none of which are entirely still. The degree to which they move determines the amount of thermal energy in an object.

While thermal energy comes into play in a wide range of phenomena, the SAT II will focus primarily on the sorts of things you might associate with words like "heat" and "temperature." We'll learn how heat is transferred from one body to another, how temperature and heat are related, and how these concepts affect solids, liquids, gases, and the phase changes between the three.

Heat and Temperature

In everyday speech, heat and temperature go hand in hand: the hotter something is, the greater its temperature. However, there is a subtle difference in the way we use the two words in everyday speech, and this subtle difference becomes crucial when studying physics.

Temperature is a property of a material, and thus depends on the material, whereas heat is a form of energy existing on its own. The difference between heat and temperature is analogous to the difference between money and wealth. For example, $200 is an amount of money: regardless of who owns it, $200 is $200. With regard to wealth, though, the significance of $200 varies from person to person. If you are ten and carrying $200 in your wallet, your friends might say you are wealthy or ask to borrow some money. However, if you are thirty-five and carrying $200 in your wallet, your friends will probably not take that as a sign of great wealth, though they may still ask to borrow your money.

Temperature

While temperature is related to thermal energy, there is no absolute correlation between the amount of thermal energy (heat) of an object and its temperature. Temperature measures the concentration of thermal energy in an object in much the same way that density measures the concentration of matter in an object. As a result, a large object will have a much lower temperature than a small object with the same amount of thermal energy. As we shall see shortly, different materials respond to changes in thermal energy with more or less dramatic changes in temperature.

Degrees Celsius

In the United States, temperature is measured in degrees Fahrenheit (°F). However, Fahrenheit is not a metric unit, so it will not show up on the SAT II. Physicists and non-Americans usually talk about temperature in terms of degrees **Celsius**, a.k.a. centigrade (°C). Water freezes at exactly 0°C and boils at 100°C. This is not a remarkable coincidence—it is the way the Celsius scale is defined.

The SAT II won't ask you to convert between Fahrenheit and Celsius, but if you have a hard time thinking in terms of degrees Celsius, it may help to know how to switch back and forth between the two. The freezing point of water is 0°C and 32°F. A change in temperature of nine degrees Fahrenheit corresponds to a change of five degrees Celsius, so that, for instance, 41°F is equivalent to 5°C. In general, we can relate any temperature of y°F to any temperature of x°C with the following equation:

$$y^\circ\text{F} = \frac{9}{5}x^\circ\text{C} + 32$$

Kelvins

In many situations we are only interested in changes of temperature, so it doesn't really matter where the freezing point of water is arbitrarily chosen to be. But in other cases, as we shall see when we study gases, we will want to do things like "double the temperature," which is meaningless if the zero point of the scale is arbitrary, as with the Celsius scale.

The **Kelvin** scale (K) is a measure of absolute temperature, defined so that temperatures expressed in Kelvins are always positive. **Absolute zero**, 0 K, which is equivalent to –273°C, is the lowest theoretical temperature a material can have. Other than the placement of the zero point, the Kelvin and Celsius scales are the same, so water freezes at 273 K and boils at 373 K.

Definition of Temperature

The temperature of a material is a measure of the average kinetic energy of the molecules that make up that material. Absolute zero is defined as the temperature at which the molecules have zero kinetic energy, which is why it is impossible for anything to be colder.

Solids are rigid because their molecules do not have enough kinetic energy to go anywhere—they just vibrate in place. The molecules in a liquid have enough energy to move around one another—which is why liquids flow—but not enough to escape each other. In a gas, the molecules have so much kinetic energy that they disperse and the gas expands to fill its container.

Heat

Heat is a measure of how much thermal energy is transmitted from one body to another. We cannot say a body "has" a certain amount of heat any more than we can say a body "has" a certain amount of work. While both work and heat can be measured in terms of joules, they are not measures of energy but rather of energy transfer. A hot water bottle has a certain amount of thermal energy; when you cuddle up with a hot water bottle, it transmits a certain amount of heat to your body.

Calories

Like work, heat can be measured in terms of joules, but it is frequently measured in terms of **calories** (cal). Unlike joules, calories relate heat to changes in temperature, making them a more convenient unit of measurement for the kinds of thermal physics problems you will encounter on the SAT II. Be forewarned, however, that a question on thermal physics on the SAT II may be expressed either in terms of calories or joules.

A calorie is defined as the amount of heat needed to raise the temperature of one gram of water by one degree Celsius. One calorie is equivalent to 4.19 J.

$$1 \text{ cal} = 1 \text{ g}/°\text{C} = 4.19 \text{ J}$$

You're probably most familiar with the word "calorie" in the context of a food's nutritional content. However, food calories are not quite the same as what we're discussing here: they are actually Calories, with a capital "C," where 1 Calorie = 1000 calories. Also, these Calories are not a measure of thermal energy, but rather a measure of the energy stored in the chemical bonds of food.

Specific Heat

Though heat and temperature are not the same thing, there is a correlation between the two, captured in a quantity called **specific heat**, c. Specific heat measures how much heat is required to raise the temperature of a certain mass of a given substance. Specific heat is measured in units of J/kg · °C or cal/g · °C. Every substance has a different specific heat, but specific heat is a constant for that substance.

For instance, the specific heat of water, c_{water}, is 4.19×10^3 J/kg · °C or 1 cal/g · °C. That means it takes 4.19×10^3 joules of heat to raise one kilogram of water by one degree Celsius. Substances that are easily heated, like copper, have a low specific heat, while substances that are difficult to heat, like rubber, have a high specific heat.

Specific heat allows us to express the relationship between heat and temperature in a mathematical formula:

$$Q = mc\Delta T$$

where Q is the heat transferred to a material, m is the mass of the material, c is the specific heat of the material, and ΔT is the change in temperature.

Example

> 4190 J of heat are added to 0.5 kg of water with an initial temperature of 12°C. What is the temperature of the water after it has been heated?

By rearranging the equation above, we can solve for ΔT:

$$\begin{aligned} \Delta T &= \frac{Q}{mc} \\ &= \frac{4190 \text{ J}}{0.5 \text{ kg} \cdot 4190 \text{ J/kg} \cdot °\text{C}} \\ &= 2 \text{ C}° \end{aligned}$$

The temperature goes up by 2 C°, so if the initial temperature was 12°C, then the final temperature is 14°C. Note that when we talk about an absolute temperature, we write °C, but when we talk about a change in temperature, we write C°.

Thermal Equilibrium

Put a hot mug of cocoa in your hand, and your hand will get warmer while the mug gets cooler. You may have noticed that the reverse never happens: you can't make your hand colder and the mug hotter by putting your hand against the mug. What you have noticed is a general truth about the world: heat flows spontaneously from a hotter object to a colder object, but never from a colder object to a hotter object. This is one way of stating the Second Law of Thermodynamics, to which we will return later in this chapter.

Whenever two objects of different temperatures are placed in contact, heat will flow from the hotter of the two objects to the colder until they both have the same temperature. When they reach this state, we say they are in **thermal equilibrium**.

Because energy is conserved, the heat that flows out of the hotter object will be equal to the heat that flows into the colder object. With this in mind, it is possible to calculate the temperature two objects will reach when they arrive at thermal equilibrium.

Example

> 3 kg of gold at a temperature of 20°C is placed into contact with 1 kg of copper at a temperature of 80°C. The specific heat of gold is 130 J/kg · °C and the specific heat of copper is 390 J/kg · °C. At what temperature do the two substances reach thermal equilibrium?

The heat gained by the gold, $Q = mc_{\text{gold}}\Delta T_{\text{gold}}$ is equal to the heat lost by the copper, $Q = mc_{\text{copper}}\Delta T_{\text{copper}}$. We can set the heat gained by the gold to be equal to the heat lost by the copper, bearing in mind that the final temperature of the gold must equal the final temperature of the copper:

$$\begin{aligned} mc_{\text{gold}}\Delta T_{\text{gold}} &= mc_{\text{copper}}\Delta T_{\text{copper}} \\ (3\text{ kg})(130\text{ J/kg}\cdot^{\circ}\text{C})\Delta T_{\text{gold}} &= (1\text{ kg})(390\text{ J/kg}\cdot^{\circ}\text{C})\Delta T_{\text{copper}} \\ 390\Delta T_{\text{gold}} &= 390\Delta T_{\text{copper}} \\ \Delta T_{\text{gold}} &= \Delta T_{\text{copper}} \end{aligned}$$

The equality between ΔT_{gold} and ΔT_{copper} tells us that the temperature change of the gold is equal to the temperature change of the copper. If the gold heats up by 30 C° and the copper cools down by 30 C°, then the two substances will reach thermal equilibrium at 50°C.

Phase Changes

As you know, if you heat a block of ice, it won't simply get warmer. It will also melt and become liquid. If you heat it even further, it will boil and become a gas. When a substance changes between being a solid, liquid, or gas, we say it has undergone a **phase change**.

Melting Point and Boiling Point

If a solid is heated through its **melting point**, it will melt and turn to liquid. Some substances—for example, dry ice (solid carbon dioxide)—cannot exist as a liquid at certain pressures and will **sublimate** instead, turning directly into gas. If a liquid is heated through its **boiling point**, it will vaporize and turn to gas. If a liquid is cooled through its melting point, it will freeze. If a gas is cooled through its boiling point, it will condense into a liquid, or sometimes **deposit** into a solid, as in the case of carbon dioxide. These phase changes are summarized in the figure below.

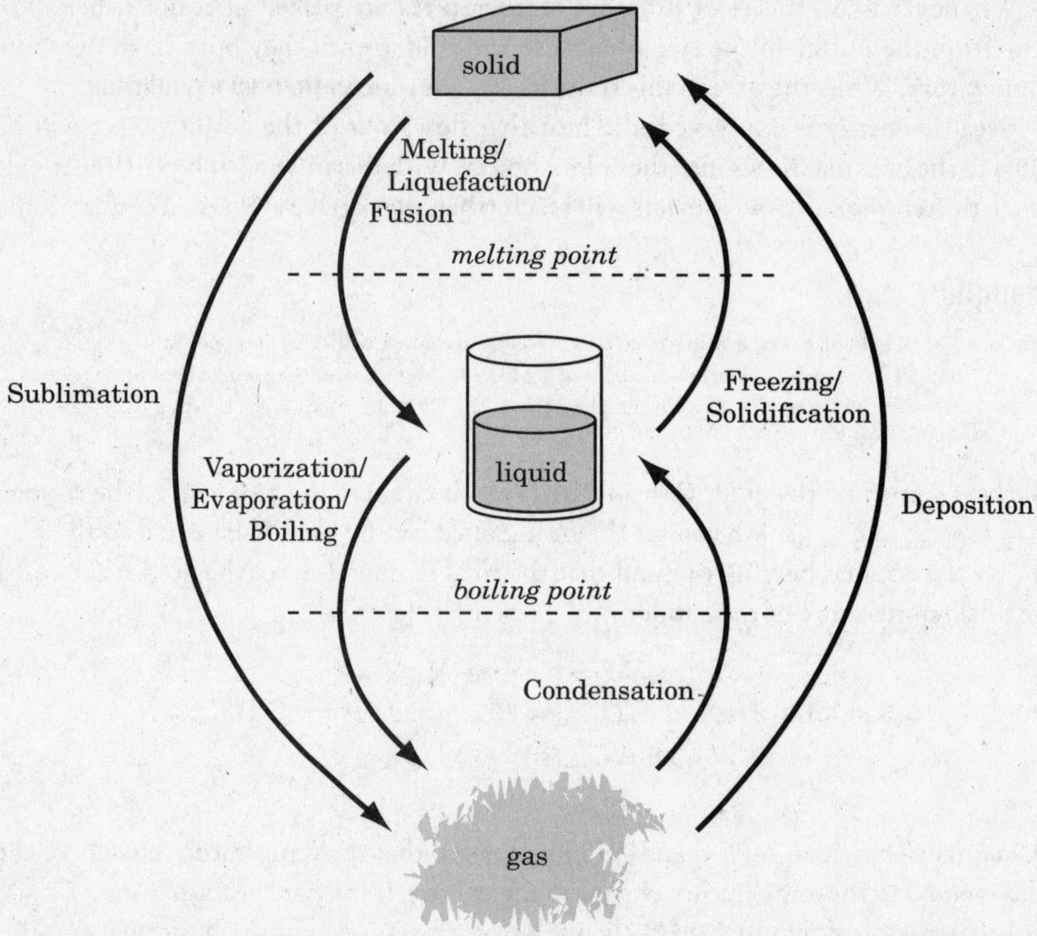

A substance requires a certain amount of heat to undergo a phase change. If you were to apply steady heat to a block of ice, its temperature would rise steadily until it reached 0°C. Then the temperature would remain constant as the block of ice slowly melted into water. Only when all the ice had become water would the temperature continue to rise.

Latent Heat of Transformation

Just as specific heat tells us how much heat it takes to increase the temperature of a substance, the **latent heat of transformation**, q, tells us how much heat it takes to change the phase of a substance. For instance, the **latent heat of fusion** of water—the latent heat gained or lost in transforming a solid into a liquid or a liquid into a solid—is 3.3×10^5 J/kg. That means that you must add 3.3×10^5 J to change one kilogram of ice into water, and remove the same amount of heat to change one kilgram of water into ice. Throughout this phase change, the temperature will remain constant at 0°C.

The **latent heat of vaporization**, which tells us how much heat is gained or lost in transforming a liquid into a gas or a gas into a liquid, is a different value from the latent heat of fusion. For instance, the latent heat of vaporization for water is 2.3×10^6 J/kg, meaning that you must add 2.3×10^6 J to change one kilogram of water into steam, or remove the same amount of heat to change one kilogram of steam into water. Throughout this phase change, the temperature will remain constant at 100°C.

To sublimate a solid directly into a gas, you need an amount of heat equal to the sum of the latent heat of fusion and the latent heat of vaporization of that substance.

Example

How much heat is needed to transform a 1 kg block of ice at –5°C to a puddle of water at 10°C?

First, we need to know how much heat it takes to raise the temperature of the ice to 0°C:

$$\begin{aligned} Q &= mc\Delta T \\ &= (1\text{kg})(4.19 \times 10^3\ \text{J/kg} \cdot {}^\circ\text{C})(5^\circ\text{C}) \\ &= 2.1 \times 10^4\ \text{J} \end{aligned}$$

Next, we need to know how much heat it takes to melt the ice into water:

$$\begin{aligned} Q &= mq_{\text{fusion}} \\ &= (1\ \text{kg})(3.3 \times 10^5\ \text{J/kg}) \\ &= 3.3 \times 10^5\ \text{J} \end{aligned}$$

Last, we need to know how much heat it takes to warm the water up to 10°C. Since the specific heat capacity of ice is the same as that of water, this would simply be twice the heat it took to warm the ice by 5 C°; that is, 4.2×10^4 J.

Now we just add the three figures together to get our answer:

$$2.1 \times 10^4 + 3.3 \times 10^5 + 4.2 \times 10^4 = 3.9 \times 10^6\ \text{J}$$

Note that far more heat was needed to melt the ice into liquid than was needed to increase the temperature.

Thermal Expansion

You may have noticed in everyday life that substances can often expand or contract with a change in temperature even if they don't change phase. If you play a brass or metal woodwind instrument, you have probably noticed that this size change creates difficulties when you're trying to tune your instrument—the length of the horn, and thus its pitch, varies with the room temperature. Household thermometers also work according to this principle: mercury, a liquid metal, expands when it is heated, and therefore takes up more space and rise in a thermometer.

Any given substance will have a **coefficient of linear expansion**, α, and a **coefficient of volume expansion**, β. We can use these coefficients to determine the change in a substance's length, L, or volume, V, given a certain change in temperature.

$$\Delta L = \alpha L_i \Delta T$$
$$\Delta V = \beta V_i \Delta T$$

Example

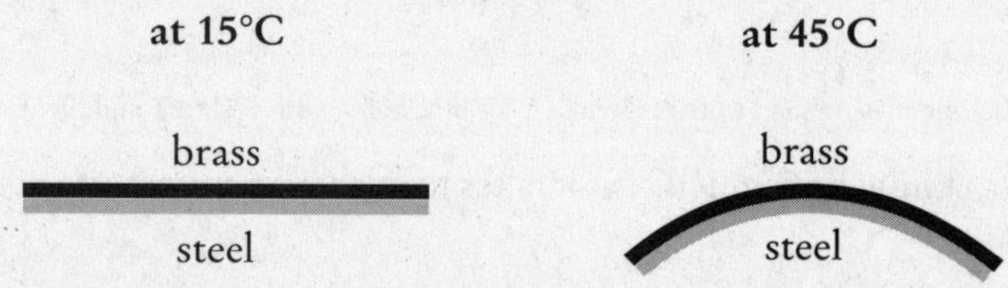

A bimetallic strip of steel and brass of length 10 cm, initially at 15°C, is heated to 45°C. What is the difference in length between the two substances after they have been heated? The coefficient of linear expansion for steel is 1.2×10^{-5} /C°, and the coefficient of linear expansion for brass is 1.9×10^{-5} /C°.

First, let's see how much the steel expands:

$$\begin{aligned}\Delta L &= \alpha L_i \Delta T \\ &= (1.2 \times 10^{-5}/\text{C}^\circ)(0.1\text{ m})(30\text{ C}^\circ) \\ &= 3.6 \times 10^{-5}\text{ m}\end{aligned}$$

Next, let's see how much the brass expands:

$$\begin{aligned}\Delta L &= \alpha L_i \Delta T \\ &= (1.9 \times 10^{-5}/\text{C}^\circ)(0.1\text{ m})(30\text{ C}^\circ) \\ &= 5.7 \times 10^{-5}\text{ m}\end{aligned}$$

The difference in length is $(5.7 \times 10^{-5}) - (3.6 \times 10^{-5}) = 2.1 \times 10^{-5}$ m. Because the brass expands more than the steel, the bimetallic strip will bend a little to compensate for the extra length of the brass.

Thermostats work according to this principle: when the temperature reaches a certain point, a bimetallic strip inside the thermostat will bend away from an electric contact, interrupting the signal calling for more heat to be sent into a room or building.

Methods of Heat Transfer

There are three different ways heat can be transferred from one substance to another or from one place to another. This material is most likely to come up on the SAT II as a question on what kind of heat transfer is involved in a certain process. You need only have a qualitative understanding of the three different kinds of heat transfer.

Conduction is the transfer of heat by intermolecular collisions. For example, when you boil water on a stove, you only heat the bottom of the pot. The water molecules at the bottom transfer their kinetic energy to the molecules above them through collisions, and this process continues until all of the water is at thermal equilibrium. Conduction is the most common way of transferring heat between two solids or liquids, or within a single solid or liquid. Conduction is also a common way of transferring heat through gases.

Convection: While conduction involves molecules passing their kinetic energy to other molecules, **convection** involves the molecules themselves moving from one place to another. For example, a fan works by displacing hot air with cold air. Convection usually takes place with gases traveling from one place to another.

Radiation: Molecules can also transform heat into electromagnetic waves, so that heat is transferred not by molecules but by the waves themselves. A familiar example is the microwave oven, which sends microwave radiation into the food, energizing the molecules in the food without those molecules ever making contact with other, hotter molecules. Radiation takes place when the source of heat is some form of electromagnetic wave, such as a microwave or sunlight.

Thermal Physics

The Kinetic Theory of Gases & the Ideal Gas Law

We said earlier that temperature is a measure of the kinetic energy of the molecules in a material, but we didn't elaborate on that remark. Because individual molecules are so small, and because there are so many molecules in most substances, it would be impossible to study their behavior individually. However, if we know the basic rules that govern the behavior of individual molecules, we can make statistical calculations that tell us roughly how a collection of millions of molecules would behave. This, essentially, is what thermal physics is: the study of the macroscopic effects of the microscopic molecules that make up the world of everyday things.

The **kinetic theory of gases** makes the transition between the microscopic world of molecules and the macroscopic world of quantities like temperature and pressure. It starts out with a few basic postulates regarding molecular behavior, and infers how this behavior manifests itself on a macroscopic level. One of the most important results of the kinetic theory is the derivation of the **ideal gas law**, which not only is very useful and important, it's also almost certain to be tested on the SAT II.

The Kinetic Theory of Gases

We can summarize the kinetic theory of gases with four basic postulates:

1. **Gases are made up of molecules:** We can treat molecules as point masses that are perfect spheres. Molecules in a gas are very far apart, so that the space between each individual molecule is many orders of magnitude greater than the diameter of the molecule.
2. **Molecules are in constant random motion:** There is no general pattern governing either the magnitude or direction of the velocity of the molecules in a gas. At any given time, molecules are moving in many different directions at many different speeds.
3. **The movement of molecules is governed by Newton's Laws:** In accordance with Newton's First Law, each molecule moves in a straight line at a steady velocity, not interacting with any of the other molecules except in a collision. In a collision, molecules exert equal and opposite forces on one another.
4. **Molecular collisions are perfectly elastic:** Molecules do not lose any kinetic energy when they collide with one another.

The kinetic theory projects a picture of gases as tiny balls that bounce off one another whenever they come into contact. This is, of course, only an approximation, but it turns out to be a remarkably accurate approximation for how gases behave in the real world.

These assumptions allow us to build definitions of temperature and pressure that are based on the mass movement of molecules.

Temperature

The kinetic theory explains why temperature should be a measure of the average kinetic energy of molecules. According to the kinetic theory, any given molecule has a certain mass, m; a certain velocity, v; and a kinetic energy of $\frac{1}{2}mv^2$. As we said, molecules in any system move at a wide variety of different velocities, but the average of these velocities reflects the total amount of energy in that system.

We know from experience that substances are solids at lower temperatures and liquids and gases at higher temperatures. This accords with our definition of temperature as average kinetic energy: since the molecules in gases and liquids have more freedom of movement, they have a higher average velocity.

Pressure

In physics, **pressure**, P, is the measure of the force exerted over a certain area. We generally say something exerts a lot of pressure on an object if it exerts a great amount of force on that object, and if that force is exerted over a small area. Mathematically:

$$P = \frac{F}{A}$$

Pressure is measured in units of **pascals** (Pa), where 1 Pa = 1 N/m^2.

Pressure comes into play whenever force is exerted on a certain area, but it plays a particularly important role with regard to gases. The kinetic theory tells us that gas molecules obey Newton's Laws: they travel with a constant velocity until they collide, exerting a force on the object with which they collide. If we imagine gas molecules in a closed container, the molecules will collide with the walls of the container with some frequency, each time exerting a small force on the walls of the container. The more frequently these molecules collide with the walls of the container, the greater the net force and hence the greater the pressure they exert on the walls of the container.

Balloons provide an example of how pressure works. By forcing more and more air into an enclosed space, a great deal of pressure builds up inside the balloon. In the meantime, the rubber walls of the balloon stretch out more and more, becoming increasingly weak. The balloon will pop when the force of pressure exerted on the rubber walls is greater than the walls can withstand.

The Ideal Gas Law

The ideal gas law relates temperature, volume, and pressure, so that we can calculate any one of these quantities in terms of the others. This law stands in relation to gases in the same way that Newton's Second Law stands in relation to dynamics: if you master this, you've mastered all the math you're going to need to know. Ready for it? Here it is:

$$PV = nRT$$

Effectively, this equation tells us that temperature, T, is directly proportional to volume, V, and pressure, P. In metric units, volume is measured in liters (L), where one liter is equal to 1000 cubic centimeters, or 1 L = 1000 cm^3.

The n stands for the number of **moles** of gas molecules. One mole (mol) is just a big number—6.023×10^{23}, to be precise—that, conveniently, is the number of hydrogen atoms in a gram of hydrogen. Because we deal with a huge number of gas molecules at any given time, it is usually a lot easier to count them in moles rather than counting them individually.

The R is a constant of proportionality called the **universal gas constant**, set at 8.31 J/mol · K. This constant effectively relates temperature to kinetic energy. If we think of RT as the kinetic energy of an average molecule, then nRT is the total kinetic energy of all the gas molecules put together.

Deriving the Ideal Gas Law

Imagine a gas in a cylinder of base A, with one moving wall. The pressure of the gas exerts a force of $F = PA$ on the moving wall of the cylinder. This force is sufficient to move the cylinder's wall back a distance L, meaning that the volume of the cylinder increases by $\Delta V = AL$. In terms of A, this equation reads $A = \Delta V / L$. If we now substitute in $\Delta V / L$ for A in the equation $F = PA$, we get $F = P\Delta V / L$, or:

$$P\Delta V = FL$$

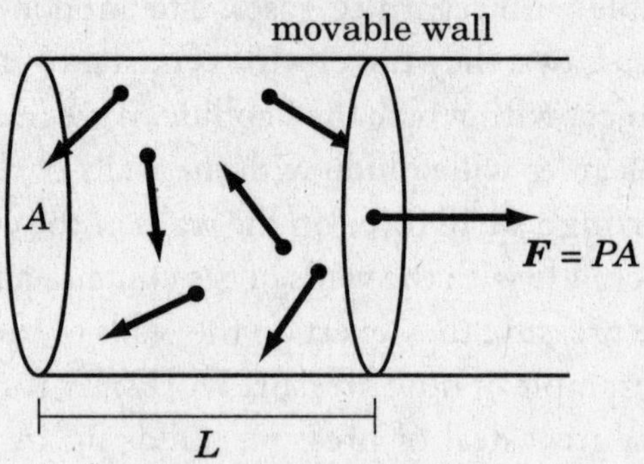

If you recall in the chapter on work, energy, and power, we defined work as force multiplied by displacement. By pushing the movable wall of the container a distance L by exerting a force F, the gas molecules have done an amount of work equal to FL, which in turn is equal to $P\Delta V$.

$$P\Delta V = FL$$

The work done by a gas signifies a change in energy: as the gas increases in energy, it does a certain amount of work on the cylinder. If a change in the value of PV signifies a change in energy, then PV itself should signify the total energy of the gas. In other words, both PV and nRT are expressions for the total kinetic energy of the molecules of a gas.

Boyle's Law and Charles's Law

The SAT II will not expect you to plug a series of numbers into the ideal gas law equation. The value of n is usually constant, and the value of R is always constant. In most problems, either T, P, or V will also be held constant, so that you will only need to consider how changes in one of those values affects another of those values. There are a couple of simplifications of the ideal gas law that deal with just these situations.

Thermal Physics

Boyle's Law

Boyle's Law deals with gases at a constant temperature. It tells us that an increase in pressure is accompanied by a decrease in volume, and vice versa: $P_iV_i = P_fV_f$. Aerosol canisters contain compressed (i.e., low-volume) gases, which is why they are marked with high-pressure warning labels. When you spray a substance out of an aerosol container, the substance expands and the pressure upon it decreases.

Charles's Law

Charles's Law deals with gases at a constant pressure. In such cases, volume and temperature are directly proportional: $V_i/T_i = V_f/T_f$. This is how hot-air balloons work: the balloon expands when the air inside of it is heated.

Gases in a Closed Container

You may also encounter problems that deal with "gases in a closed container," which is another way of saying that the volume remains constant. For such problems, pressure and temperature are directly proportional: $P_i/T_i = P_f/T_f$. This relationship, however, apparently does not deserve a name.

Example 1

> A gas in a cylinder is kept at a constant temperature while a piston compresses it to half its original volume. What is the effect of this compression on the pressure the gas exerts on the walls of the cylinder?

Questions like this come up all the time on the SAT II. Answering it is a simple matter of applying Boyle's Law, or remembering that pressure and volume are inversely proportional in the ideal gas law. If volume is halved, pressure is doubled.

Example 2

> A gas in a closed container is heated from 0ºC to 273ºC. How does this affect the pressure of the gas on the walls of the container?

First, we have to remember that in the ideal gas law, temperature is measured in Kelvins. In those terms, the temperature goes from 273 K to 546 K; in other words, the temperature doubles. Because we are dealing with a closed container, we know the volume remains constant. Because pressure and temperature are directly proportional, we know that if the temperature is doubled, then the pressure is doubled as well. This is why it's a really bad idea to heat an aerosol canister.

The Laws of Thermodynamics

Dynamics is the study of why things move the way they do. For instance, in the chapter on dynamics, we looked at Newton's Laws to explain what compels bodies to accelerate, and how. The prefix "thermo" denotes heat, so thermodynamics is the study of what compels heat to move in the way that it does. The Laws of Thermodynamics give us the whats and whys of heat flow.

The laws of thermodynamics are a bit strange. There are four of them, but they are ordered zero to three, and not one to four. They weren't discovered in the order in which they're numbered, and some—particularly the Second Law—have many different formulations, which seem to have nothing to do with one another.

There will almost certainly be a question on the **Second Law** on the SAT II, and quite possibly something on the **First Law**. The **Zeroth Law** and **Third Law** are unlikely to come up, but we include them here for the sake of completion. Questions on the Laws of Thermodynamics will probably be qualitative: as long as you understand what these laws mean, you probably won't have to do any calculating.

Zeroth Law

If system A is at thermal equilibrium with system B, and B is at thermal equilibrium with system C, then A is at thermal equilibrium with C. This is more a matter of logic than of physics. Two systems are at thermal equilibrium if they have the same temperature. If A and B have the same temperature, and B and C have the same temperature, then A and C have the same temperature.

The significant consequence of the Zeroth Law is that, when a hotter object and a colder object are placed in contact with one another, heat will flow from the hotter object to the colder object until they are in thermal equilibrium.

First Law

Consider an isolated system; that is, one where heat and energy neither enter nor leave the system. Such a system is doing no work, but we associate with it a certain **internal energy**, U, which is related to the kinetic energy of the molecules in the system, and therefore to the system's temperature. Internal energy is similar to potential energy in that it is a property of a system that is doing no work, but has the potential to do work.

The First Law tells us that the internal energy of a system increases if heat is added to the system or if work is done on the system, and decreases if the system gives off heat or does work. We can express this law as an equation:

$$\Delta U = \Delta Q + \Delta W$$

where U signifies internal energy, Q signifies heat, and W signifies work.

The First Law is just another way of stating the law of conservation of energy. Both heat and work are forms of energy, so any heat or work that goes into or out of a system must affect the internal energy of that system.

Example

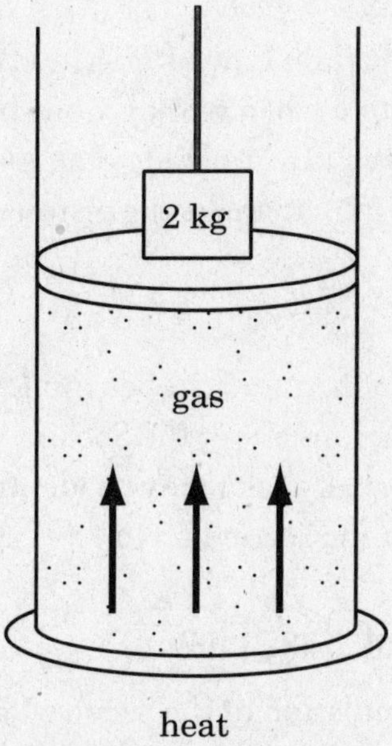

Some heat is added to a gas container that is topped by a movable piston. The piston is weighed down with a 2 kg mass. The piston rises a distance of 0.2 m at a constant velocity. Throughout this process, the temperature of the gas in the container remains constant. How much heat was added to the container?

The key to answering this question is to note that the temperature of the container remains constant. That means that the internal energy of the system remains constant ($\Delta U = 0$), which means that, according to the First Law, $\Delta Q + \Delta W = 0$. By pushing the piston upward, the system does a certain amount of work, ΔW, and this work must be equal to the amount of heat added to the system, ΔQ.

The amount of work done by the system on the piston is the product of the force exerted on the piston and the distance the piston is moved. Since the piston moves at a constant velocity, we know that the net force acting on the piston is zero, and so the force the expanding gas exerts to push the piston upward must be equal and opposite to the force of gravity pushing the piston downward. If the piston is weighed down by a two kilogram mass, we know that the force of gravity is:

$$\begin{aligned} F = mg &= (2\text{ kg})(9.8\text{ m/s}^2) \\ &= 19.6\text{ N} \end{aligned}$$

Since the gas exerts a force that is equal and opposite to the force of gravity, we know that it exerts a force of 19.6 N upward. The piston travels a distance of 0.2 m, so the total work done on the piston is:

$$\begin{aligned} W = Fd &= (19.6\text{ N})(0.2\text{ m}) \\ &= 3.92\text{ J} \end{aligned}$$

Since ΔW in the equation for the First Law of Thermodynamics is positive when work is done on the system and negative when work is done by the system, the value of ΔW is -3.92 J. Because $\Delta U = 0$, we can conclude that $\Delta Q = 3.92$ J, so 3.92 J of heat must have been added to the system to make the piston rise as it did.

Second Law

There are a number of equivalent forms of the Second Law, each of which sounds quite different from the others. Questions about the Second Law on the SAT II will invariably be qualitative. They will usually ask that you identify a certain formulation of the Second Law as an expression of the Second Law.

The Second Law in Terms of Heat Flow

Perhaps the most intuitive formulation of the Second Law is that heat flows spontaneously from a hotter object to a colder one, but not in the opposite direction. If you leave a hot dinner on a table at room temperature, it will slowly cool down, and if you leave a bowl of ice cream on a table at room temperature, it will warm up and melt. You may have noticed that hot dinners do not spontaneously get hotter and ice cream does not spontaneously get colder when we leave them out. More's the pity: we could save a fortune on ovens and freezers.

The Second Law in Terms of Heat Engines

One consequence of this law, which we will explore a bit more in the section on **heat engines**, is that no machine can work at 100% efficiency: all machines generate some heat, and some of that heat is always lost to the machine's surroundings.

The Second Law in Terms of Entropy

The Second Law is most famous for its formulation in terms of **entropy**. The word "entropy" was coined in the 19th century as a technical term for talking about disorder. The same principle that tells us that heat spontaneously flows from hot to cold but not in the opposite direction also tells us that, in general, ordered systems are liable to fall into disorder, but disordered systems are not liable to order themselves spontaneously.

Imagine pouring a tablespoon of salt and then a tablespoon of pepper into a jar. At first, there will be two separate heaps: one of salt and one of pepper. But if you shake up the mixture, the grains of salt and pepper will mix together. No amount of shaking will then help you separate the mixture of grains back into two distinct heaps. The two separate heaps of salt and pepper constitute a more ordered system than the mixture of the two.

Next, suppose you drop the jar on the floor. The glass will break and the grains of salt and pepper will scatter across the floor. You can wait patiently, but you'll find that, while the glass could shatter and the grains could scatter, no action as simple as dropping a jar will get the glass to fuse back together again or the salt and pepper to gather themselves up. Your system of salt and pepper in the jar is more ordered than the system of shattered glass and scattered condiments.

Entropy and Time

You may have noticed that Newton's Laws and the laws of kinematics are time-invariant. That is, if you were to play a videotape of kinematic motion in reverse, it would still obey the laws of kinematics. Videotape a ball flying up in the air and watch it drop. Then play the tape backwards: it goes up in the air and drops in just the same way.

By contrast, you'll notice that the Second Law is not time-invariant: it tells us that, over time, the universe tends toward greater disorder. Physicists suggest that the Second Law is what gives time a direction. If all we had were Newton's Laws, then there would be no difference between time going forward and time going backward. So we were a bit inaccurate when we said that entropy increases over time. We would be more accurate to say that time moves in the direction of entropy increase.

Third Law

Though the universe tends toward maximum entropy, this entropy state is never actually reached. This law is irrelevant as far as the SAT II is concerned, but we have included it for the sake of completeness.

Heat Engines

A heat engine is a machine that converts heat into work. Heat engines are important not only because they come up on the SAT II, but also because a large number of the machines we use—most notably our cars—employ heat engines.

A heat engine operates by taking heat from a hot place, converting some of that heat into work, and dumping the rest in a cooler heat reservoir. For example, the engine of a car generates heat by combusting gasoline. Some of that heat drives pistons that make the car do work on the road, and some of that heat is dumped out the exhaust pipe.

Assume that a heat engine starts with a certain internal energy U, intakes heat ΔQ_{in} from a heat source at temperature T_{in}, does work ΔW, and exhausts heat ΔQ_{out} into a the cooler heat reservoir with temperature T_{out}. With a typical heat engine, we only want to use the heat intake, not the internal energy of the engine, to do work, so $\Delta U = 0$. The First Law of Thermodynamics tells us:

$$\Delta U = 0 = \Delta Q_{in} - \Delta Q_{out} - \Delta W$$

To determine how effectively an engine turns heat into work, we define the **efficiency**, e, as the ratio of work done to heat input:

$$\begin{aligned} e \equiv \frac{\Delta W}{\Delta Q_{in}} &= \frac{\Delta Q_{in} - \Delta Q_{out}}{\Delta Q_{in}} \\ &= 1 - \frac{\Delta Q_{out}}{\Delta Q_{in}} \end{aligned}$$

Because the engine is doing work, we know that $\Delta W > 0$, so we can conclude that $\Delta Q_{in} < \Delta Q_{out}$. Both ΔQ_{in} and ΔQ_{out} are positive, so the efficiency is always between 0 and 1:

$$0 \leq e < 1$$

Efficiency is usually expressed as a percentage rather than in decimal form. That the efficiency of a heat engine can never be 100% is a consequence of the Second Law of Thermodynamics. If there were a 100% efficient machine, it would be possible to create perpetual motion: a machine could do work upon itself without ever slowing down.

Example

80 J of heat are injected into a heat engine, causing it to do work. The engine then exhausts 20 J of heat into a cool reservoir. What is the efficiency of the engine?

If we know our formulas, this problem is easy. The heat into the system is $\Delta Q_{in} = 80$ J and the heat out of the system is $\Delta Q_{out} = 20$ J. The efficiency, then, is: $1 - 20/80 = 0.75 = 75\%$.

Thermal Physics

Key Formulas

Conversion between Fahrenheit and Celsius	$y°\text{F} = \frac{9}{5}x°\text{C} + 32$
Conversion between Celsius and Kelvin	$0°\text{C} = 273\ \text{K}$ $\text{K} = °\text{C} + 273$
Relationship between Heat and Temperature	$Q = mc\Delta T$
Coefficient of Linear Expansion	$\Delta L = \alpha L_i \Delta T$
Coefficient of Volume Expansion	$\Delta V = \beta V_i \Delta T$
Ideal Gas Law	$PV = nRT$
Boyle's Law	$P_i V_i = P_f V_f$
Charles's Law	$\frac{V_i}{T_i} = \frac{V_f}{T_f}$
First Law of Thermodynamics	$\Delta U = \Delta Q + \Delta W$
Efficiency of a Heat Engine	$e = 1 - \frac{\Delta Q_{\text{out}}}{\Delta Q_{\text{in}}}$
Theoretical Limits on Heat Engine Efficiency	$0 \leq e < 1$

Practice Questions

1. 1 kg of cold water at 5°C is added to a container of 5 kg of hot water at 65° C. What is the final temperature of the water when it arrives at thermal equilibrium?

   (A) 10°C
   (B) 15°C
   (C) 35°C
   (D) 55°C
   (E) 60°C

2. Which of the following properties must be known in order to calculate the amount of heat needed to melt 1.0 kg of ice at 0°C?

   I. The specific heat of water
   II. The latent heat of fusion for water
   III. The density of water

   (A) I only
   (B) I and II only
   (C) I, II, and III
   (D) II only
   (E) I and III only

3. Engineers design city sidewalks using blocks of asphalt separated by a small gap to prevent them from cracking. Which of the following laws best explains this practice?

   (A) The Zeroth Law of Thermodynamics
   (B) The First Law of Thermodynamics
   (C) The Second Law of Thermodynamics
   (D) The law of thermal expansion
   (E) Conservation of charge

4. Which of the following is an example of convection?

   (A) The heat of the sun warming our planet
   (B) The heat from a stove warming a frying pan
   (C) Ice cubes cooling a drink
   (D) A microwave oven cooking a meal
   (E) An overhead fan cooling a room

5. An ideal gas is enclosed in a sealed container. Upon heating, which property of the gas does not change?

   (A) Volume
   (B) Pressure
   (C) The average speed of the molecules
   (D) The rate of collisions of the molecules with each other
   (E) The rate of collisions of the molecules with the walls of the container

6. A box contains two compartments of equal volume separated by a divider. The two compartments each contain a random sample of n moles of a certain gas, but the pressure in compartment A is twice the pressure in compartment B. Which of the following statements is true?

   (A) The temperature in A is twice the temperature in B
   (B) The temperature in B is twice the temperature in A
   (C) The value of the ideal gas constant, R, in A is twice the value of R in B
   (D) The temperature in A is four times as great as the temperature in B
   (E) The gas in A is a heavier isotope than the gas in B

7. An ideal gas is heated in a closed container at constant volume. Which of the following properties of the gas increases as the gas is heated?

   (A) The atomic mass of the atoms in the molecules
   (B) The number of molecules
   (C) The density of the gas
   (D) The pressure exerted by the molecules on the walls of the container
   (E) The average space between the molecules

8. 24 J of heat are added to a gas in a container, and then the gas does 6 J of work on the walls of the container. What is the change in internal energy for the gas?

   (A) –30 J
   (B) –18 J
   (C) 4 J
   (D) 18 J
   (E) 30 J

9. When water freezes, its molecules take on a more structured order. Why doesn't this contradict the Second Law of Thermodynamics?

   (A) Because the density of the water is decreasing
   (B) Because the water is gaining entropy as it goes from liquid to solid state
   (C) Because the water's internal energy is decreasing
   (D) Because the surroundings are losing entropy
   (E) Because the surroundings are gaining entropy

10. A heat engine produces 100 J of heat, does 30 J of work, and emits 70 J into a cold reservoir. What is the efficiency of the heat engine?

    (A) 100%
    (B) 70%
    (C) 42%
    (D) 40%
    (E) 30%

Explanations

1. (D)

The amount of heat lost by the hot water must equal the amount of heat gained by the cold water. Since all water has the same specific heat capacity, we can calculate the change in temperature of the cold water, ΔT_c, in terms of the change in temperature of the hot water, ΔT_h:

$$m_c c \Delta T_c = m_h c \Delta T_h$$
$$\Delta T_c = \frac{m_h}{m_c} \Delta T_h$$
$$= 5 \Delta T_h$$

At thermal equilibrium, the hot water and the cold water will be of the same temperature. With this in mind, we can set up a formula to calculate the value of ΔT_h:

$$65°\text{C} - \Delta T_h = 5°\text{C} + \Delta T_c$$
$$6 \Delta T_h = 60°\text{C}$$
$$\Delta T_h = 10°\text{C}$$

Since the hot water loses 10 C°, we can determine that the final temperature of the mixture is 65°C – 10 C° = 55°C.

2. (D)

If a block of ice at 0°C is heated, it will begin to melt. The temperature will remain constant until the ice is completely transformed into liquid. The amount of heat needed to melt a certain mass of ice is given by the latent heat of fusion for water. The specific heat of water is only relevant when the temperature of the ice or water is changing, and the density of the water is not relevant.

3. (D)

Asphalt, like most materials, has a positive coefficient of linear expansion, meaning that it expands as temperatures rise in summer and shrinks as temperatures fall in winter. This effect is called the law of thermal expansion, (D). The gaps in the sidewalk allow the blocks to expand without pushing against each other and cracking.

4. (E)

Convection is a form of heat transfer where a large number of molecules move from one place to another. An overhead fan works precisely by this method: it sends cooler air molecules down into a hot room, cooling the temperature of the room. The heat of the sun and the cooking action of a microwave are both forms of radiation, while the heat on a frying pan and the cooling action of ice cubes are both forms of conduction.

5. (A)
Since the gas is in a closed container, its volume remains constant, so the correct answer is (A).

When the gas is heated, its temperature increases, meaning that the average speed of the gas molecules increases. An increase in temperature also means there are more collisions between molecules.

According to the ideal gas law, when volume is constant and temperature is increased, then pressure will also increase. Pressure is determined by the rate of collisions of the gas molecules with the walls of the container.

6. (A)
According to the ideal gas law, temperature is directly proportional to volume and pressure. Since the volume of the container is constant, that means that doubling the temperature will double the pressure.

R is a constant: it doesn't vary under different circumstances, so (C) is wrong. Also, we are looking at a random sample of the gas, so there won't be a heavier isotope in one or the other of the containers: (E) is also wrong.

7. (D)
The ideal gas law states that temperature is directly proportional to pressure and volume. Since the gas is in a closed container, the volume is fixed, so an increase in temperature leads to an increase in pressure. The correct answer is (D).

The atomic mass and the number of molecules are fixed properties of the gas sample, and cannot change with heat. The density depends on the mass and the volume. The mass is also a fixed property of the gas sample, and the volume is being held constant, since we are dealing with a closed container. Therefore, the density must also remain constant. Because the number of molecules and the volume are constant, the average space between the molecules must remain constant.

8. (D)
The First Law of Thermodynamics tells us that $\Delta U = \Delta Q + \Delta W$: the change in internal energy is equal to the change in heat plus the work done on the system. The value of ΔQ is 24 J, since that much heat is added to the system, and the value of ΔW is –6 J, since the system *does* work rather than has work *done on it*. With this in mind, calculating ΔU is a simple matter of subtraction:

$$\begin{aligned}\Delta U &= \Delta Q + \Delta W = 24\text{ J} - 6\text{ J}\\ &= 18\text{ J}\end{aligned}$$

9. (E)

The Second Law of Thermodynamics tells us that the total amount of disorder, or entropy, in the universe is increasing. The entropy in a particular system can decrease, as with water molecules when they turn to ice, but only if the entropy in the surroundings of that system increases to an equal or greater extent. The Second Law of Thermodynamics holds, but only because the surroundings are gaining entropy, so the correct answer is (E). Answer (D) refers to the key part of the answer, but gives the wrong information about the change in entropy of the surroundings.

Be careful not to fall for answer (C). This is an explanation for why the water does not lose heat when it freezes: it is, in fact, losing internal energy. This is an instance of the *First* Law of Thermodynamics, which states that the change in a system's internal energy is equal to the value of the heat transfer in the system minus the work done by the system.

10. (E)

The efficiency of a heat engine is defined as $e = 1 - \Delta Q_{out}/\Delta Q_{in}$, where ΔQ_{out} is the amount of heat output into the cold reservoir and ΔQ_{in} is the amount of heat produced by the heat engine. Plugging the numbers in the question into this formula, we find that:

$$e = 1 - \frac{70 \text{ J}}{100 \text{ J}} = 0.3$$

An efficiency of 0.3 is the same thing as 30%.

Electric Forces, Fields, and Potential

Chapter Contents

Electric Charge 228
Electric Force 229
Electric Field 232
Electric Potential 234
Conductors and Insulators 236
Key Formulas 237
Review Questions 237
Explanations 240

DEMOCRITUS, A GREEK PHILOSOPHER of the 5th century B.C., was the first to propose that all things are made of indivisible particles called **atoms**. As is not unusual for the ancient Greeks, he had it only half right. The things we call atoms today are in fact made up of three different kinds of particles: **protons**, **neutrons**, and **electrons**. Electrons are much smaller than the other two particles. The electrons orbit the **nucleus** of the atom, which contains protons and neutrons.

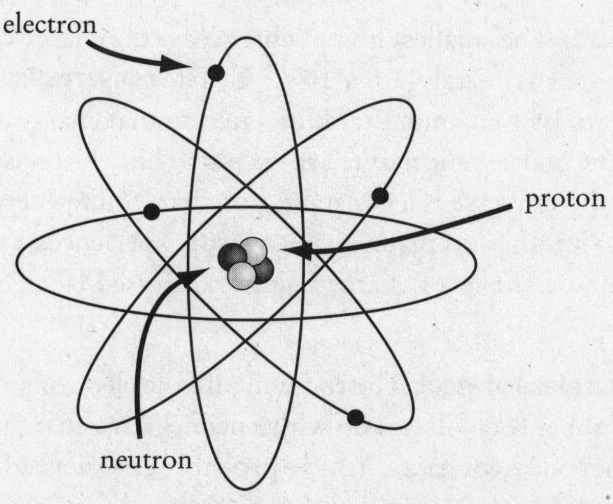

Electrons orbit the nucleus under the influence of the **electric force**. Protons and electrons both carry **electric charge**, which causes them to be attracted to one another. In most atoms, there are as many electrons as there are protons, and the opposite charges of these two kinds of particle balance out. However, it is possible to break electrons free from their orbits about the nucleus, causing an imbalance in charge. The movement of free electrons is the source of everything that we associate with electricity, a phenomenon whose power we have learned to harness over the past few hundred years to revolutionary effect.

Electric Charge

It is very difficult, if not impossible, to understand fully what electric charge, q, is. For the SAT II, you need only remember the old phrase: opposites attract. Protons carry a positive charge and electrons carry a negative charge, so you can just remember these three simple rules:

- Two positive charges will repel one another.
- Two negative charges will repel one another.
- A positive charge and a negative charge will attract one another.

The amount of positive charge in a proton is equal to the amount of negative charge in an electron, so an atom with an equal number of protons and electrons is electrically neutral, since the positive and negative charges balance out. Our focus will be on those cases when electrons are liberated from their atoms so that the atom is left with a net positive charge and the electron carries a net negative charge somewhere else.

Conservation of Charge

The SI unit of charge is the **coulomb** (C). The smallest unit of charge, e—the charge carried by a proton or an electron—is approximately 1.6×10^{-19} C. The **conservation of charge**—a hypothesis first put forward by Benjamin Franklin—tells us that charge can be neither created nor destroyed. The conservation of charge is much like the conservation of energy: the net charge in the universe is a constant, but charge, like energy, can be transferred from one place to another, so that a given system experiences a net gain or loss of charge. Two common examples of charge being transferred from one place to another are:

1. **Rubbing a rubber rod with a piece of wool:** The rod will pull the electrons off the wool, so that the rubber rod will end up with a net negative charge and the wool will have a net positive charge. You've probably experienced the "shocking" effects of rubbing rubber-soled shoes on a wool carpet.
2. **Rubbing a glass rod with a piece of silk:** The silk will pull the electrons off the glass, so that the glass rod will end up with a net positive charge and the silk will have a net negative charge.

Remember, net charge is always conserved: the positive charge of the wool or glass rod will balance out the negative charge of the rubber rod or silk.

The Electroscope

The **electroscope** is a device commonly used—and sometimes included on the SAT II—to demonstrate how electric charge works. It consists of a metal bulb connected to a rod, which in turn is connected to two thin leaves of metal contained within an evacuated glass chamber. When a negatively charged object is brought close to the metal bulb, the electrons in the bulb are repelled by the charge in the object and move down the rod to the two thin leaves. As a result, the bulb at the top takes on a positive charge and the two leaves take on a negative charge. The two metal leaves then push apart, as they are both negatively charged, and repel one another.

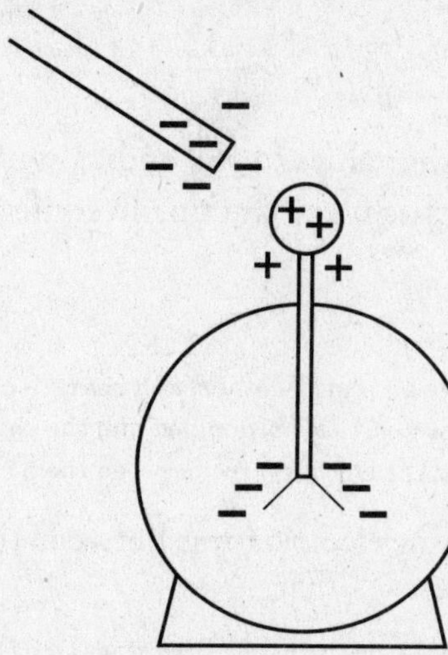

When a positively charged object approaches the metal bulb, the exact opposite happens, but with the same result. Electrons are drawn up toward the bulb, so that the bulb takes on a negative charge and the metal leaves have a positive charge. Because both leaves still have the same charge, they will still push apart.

Electric Force

There is a certain force associated with electric charge, so when a net charge is produced, a net electric force is also produced. We find electric force at work in anything that runs on batteries or uses a plug, but that isn't all. Almost all the forces we examine in this book come from electric charges. When two objects "touch" one another—be it in a car crash or a handshake—the atoms of the two objects never actually come into contact. Rather, the atoms in the two objects repel each other by means of an electric force.

Coulomb's Law

Electric force is analogous to gravitational force: the attraction or repulsion between two particles is directly proportional to the charge of the two particles and inversely proportional to the square of the distance between them. This relation is expressed mathematically as **Coulomb's Law**:

$$F = k\frac{q_1 q_2}{r^2}$$

In this equation, q_1 and q_2 are the charges of the two particles, r is the distance between them, and k is a constant of proportionality. In a vacuum, this constant is **Coulumb's constant**, k_0, which is approximately 9×10^9 N · m² / C². Coulomb's constant is often expressed in terms of a more fundamental constant—the **permittivity of free space**, ε_0, which has a value of 8.85×10^{-12} C²/ N · m²:

$$k_0 = \frac{1}{4\pi\varepsilon_0}$$

If they come up on the SAT II, the values for k_0 and ε_0 will be given to you, as will any other values for k when the electric force is acting in some other medium.

Example

> Two particles, one with charge $+q$ and the other with charge $-q$, are a distance r apart. If the distance between the two particles is doubled and the charge of one of the particles is doubled, how does the electric force between them change?

According to Coulomb's Law, the electric force between the two particles is initially

$$F = \frac{k(q)(q)}{r^2} = \frac{kq^2}{r^2}$$

If we double one of the charges and double the value of r, we find:

$$F = \frac{k(2q)(q)}{(2r)^2} = \frac{2kq^2}{4r^2} = \frac{kq^2}{2r^2}$$

Doubling the charge on one of the particles doubles the electric force, but doubling the distance between the particles divides the force by four, so in all, the electric force is half as strong as before.

Superposition

If you've got the hang of vectors, then you shouldn't have too much trouble with the law of **superposition** of electric forces. The net force acting on a charged particle is the vector sum of all the forces acting on it. For instance, suppose we have a number of charged particles, q_1, q_2, and q_3. The net force acting on q_1 is the force exerted on it by q_2 added to the force exerted on it by q_3. More generally, in a system of n particles:

$$F_1 = F_{2\to1} + F_{3\to1} + \cdots + F_{n\to1}$$

where $F_{n\to1}$ is the force exerted on particle 1 by particle n and F_1 is the net force acting on particle 1. The particle in the center of the triangle in the diagram below has no net force acting upon it, because the forces exerted by the three other particles cancel each other out.

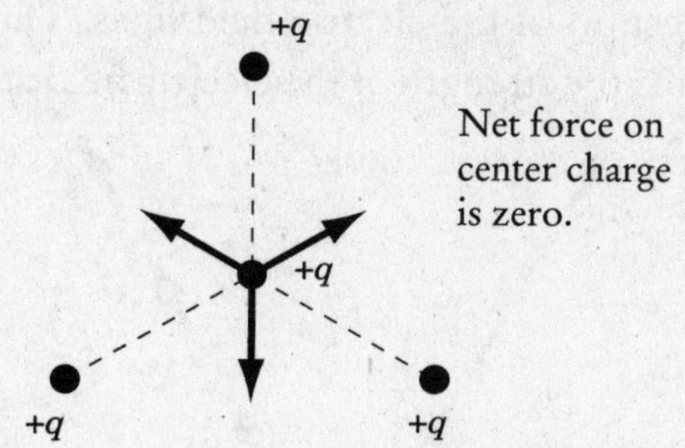

Example

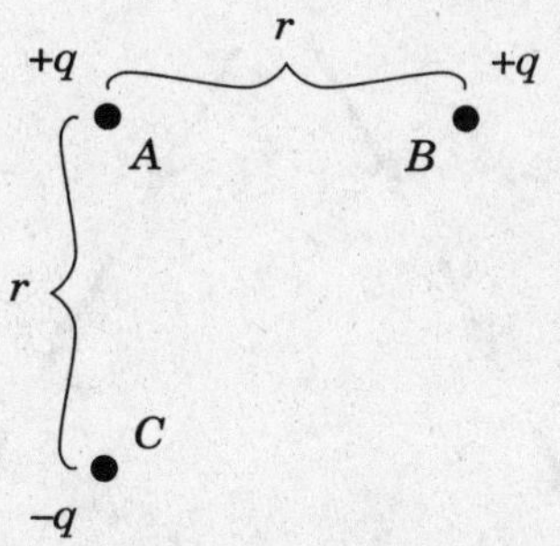

In the figure above, what is the direction of the force acting on particle *A*?

The net force acting on A is the vector sum of the force of B acting on A and the force of C acting on A. Because they are both positive charges, the force between A and B is repulsive, and the force of B on A will act to push A toward the left of the page. C will have an attractive force on A and will pull it toward the bottom of the page. If we add the effects of these two forces together, we find that the net force acting on A is diagonally down and to the left.

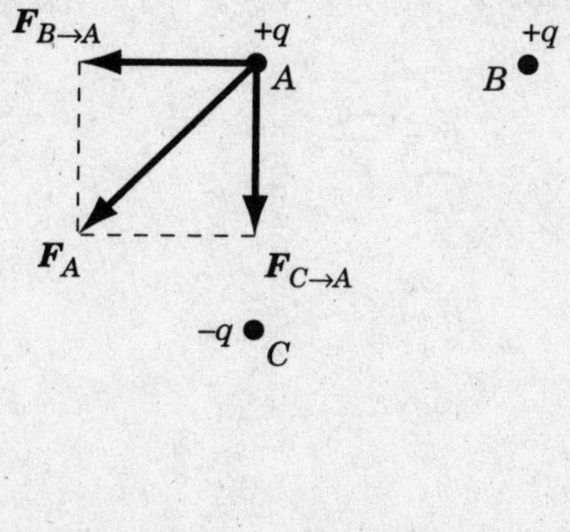

Electric Field

An electric charge, q, can exert its force on other charged objects even though they are some distance away. Every charge has an **electric field** associated with it, which exerts an electric force over all charges within that field. We can represent an electric field graphically by drawing vectors representing the force that would act upon a positive point charge placed at that location. That means a positive charge placed anywhere in an electric field will move in the direction of the electric field lines, while a negative charge will move in the opposite direction of the electric field lines. The density of the resulting electric field lines represents the strength of the electric field at any particular point.

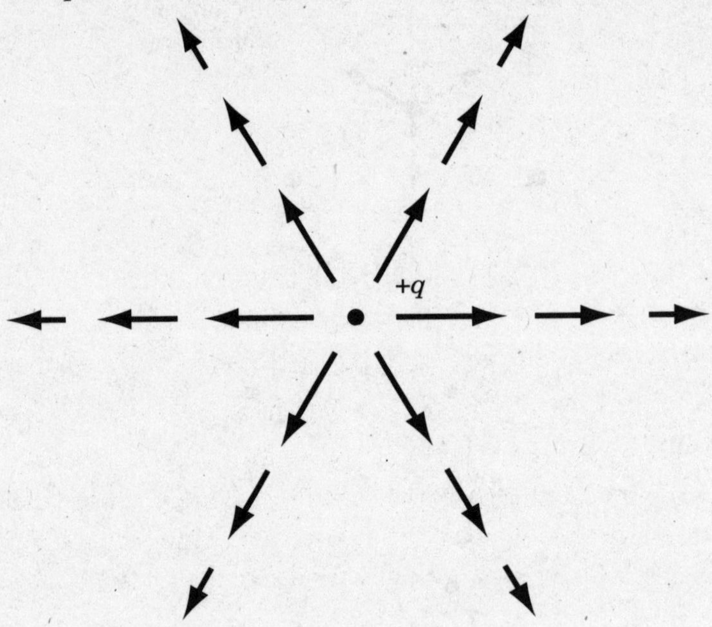

Calculating Electric Field

The electric field is a vector field: at each point in space, there is a vector corresponding to the electric field. The force $\boldsymbol{F}$ experienced by a particle q in electric field $\boldsymbol{E}$ is:

$$\boldsymbol{F}_q = q\boldsymbol{E}$$

Combining this equation with Coulomb's Law, we can also calculate the magnitude of the electric field created by a charge q at any point in space. Simply substitute Coulomb's Law in for $\boldsymbol{F}_q$, and you get:

$$E = \frac{kq}{r^2}$$

Drawing Electric Field Lines

The SAT II may ask a question about electric fields that involves the graphical representation of electric field lines. We saw above how the field lines of a single point charge are represented. Let's now take a look at a couple of more complicated cases.

Electric Fields for Multiple Charges

Just like the force due to electric charges, the electric field created by multiple charges is the sum of the electric fields of each charge. For example, we can sketch the electric field due to two charges, one positive and one negative:

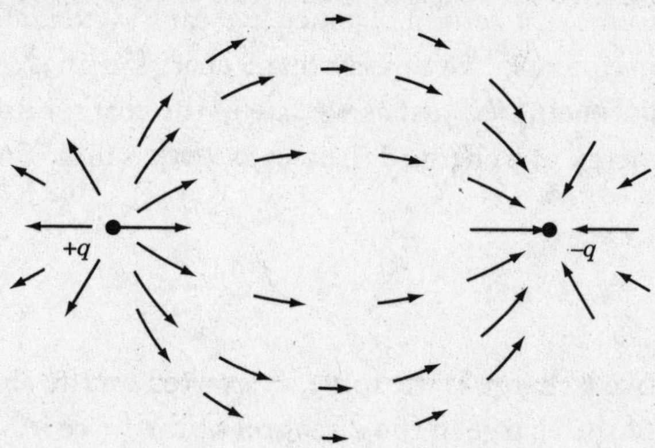

Line Charges and Plane Charges

Suppose we had a line of charge, rather than just a point charge. The electric field strength then decreases linearly with distance, rather than as the square of the distance. For a plane of charge, the field is constant with distance.

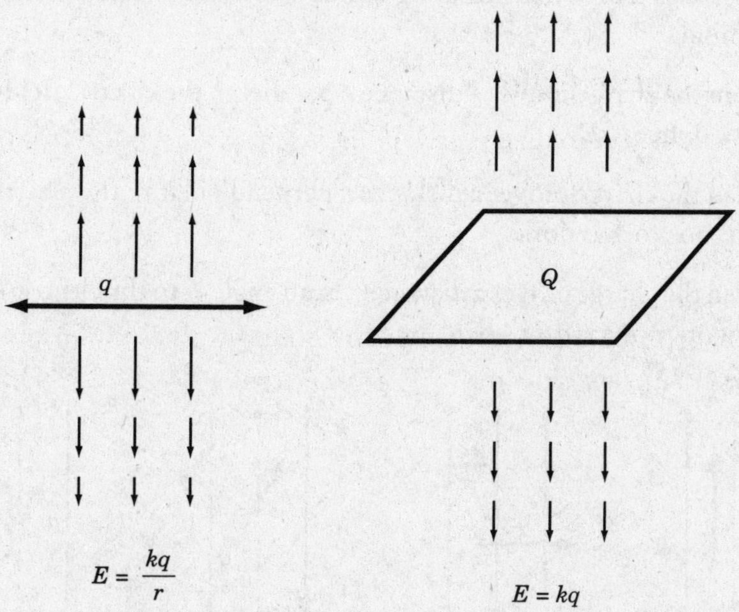

$E = \frac{kq}{r}$ $\qquad$ $E = kq$

Electric Potential

Because the electric force can displace charged objects, it is capable of doing work. The presence of an electric field implies the potential for work to be done on a charged object. By studying the **electric potential** between two points in an electric field, we can learn a great deal about the work and energy associated with electric force.

Electric Potential Energy

Because an electric field exerts a force on any charge in that field, and because that force causes charges to move a certain distance, we can say that an electric field does work on charges. Consequently, we can say that a charge in an electric field has a certain amount of **potential energy**, U. Just as we saw in the chapter on work, energy, and power, the potential energy of a charge decreases as **work** is done on it:

$$\Delta U = -W$$

Work

The work done to move a charge is the force, F, exerted on the charge, multiplied by the displacement, d, of the charge in the direction of the force. As we saw earlier, the magnitude of the force exerted on a charge q in an electric field E is $F_q = qE$. Thus, we can derive the following equation for the work done on a charge:

$$W = qEd$$

Remember that d is not simply the displacement; it is the displacement in the direction that the force is exerted. When thinking about work and electric fields, keep these three rules in mind:

1. When the charge moves a distance r parallel to the electric field lines, the work done is qEr.
2. When the charge moves a distance r perpendicular to the electric field lines, no work is done.
3. When the charge moves a distance r at an angle θ to the electric field lines, the work done is $qEr \cos\theta$.

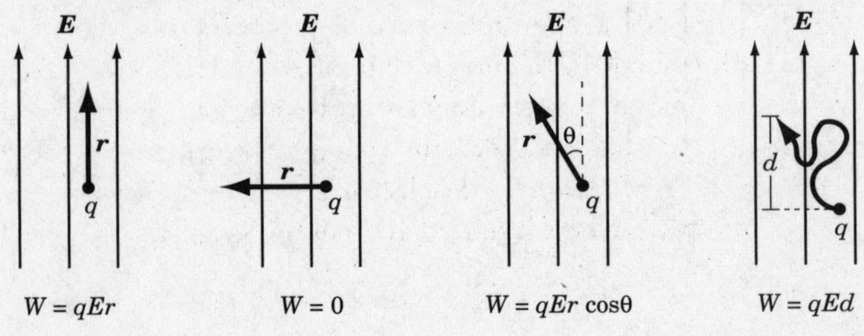

Electric Forces & Fields

Example

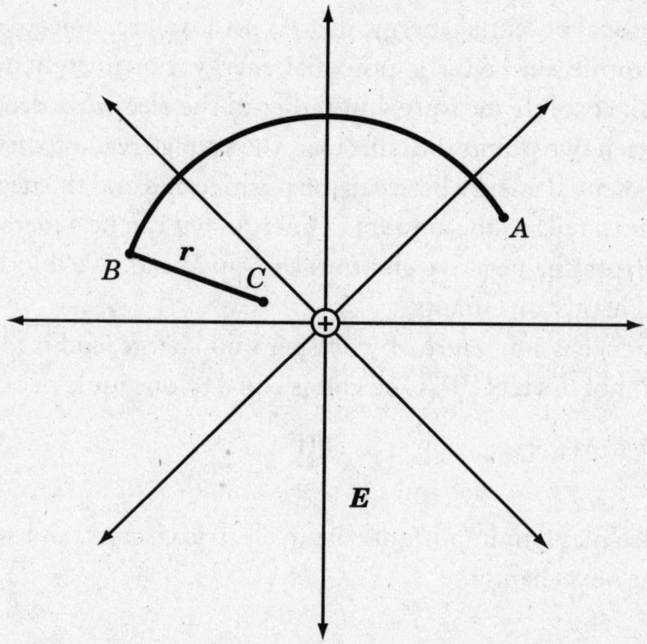

In an electric field, E, a positive charge, q, is moved in the circular path described above, from point A to point B, and then in a straight line of distance r toward the source of the electric field, from point B to point C. How much work is done by the electric field on the charge? If the charge were then made to return in a straight line from point C to point A, how much work would be done?

How much work is done moving the charge from point *A* to point *B* to point *C*? The path from point A to point B is perpendicular to the radial electric field throughout, so no work is done. Moving the charge from point B to point C requires a certain amount of work to be done against the electric field, since the positive charge is moving against its natural tendency to move in the direction of the electric field lines. The amount of work done is:

$$W = -qEr$$

The negative sign in the equation reflects the fact that work was done against the electric field.

How much work is done moving the charge directly from point *C* back to point *A*? Like all other forces in nature, the electric force is a conservative force, meaning that the path taken from one point in the electric field to another is irrelevant. The charge could move in a straight line from point C to point A or in a complex series of zig-zags: either way, the amount of work done by the electric field on the charge would be the same. The only thing that affects the amount of work done is the displacement of the charge in the direction of the electric field lines. Because we are simply moving the charge back to where it started, the amount of work done is $W = qEr$.

Potential Difference

Much like gravitational potential energy, there is no absolute, objective point of reference from which to measure electric potential energy. Fortunately, we are generally not interested in an absolute measure, but rather in the electric potential, or **potential difference**, V, between two points. For instance, the voltage reading on a battery tells us the difference in potential energy between the positive end and the negative end of the battery, which in turn tells us the amount of energy that can be generated by allowing electrons to flow from the negative end to the positive end. We'll look at batteries in more detail in the chapter on circuits.

Potential difference is a measure of work per unit charge, and is measured in units of joules per coulomb, or **volts** (V). One volt is equal to one joule per coulomb.

$$V = \frac{W}{q}$$

Potential difference plays an important role in electric circuits, and we will look at it more closely in the next chapter.

Conductors and Insulators

Idealized point charges and constant electric fields may be exciting, but, you may ask, what about the real world? Well, in some materials, such as copper, platinum, and most other metals, the electrons are only loosely bound to the nucleus and are quite free to flow, while in others, such as wood and rubber, the electrons are quite tightly bound to the nucleus and cannot flow. We call the first sort of materials **conductors** and the second **insulators**. The behavior of materials in between these extremes, called **semiconductors**, is more complicated. Such materials, like silicon and germanium, are the basis of all computer chips.

In a conductor, vast numbers of electrons can flow freely. If a number of electrons are transmitted to a conductor, they will quickly distribute themselves across the conductor so that the forces between them cancel each other out. As a result, the electric field within a conductor will be zero. For instance, in the case of a metal sphere, electrons will distribute themselves evenly so that there is no charge within the sphere, but only on the surface of the sphere.

Key Formulas

Coulomb's Law	$F = k\frac{q_1 q_2}{r^2}$
The Law of Superposition	$F_1 = F_{2\to1} + F_{3\to1} + \cdots + F_{n\to1}$
Definition of the Electric Field	$\boldsymbol{F}_q = q\boldsymbol{E}$
Electric Potential Energy	$\Delta U = -W$
Work Done by an Electric Field	$W = qEd$
Electric Potential	$V = \frac{W}{q}$

Review Questions

1. When a long-haired woman puts her hands on a Van de Graaff generator—a large conducting sphere with charge being delivered to it by a conveyer belt—her hair stands on end. Which of the following explains this phenomenon?

    (A) Positive charges are being brought to her
    (B) Negative charges are being brought to her
    (C) Positive charges are being taken from her
    (D) Negative charges are being taken from her
    (E) Her hair won't stand on end at all

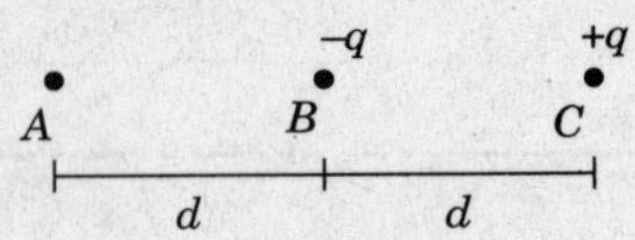

2. Three particles, A, B, and C, are set in a line, with a distance of d between each of them, as shown above. If particle B is attracted to particle A, what can we say about the charge, q_A, of particle A?

(A) $q_A < -q$
(B) $-q < q_A < 0$
(C) $q_A = 0$
(D) $0 < q_A < +q$
(E) $q_A > +q$

3. A particle of charge $+2q$ exerts a force $\boldsymbol{F}$ on a particle of charge $-q$. What is the force exerted by the particle of charge $-q$ on the particle of charge $+2q$?

(A) ½ $\boldsymbol{F}$
(B) 0
(C) 2$\boldsymbol{F}$
(D) $\boldsymbol{F}$
(E) $-\boldsymbol{F}$

4. Two charged particles exert a force of magnitude F on one another. If the distance between them is doubled and the charge of one of the particles is doubled, what is the new force acting between them?

(A) ¼ F
(B) ½ F
(C) F
(D) 2F
(E) 4F

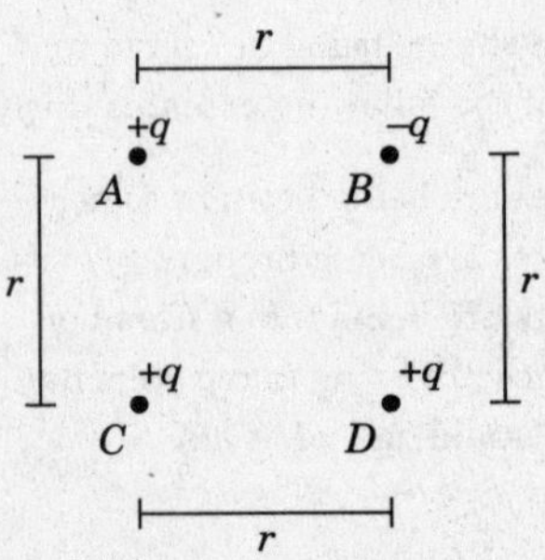

5. Four charged particles are arranged in a square, as shown above. What is the direction of the force acting on particle A?

(A) 
(B) 
(C) 
(D) 
(E) 

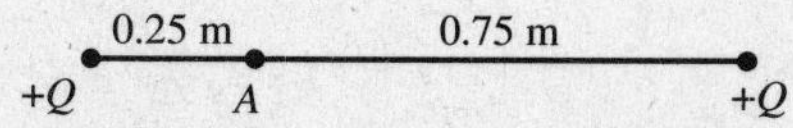

6. Two identical positive charges of $+Q$ are 1 m apart. What is the magnitude and direction of the electric field at point A, 0.25 m to the right of the left-hand charge?

(A) $\frac{3}{4}kQ$ to the right
(B) $\frac{128}{9}kQ$ to the left
(C) $\frac{160}{9}kQ$ to the left
(D) $\frac{160}{9}kQ$ to the right
(E) $\frac{128}{9}kQ$ to the right

7. A particle of charge $+q$ is a distance r away from a charged flat surface and experiences a force of magnitude F pulling it toward the surface. What is the magnitude of the force exerted on a particle of charge $+q$ that is a distance $2r$ from the surface?

(A) $\frac{1}{8}F$
(B) $\frac{1}{4}F$
(C) $\frac{1}{2}F$
(D) F
(E) $2F$

8. What is the change in potential energy of a particle of charge $+q$ that is brought from a distance of $3r$ to a distance of $2r$ by a particle of charge $-q$?

(A) kq^2/r
(B) $-kq^2/r$
(C) $kq^2/4r^2$
(D) $-kq^2/4r^2$
(E) kq^2/r^2

9. Two charges are separated by a distance d. If the distance between them is doubled, how does the electric potential between them change?

(A) It is doubled
(B) It is halved
(C) It is quartered
(D) It is quadrupled
(E) It is unchanged

10. A solid copper sphere has a charge of $+Q$ on it. Where on the sphere does the charge reside?

(A) $+Q$ at the center of the sphere
(B) $Q/2$ at the center of the sphere and $Q/2$ on the outer surface
(C) $-Q$ at the center of the sphere and $+2Q$ on the outer surface
(D) $+Q$ on the outer surface
(E) The charge is spread evenly throughout the sphere

Explanations

1. (B)

The conveyer belt brings electrons to the conducting sphere, which are then transferred to the woman's body. The charges distribute themselves evenly throughout the woman's body so that she has a net negative charge. Because her hair becomes negatively charged, each strand repels each other strand, and all the strands stand on end to get as far away from each other as possible. Thus, the answer is (B)

2. (E)

Particle C exerts an attractive force on the negatively charged particle B. If B is to be pulled in the direction of A, A must exert an even stronger attractive force than particle C. That means that particle A must have a stronger positive charge than particle C, which is $+q$.

3. (E)

The electric force exerted by one charged particle on another is proportional to the charge on both particles. That is, the force exerted by the $+2q$ particle on the $-q$ particle is of the same magnitude as the force exerted by the $-q$ particle on the $+2q$ particle, because, according to Coulomb's Law, both forces have a magnitude of:

$$F = k\frac{(+2q)(-q)}{r^2}$$

Since one particle is positive and the other is negative, this force is attractive: each particle is pulled toward the other. Since the two particles are pulled toward each other, the forces must be acting in opposite directions. If one particle experiences a force of $\mathbf{F}$, then the other particle must experience a force of $-\mathbf{F}$.

4. (B)

Coulomb's Law tells us that $F = kq_1q_2/r^2$: the force between two particles is directly proportional to their charges and inversely proportional to the square of the distance between them. If the charge of one of the particles is doubled, then the force is doubled. If the distance between them is doubled, then the force is divided by four. Since the force is multiplied by two and divided by four, the net effect is that the force is halved.

5. (C)

Particles C and D exert a repulsive force on A, while B exerts an attractive force. The force exerted by D is somewhat less than the other two, because it is farther away. The resulting forces are diagrammed below:

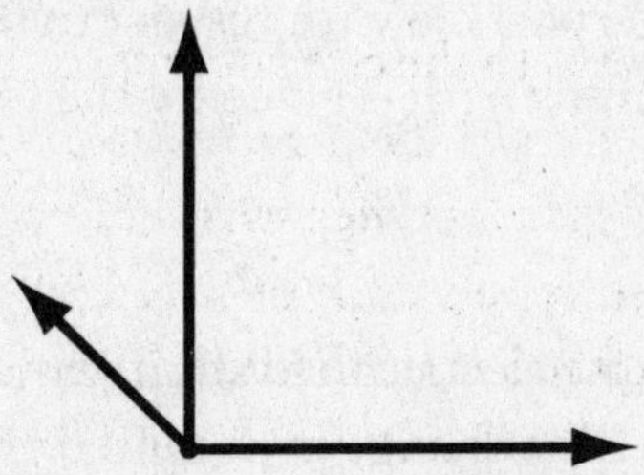

The vector sum of the three vectors will point diagonally up and to the right, as does the vector in (C).

6. (E)

The vector for electric field strength at any point has a magnitude of $E = kQ/r^2$ and points in the direction that a positive point charge would move if it were at that location. Because there are two different point charges, Q_1 and Q_2, there are two different electric fields acting at point A. The net electric field at A will be the vector sum of those two fields. We can calculate the magnitude of the electric field of each charge respectively:

$$E_{Q_1} = \frac{kQ}{(0.25)^2} = 16kQ$$

$$E_{Q_2} = \frac{kQ}{(0.75)^2} = \frac{16}{9}kQ$$

Since both Q_1 and Q_2 would exert a repulsive force on a positive point charge, $\boldsymbol{E}_{Q_1}$ points to the right and $\boldsymbol{E}_{Q_2}$ points to the left. The net electric field is:

$$E_{Q_1} - E_{Q_2} = 16kQ - \frac{16}{9}kQ = \frac{128}{9}kQ$$

Because Q_1 is closer to A than Q_2, the electric field from Q_1 will be stronger than the electric field from Q_2, and so the net electric field will point to the right.

7. (D)

The charged surface is a plane charge, and the electric field exerted by a plane charge is $E = kq$. That is, the magnitude of the electric field strength does not vary with distance, so a particle of charge $+q$ will experience the same attractive force toward the charged surface no matter how far away it is.

8. (B)

The change in potential energy of a charged particle is related to the work done on that particle by the formula $\Delta U = -W$. In turn, the work done on a particle is given by the formula $W = qEd$, where q is the charge of the particle, E is the electric field strength, and d is the displacement of the particle with respect to the electric field.

In this case, the electric field strength is given by the formula:

$$E = \frac{kq}{r^2}$$

Since we know the charge and the electric field strength, and can calculate that the value of d is $3r - 2r$, we can calculate the change in potential energy:

$$\Delta U = -W = -qEd = (-q)\frac{kq}{r^2}(r)$$

$$= -\frac{kq^2}{r}$$

9. (B)

The electric potential of a charge is given by the equation $V = kq/r$. In other words, distance is inversely proportional to electric potential. If the distance is doubled, then the electric potential must be halved.

10. (D)

Excess charges always reside on the surface of a conductor because they are free to move, and feel a repulsive force from each other.

DC Circuits

Chapter Contents

Voltage ... *243*
Current ... *244*
Resistance ... *245*
Energy, Power, and Heat ... *247*
Circuits ... *249*
Capacitors ... *262*
Key Formulas ... *265*
Practice Questions ... *266*
Explanations ... *268*

In the previous chapter, we looked at the movement of charges, showing that a net charge creates an electric field with differences in electric potential energy at different points in the field. When two points in a field with a potential difference are connected by a conducting material, electrons will flow spontaneously from one point to another. For instance, when the two terminals of a battery (a source of potential difference) are connected by a copper wire (a conducting material), electrons flow spontaneously from the negative terminal of the battery toward the positive terminal. This mass flow of electrons in a particular direction creates a **current**, which is the source of the circuits that we will examine in this chapter.

As fans of hard rock know, there are two kinds of circuits, AC and DC. "AC" stands for "alternating current": an electromagnetic generator induces a current that alternates in direction. AC circuits can be quite complicated, so you'll be relieved to know this is the last you'll hear of them: they don't appear on the SAT II. However, you should expect a good number of questions on DC, or "direct current," circuits. These are the more familiar circuits, where a current flows steadily in a single direction.

DC Circuits

Voltage

The batteries we use in flashlights and clock radios operate on chemical energy. This chemical energy—which you may learn more about in chemistry class—separates charges, creating a potential difference. To separate charges and create a positive and negative terminal, the battery must do a certain amount of work on the charges. This work per unit charge is called the **voltage**, *V*, or **electromotive force**, emf, and is measured in volts (V). Remember, one volt is equal to one joule per coulomb.

You'll notice that voltage is measured in the same units as potential difference. That's because they are essentially the same thing. The voltage of a battery is a measure of the work that has been done to set up a potential difference between the two terminals. We could draw an analogy to the amount of work required to lift an object in the air, giving it a certain amount of gravitational potential energy: both work and gravitational potential energy are measured in joules, and the amount of work done on the object is exactly equal to the amount of gravitational potential energy it acquires.

When a current flows about a circuit, we say there is a certain "voltage drop" or "drop in potential" across the circuit. An electric current converts potential energy into work: the electric field in the circuit does work on the charges to bring them to a point of lower potential. In a circuit connected to a 30 V battery, the current must drop 30 volts to send the electrons from the negative terminal to the positive terminal.

Current

When a wire is connected between the terminals of a battery, the potential difference in the battery creates an electric field in the wire. The electrons at the negative terminal move through the wire to the positive terminal.

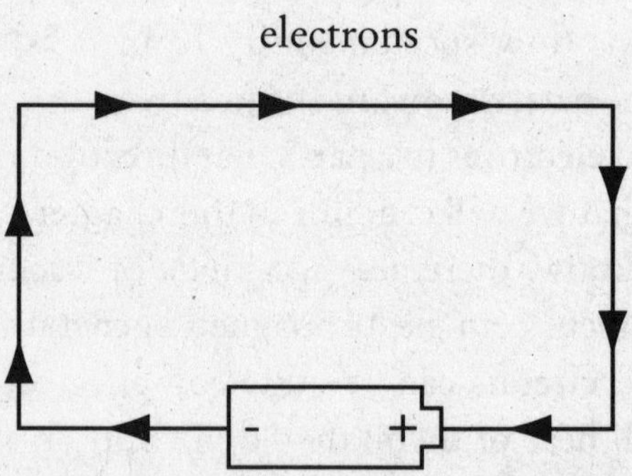

Although the electrons in the wire move quickly, they go in random directions and collide with other electrons and the positive charges in the wire. Each electron moves toward the positive terminal at a speed v_d, called the **drift speed**, which is only about one millimeter per second. However, when we study circuits, we do not follow individual electrons as they move along the wire, but rather we look at the current, I, that they create. Current is the charge per unit time across an imaginary plane in the wire:

$$I = \frac{Q}{t}$$

The unit of current is the coulomb per second, which is called an **ampere** (A): 1 A = 1 C/s.

DC Circuits

Direction of Current

Although the electrons are the charge carriers and move from the negative terminal to the positive terminal of the battery, the current flows in the opposite direction, from the positive terminal to the negative terminal. This may seem odd, but we can draw an analogous example from everyday life. Suppose you arrange 12 chairs in a circle, and get 11 people to sit down, leaving one chair empty. If each person in turn were to shift over in the clockwise direction to fill the vacant spot, the vacant spot would appear to move in the counterclockwise direction. If we think of the electrons in a circuit as the people, then the current moves in the direction of the vacant spot.

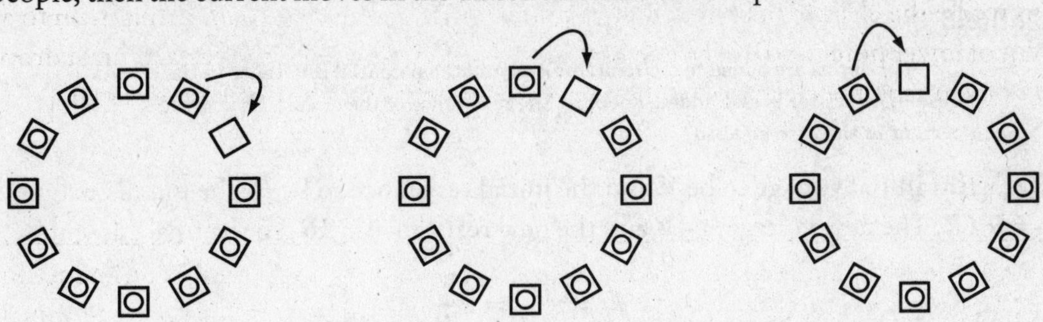

People move clockwise; vacant spot moves counterclockwise

Resistance

Some materials conduct current better than others. If we had a copper wire and a glass wire with the same length and cross section, and put the same potential difference across them, the current in the copper wire would be much larger than the current in the glass wire. The structure of copper, a conductor, is such that it permits electrons to move about more freely than glass, an insulator. We say that the glass wire has a higher **resistance**, R, than the copper wire.

We can express resistance in terms of the potential difference, ΔV, and the current, I:

$$R = \frac{\Delta V}{I}$$

Generally, the Δ is omitted. For a given voltage, the larger the current, the smaller the resistance. The unit of resistance is the **ohm** (Ω). One ohm is equal to one volt per ampere: $1\ \Omega = 1\ \text{V/A}$.

DC Circuits

Ohm's Law

Ohm's Law relates the three important quantities of current, voltage, and resistance:

$$I = \frac{V}{R}$$

This equation tells us that we can maximize the current by having a large voltage drop and a small resistance. This is one of the most important equations dealing with electromagnetism, and the SAT II is bound to call upon you to remember it.

Example

> **Three batteries are added to a circuit, multiplying the potential difference in the circuit by four. A resistor is also added, doubling the resistance of the circuit. How is the current in the wire affected?**

Taking the initial voltage to be V and the initial resistance to be R, the initial current is $I_i = V/R$. The new voltage is $4V$ and the new resistance is $2R$, so the final current is:

$$I_f = \frac{4V}{2R} = 2\frac{V}{R}$$

These changes double the current.

Resistivity

Resistivity, ρ, is a property of a material that affects its resistance. The higher the resistivity, the higher the resistance. Resistance also depends on the dimensions of the wire—on its length, L, and cross-sectional area, A:

$$R = \rho\frac{L}{A}$$

A longer wire provides more resistance because the charges have farther to go. A larger cross sectional area reduces the resistance because it is easier for the charges to move. The unit of resistivity is the ohm-meter, $\Omega \cdot$ m. The resistivity of copper is about 10^{-8} $\Omega \cdot$ m and the resistivity of glass is about 10^{12} $\Omega \cdot$ m. At higher temperatures, the resistivity of most metals increases.

Example

> **A copper wire of length 4 m and cross-sectional area 4 mm^2 is connected to a battery with a potential difference of 9 V. What is the current that runs through the wire? Approximate the resistivity for copper to be 10^{-8} $\Omega \cdot$ m.**

As we know, the current in a wire is a measure of voltage divided by resistance. We know that the voltage for the circuit is 9 V, but we don't know the resistance. However, since we know that the resistivity for copper is 10^{-8} $\Omega \cdot$ m, we can use the formula for resistivity to calculate the resistance in the wire.

First, we need to remember that area is measured in m^2, not mm^2. If 1 mm = 1×10^{-3} m, then 4 $mm^2 = 4 \times (10^{-3}\text{ m})^2 = 4 \times 10^{-6}\text{ m}^2$.

Now we can plug the values for the resistivity of copper and the length and cross-sectional area of the wire into the equation for resistivity:

$$R = \rho\frac{L}{A} = (10^{-8}\ \Omega \cdot \text{m})\frac{4\text{m}}{4 \times 10^{-6}\text{ m}^2}$$
$$= 1 \times 10^{-2}\ \Omega$$

Once we know the resistance of the circuit, calculating the current involves a simple application of Ohm's Law:

$$I = \frac{V}{R} = \frac{9\text{V}}{1 \times 10^{-2}\ \Omega} = 900\text{ A}$$

Conductivity

Infrequently, you may come across talk of **conductivity** and conductance rather than resistivity and resistance. As the names suggest, these are just the inverse of their resistant counterparts. Saying a material has high conductivity is another way of saying that material has a low resistivity. Similarly, a circuit with high conductance has low resistance. Someone with half a sense of humor named the unit of conductance the mho (℧), where 1 ℧ = 1 Ω^{-1}.

Energy, Power, and Heat

As a charge carrier moves around a circuit and drops an amount of potential, V, in time t, it loses an amount of potential energy, qV. The power, or the rate at which it loses energy, is qV/t. Since the current, I, is equal to q/t, the power can be expressed as:

$$P = IV$$

The unit of power is the **watt** (W). As you learned in Chapter 4, one watt is equal to one joule per second.

VIR and PIV Triangles

Ohm's Law and the formula for power express fundamental relationships between power, current, and voltage, and between voltage, current, and resistance. On occasion, you may be asked to calculate any one of the three variables in these equations, given the other two. As a result, good mnemonics to remember are the VIR and PIV triangles:

DC Circuits

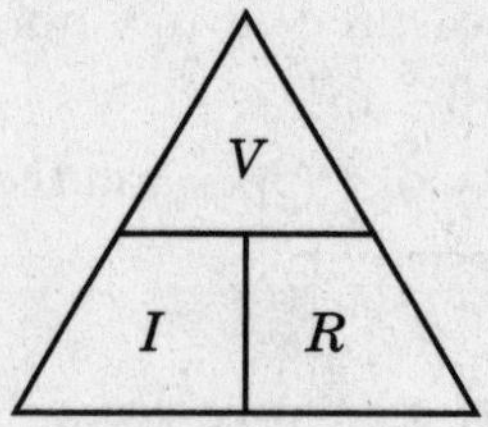

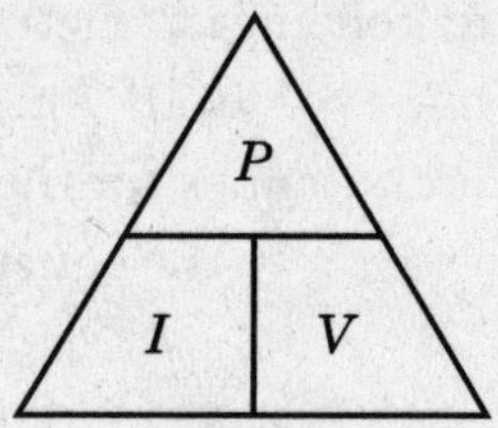

$$V = IR$$

$$I = \frac{V}{R}$$

$$R = \frac{V}{I}$$

$$P = IV$$

$$I = \frac{P}{V}$$

$$V = \frac{P}{I}$$

If the two variables you know are across from one another, then multiplying them will get you the third. If the two variables you know are above and below one another, then you can get the third variable by dividing the one above by the one below. For instance, if you know the power and the voltage in a given circuit, you can calculate the current by dividing the power by the voltage.

Power and Resistance

We can combine the equations for power and Ohm's Law to get expressions for power in terms of resistance:

$$P = I^2R = \frac{V^2}{R}$$

Heat

As current flows through a resistor, the resistor heats up. The heat in joules in given by:

$$H = I^2Rt = Pt$$

where t is the time in seconds. In other words, a resistor heats up more when there is a high current running through a strong resistor over a long stretch of time.

Example

> **A circuit with a potential difference of 10 V is hooked up to a light bulb whose resistance is 20 Ω. The filament in the light bulb heats up, producing light. If the light bulb is left on for one minute, how much heat is produced?**

We are being asked for the amount of heat that is dissipated, which is the product of power and time. We have learned to express power in terms of voltage and resistance in the formula $P = V^2/R$. Applying that formula to the problem at hand, we find:

$$P = \frac{V^2}{R} = \frac{(10\text{ V})^2}{20\ \Omega}$$
$$= 5\text{ W}$$

Then, plugging the appropriate numbers into the equation for heat, we find:

$$H = Pt = (5\text{ W})(60\text{ s})$$
$$= 300\text{ J}$$

Every minute, the filament produces 300 J of heat.

Kilowatt-Hours

When electric companies determine how much to charge their clients, they measure the power output and the amount of time in which this power was generated. Watts and seconds are relatively small units, so they measure in kilowatt-hours, where one kilowatt is equal to 1000 watts. Note that the kilowatt-hour, as a measure of power multiplied by time, is a unit of energy. A quick calculation shows that:

$$1\text{ kW}\cdot\text{h}\left(\frac{1000\text{ W}}{1\text{ kW}}\right)\left(\frac{3600\text{ s}}{1\text{ h}}\right) = 3.6\times10^6\text{ J}$$

Circuits

Most SAT II questions on circuits will show you a circuit diagram and ask you questions about the current, resistance, or voltage at different points in the circuit. These circuits will usually consist of a power source and one or more resistors arranged in parallel or in series. You will occasionally encounter other circuit elements, such as a voltmeter, an ammeter, a fuse, or a capacitor. Reading the diagrams is not difficult, but since there will be a number of questions on the test that rely on diagrams, it's important that you master this skill. Here's a very simple circuit diagram:

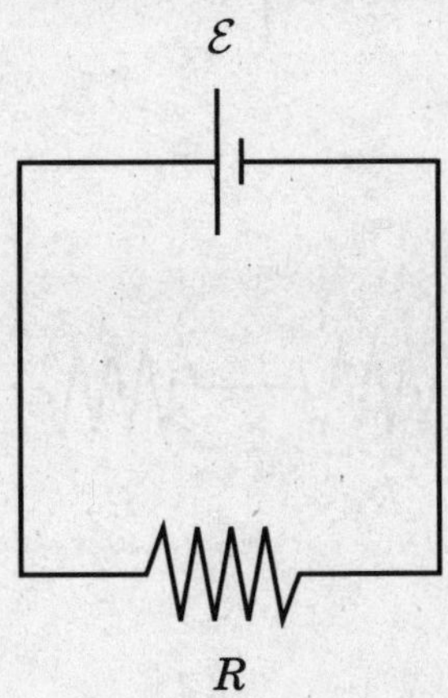

DC Circuits

Zigzags represent resistors, and a pair of parallel, unequal lines represents a battery cell. The longer line is the positive terminal and the shorter line is the negative terminal. That means the current flows from the longer line around the circuit to the shorter line. In the diagram above, the current flows counterclockwise. Often, more than one set of unequal parallel lines are arranged together; this just signifies a number of battery cells arranged in series.

Example

> **In the diagram above, $\mathcal{E}$ = 6 V and R = 12 Ω. What is the current in the circuit and what is the power dissipated in the resistor?**

You don't really need to refer to the diagram in order to solve this problem. As long as you know that there's a circuit with a six-volt battery and a 12-ohm resistor, you need only apply Ohm's Law and the formula for power.

Since $I = V/R$, the current is:

$$I = \frac{6\text{ V}}{12\ \Omega} = 0.5\text{ A}$$

The power is:

$$P = I^2R = (0.5\text{ A})^2(12\ \Omega) = 3\text{ W}$$

Resistors in Series

Two resistors are in **series** when they are arranged one after another on the circuit, as in the diagram below. The same amount of current flows first through one resistor and then the other, since the current does not change over the length of a circuit.

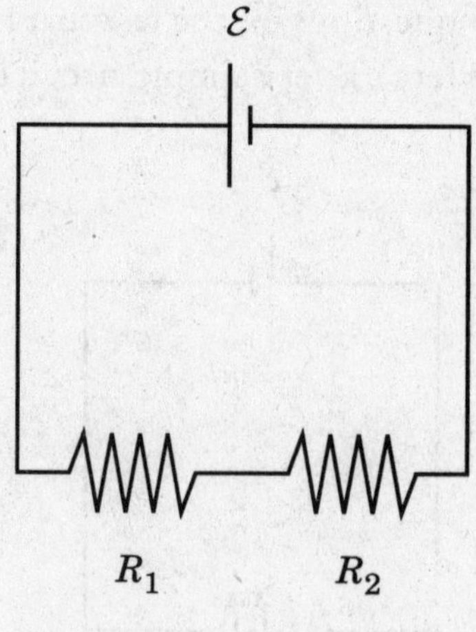

DC Circuits

However, each resistor causes a voltage drop, and if there is more than one resistor in the circuit, the sum of the voltage drops across each resistor in the circuit is equal to the total voltage drop in the circuit. The total resistance in a circuit with two or more resistors in series is equal to the sum of the resistance of all the resistors: a circuit would have the same resistance if it had three resistors in series, or just one big resistor with the resistance of the original three resistors put together. In equation form, this principle is quite simple. In a circuit with two resistors, R_1 and R_2, in series, the total resistance, R_t is:

$$R_t = R_1 + R_2$$

Example

In the figure above, a battery supplies 30 V to a circuit with a 10 Ω resistor and a 20 Ω resistor. What is the current in the circuit, and what is the voltage drop across each resistor?

What is the current in the circuit? We can determine the current in the circuit by applying Ohm's Law: $I = V/R$. We know what V is, but we need to calculate the total resistance in the circuit by adding together the individual resistances of the two resistors in series.:

$$\begin{aligned} R_t &= R_1 + R_2 = 10\ \Omega + 20\ \Omega \\ &= 30\ \Omega \end{aligned}$$

When we know the total resistance in the circuit, we can determine the current through the circuit with a simple application of Ohm's Law:

$$\begin{aligned} I &= \frac{V}{R} = \frac{30\ \text{V}}{30\ \Omega} \\ &= 1\ \text{A} \end{aligned}$$

What is the voltage drop across each resistor? Determining the voltage drop across an individual resistor in a series of resistors simply requires a reapplication of Ohm's Law. We know the current through the circuit, and we know the resistance of that individual resistor, so the voltage drop across that resistor is simply the product of the current and the resistance. The voltage drop across the two resistors is:

$$\begin{aligned} V_1 &= IR_1 = (1\ \text{A})(10\ \Omega) = 10\ \text{V} \\ V_2 &= IR_2 = (1\ \text{A})(20\ \Omega) = 20\ \text{V} \end{aligned}$$

Note that the voltage drop across the two resistors is 10 V + 20 V = 30 V, which is the total voltage drop across the circuit.

Resistors in Parallel

Two resistors are in **parallel** when the circuit splits in two and one resistor is placed on each of the two branches.

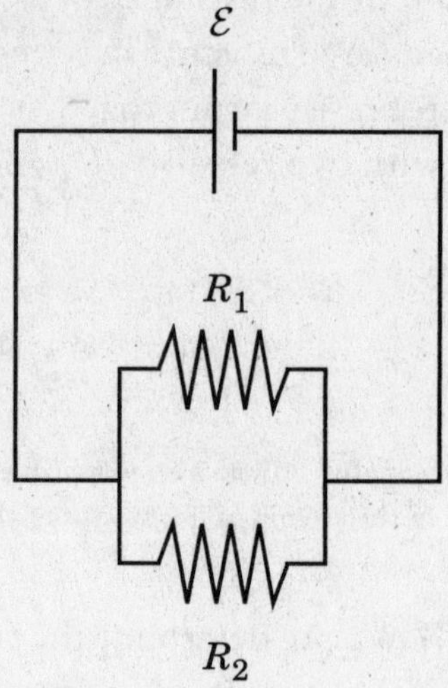

In this circumstance, it is often useful to calculate the equivalent resistance as if there were only one resistor, rather than deal with each resistor individually. Calculating the equivalent resistance of two or more resistors in parallel is a little more complicated than calculating the total resistance of two or more resistors in series. Given two resistors, R_1 and R_2, in parallel, the equivalent resistance, R_t, is:

$$\frac{1}{R_t} = \frac{1}{R_1} + \frac{1}{R_2}$$

When a circuit splits in two, the current is divided between the two branches, though the current through each resistor will not necessarily be the same. The voltage drop must be the same across both resistors, so the current will be stronger for a weaker resistor, and vice versa.

Example

> Two resistors, $R_1 = 5\ \Omega$ and $R_2 = 20\ \Omega$, are set up in parallel, as in the diagram above. The battery produces a potential difference of $\mathcal{E} = 12$ V. What is the total resistance in the circuit? What is the current running through R_1 and R_2? What is the power dissipated in the resistors?

What is the total resistance in the circuit? Answering this question is just a matter of plugging numbers into the formula for resistors in parallel.

$$\frac{1}{R_t} = \frac{1}{5\ \Omega} + \frac{1}{20\ \Omega} = \frac{1}{4\ \Omega}$$

So $R_t = 4\ \Omega$.

What is the current running through R_1 and R_2? We know that the total voltage drop is 12 V, and since the voltage drop is the same across all the branches of a set of resistors in parallel, we know that the voltage drop across both resistors will be 12 V. That means we just need to apply Ohm's Law twice, once to each resistor:

$$I_1 = \frac{12\text{ V}}{5\ \Omega} = 2.4\text{ A}$$
$$I_2 = \frac{12\text{ V}}{20\ \Omega} = 0.6\text{ A}$$

If we apply Ohm's Law to the total resistance in the system, we find that $I_t = (12\text{ V})/(4\ \Omega) = 3$ A. As we might expect, the total current through the system is the sum of the current through each of the two branches. The current is split into two parts when it branches into the resistors in parallel, but the total current remains the same throughout the whole circuit. This fact is captured in the junction rule we will examine when we look at Kirchhoff's Rules.

What is the power dissipated in the resistors? Recalling that $P = I^2R$, we can solve for the power dissipated through each resistor individually, and in the circuit as a whole. Let P_1 be the power dissipated in R_1, P_2 the power dissipated in R_2, and P_t the power dissipated in R_t.

$$P_1 = (2.4\text{A})^2(5\Omega) = 28.8\text{W}$$
$$P_2 = (0.6\text{A})^2(20\Omega) = 7.2\text{W}$$
$$P_t = (3\text{A})^2(4\Omega) = 36\text{W}$$

Note that $P_1 + P_2 = P_t$.

Circuits with Resistors in Parallel and in Series

Now that you know how to deal with resistors in parallel and resistors in series, you have all the tools to approach a circuit that has resistors both in parallel and in series. Let's take a look at an example of such a circuit, and follow two important steps to determine the total resistance of the circuit.

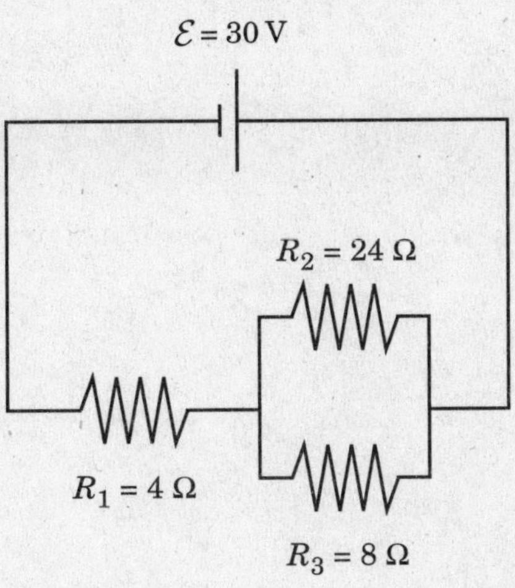

DC Circuits

1. **Determine the equivalent resistance of the resistors in parallel.** We've already learned to make such calculations. This one is no different:

$$\frac{1}{R_{2+3}} = \frac{1}{R_2} + \frac{1}{R_3} = \frac{1}{24\ \Omega} + \frac{1}{8\ \Omega}$$
$$= \frac{1}{6\ \Omega}$$

So the equivalent resistance is 6 Ω. In effect, this means that the two resistors in parallel have the same resistance as if there were a single 6 Ω resistor in their place. We can redraw the diagram to capture this equivalence:

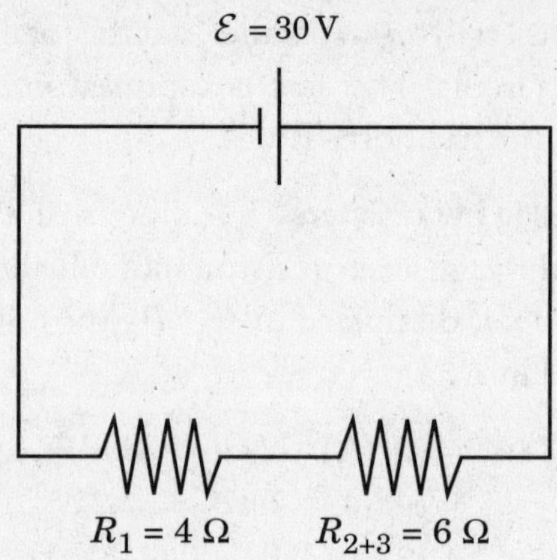

2. **Treating the equivalent resistance of the resistors in parallel as a single resistor, calculate the total resistance by adding resistors in series.** The diagram above gives us two resistors in series. Calculating the total resistance of the circuit couldn't be easier:

$$R_t = R_1 + R_{2+3} = 4\ \Omega + 6\ \Omega$$
$$= 10\ \Omega$$

Now that you've looked at this two-step technique for dealing with circuits in parallel and in series, you should have no problem answering a range of other questions.

Example

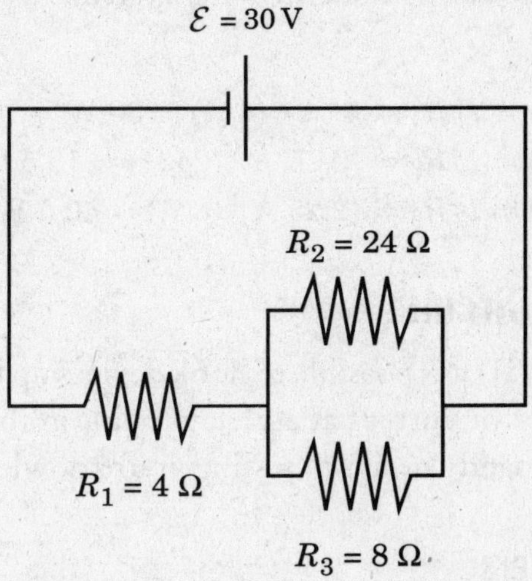

Consider again the circuit whose total resistance we have calculated. What is the current through each resistor? What is the power dissipated in each resistor?

What is the current running through each resistor? We know that resistors in series do not affect the current, so the current through R_1 is the same as the total current running through the circuit. Knowing the total resistance of the circuit and the voltage drop through the circuit, we can calculate the circuit's total current by means of Ohm's Law:

$$\begin{aligned} I &= \frac{V}{R} = \frac{30\ \text{V}}{30\ \Omega} \\ &= 1\ \text{A} \end{aligned}$$

Therefore, the current through R_1 is 3 A.

But be careful before you calculate the current through R_2 and R_3: the voltage drop across these resistors is not the total voltage drop of 30 V. The sum of the voltage drops across R_1 and the two resistors in parallel is 30 V, so the voltage drop across just the resistors in parallel is less than 30 V.

If we treat the resistors in parallel as a single equivalent resistor of 6 Ω, we can calculate the voltage drop across the resistors by means of Ohm's Law:

$$V = IR_{2+3} = (3\ \text{A})(6\ \Omega) = 18\ \text{V}$$

Now, recalling that current is divided unevenly between the branches of a set of resistors in parallel, we can calculate the current through R_2 and R_3 in the familiar way:

$$\begin{aligned} I_2 &= \frac{V}{R_2} = \frac{18\ \text{V}}{24\ \Omega} = 0.75\ \text{A} \\ I_3 &= \frac{V}{R_3} = \frac{18\ \text{V}}{8\ \Omega} = 2.25\ \text{A} \end{aligned}$$

What is the power dissipated across each resistor? Now that we know the current across each resistor, calculating the power dissipated is a straightforward application of the formula $P = I^2R$:

$$P_1 = I_1^2 R_1 = (3\text{ A})^2(4\,\Omega) = 36\text{ W}$$
$$P_2 = I_2^2 R_2 = (0.75\text{ A})^2(24\,\Omega) = 13.5\text{ W}$$
$$P_3 = I_3^2 R_3 = (2.25\text{ A})^2(8\,\Omega) = 40.5\text{ W}$$

Common Devices in Circuits

In real life (and on the SAT II) it is possible to hook devices up to a circuit that will read off the potential difference or current at a certain point in the circuit. These devices provide the SAT II with a handy means of testing your knowledge of circuits.

Voltmeters and Ammeters

A **voltmeter**, designated by a (V), measures the voltage across a wire. It is connected in parallel with the stretch of wire whose voltage is being measured, since an equal voltage crosses both branches of two wires connected in parallel.

An **ammeter**, designated by an (A), is connected in series. It measures the current passing through that point on the circuit.

Example

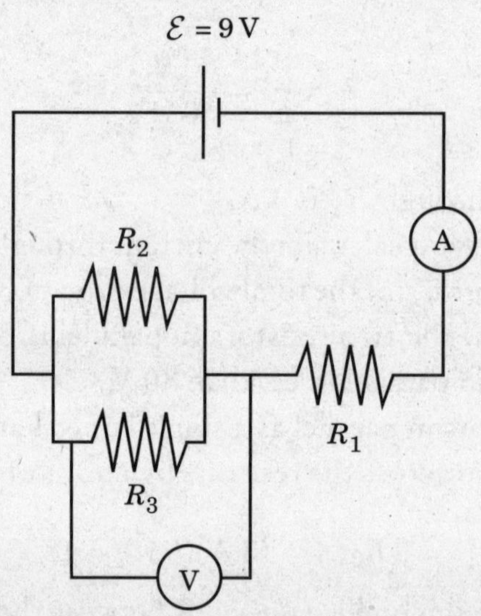

In the diagram above, $\mathcal{E}$ = 9 V, R_1 = 5 Ω, R_2 = 5 Ω, and R_3 = 20 Ω. What are the values measured by the ammeter and the voltmeter?

What does the ammeter read? Since the ammeter is not connected in parallel with any other branch in the circuit, the reading on the ammeter will be the total current in the circuit. We can use Ohm's Law to determine the total current in the circuit, but only if we first determine the total resistance in the circuit.

This circuit consists of resistors in parallel and in series, an arrangement we have looked at before. Following the same two steps as we did last time, we can calculate the total resistance in the circuit.

1. **Determine the equivalent resistance of the resistors in parallel.**

$$\frac{1}{R_{2+3}} = \frac{1}{5\ \Omega} + \frac{1}{20\ \Omega} = \frac{1}{4\ \Omega}$$

   We can conclude that $R_{2+3} = 4\ \Omega$.

2. **Treating the equivalent resistance of the resistors in parallel as a single resistor, calculate the total resistance by adding resistors in series.**

$$\begin{aligned} R_t &= R_1 + R_{2+3} = 5\ \Omega + 4\ \Omega \\ &= 9\ \Omega \end{aligned}$$

   Given that the total resistance is 9 Ω and the total voltage is 9 V, Ohm's Law tells us that the total current is:

$$\begin{aligned} I &= \frac{V}{R} = \frac{9\text{ V}}{9\ \Omega} \\ &= 1\text{ A} \end{aligned}$$

The ammeter will read 1 A.

What does the voltmeter read? The voltmeter is connected in parallel with R_2 and R_3, so it will register the voltage drop across these two resistors. Recall that the voltage drop across resistors in series is the same for each resistor.

We know that the total voltage drop across the circuit is 9 V. Some of this voltage drop will take place across R_1, and the rest of the voltage drop will take place across the resistors in parallel. By calculating the voltage drop across R_1 and subtracting from 9 V, we will have the voltage drop across the resistors in parallel, which is what the voltmeter measures.

$$\begin{aligned} V_1 &= I_1 R_1 = (1\text{ A})(5\ \Omega) \\ &= 5\text{ V} \end{aligned}$$

If the voltage drop across R_1 is 5 V, then the voltage drop across the resistors in parallel is 9 V – 5 V = 4 V. This is what the voltmeter reads.

Fuses

A **fuse** burns out if the current in a circuit is too large. This prevents the equipment connected to the circuit from being damaged by the excess current. For example, if the ammeter in the previous problem were replaced by a half-ampere fuse, the fuse would blow and the circuit would be interrupted.

Fuses rarely come up on the SAT II. If a question involving fuses appears, it will probably ask you whether or not the fuse in a given circuit will blow under certain circumstances.

Kirchhoff's Rules

Gustav Robert Kirchhoff came up with two simple rules that simplify many complicated circuit problems. The **junction rule** helps us to calculate the current through resistors in parallel and other points where a circuit breaks into several branches, and the **loop rule** helps us to calculate the voltage at any point in a circuit. Let's study Kirchhoff's Rules in the context of the circuit represented below:

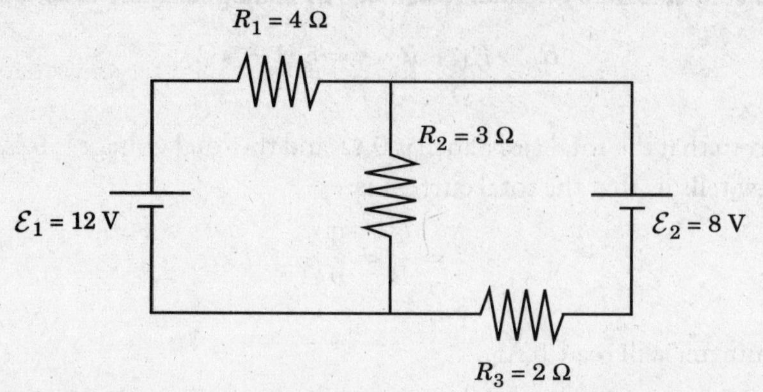

Before we can apply Kirchhoff's Rules, we have to draw arrows on the diagram to denote the direction in which we will follow the current. You can draw these arrows in any direction you please—they don't have to denote the actual direction of the current. As you'll see, so long as we apply Kirchhoff's Rules correctly, it doesn't matter in what directions the arrows point. Let's draw in arrows and label the six vertices of the circuit:

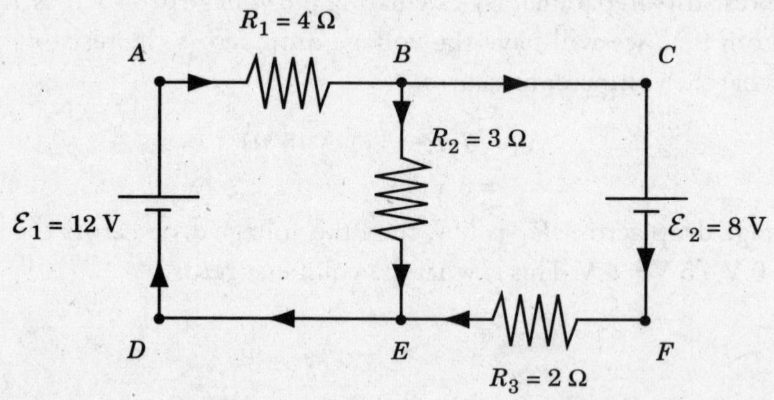

We repeat, these arrows do not point in the actual direction of the current. For instance, we have drawn the current flowing into the positive terminal and out of the negative terminal of $\mathcal{E}_2$, contrary to how we know the current must flow.

The Junction Rule

The junction rule deals with "junctions," where a circuit splits into more than one branch, or when several branches reunite to form a single wire. The rule states:

The current coming into a junction equals the current coming out.

This rule comes from the conservation of charge: the charge per unit time going into the junction must equal the charge per unit time coming out. In other words, when a circuit separates into more than one branch—as with resistors in parallel—then the total current is split between the different branches.

The junction rule tells us how to deal with resistors in series and other cases of circuits branching in two or more directions. If we encounter three resistors in series, we know that the sum of the current through all three resistors is equal to the current in the wire before it divides into three parallel branches.

Let's apply the junction rule to the junction at B in the diagram we looked at earlier.

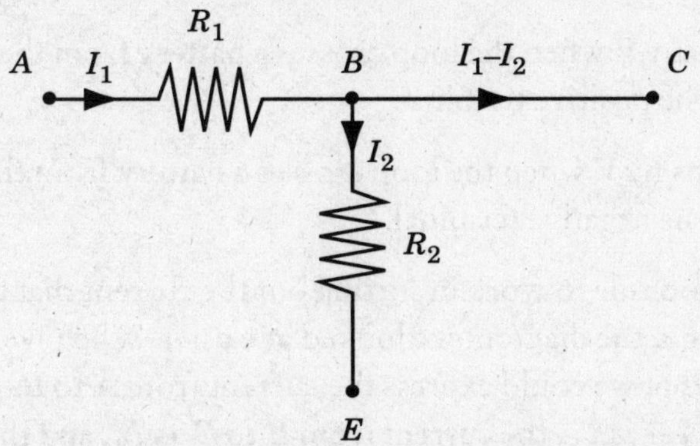

According to the arrows we've drawn, the current in the diagram flows from A into B across R_1 and flows out of B in two branches: one across R_2 toward E and the other toward C. According to the junction rule, the current flowing into B must equal the current flowing out of B. If we label the current going into B as I_1 and the current going out of B toward E as I_2, we can conclude that the current going out of B toward C is $I_1 - I_2$. That way, the current flowing into B is I_1 and the current flowing out of B is $I_2 + (I_1 - I_2) = I_1$.

The Loop Rule

The loop rule addresses the voltage drop of any closed loop in the circuit. It states:

The sum of the voltage drops around a closed loop is zero.

DC Circuits

This is actually a statement of conservation of energy: every increase in potential energy, such as from a battery, must be balanced by a decrease, such as across a resistor. In other words, the voltage drop across all the resistors in a closed loop is equal to the voltage of the batteries in that loop.

In a normal circuit, we know that when the current crosses a resistor, R, the voltage drops by IR, and when the current crosses a battery, V, the voltage rises by V. When we trace a loop—we can choose to do so in the clockwise direction or the counterclockwise direction—we may sometimes find ourselves tracing the loop against the direction of the arrows we drew. If we cross a resistor against the direction of the arrows, the voltage rises by IR. Further, if our loop crosses a battery in the wrong direction—entering in the positive terminal and coming out the negative terminal—the voltage drops by V. To summarize:

- Voltage drops by IR when the loop crosses a resistor in the direction of the current arrows.
- Voltage rises by IR when the loop crosses a resistor against the direction of the current arrows.
- Voltage rises by V when the loop crosses a battery from the negative terminal to the positive terminal.
- Voltage drops by V when the loop crosses a battery from the positive terminal to the negative terminal.

Let's now put the loop rule to work in sorting out the current that passes through each of the three resistors in the diagram we looked at earlier. When we looked at the junction rule, we found that we could express the current from A to B—and hence the current from E to D to A—as I_1, the current from B to E as I_2, and the current from B to C—and hence the current from C to F to E—as $I_1 - I_2$. We have two variables for describing the current, so we need two equations in order to solve for these variables. By applying the loop rule to two different loops in the circuit, we should be able to come up with two different equations that include the variables we're looking for. Let's begin by examining the loop described by $ABED$.

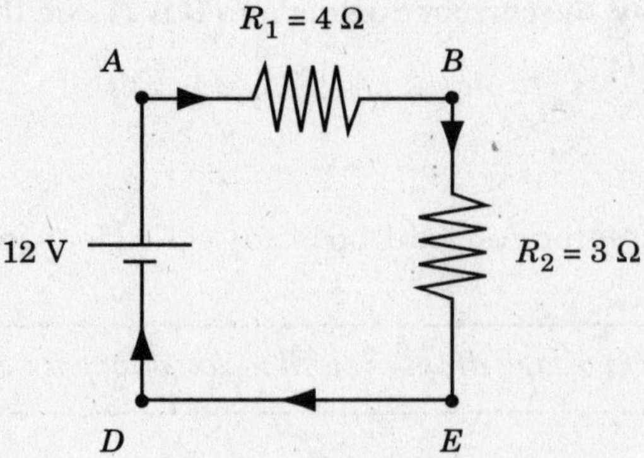

DC Circuits

Remember that we've labeled the current between A and B as I_1 and the current between B and E as I_2. Because the current flowing from E to A is the same as that flowing from A to B, we know this part of the circuit also has a current of I_1.

Tracing the loop clockwise from A, the current first crosses R_1 and the voltage drops by I_1R_1. Next it crosses R_2 and the voltage drops by I_2R_2. Then the current crosses $\mathcal{E}_1$, and the voltage rises by 12 V. The loop rule tells us that the net change in voltage is zero across the loop. We can express these changes in voltage as an equation, and then substitute in the values we know for R_1, R_2, and $\mathcal{E}_1$:

$$-(I_1R_1) - I_2R_2 + \mathcal{E}_1 = 0$$
$$-4I_1 - 3I_2 + 12 = 0$$
$$4I_1 + 3I_2 = 12$$

Now let's apply the loop rule to the loop described by $BCFE$.

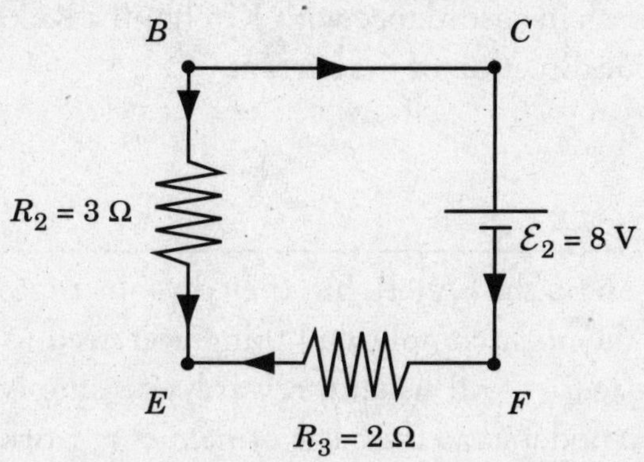

Tracing the loop clockwise from B, the arrows cross $\mathcal{E}_2$,but in the wrong direction, from positive to negative, meaning that the voltage drops by 8 V. Next, the current crosses R_3, with an additional voltage drop of $(I_1 - I_2)R_3$. Finally, it crosses R_2, but in the opposite direction of the arrows, so the current goes up by I_2R_2. Now we can construct a second equation:

$$-\mathcal{E}_2 - (I_1 - I_2)R_3 + I_2R_2 = 0$$
$$-8 - 2(I_1 - I_2) + 3I_2 = 0$$
$$I_1 = \frac{5}{2}I_2 - 4$$

Plugging this solution for I_1 into the earlier equation of $4I_1 + 3I_2 = 12$, we get:

$$4(\frac{5}{2}I_2 - 4) + 3I_2 = 12$$
$$13I_2 = 28$$
$$I_2 = \frac{28}{13}$$

So the current across R_2 is 28/13 A. With that in mind, we can determine the current across R_1 and R_3 by plugging the value for I_2 into the equations we derived earlier:

$$I_{R_1} = I_1 = \frac{5}{2}I_2 - 4 = \frac{14}{13}\text{ A}$$
$$I_{R_3} = I_1 - I_2 = -\frac{14}{13}\text{ A}$$

The negative value for the current across R_3 means that the current actually flows in the opposite direction of the arrow we drew. This makes perfect sense when we consider that current should normally flow out of the positive terminal and into the negative terminal of battery $\mathcal{E}_2$.

It doesn't matter how you draw the current arrows on the diagram, because if you apply Kirchhoff's Rules correctly, you will come up with negative values for current wherever your current arrows point in the opposite direction of the true current. Once you have done all the math in accordance with Kirchhoff's Rules, you will quickly be able to determine the true direction of the current.

Capacitors

Capacitors rarely come up on the SAT II, but they do sometimes make an appearance. Because capacitance is the most complicated thing you need to know about DC circuits, questions on capacitors will usually reward you simply for knowing what's going on. So long as you understand the basic principles at work here, you're likely to get a right answer on a question most students will answer wrong.

A capacitor is a device for storing charge, made up of two parallel plates with a space between them. The plates have an equal and opposite charge on them, creating a potential difference between the plates. A capacitor can be made of conductors of any shape, but the **parallel-plate capacitor** is the most common kind. In circuit diagrams, a capacitor is represented by two equal parallel lines.

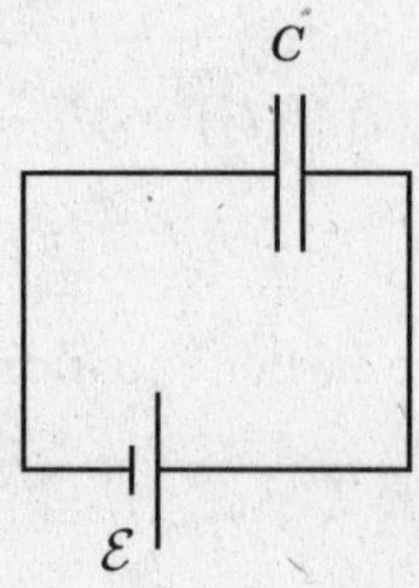

For any capacitor, the ratio of the charge to the potential difference is called the **capacitance**, C:

$$C = \frac{Q}{\Delta V}$$

For a parallel-plate capacitor, C is directly proportional to the area of the plates, A, and inversely proportional to the distance between them, d. That is, if the area of the plates is doubled, the capacitance is doubled, and if the distance between the plates is doubled, the capacitance is halved. The proportionality constant between C and A/d is ε_0, called the permittivity of free space, which we encountered in the previous chapter in relation to Coulomb's constant. In case you forgot, $\varepsilon_0 = 8.85 \times 10^{-12}\ C^2/N \cdot m^2$.

$$C = \epsilon_0 \frac{A}{d}$$

The unit of capacitance is the **farad** (F). One farad is equal to one coulomb per volt. Most capacitors have very small capacitances, which are usually given in microfarads, where $1\ \mu F = 10^{-6}\ F$.

Energy

To move a small amount of negative charge from the positive plate to the negative plate of a capacitor, an external agent must do work. This work is the origin of the energy stored by the capacitor.

If the plates have a charge of magnitude q, the potential difference is $\Delta V = q/C$. If $q = 0$, and work is done to add charge until $q = Q$, the total work required is:

$$U = \frac{1}{2} Q \Delta V$$

This is the energy stored by the capacitor. Manipulating this equation and the equation for capacitance, $C = Q/\Delta V$, we can derive a number of equivalent forms:

$$U = \frac{1}{2} C V^2 = \frac{Q^2}{C} = \frac{1}{2} Q V$$

Equivalent Capacitance

Like resistors, capacitors can be arranged in series or in parallel. The rule for adding capacitance is the reverse of adding resistance:

Capacitors in series add like resistors in parallel, and capacitors in parallel add like resistors in series.

For two capacitors in series:

$$\frac{1}{C_t} = \frac{1}{C_1} + \frac{1}{C_2}$$

For two capacitors in parallel:

$$C_t = C_1 + C_2$$

Example

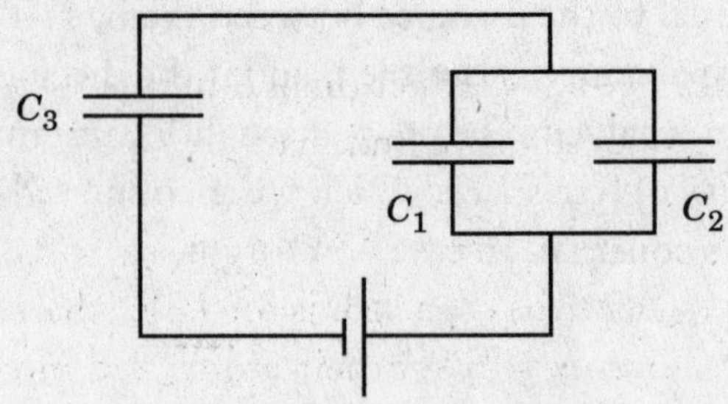

Given $C_1 = 2$ µF, $C_2 = 6$ µF, and $C_3 = 3$ µF, what is the total capacitance of the circuit in the figure above?

First, we find the equivalent capacitance of C_1 and C_2. Since they are in parallel, $C_{1+2} = C_1 + C_2 = 8$ µF. Then C_t is given by:

$$\begin{aligned}\frac{1}{C_t} &= \frac{1}{C_{1+2}} + \frac{1}{C_3} \\ &= \frac{1}{8\ \mu\text{F}} + \frac{1}{3\ \mu\text{F}} \\ &= \frac{11}{24}\ \mu\text{F}\end{aligned}$$

Dielectrics

One way to keep the plates of a capacitor apart is to insert an insulator called a **dielectric** between them. A dielectric increases the capacitance. There is an electric field between the plates of a capacitor. This field polarizes the molecules in the dielectric; that is, some of the electrons in the molecules move to the end of the molecule, near the positive plate:

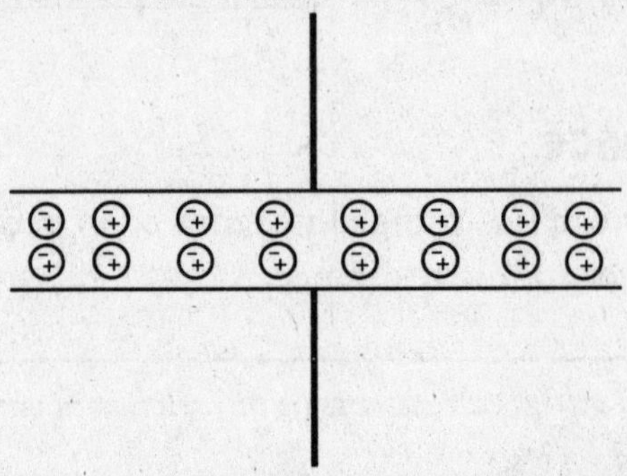

The movement of electrons creates a layer of negative charge by the positive plate and a layer of positive charge by the negative plate. This separation of charge, in turn, creates an electric field in the dielectric that is in the opposite direction of the original field of the capacitor. This reduces the total electric field:

$$E_{\text{new}} = E_{\text{old}} - E_{\text{induced}} = \frac{E}{\kappa}$$

The Greek letter κ is called the **dielectric constant**, and it varies from material to material. For all materials, $\kappa > 1$.

For a parallel-plate capacitor, the reduction in E means that ΔV is also reduced by a factor of κ. Then, since $C = Q/\Delta V$, we find that:

$$C_{\text{new}} = \kappa C_{\text{old}}$$

If the potential difference across the capacitor is too large, then the electric field will be so strong that the electrons escape from their atoms and move toward the positive plate. This **dielectric breakdown** not only discharges the capacitor, but also burns a hole in the dielectric and ruins the capacitor.

Key Formulas

Ohm's Law	$I = \frac{V}{R}$
Resistance	$R = \rho\frac{L}{A}$
Power Dissipated in a Resistor	$P = IV = I^2R = \frac{V^2}{R}$
Heat Dissipated in a Resistor	$H = I^2Rt$
Equivalent Resistance of Two Resistors in Series	$R_t = R_1 + R_2$
Equivalent Resistance of Two Resistors in Parallel	$\frac{1}{R_t} = \frac{1}{R_1} + \frac{1}{R_2}$
Stored Energy of a Capacitor	$U = \frac{1}{2}Q\Delta V = \frac{1}{2}CV^2 = \frac{Q^2}{C} = \frac{1}{2}QV$
Equivalent Capacitance of Two Capacitors in Series	$\frac{1}{C_t} = \frac{1}{C_1} + \frac{1}{C_2}$
Equivalent Capacitance of Two Capacitors in Parallel	$C_t = C_1 + C_2$

Practice Questions

The following choices refer to a circuit consisting of a resistor and a battery.

(A) It is doubled.
(B) It is quadrupled.
(C) It is halved.
(D) It is quartered.
(E) It remains the same.

1. What happens to the current in the resistor when the voltage is doubled and the resistance remains the same?

2. What happens to the power dissipated in the resistor when the resistance is quadrupled and the voltage remains constant?

3. Which of the following affects the resistance of a wire?

I. The material from which it is made
II. The length of the wire
III. The diameter of the wire

(A) I only
(B) II only
(C) I and II only
(D) I and III only
(E) I, II, and III

4. Two resistors, R_1 and R_2, are identical, but the potential difference across R_1 is half the potential difference across R_2. What is the ratio of the current in R_1 to the current in R_2?

(A) $\frac{1}{4}$ (B) $\frac{1}{2}$ (C) 1 (D) 2 (E) 4

Questions 5 and 6 refer to two identical resistors, arranged in parallel.

5. If a third identical resistor is added in parallel, what is the ratio of the new equivalent resistance to the old?

(A) $\frac{4}{9}$ (B) $\frac{2}{3}$ (C) 1 (D) $\frac{3}{2}$ (E) $\frac{9}{4}$

6. Assuming the voltage is kept constant, what is the ratio between the new current and the old current when a third identical resistor is added in parallel with the earlier two?

(A) $\frac{4}{9}$ (B) $\frac{2}{3}$ (C) 1 (D) $\frac{3}{2}$ (E) $\frac{9}{4}$

7. How much heat is produced in a 5 Ω resistor in 10 s when a current of 2 A flows through it?

(A) 2 J (B) 10 J (C) 20 J (D) 100 J (E) 200 J

8. Two identical capacitors are arranged in a circuit. What is the ratio of the equivalent capacitance of the circuit when the capacitors are in series to that when they are in parallel?

(A) $\frac{1}{4}$ (B) $\frac{1}{2}$ (C) 1 (D) 2 (E) 4

9. A potential difference of ΔV exists between two plates of a parallel-plate capacitor with capacitance C. A dielectric with a dielectric constant of κ is then placed between the plates of the capacitor. What is the energy stored in the capacitor?

(A) $\frac{1}{2}(C/\kappa)(\Delta V)^2$

(B) $\frac{1}{2}(\kappa/C)(\Delta V)^2$

(C) $\frac{1}{2}\kappa C(\Delta V)^2$

(D) $\frac{1}{2}\kappa C\Delta V$

(E) $\frac{1}{2}(C/\kappa)\Delta V$

10. A dielectric is inserted into a capacitor while the charge on it is kept constant. What happens to the potential difference and the stored energy?

(A) The potential difference decreases and the stored energy increases.
(B) Both the potential difference and the stored energy increase.
(C) The potential difference increases and the stored energy decreases.
(D) Both the potential difference and the stored energy decrease.
(E) Both the potential difference and the stored energy remain the same.

Explanations

1. (A)

Ohm's Law tells us that current and voltage are directly proportional: doubling the voltage will also double the current.

2. (D)

The power dissipated in a resistor is given by the formula $P = V^2/R$. Since P and R are inversely proportional, multiplying the resistance by four will divide the power by four.

3. (E)

The resistance for a wire is given by the formula $R = \rho\, L/A$, where ρ is the resistivity of the material the wire is made of, L is the length of the wire, and A is the cross-sectional area of the wire.

The value of ρ varies from material to material, so the material the wire is made of does affect the resistance in the wire, which is why we don't wire our houses with glass or wooden wires. The length of the wire, L, also affects the resistance, since the longer a wire gets, the farther the electrons in the wire have to travel. The cross-sectional area, A, and hence the diameter of the wire affects the resistance, since charges have more room to move in a wider wire. Since all three of the statements are true, the answer is (E).

4. (B)

According to Ohm's Law, $V = IR$: current is directly proportional to potential difference. If the potential difference across R_1 is half the potential difference across R_2, and if R_1 and R_2 have the same resistance, then the current through R_1 is half the current through R_2.

5. (B)

The equivalent resistance, R_t, of two identical resistors in parallel is given by the formula:

$$\frac{1}{R_t} = \frac{1}{R} + \frac{1}{R} = \frac{2}{R}$$

$$R_t = \frac{R}{2}$$

The equivalent resistance of three identical resistors in parallel is given by the formula:

$$\frac{1}{R_t} = \frac{1}{R} + \frac{1}{R} + \frac{1}{R} = \frac{3}{R}$$

The ratio, then, between the new resistance and the old is:

$$\frac{R/3}{R/2} = \frac{2}{3}$$

6. (D)
According to Ohm's Law, $V = IR$, current and resistance are inversely proportional. In the previous question, we saw that the new resistance is ⅔ the old resistance. That means that, inversely, the new current is $^3/_2$ times the old resistance.

7. (E)
The power dissipated in a resistor is given by the formula $P = I^2R$, which in this case has a value of 20 W. The heat dissipated in a resistor is given by the formula $H = Pt$: every second, the resistor dissipates 20 J of heat. Since we are looking at a 10 second period, the total heat dissipated is 200 J.

8. (A)
The equivalent capacitance of two capacitors in series is:

$$\frac{1}{C_t} = \frac{1}{C_1} + \frac{1}{C_2} = \frac{2}{C}$$
$$C_t = \frac{C}{2}$$

The equivalent capacitance of two capacitors in parallel is simply the sum of the two capacitors, so $C_t = C_1 + C_2 = 2C$. The ratio between the equivalent capacitance of the two capacitors in series and the two capacitors in parallel is therefore:

$$\frac{C/2}{2C} = \frac{1}{4}$$

9. (C)
The energy stored in a capacitor is $U = \frac{1}{2}C(\Delta V)^2$. When a dielectric with a dielectric constant of κ is inserted between the plates of a capacitor with capacitance C, the new capacitance is $\kappa\ C$. So (C) is the correct answer.

10. (D)
When the dielectric is inserted, the electrons in it create an electric field that opposes the field between the plates of the capacitor. Since electric field and potential difference are directly proportional, this decrease in the electric field causes a decrease in the potential difference.

The energy stored in a capacitor is given by the equation $U = (1/2)Q\Delta V$, so a decrease in the potential difference also leads to a decrease in the stored energy.

Magnetic Force and Fields

Chapter Contents

Permanent Magnets *271*
Magnetic Force on Charges *273*
Magnetic Force on Current-Carrying Wires *278*
The Magnetic Field Due to a Current . . . *279*
Key Formulas *281*
Practice Questions *282*
Explanations *286*

When we think "magnet," we might envision those things we stick on our fridge door. It may be a bit confusing, then, to discover that magnetism is closely related to electricity. In fact, there is a single force—the **electromagnetic force**—that governs the behavior of both magnets and electric charges.

We have seen that there is a reciprocal relationship between electric charges and electric fields: electric charges generate electric fields and electric fields exert a force on electric charges. Similarly, there is a reciprocal relationship between a *moving* electric charge and a **magnetic field**: a moving electric charge creates a magnetic field, and magnetic fields exert a force that compels charges to move.

Bearing this reciprocal relationship in mind, we can make sense of **electromagnets**, the on-off magnets you see, for instance, lifting and dropping cars at the junkyard. The magnetism in these electromagnets is generated by a current running through the magnet that can be turned on and off at will. However, we still haven't explained how any of this connects with the **permanent magnets** we stick to our fridge door.

Permanent Magnets

Like all other materials, permanent magnets are made up of atoms that have electrons orbiting a nucleus of protons and neutrons. In moving around the nucleus, these electrons create miniscule magnetic fields. In most materials, these tiny fields all point in different random directions, so the bulk material does not have a magnetic field. But in permanent magnets, the fields are all lined up together, and so the material is

magnetized. Materials, like iron, that can be magnetized, are called **ferromagnetic**. Other materials, like wood or plastic, which have atoms that won't line up in this fashion and can't create magnetic fields, are called **paramagnetic**. In the same way that semiconductors straddle the fence between conductors and insulators, there are **diamagnetic** materials that are not quite ferromagnetic or paramagnetic. Diamagnetic materials produce weak magnetic fields.

Magnetic Field Lines

Permanent magnets—and electromagnets—have positive and negative poles, often called "north" and "south," respectively. Like electric field lines, magnetic field lines go from the positive, or north, pole, toward the negative, or south, pole. For example, the magnetic field of a bar magnet looks like this:

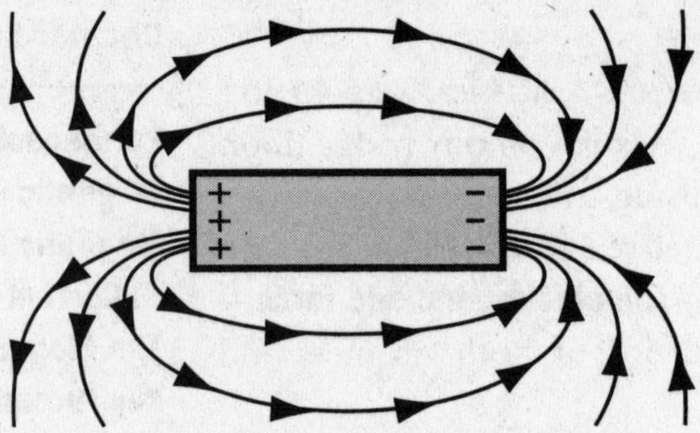

A horseshoe-shaped magnet creates a magnetic field like this:

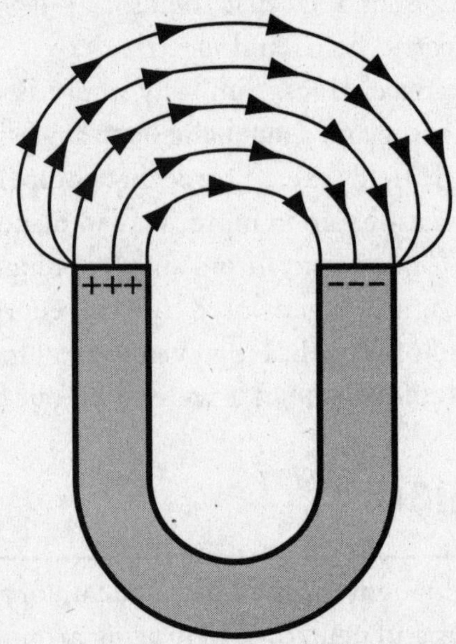

It is possible to do a nifty experiment to see these magnetic field lines by scattering iron fillings around a permanent magnet—the filings will move to trace the lines.

The Earth's Magnetic Field

The Earth itself acts like a huge bar magnet. The iron in the Earth's crust has become magnetically **polarized** over the years. The presence of a magnetic field about the Earth allows us to use compasses that point northward, and creates a spectacular aurora over the northern and southern skies. But the magnetism of the Earth is quite complicated, and is still an active subject of research for geologists, so let us turn to the simpler cases of idealized charges and constant magnetic fields.

Magnetic Force on Charges

The questions on magnetism that you'll find on the SAT II will deal for the most part with the reciprocal relationship between magnetic fields and moving charges. Generally, these questions will expect you to predict the motion of a charge through a magnetic field, or to calculate the magnitude of the magnetic force or magnetic field strength necessary to move a charge in a certain manner.

Calculating Magnetic Force

A magnetic field exerts a force on a moving charge. Given a magnetic field, $\boldsymbol{B}$, and a charge, q, moving with velocity, $\boldsymbol{v}$, the force, $\boldsymbol{F}$, on the charge is:

$$\boldsymbol{F} = q(\boldsymbol{v} \times \boldsymbol{B})$$

Magnetic field strength is measured in **teslas** (T), where $1 \text{ T} = 1 \text{ N/A} \cdot \text{m}$.

You'll notice that the force on a moving particle is calculated as a cross product of the particle's velocity and the magnetic field's strength. You can determine the direction of the $\boldsymbol{v} \times \boldsymbol{B}$ vector by using the right-hand rule as follows: point the fingers of your right hand in the direction of the velocity vector and then curl them around to point in the direction of the magnetic field vector. The direction in which your thumb points gives you the direction of the $\boldsymbol{v} \times \boldsymbol{B}$ vector.

However, though q is a scalar quantity, it can affect the direction of the force vector. If q has a negative value, then $q(\boldsymbol{v} \times \boldsymbol{B})$ has a negative value, and so the force vector will point in a direction opposite from what the right-hand rule might tell you.

You can calculate the magnitude of the magnetic force without using the right-hand rule, so long as you know the angle, θ, between the velocity vector and the magnetic field vector:

$$F = qvB \sin \theta$$

The $\sin \theta$ term is important, because it lets us see very quickly that there is no force if a charge moves parallel to a magnetic field, and that the greatest force occurs when a charge moves perpendicular to the magnetic field.

Example

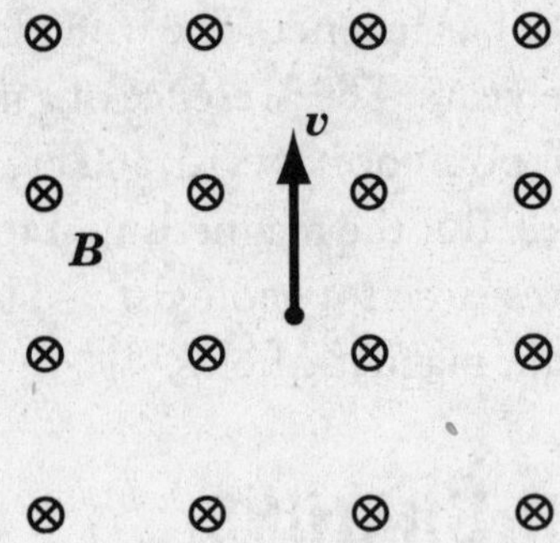

A charge of 5 C moves upward at 2 m/s in a magnetic field of 0.3 T that points into the page. What is the magnitude and direction of the force that the charge experiences?

The cross product of $\boldsymbol{v} \times \boldsymbol{B}$ is a vector of magnitude $qvB \sin \theta = 3$ N. Following the right-hand rule, point your fingers toward the top of the page, and then curl them around so that they point into the page. You'll find that your thumb is pointing left, which is the direction of the $\boldsymbol{v} \times \boldsymbol{B}$ vector. Because the value of q is positive, the force acting on the particle will also be in the leftward direction.

A Quick Note on Vectors Going In and Out of the Page

The magnetic field lines illustrated in this example that are going into the page are represented by circles with an "x" inscribed in them. Vector lines pointing out of the page are represented by circles with a dot in them. You can think about these symbols as arrows appearing from in front or behind: from in front, you see the conical tip of the arrow, and from behind you see the fletching of the four feathers in an "x" shape.

Trajectory of Charges in a Magnetic Field

The direction of the force on a moving charge depends on the direction of its velocity. As its velocity changes, so will its direction. *The magnitude of the velocity will not change*, but charged particles moving in a magnetic field experience nonlinear trajectories.

When the Velocity Vector and Magnetic Field Lines are Perpendicular

In the example above, we saw that a force of 3 N would pull the charged particle to the left. However, as soon as the particle begins to move, the velocity vector changes, and so must the force acting on the particle. As long as the particle's velocity vector is at a right angle to the magnetic field lines, the force vector will be at right angles to both the velocity vector and the magnetic field. As we saw in the chapter on circular motion and gravitation, a force that always acts perpendicular to the velocity of an object causes that object to move in circular motion.

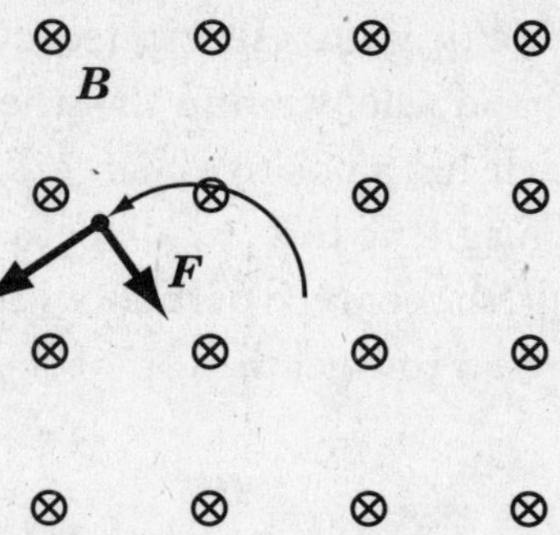

Because the velocity vector and the magnetic field lines are at right angles to one another, the magnitude of the magnetic force is $F = qvB$. Further, because the magnetic force pulls the particle in a circular path, it is a centripetal force that fits the equation $F = mv^2/r$. Combining these two equations, we can solve for r to determine the radius of the circle of the charged particle's orbit:

$$F = qvB = \frac{mv^2}{r}$$
$$r = \frac{mv}{qB}$$

When the Velocity Vector and Magnetic Field Lines are Parallel

The magnetic force acting on a moving charged particle is the cross product of the velocity vector and the magnetic field vector, so when these two vectors are parallel, the magnetic force acting on them is zero.

When the Velocity Vector and Magnetic Field Lines are Neither Perpendicular nor Parallel

The easiest way to deal with a velocity vector that is neither parallel nor perpendicular to a magnetic field is to break it into components that are perpendicular and parallel to the magnetic field.

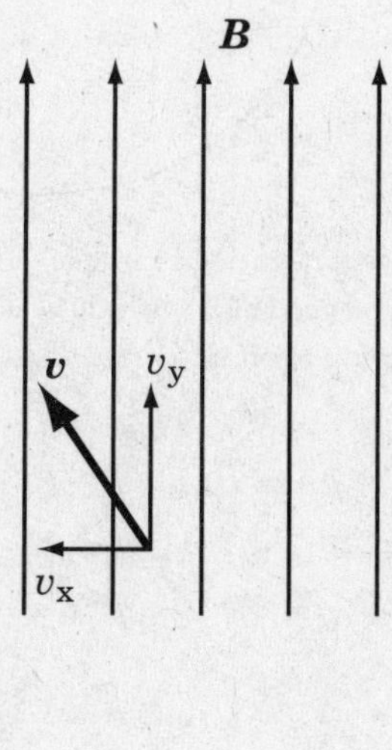

The x-component of the velocity vector illustrated above will move with circular motion. Applying the right-hand rule, we find that the force will be directed downward into the page if the particle has a positive charge. The y-component of the velocity vector will experience no magnetic force at all, because it is moving parallel to the magnetic field lines. As a result, the charged particle will move in a helix pattern, spiraling around while also moving up toward the top of the page. Its trajectory will look something like this:

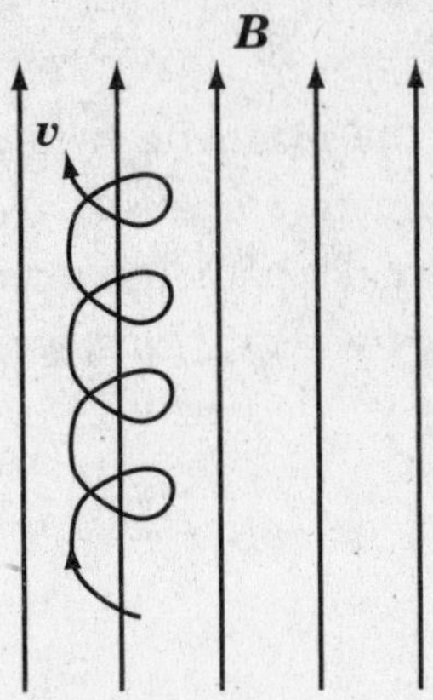

If the particle has a positive charge it will move in a counterclockwise direction, and if it has a negative charge it will move in a clockwise direction.

Example

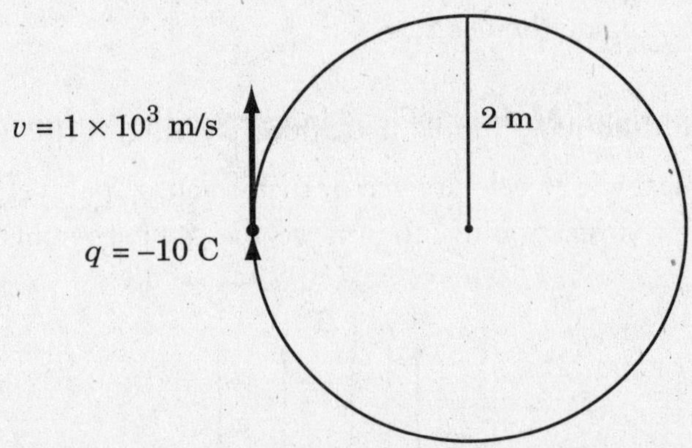

A particle of mass 1×10^{-5} kg has a negative charge of –10 C. It moves in a clockwise circular pattern of radius 2 m at a speed of 1×10^{3} m/s. What is the magnitude and direction of the magnetic field acting upon it?

We know the velocity, mass, charge, and radius of the orbit of the particle. These four quantities are related to magnetic field strength, B, in the equation $r = mv/qB$. By rearranging this equation, we can solve for B:

$$\begin{aligned} B &= \frac{mv}{qr} \\ &= \frac{(1 \times 10^{-5}\ \text{kg})(1 \times 10^{3}\ \text{m/s})}{(-10\ \text{C})(2\ \text{m})} \\ &= -5 \times 10^{-4}\ \text{T} \end{aligned}$$

Now we just need to determine the direction of the magnetic field. To find the direction, apply the right-hand rule in reverse: point your thumb in the direction of the force—toward the center of the circle—and then stretch your fingers in the direction of the velocity. When you curl your fingers around, they will point out of the page. However, because the particle has a negative charge, the magnetic field has the opposite direction—into the page.

Magnetic Fields and Electric Fields Overlapping

There's no reason why a magnetic field and an electric field can't operate in the same place. Both will exert a force on a moving charge. Figuring out the total force exerted on the charge is pretty straightforward: you simply add the force exerted by the magnetic field to the force exerted by the electric field. Let's look at an example.

Example

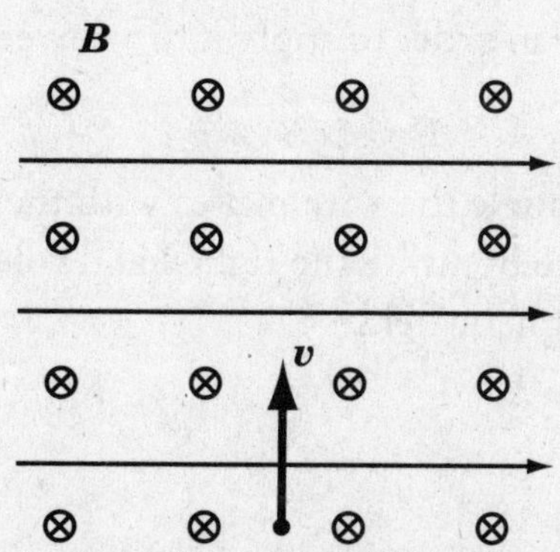

A particle with a positive charge of 3 C moves upward at a speed of 10 m/s. It passes simultaneously through a magnetic field of 0.2 T directed into the page and an electric field of 2 N/C directed to the right. How is the motion of the particle affected?

Answering this question is a matter of calculating the force exerted by the magnetic field and the force exerted by the electric field, and then adding them together. The force exerted by the magnetic field is:

$$\begin{aligned} F = qvB &= (3\ \text{C})(10\ \text{m/s})(0.2\ \text{T}) \\ &= 6\ \text{N} \end{aligned}$$

Using the right-hand-rule, we find that this force is directed to the left. The force exerted by the electric field is:

$$\begin{aligned} F = qE &= (3\ \text{C})(2\ \text{N/C}) \\ &= 6\ \text{N} \end{aligned}$$

This force is directed to the right. In sum, we have one force of 6 N pushing the particle to the left and one force of 6 N pushing the particle to the right. The net force on the particle is zero, so it continues toward the top of the page with a constant velocity of 10 m/s.

Magnetic Force on Current-Carrying Wires

Since an electric current is just a bunch of moving charges, wires carrying current will be subject to a force when in a magnetic field. When dealing with a current in a wire, we obviously can't use units of q and v. However, qv can equally be expressed in terms of $I\ell$, where I is the current in a wire, and ℓ is the length, in meters, of the wire—both qv and $I\ell$ are expressed in units of C · m/s. So we can reformulate the equation for the magnitude of a magnetic force in order to apply it to a current-carrying wire:

$$F = I\ell B \sin\theta$$

In this formulation, θ is the angle the wire makes with the magnetic field. We determine the direction of the force by using the right-hand rule between the direction of the current and that of the magnetic field.

Example

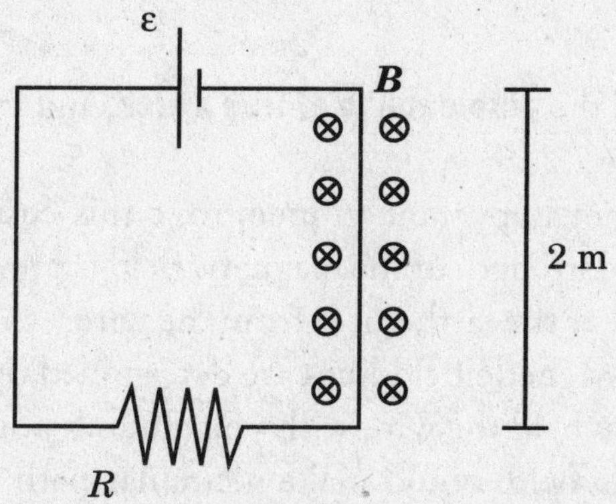

In the figure above, a magnetic field of 4.0 × 10^{-5} T is applied locally to one part of an electric circuit with a 5 Ω resistor and a voltage of 30 V. The length of wire to which the magnetic field is applied is 2 m. What is the magnetic force acting on that stretch of wire?

We are only interested in the stretch of wire on the right, where the current flows in a downward direction. The direction of current is perpendicular to the magnetic field, which is directed into the page, so we know the magnetic force will have a magnitude of $F = I\ell B$, and will be directed to the right.

We have been told the magnetic field strength and the length of the wire, but we need to calculate the current in the wire. We know the circuit has a voltage of 30 V and a resistance of 5 Ω, so calculating the current is just a matter of applying Ohm's Law:

$$I = \frac{V}{R} = \frac{30\ \text{V}}{5\ \Omega}$$
$$= 6\ \text{A}$$

Now that we know the current, we can simply plug numbers into the equation for the force of a magnetic field on a current-carrying wire:

$$F = I\ell B = (6\ \text{A})(2\ \text{m})(4.0 \times 10^{-5}\ \text{T})$$
$$= 4.8 \times 10^{-4}\ \text{N}$$

The Magnetic Field Due to a Current

So far we have discussed the effect a magnetic field has on a moving charge, but we have not discussed the reverse: the fact that a moving charge, or current, can generate a magnetic field. There's no time like the present, so let's get to it.

The magnetic field created by a single moving charge is actually quite complicated, and is not covered by the SAT II. However, the magnetic field created by a long straight wire carrying a current, I, is relatively simple, and is fair game for the SAT II.

The magnetic field strength is given by:

$$B = \frac{\mu_0}{2\pi}\frac{I}{r}$$

The constant μ_0 is called the **permeability of free space**, and in a vacuum it has a value of about $4\pi \times 10^{7}$ N/A^2.

For the SAT II, it's not important to memorize this equation exactly. It's more important to note that the strength of the magnetic field is proportional to the strength of the current and is weaker the farther it is from the wire.

The direction of the magnetic field lines are determined by an alternate version of the right-hand rule: if you held the wire with your thumb pointing in the direction of the current, the magnetic field would make a circular path around the wire, in the direction that your fingers curl.

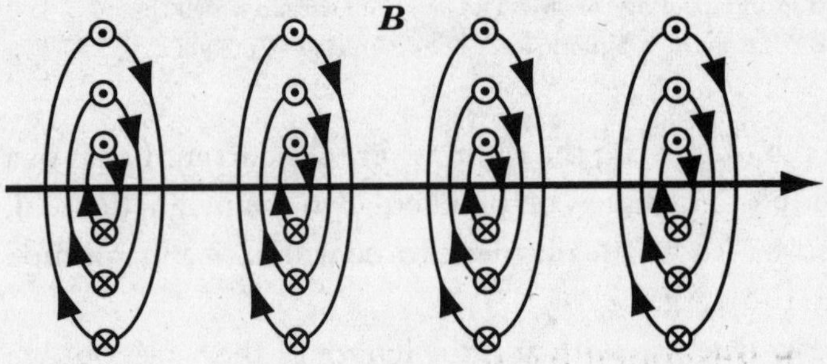

wire seen from above

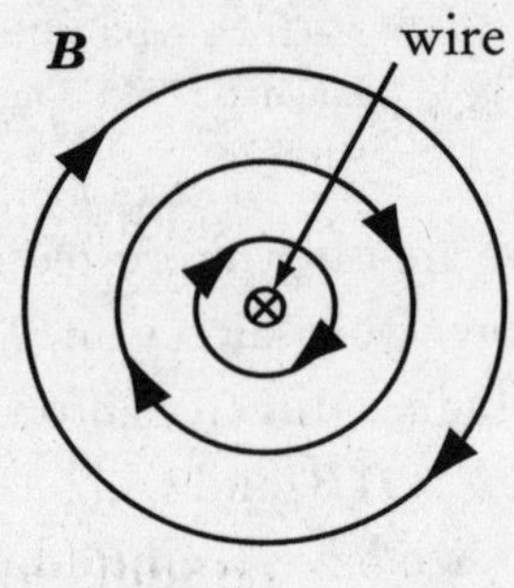

wire seen from behind

Example

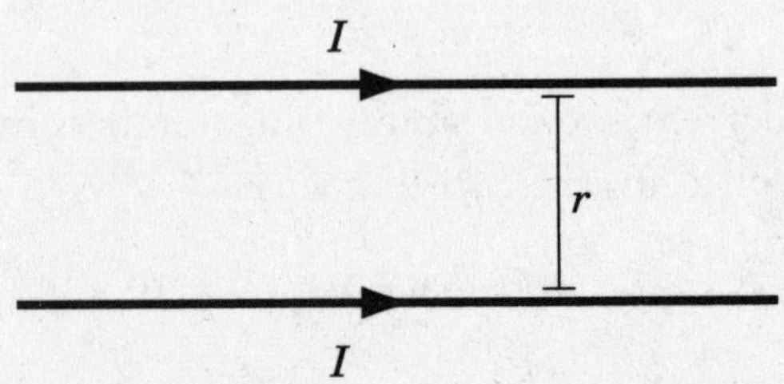

Two parallel long straight wires carrying a current I stand a distance r apart. What force does one wire exert on the other?

Consider the magnetic field created by the bottom wire as it affects the top wire. According to the right-hand rule, the magnetic field will point out of the page, and will have a strength of $B = (\mu_0 I)/(2\pi r)$.

The force exerted by the bottom wire on the top wire is $F = I\ell B$. If we substitute in for B the equation we derived above, we find the force per unit length is:

$$\frac{F}{\ell} = IB = \frac{I^2 \mu_0}{2\pi r}$$

Using the right-hand rule once more, we find that the force pulls the top wire down toward the bottom wire.

We can apply the same equations to find that the top wire pulls the bottom wire up. In other words, the two wires generate magnetic fields that pull one another toward each other. Interestingly, the fact that each wire exerts an opposite force on the other is further evidence of Newton's Third Law.

Key Formulas

Magnetic Force on a Moving Charge	$\boldsymbol{F} = q(\boldsymbol{v} \times \boldsymbol{B})$
Magnitude of the Magnetic Force on a Moving Charge	$F = qvB \sin\theta$
Radius of the Circle Described by a Charged Particle Moving Perpendicular to a Magnetic Field	$r = \frac{mv}{qB}$
Magnetic Force on a Current	$F = I\ell B \sin\theta$
Magnetic Field Created by a Current	$B = \frac{\mu_0}{2\pi}\frac{I}{r}$

Practice Questions

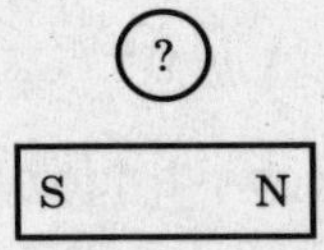

1. The pointer on a compass is the south pole of a small magnet. If a compass were placed next to a bar magnet, as shown above, in what direction would the pointer point?

(A) 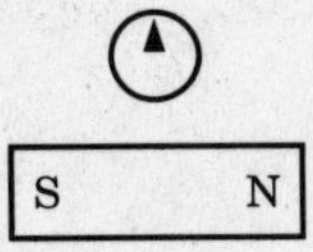

(B) 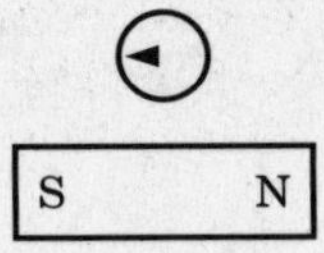

(C) 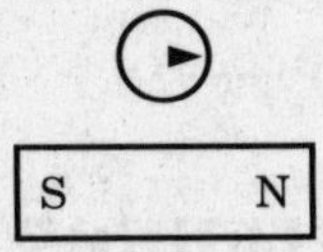

(D) 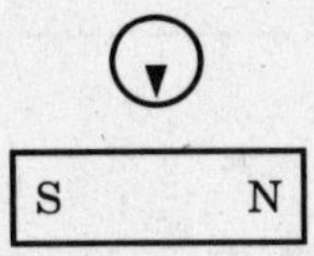

(E) 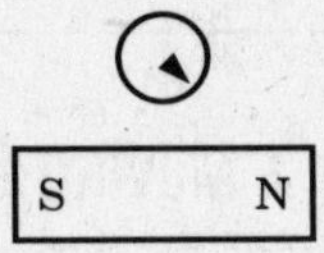

2. A positively charged particle in a uniform magnetic field moves in a circular path in the clockwise direction, parallel to the plane of the page. In what direction do the magnetic field lines point?

(A) Out of the page
(B) Into the page
(C) To the left
(D) To the right
(E) In a clockwise pattern parallel to the plane of the page

3. What should one do to maximize the magnitude of the magnetic force acting on a charged particle?

I. Maximize the magnetic field the charged particle is in
II. Minimize the particle's velocity
III. Ensure that the particle is moving in the same direction as the magnetic field lines

(A) I only
(B) I and II only
(C) I and III only
(D) II and III only
(E) I, II, and III

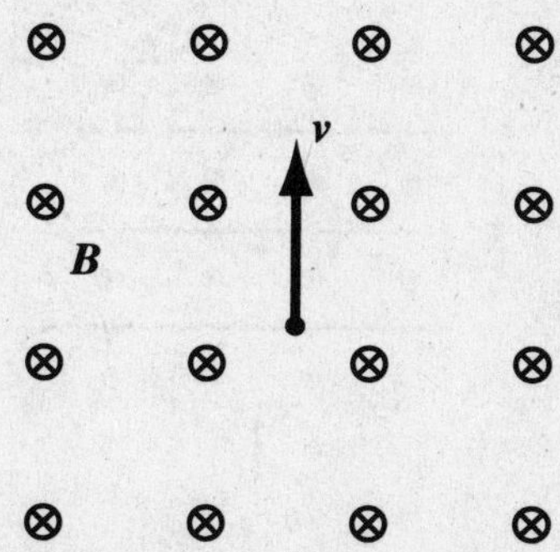

4. What is the magnetic force experienced by a negatively charged particle of 1.0 C that is moving upward at a velocity of 2.0×10^3 m/s in a magnetic field of strength 4.0×10^{-4} T, directed into the page?

(A) 0.8 N to the left
(B) 0.8 N to the right
(C) 2.0×10^{-7} N to the left
(D) 2.0×10^{-7} N to the right
(E) 5.0×10^6 N to the left

5. A charged particle is moving in a circular orbit in a magnetic field. If the strength of the magnetic field doubles, how does the radius of the particle's orbit change?

(A) It is quartered.
(B) It is halved.
(C) It is unchanged.
(D) It is doubled.
(E) It is quadrupled.

6. Which of the following is not a possible trajectory of a charged particle in a uniform magnetic field?

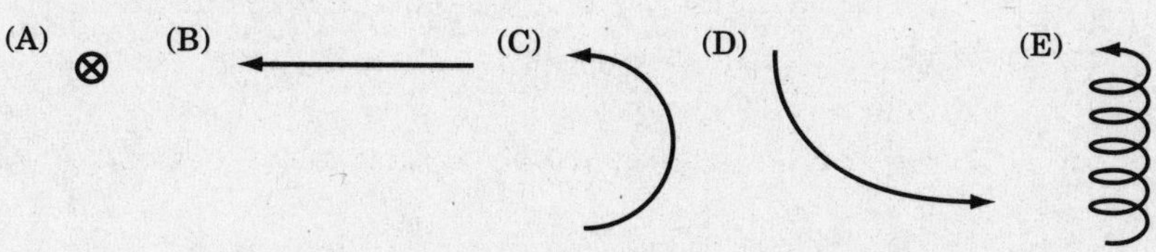

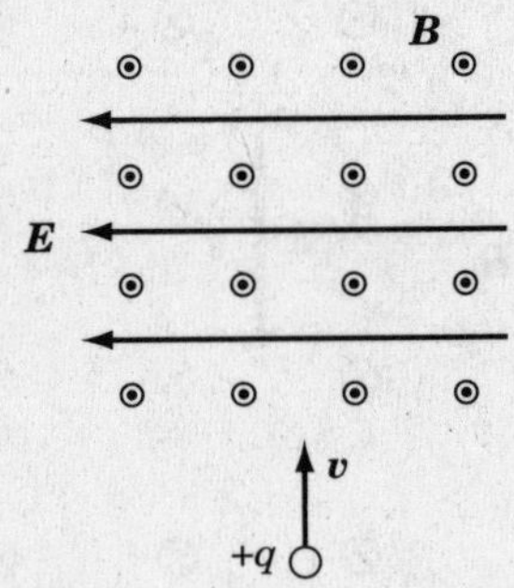

7. A positively charged particle of 2.0 C moves upward into an area where both a magnetic field and an electric field are acting. The magnetic field has a magnitude of 4.0×10^{-4} T and the electric field has a magnitude of 0.1 N/C. At what velocity must the particle be moving if it is not deflected when it enters this area?

   (A) 4.0×10^{-3} m/s
   (B) 125 m/s
   (C) 250 m/s
   (D) 500 m/s
   (E) The particle will be deflected to the left regardless of its velocity

8. A current-carrying wire in a magnetic field is subject to a magnetic force. If the current in the wire is doubled, what happens to the magnetic force acting on the wire?

   (A) It is quartered.
   (B) It is halved.
   (C) It is unchanged.
   (D) It is doubled.
   (E) It is quadrupled.

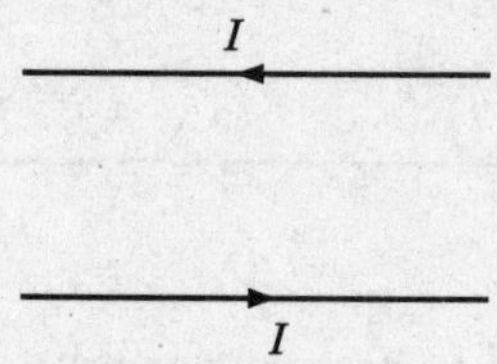

9. Two wires carry current in opposite directions. Which of the following graphs represents the magnetic force acting on each wire?

(A) 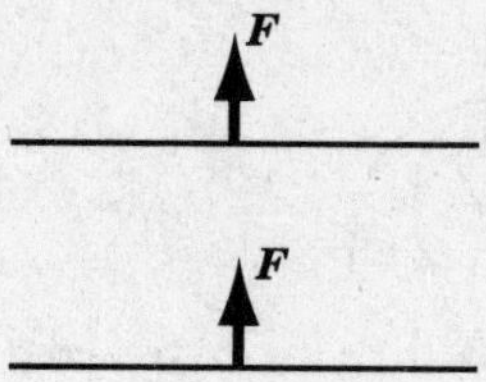

(B) 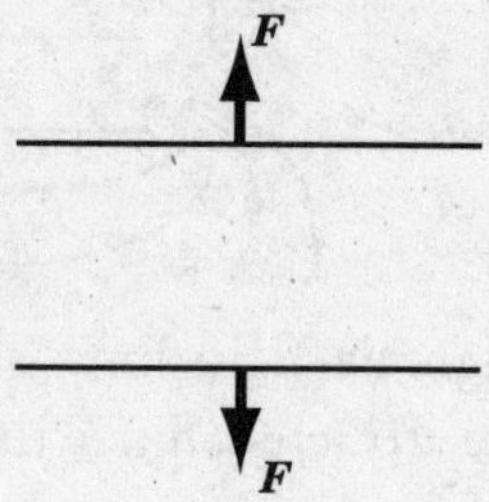

(C) 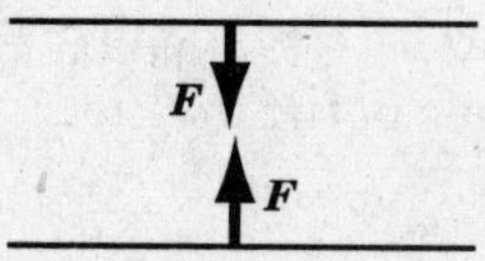

(D) 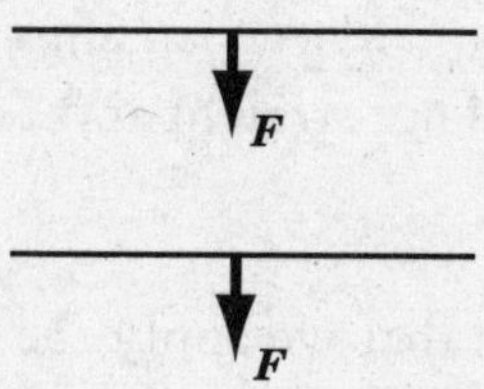

(E) There is no net force acting on either wire.

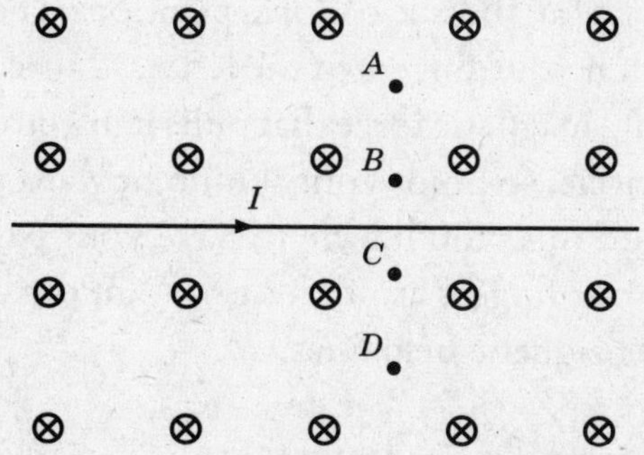

10. A current carrying wire passes through a uniform magnetic field, as shown above. At which point is the magnetic field the strongest?

(A) *A*
(B) *B*
(C) *C*
(D) *D*
(E) The magnetic field strength is uniform throughout.

Explanations

1. (C)

To solve this problem, it is helpful to remember how the magnetic field lines around a bar magnet look:

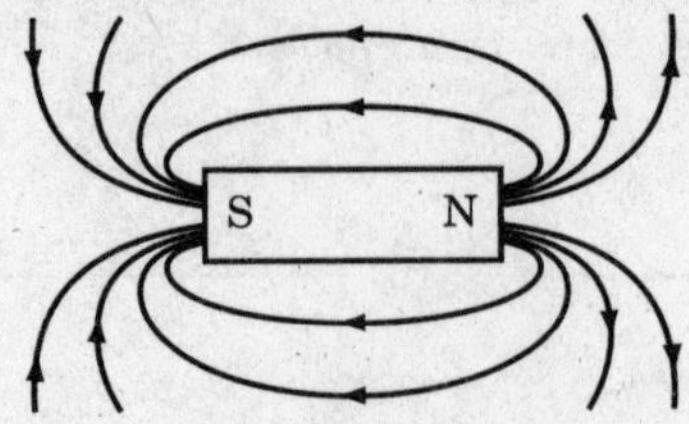

The arrows of the magnetic field lines show the direction toward which a north magnetic pole would be attracted. Since the compass needle is a south magnetic pole, it's attracted in the opposite direction of the field lines.

Note that the correct answer is (C), and not (E). The magnet points along the magnetic field lines, and not straight at the north pole of the magnet.

2. (A)

This question demands that we apply the right-hand rule backward. Force, velocity, and magnetic strength are related by the formula $\boldsymbol{F} = q(\boldsymbol{v} \times \boldsymbol{B})$. Since the particle is positively charged, q is positive, and the $\boldsymbol{F}$ vector will point in the same direction as the $\boldsymbol{v} \times \boldsymbol{B}$ vector.

Let's imagine the particle at the six o'clock position. That means the particle is moving to the left, so stretch your fingers in the leftward direction. It's moving under the influence of a centripetal magnetic force that pulls it in a circle. This force is directed toward the center of the circle, so point your thumb upward toward the center of the imaginary clock face. To do this, you'll have to have your palm facing up, and you'll find that when you curl your fingers around, they point out of the plane of the page. That's the direction of the magnetic field lines.

3. (B)

The magnetic force experienced by a moving particle is given by the formula $\boldsymbol{F} = q(\boldsymbol{v} \times \boldsymbol{B})$. Since $\boldsymbol{F}$ is proportional to the cross product of $\boldsymbol{v}$ and $\boldsymbol{B}$, we can maximize $\boldsymbol{F}$ by maximizing $\boldsymbol{v}$ and $\boldsymbol{B}$, and by ensuring that $\boldsymbol{v}$ and $\boldsymbol{B}$ are perpendicular to one another. According to these requirements, only statement I will maximize the magnetic force: both statements II and III will serve to minimize the magnetic force.

4. (B)

Magnetic force is related to charge, velocity, and magnetic field strength by the formula $\boldsymbol{F} = q(\boldsymbol{v} \times \boldsymbol{B})$. Since the velocity vector and the magnetic field strength vector are perpendicular, we can calculate the magnitude of the magnetic force quite easily:

$$F = qvB = (-1.0\text{ C})(2.0 \times 10^3\text{ m/s})(4.0 \times 10^{-4}\text{ T}) \\ = -0.8\text{ N}$$

The minus sign in the answer signifies the fact that we are dealing with a negatively charged particle. That means that the force is in the opposite direction of the $\boldsymbol{v} \times \boldsymbol{B}$ vector. We can determine the direction of this vector using the right-hand rule: point your fingers upward in the direction of the $\boldsymbol{v}$ vector and curl them downward in the direction of the $\boldsymbol{B}$ vector; your thumb will be pointing to the left. Since we're dealing with a negatively charged particle, it will experience a force directed to the right.

5. (B)
If the particle is moving in a circular orbit, its velocity is perpendicular to the magnetic field lines, and so the magnetic force acting on the particle has a magnitude given by $F = qvB$. Since this force pulls the particle in a circular orbit, we can also describe the force with the formula for centripetal force: $F = mv^2/r$. By equating these two formulas, we can get an expression for orbital radius, r, in terms of magnetic field strength, B:

$$qvB = \frac{mv^2}{r}$$
$$B = \frac{mv}{qr}$$

Since magnetic field strength is inversely proportional to orbital radius, doubling the magnetic field strength means halving the orbital radius.

6. (D)
When a charged particle moves in the direction of the magnetic field lines, it experiences no magnetic force, and so continues in a straight line, as depicted in (A) and (B). When a charged particle moves perpendicular to the magnetic field lines, it moves in a circle, as depicted in (C). When a charged particle has a trajectory that is neither perfectly parallel nor perfectly perpendicular to the magnetic field lines, it moves in a helix pattern, as depicted in (E). However, there are no circumstances in which a particle that remains in a uniform magnetic field goes from a curved trajectory to a straight trajectory, as in (D).

7. (C)
The electric field will pull the charged particle to the left with a force of magnitude $F = qE$. The magnetic field will exert a force of magnitude $F = qvB$. The direction of this force can be determined using the right-hand rule: extend your fingers upward in the direction of the velocity vector, then point them out of the page in the direction of the magnetic field vector. You will find your thumb is pointing to the right, and so a positively charged particle will experience a magnetic force to the right.

If the particle is to move at a constant velocity, then the leftward electric force must be equal in magnitude to the rightward magnetic force, so that the two cancel each other out:

$$qvB = qE$$
$$v = \frac{E}{B} = \frac{0.1 \text{ N/C}}{4.0 \times 10^{-4} \text{ T}}$$
$$= 250 \text{ m/s}$$

8. (D)

The magnetic force, F, due to a magnetic field, B, on a current-carrying wire of current I and length ℓ has a magnitude $F = I\ell B$. Since F is directly proportional to I, doubling the current will also double the force.

9. (B)

Each wire exerts a magnetic force on the other wire. Let's begin by determining what force the lower wire exerts on the upper wire. You can determine the direction of the magnetic field of the lower wire by pointing the thumb of your right hand in the direction of the current, and wrapping your fingers into a fist. This shows that the magnetic field forms concentric clockwise circles around the wire, so that, at the upper wire, the magnetic field will be coming out of the page. Next, we can use the right-hand rule to calculate the direction of the force on the upper wire. Point your fingers in the direction of the current of the upper wire, and then curl them upward in the direction of the magnetic field. You will find you thumb pointing up, away from the lower wire: this is the direction of the force on the upper wire.

If you want to be certain, you can repeat this exercise with the lower wire. The easiest thing to do, however, is to note that the currents in the two wires run in opposite directions, so whatever happens to the upper wire, the reverse will happen to the lower wire. Since an upward force is exerted on the upper wire, downward force will be exerted on the lower wire. The resulting answer, then, is (B).

10. (C)

There are two magnetic fields in this question: the uniform magnetic field and the magnetic field generated by the current-carrying wire. The uniform magnetic field is the same throughout, pointing downward. The magnetic field due to the current-carrying wire forms concentric clockwise circles around the wire, so that they point out of the page above the wire and into the page below the wire. That means that at points *A* and *B*, the upward magnetic field of the current-carrying wire will counteract the downward uniform magnetic field. At points *C* and *D*, the downward magnetic field of the current-carrying wire will complement the downward uniform magnetic field. Since the magnetic field due to a current-carrying wire is stronger at points closer to the wire, the magnetic field will be strongest at point *C*.

Electromagnetic Induction

Chapter Contents

Motional Emf *289*
Faraday's Law *291*
Applications *296*
Key Formulas *297*
Practice Questions *297*
Explanations *299*

CHARGES MOVING IN A MAGNETIC FIELD create an electric field, just as charges moving in an electric field create a magnetic field. This is called **electromagnetic induction**. Induction provides the basis of everyday technology like transformers on power lines and electric generators.

On average, the SAT II asks only one question about electromagnetic induction. However, less than half of test-takers usually get this question right, so if you get the hang of this material, you'll be separating yourself from the crowd. On the whole, this question will be qualitative, with only a minimum of calculation involved.

Motional Emf

Consider the bar in the figure below. It has length ℓ and moves at speed v to the right in magnetic field $\boldsymbol{B}$, which is directed into the page.

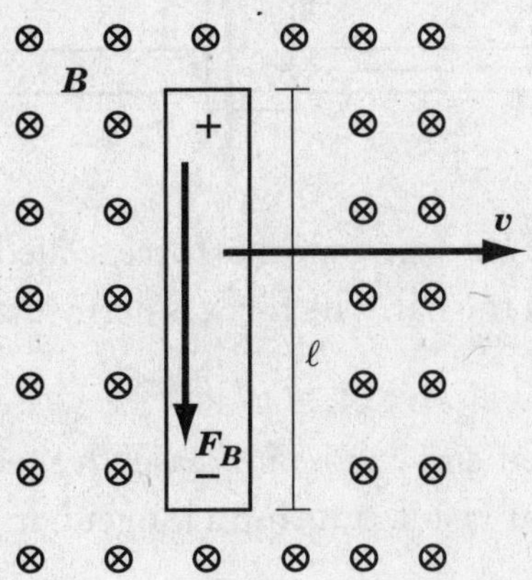

Electromagnetic Induction

The field exerts a magnetic force on the free electrons in the bar. That force is $\boldsymbol{F}_B = q(\boldsymbol{v} \times \boldsymbol{B})$: using the right-hand rule, you will find that the $\boldsymbol{v} \times \boldsymbol{B}$ vector is directed upward along the bar, but since electrons are negatively charged, the magnetic force acting upon them is directed downward. As a result, electrons flow to the bottom of the bar, and the bottom becomes negatively charged while the top becomes positively charged.

The separation of charge in the rod creates an electric field within the bar in the downward direction, since the top of the bar is positively charged and the bottom of the bar is negatively charged. The force from the electric field, $\boldsymbol{F}_E = q\boldsymbol{E}$, pulls negative charges upward while the force from the magnetic field pulls negative charges downward. Initially, the magnetic field is much stronger than the electric field, but as more electrons are drawn to the bottom of the bar, the electric field becomes increasingly stronger. When the two fields are of equal strength, the forces balance one another out, halting the flow of electrons in the bar. This takes place when:

$$\boldsymbol{F}_E = -\boldsymbol{F}_B$$
$$q\boldsymbol{E} = -q(\boldsymbol{v} \times \boldsymbol{B})$$
$$\boldsymbol{E} = -(\boldsymbol{v} \times \boldsymbol{B})$$

Induced Current and Motional Emf

The electric field in the metal bar causes a potential difference of $V = E\ell = vB\ell$. If the bar slides along metal rails, as in the figure below, a closed circuit is set up with current flowing in the counterclockwise direction, up the bar and then around the metal rail back to the bottom of the bar. This is called an **induced current**.

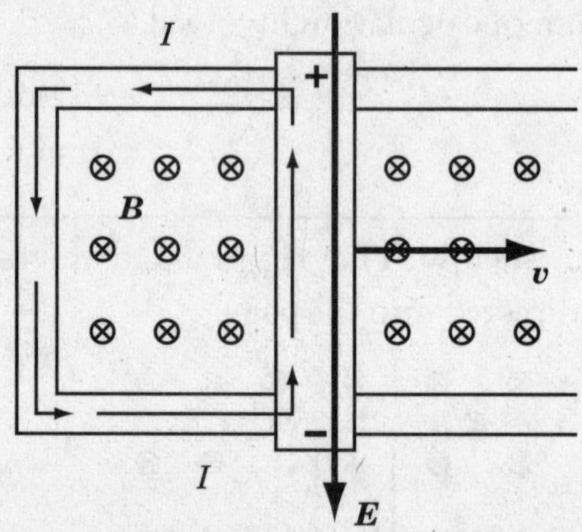

The moving bar is a source of an electromotive force, called **motional emf**, since the emf is generated by the motion of the bar. The force is defined as:

$$\mathcal{E} = vB\ell$$

The magnitude of the induced emf can be increased by increasing the strength of the magnetic field, moving the bar faster, or using a longer bar.

Example

A bar of length 10 cm slides along metal rails at a speed of 5 m/s in a magnetic field of 0.1 T. What is the motional emf induced in the bar and rails?

Now that we've defined motional emf, solving this problem is simply a matter of plugging numbers into the appropriate equation:

$$\begin{aligned}\mathcal{E} &= vB\ell = (5\ \text{m/s})(0.1\ \text{T})(0.1\ \text{m}) \\ &= 5 \times 10^{-2}\ \text{V}\end{aligned}$$

Faraday's Law

Moving a conductor through a magnetic field is just one way of inducing an electric current. A more common way of inducing current, which we will examine now, is by changing the magnetic flux through a circuit.

Magnetic Flux

The **magnetic flux**, Φ, through an area, $\boldsymbol{A}$, is the product of the area and the magnetic field perpendicular to it:

$$\Phi = \boldsymbol{B} \cdot \boldsymbol{A} = BA\cos\theta$$

The $\boldsymbol{A}$ vector is perpendicular to the area, with a magnitude equal to the area in question. If we imagine flux graphically, it is a measure of the number and length of flux lines passing through a certain area.

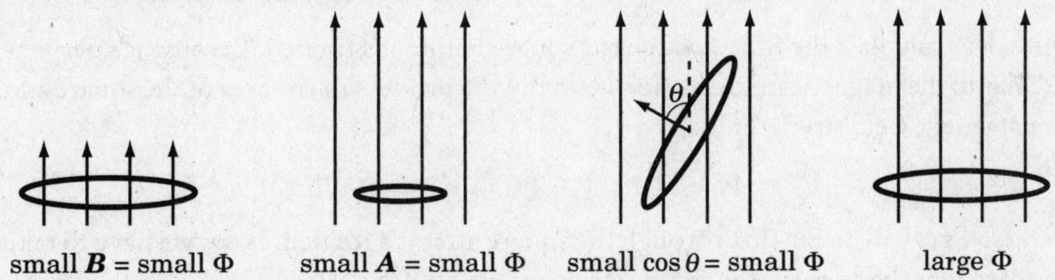

The unit of flux is the **weber** (Wb), where 1 Wb = 1 T · m².

Changing Magnetic Flux

As we will see shortly, $\Delta\Phi$ is more important than Φ: our interest is in how flux changes, not in its fixed value. The formula for magnetic flux suggests that there are three ways of changing magnetic flux:

1. **Change the magnetic field strength:** By sliding a permanent magnet back and forth, the magnetic field in a certain area will fluctuate. We will look at this phenomenon a bit later in this chapter.
2. **Change the area:** When a bar slides on rails in a magnetic field, as in our discussion of motional emf, the square bounded by the bar and the rails gets larger. As it grows, the number of field lines passing through it increases, and thus the flux increases as the bar moves.
3. **Rotate the area, changing the angle between the area and the magnetic field:** When the area is perpendicular to the magnetic field, the magnetic flux will simply be the product of the magnitudes of the area and the magnetic field strength. However, as you rotate the area so that it is at an angle to the magnetic field, fewer field lines will pass through it, and so the magnetic flux will decrease.

Example

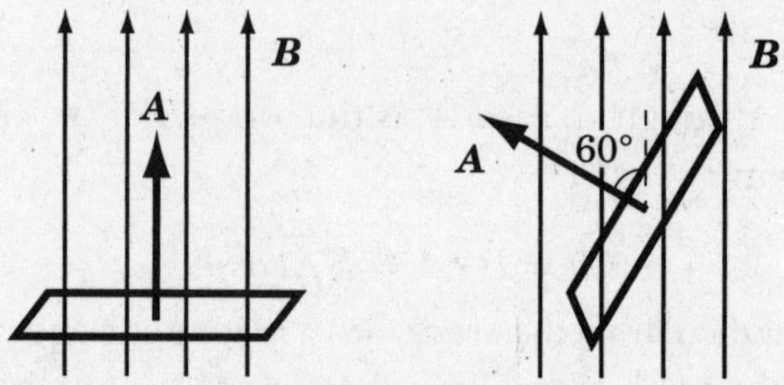

A square with sides of length 2 m is perpendicular to a magnetic field of strength 10 T. If the square is rotated by 60°, what is the change in magnetic flux through the square?

First, let's calculate the flux through the square before it's rotated. Because it's perpendicular to the magnetic field, the flux is simply the product of the area of the square and the magnetic field strength:

$$\Phi_i = BA = (10\ \text{T})(4\ \text{m}^2) = 40\ \text{Wb}$$

Next, let's calculate the flux through the square after it's rotated. Now we have to take into account the fact that the square is at an angle of 60°:

$$\Phi_f = BA\cos 60^\circ = (10\ \text{T})(4\ \text{m}^2)(0.5) = 20\ \text{Wb}$$

So the change in magnetic flux is :

$$\Phi_f - \Phi_i = 20\ \text{Wb} - 40\ \text{Wb} = -20\text{Wb}$$

The magnetic flux decreases because, as the square is rotated, fewer magnetic field lines can pass through it.

Faraday's Law

We have seen earlier that a bar sliding along rails is a source of induced emf. We have also seen that it is a source of changing magnetic flux: as it moves, it changes the area bounded by the bar and the rails. The English scientist Michael Faraday discovered that this is no coincidence: induced emf is a measure of the change in magnetic flux over time.

$$|\mathcal{E}| = \frac{\Delta\Phi}{\Delta t}$$

This formula is called **Faraday's Law.**

Equivalence of Faraday's Law with $\mathcal{E} = vB\ell$

The earlier example of a metal bar rolling along tracks to induce a current is just a particular case of the more general Faraday's Law. If the bar is moving at a constant velocity $\boldsymbol{v}$, at which it covers a distance Δx in a time Δt, then:

$$\begin{aligned}\mathcal{E} &= vB\ell = (\frac{\Delta x}{\Delta t})B\ell \\ &= \frac{\Delta(xB\ell)}{\Delta t}\end{aligned}$$

Because $\Delta(x\ell)$ is the same thing as ΔA, we get:

$$\frac{\Delta(xB\ell)}{\Delta t} = \frac{\Delta(BA)}{\Delta t} = \frac{\Delta\Phi}{\Delta t}$$

Lenz's Law

Faraday's Law tells us that a change in magnetic flux induces a current in a loop of conducting material. However, it doesn't tell us in what direction that current flows. According to **Lenz's Law**, the current flows so that it opposes the change in magnetic flux by creating its own magnetic field. Using the right-hand rule, we point our thumb in the opposite direction of the change in magnetic flux, and the direction in which our fingers wrap into a fist indicates the direction in which current flows.

Lenz's Law is included in Faraday's Law by introducing a minus sign:

$$\mathcal{E} = -\frac{\Delta\Phi}{\Delta t}$$

Example

The square in the previous example, with sides of length 2 m and in a magnetic field of strength 10 T, is rotated by 60° in the course of 4 s. What is the induced emf in the square? In what direction does the current flow?

We established in the previous example that the change in flux as the square is rotated is –20 Wb. Knowing that it takes 4 seconds to rotate the square, we can calculate the induced emf using Lenz's Law:

$$\begin{aligned}\mathcal{E} &= -\frac{\Delta\Phi}{\Delta t} = -\frac{-20\text{ Wb}}{4\text{ s}} \\ &= 5\text{ V}\end{aligned}$$

As for determining the direction of the current, we first need to determine the direction of the change in magnetic flux. From the diagram we saw in the previous example, we see that the magnetic field lines, $\boldsymbol{B}$, move in the upward direction. Because we rotated the square so that it is no longer perpendicular to the field lines, we decreased the magnetic flux. Saying that the magnetic flux changed by –20 Wb is equivalent to saying that the flux changed by 20 Wb in the downward direction.

The direction of the current must be such that it opposes the downward change in flux. In other words, the current must have an "upward" direction. Point the thumb of your right hand upward and wrap your fingers into a fist, and you will find that they curl in a counterclockwise direction. This is the direction of the current flow.

Conservation of Energy

Lenz's Law is really a special case of the conservation of energy. Consider again the bar sliding on rails. What would happen if the induced current did not oppose the change in flux?

Since the current flows counterclockwise, the current in the bar flows toward the top of the page. Thus, the magnetic field exerts a leftward force on the bar, opposing the external force driving it to the right. If the current flowed in the other direction, the force on the bar would be to the right. The bar would accelerate, increasing in speed and kinetic energy, without any input of external energy. Energy would not be conserved, and we know this cannot happen.

Changing the Flux by Changing the Magnetic Field

So far, we have changed the magnetic flux in two ways: by increasing the size of the circuit and by rotating the circuit in a constant magnetic field. A third way is to keep the circuit still and change the field. If a permanent magnet moves toward a loop of wire, the magnetic field at the loop changes.

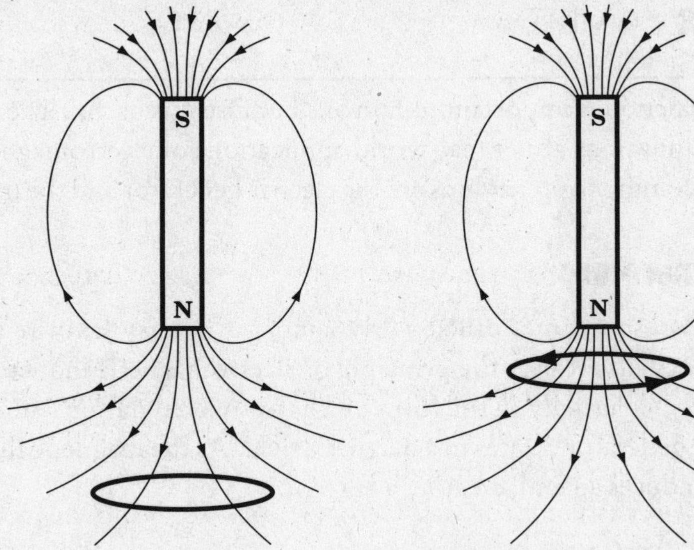

downward flux increases

Remember that field lines come *out* of the north (N) pole of a magnet. As the magnet moves closer to the loop, the flux in the downward direction increases. By Lenz's Law, the current must then be in the upward direction. Using the right-hand rule, we find that the current will flow counterclockwise as viewed from above.

As the middle of the magnet passes through the loop, the flux decreases in the downward direction. A decrease in the magnitude of the downward flux is the same as a change in flux in the upward direction, so at this point the change in flux is upward, and the current will change direction and flow clockwise.

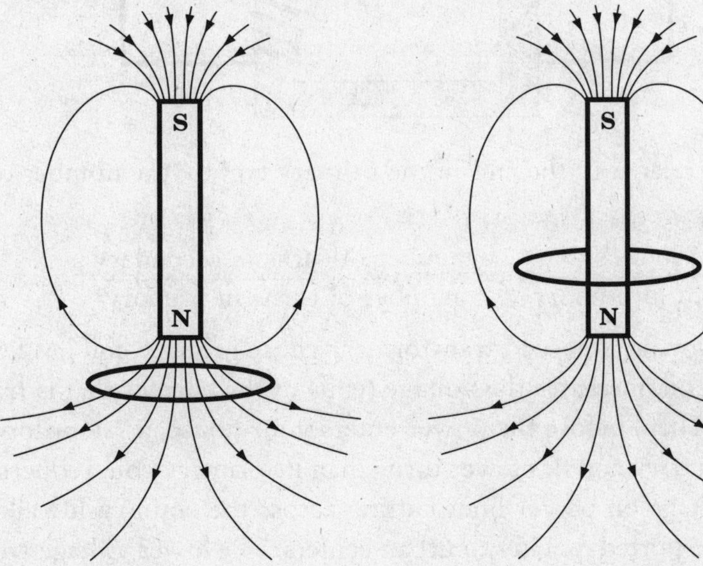

downward flux decreases

It doesn't matter whether the magnet or the loop is moving, so long as one is moving relative to the other.

Applications

Electromagnetic induction is important to humans because it is useful. The SAT II has been known to ask questions about real-world applications of electromagnetic induction. The two most common applications are the electric generator and the transformer.

The Electric Generator

The **electric generator**, sometimes called a "dynamo," is a noisy favorite at outdoor events that need electricity. It uses the principle of electromagnetic induction to convert mechanical energy—usually in the form of a gas-powered motor—into electrical energy. A coil in the generator rotates in a magnetic field. As the magnetic flux through the coil changes, it induces an emf, creating a current.

The Transformer

The **transformer** converts current of one voltage to current of another voltage. A simple transformer consists of two coils wrapped around an iron core. Transformers rely on the property of **mutual induction**: the change in current in one coil induces an emf in another coil. The coil with the applied current is called the primary coil, and the coil with the induced emf is called the secondary coil.

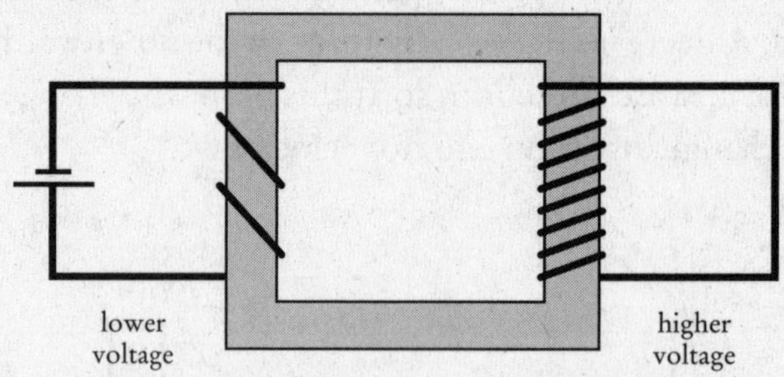

The induced emf is related to the emf in the primary coil by the number of turns in each coil:

$$\frac{\text{emf in secondary}}{\text{emf in primary}} = \frac{\text{number of turns in secondary}}{\text{number of turns in primary}}$$

Outside a power plant, a "step-up" transformer, whose primary coil has fewer turns than its secondary coil, increases the voltage (emf) of the current that is transported along power lines. Then, before the power enters your house, a "step-down" transformer, whose secondary coil has fewer turns than its primary coil, reduces the voltage. The higher voltage on power lines cutting across the countryside allows more electricity to be transported quickly to urban centers. The lower voltage within your house renders the electricity safer: children get a shock when they stick a paper clip in an electric socket, but they don't burn to a crisp.

Key Formulas

Motional Emf	$\mathcal{E} = vB\ell$
Magnetic Flux	$\Phi = \boldsymbol{B} \cdot \boldsymbol{A} = BA\cos\theta$
Faraday's Law / Lenz's Law	$\mathcal{E} = -\dfrac{\Delta\Phi}{\Delta t}$
Emf Induced in a Transformer	$\dfrac{\text{emf in secondary}}{\text{emf in primary}} = \dfrac{\text{number of turns in secondary}}{\text{number of turns in primary}}$

Practice Questions

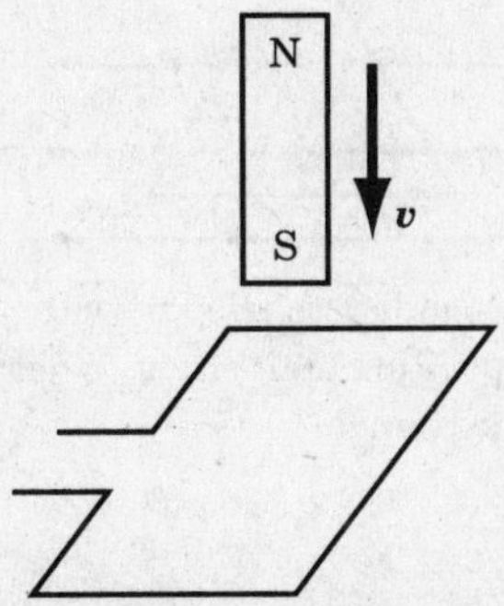

1. A bar magnet is moving downward, south pole first, toward a loop of wire. Which of the following best describes the current induced in the wire?

   (A) Clockwise, as viewed from above
   (B) Counterclockwise, as viewed from above
   (C) The current alternates
   (D) There is no current induced in the wire
   (E) The direction of the current cannot be determined from the information given here

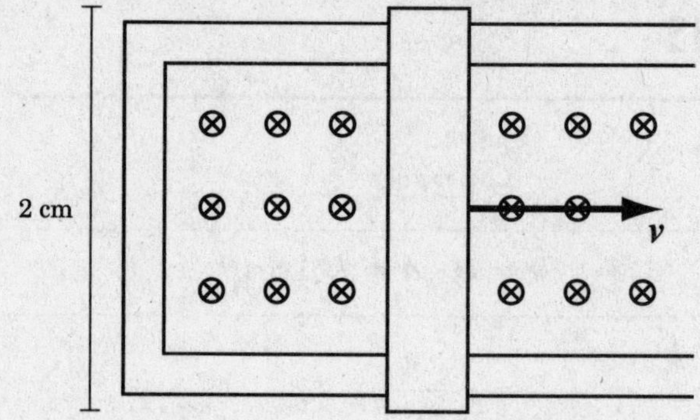

2. A bar of length 2 cm slides along metal rails at a speed of 1 cm/s. The bar and rails are in a magnetic field of 2 T, pointing out of the page. What is the induced emf in the bar and rails?

(A) 2×10^{-5} V
(B) 2×10^{-4} V
(C) 4×10^{-4} V
(D) 2×10^{-3} V
(E) 4×10^{-3} V

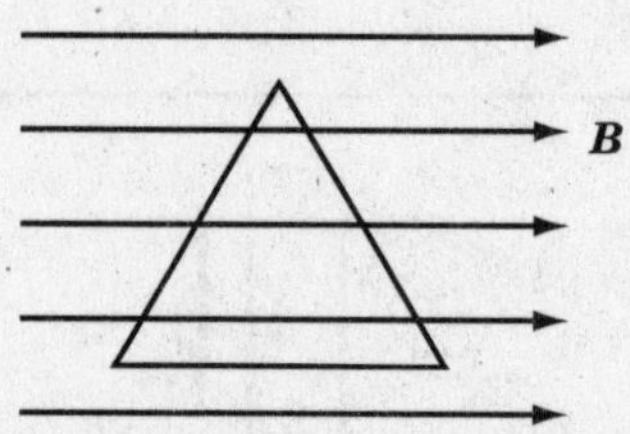

3. A wire in the shape of an equilateral triangle with sides of length 1.00 m sits in a magnetic field of 2.00 T, pointing to the right. What is the magnitude of the magnetic flux through the triangle?

(A) 0 Wb
(B) 1.00 Wb
(C) 1.73 Wb
(D) 2.00 Wb
(E) 3.46 Wb

4. A device that transforms mechanical energy into electrical energy is called a:

(A) Transformer
(B) Inductor
(C) Motor
(D) Galvanometer
(E) Generator

5. A wire carrying 5.0 V is applied to a transformer. The primary coil has 5 turns and the secondary coil has 10 turns. What is the emf induced in the secondary coil?

(A) 0.50 V (B) 5.0 V (C) 10 V (D) 50 V (E) 100 V

Explanations

1. (A)

The magnet moving downward creates a downward magnetic flux. Using the right-hand rule, we find that the current related to a downward flux flows clockwise.

2. (C)

The induced emf, $\mathcal{E}$, from a bar of length ℓ moving along rails at a speed v in a magnetic field of magnitude B is given by the formula $\mathcal{E} = vB\ell$. Since we are given the values for v, B, and ℓ, this is simply a matter of plugging numbers into a formula. Remember that we need to convert to units of meters:

$$\begin{aligned}\mathcal{E} &= vB\ell = (0.01\ \text{m/s})(2\ \text{T})(0.02\ \text{m}) \\ &= 4 \times 10^{-4}\ \text{V}\end{aligned}$$

3. (A)

Magnetic flux is given by the formula $\Phi = BA\cos\theta$, where B is the magnetic field strength, A is the area, and θ is the angle between the magnetic field vector and a vector pointing perpendicular to the area. In this case, the value of θ is 90°, and since cos 90° = 0, the magnetic flux through the area is zero.

A more intuitive way of thinking about this problem is to see that, since the magnetic field lines pass across the triangle rather than through it, there are no magnetic field lines passing through the area, and so the flux is equal to zero.

4. (E)

A generator, also called a dynamo, is normally run by a gas-powered motor that rotates a coil in a magnetic field, thereby inducing emf and generating an electric current.

5. (C)

The relationship between the voltage in a primary coil and in a secondary coil is given by the formula:

$$\frac{\text{emf in secondary}}{\text{emf in primary}} = \frac{\text{number of turns in secondary}}{\text{number of turns in primary}}$$

Since the primary has an emf of 5.0 V, and the secondary has twice as many turns as the primary, the secondary has an emf of 10 V.

Wave Phenomena

Chapter Contents

Periodic Motion 301
Wave Motion 302
Transverse and Longitudinal Waves . . . 305
Superposition. 308
Standing Waves and Resonance 310
The Doppler Effect 313
Key Formulas 314
Practice Questions 315
Explanations 317

WAVE PHENOMENA OCCUR ALMOST anywhere there is periodic motion. We have already encountered such periodic motion in the back-and-forth movement of pendulums and masses on a spring and with the cyclic orbits of objects in a gravitational field. The physics of waves is also central in explaining how light and sound work. Anything from a violin string to a drum skin to a wine glass can make a sound, suggesting that there are few things in the world that cannot produce wave phenomena. We find waves in the air, in our bodies, in earthquakes, in computers—and, if we're surfers, at the beach.

Periodic Motion

We've already covered some of the basics of periodic motion with our discussion of a mass on a spring back in Chapter 5. When the end of a spring is stretched or compressed, the spring exerts a force so as to return the mass at its end to its **equilibrium position**. The maximum displacement of the mass from its equilibrium position during each cycle is the **amplitude** of the oscillation. One **cycle** of periodic motion is completed each time the spring returns to its starting point, and the time it takes to complete one cycle is the **period**, T, of oscillation. The **frequency**, f, of the spring's motion is the number of cycles it completes per second. A high frequency means each period is relatively short, so frequency and period are inversely proportional:

$$f = \frac{1}{T}$$

Frequency is measured in units of **hertz** (Hz), where 1 Hz = 1 cycle. The unit of hertz is technically defined as an inverse second (s^{-1}) and can be applied to any process that measures how frequently a certain event recurs.

We can summarize all of these concepts in an equation describing the position of the mass at the end of a spring, x, as a function of time, t:

$$x(t) = A\sin(2\pi ft) = A\sin(2\pi\frac{t}{T})$$

In this equation, A is the amplitude, f is the frequency, and T is the period of the oscillation. It is useful to think of each of these quantities in terms of a graph plotting the mass's displacement over time.

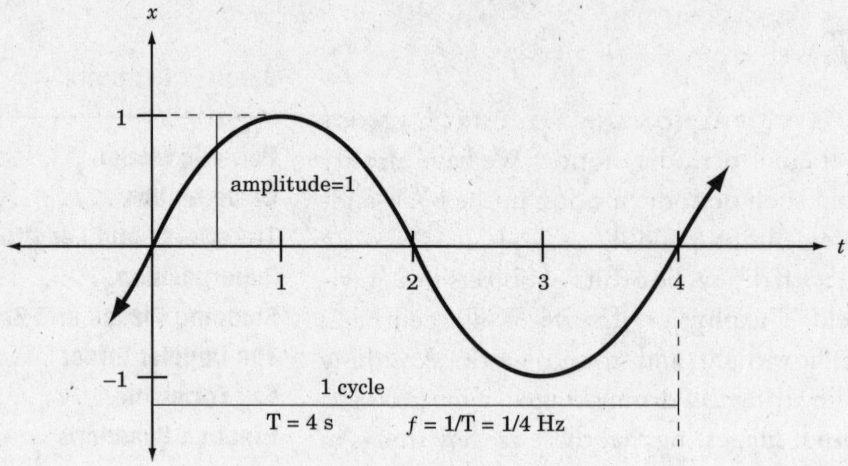

The graph shows us an object moving back and forth withina distance of 1 m from its equilibrium position. It reaches its equilibrium position of $x = 0$ at $t = 0$, $t = 2$, and $t = 4$.

Note that one cycle is completed not at $t = 2$ but at $t = 4$. Though the object is at the same position, $x = 0$, at $t = 2$ as it was at $t = 0$, it is moving in the opposite direction. At the beginning of a new cycle, both the position and the velocity must be identical to the position and velocity at the beginning of the previous cycle.

Wave Motion

Because both masses suspended on a spring and waves at the beach exhibit periodic motion, we can use much of the same vocabulary and mathematical tools to describe both. However, there is a significant difference: waves are extended in space, while a mass on a spring just oscillates back and forth in one place.

The Basics

A familiar and concrete example of wave motion is the "wave" spectators create at sporting events by standing up and sitting down at appropriate intervals. Each person stands up just as that person's neighbor stands up, transmitting a form of energy all the way around the stadium. There are two things worth noting about how this works:

1. **Waves are transmitted through a medium:** The energy and the "wave" are both created by the successive action of people standing up and down. If there were no people in the stadium, no wave could exist and no energy could be transmitted. We call the people at the stadium, the water at the beach, the air molecules transmitting sound, etc., the medium through which these waves are transmitted.
2. **The medium itself is not propagated:** For the "wave" to work, each person in the stadium only needs to stand up and sit back down. The "wave" travels around the stadium, but the people do not.

Think of waves as a means of transmitting energy over a distance. One object can transmit energy to another object without either object, or anything in between them, being permanently displaced. For instance, if a friend shouts to you across a room, the sound of your friend's voice is carried as a wave of agitated air particles. However, no air particle has to travel the distance between your friend and your ear for you to hear the shout. The air is a medium, and it serves to propagate sound energy without itself having to move. Waves are so widespread and important because they transmit energy through matter without permanently displacing the matter through which they move.

Crests, Troughs, and Wavelength

Waves travel in **crests** and **troughs**, although, for reasons we will discuss shortly, we call them **compressions** and **rarefactions** when dealing with **longitudinal waves**. The terms "crest" and "trough" are used in physics just as you would use them to refer to waves on the sea: the crest of a wave is where the wave is at its maximum positive displacement from the equilibrium position, and the trough is where it is at its maximum negative displacement. Therefore, the displacement at the crest is the wave's amplitude, while the displacement at the trough is the negative amplitude. There is one crest and one trough in every cycle of a wave. The **wavelength**, λ, of a traveling wave is the distance between two successive crests or two successive troughs.

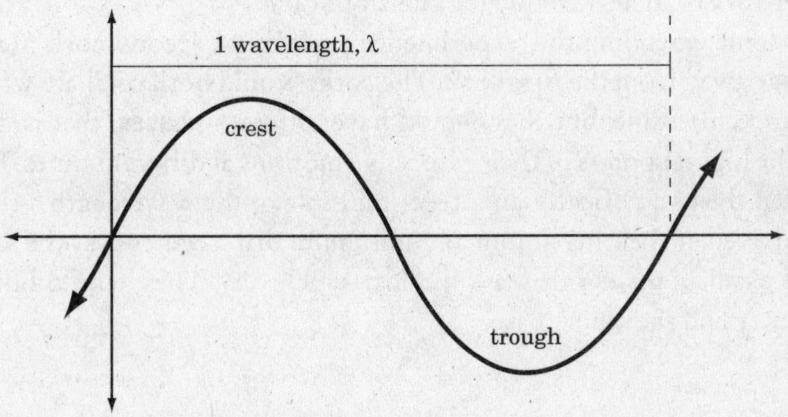

Wave Speed

The period of oscillation, T, is simply the time between the arrival of successive wave crests or wave troughs at a given point. In one period, then, the crests or troughs travel exactly one wavelength. Therefore, if we are given the period and wavelength, or the frequency and wavelength, of a particular wave, we can calculate the **wave speed**, v:

$$v = \frac{\lambda}{T} = \lambda f$$

Example

Ernst attaches a stretched string to a mass that oscillates up and down once every half second, sending waves out across the string. He notices that each time the mass reaches the maximum positive displacement of its oscillation, the last wave crest has just reached a bead attached to the string 1.25 m away. What are the frequency, wavelength, and speed of the waves?

Determining frequency: The oscillation of the mass on the spring determines the oscillation of the string, so the period and frequency of the mass's oscillation are the same as those of the string. The period of oscillation of the string is $T = 0.5$ s, since the string oscillates up and down once every half second. The frequency is just the reciprocal of the period: $f = 1/T = 2$ Hz.

Determining wavelength: The maximum positive displacement of the mass's oscillation signifies a wave crest. Since each crest is 1.25 m apart, the wavelength, λ, is 1.25 m.

Determining wave speed: Given the frequency and the wavelength, we can also calculate the wave speed: $v = f\lambda = (2 \text{ Hz})(1.25 \text{ m}) = 2.5 \text{ m/s}$.

Phase

Imagine placing a floating cork in the sea so that it bobs up and down in the waves. The up-and-down oscillation of the cork is just like that of a mass suspended from a spring: it oscillates with a particular frequency and amplitude.

Now imagine extending this experiment by placing a second cork in the water a small distance away from the first cork. The corks would both oscillate with the same frequency and amplitude, but they would have different **phases**: that is, they would each reach the highest points of their respective motions at different times. If, however, you separated the two corks by an integer multiple of the wavelength—that is, if the two corks arrived at their maximum and minimum displacements at the same time—they would oscillate up and down in perfect synchrony. They would both have the same frequency and the same phase.

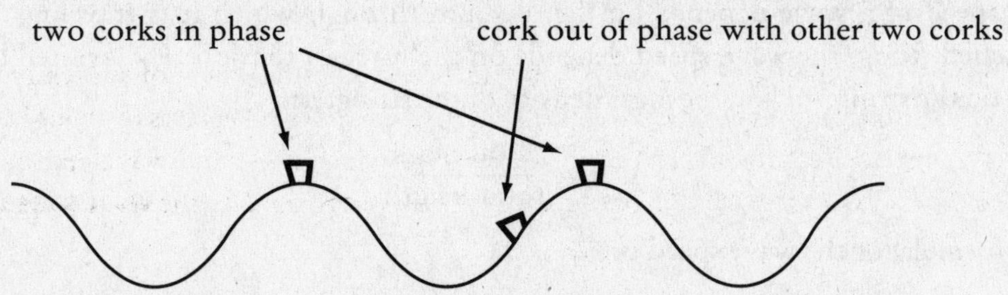

Transverse Waves and Longitudinal Waves

There are two major kinds of waves: **transverse waves** and **longitudinal waves**. The medium transmitting transverse waves oscillates in a direction perpendicular to the direction the wave is traveling. A good example is waves on water: the water oscillates up and down while transmitting a wave horizontally. Other common examples include a wave on a string and electromagnetic waves. By contrast, the medium transmitting longitudinal waves oscillates in a direction parallel to the direction the wave is traveling. The most commonly discussed form of longitudinal waves is sound.

Transverse Waves: Waves on a String

Imagine—or better yet, go grab some twine and set up—a length of string stretched between two posts so that it is taut. Each point on the string is just like a mass on a spring: its equilibrium position lies on the straight line between the two posts, and if it is plucked away from its resting position, the string will exert a force to restore its equilibrium position, causing periodic oscillations. A string is more complicated than a simple mass on a spring, however, since the oscillation of each point influences nearby points along the string. Plucking a string at one end causes periodic vibrations that eventually travel down the whole length of the string. Now imagine detaching one end of the string from the pole and connecting it to a mass on a spring, which oscillates up and down, as in the figure below. The oscillation at one end of the string creates waves that propagate, or travel, down the length of the string. These are called, appropriately, **traveling waves**. Don't let this name confuse you: the string itself only moves up and down, returning to its starting point once per cycle. The wave travels, but the medium—the string, in this case—only oscillates up and down.

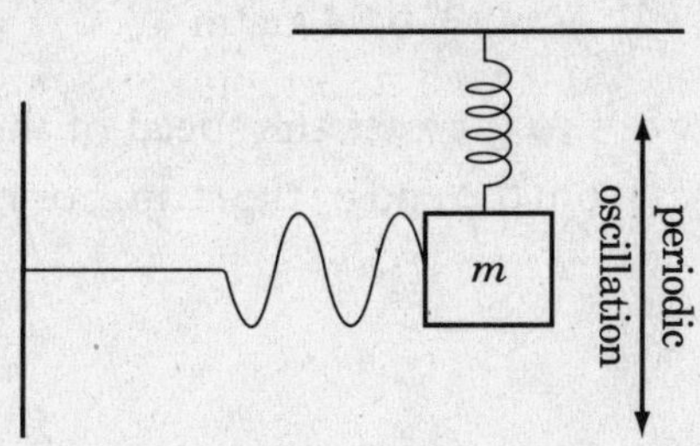

The speed of a wave depends on the medium through which it is traveling. For a stretched string, the wave speed depends on the force of tension, F_T, exerted by the pole on the string, and on the mass density of the string, μ:

$$\mu = \frac{\text{total mass}}{\text{total length}}$$

The formula for the wave speed is:

$$v = \sqrt{\frac{F_T}{\mu}}$$

Example

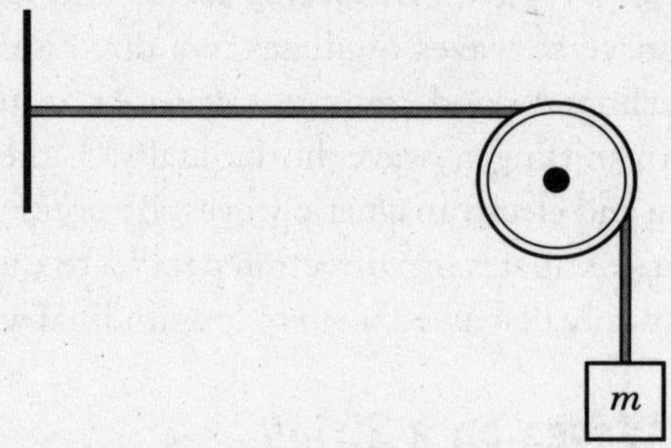

A string is tied to a pole at one end and 100 g mass at the other, and wound over a pulley. The string's mass is 100 g, and it is 2.5 m long. If the string is plucked, at what speed do the waves travel along the string? How could you make the waves travel faster? Assume the acceleration due to gravity is 10 m/s².

Since the formula for the speed of a wave on a string is expressed in terms of the mass density of the string, we'll need to calculate the mass density before we can calculate the wave speed.

$$\mu = \frac{0.1\text{ kg}}{2.5\text{ m}} = 0.04\text{ kg/m}$$

The tension in the string is the force of gravity pulling down on the weight, $F_T = mg = (0.1\text{ kg})(10\text{ m/s}^2) = 1\text{ N}$. The equation for calculating the speed of a wave on a string is:

$$v = \sqrt{\frac{F_T}{\mu}} = \sqrt{\frac{1\text{ N}}{0.04\text{ kg/m}}} = 5\text{ m/s}$$

This equation suggests two ways to increase the speed of the waves: increase the tension by hanging a heavier mass from the end of the string, or replace the string with one that is less dense.

Longitudinal Waves: Sound

While waves on a string or in water are transverse, sound waves are longitudinal. The term "longitudinal" means that the medium transmitting the waves—air, in the case of sound waves—oscillates back and forth, parallel to the direction in which the wave is moving. This back-and-forth motion stands in contrast to the behavior of transverse waves, which oscillate up and down, perpendicular to the direction in which the wave is moving.

Imagine a slinky. If you hold one end of the slinky in each of your outstretched arms and then jerk one arm slightly toward the other, you will send a pulse across the slinky toward the other arm. This pulse is transmitted by each coil of the slinky oscillating back and forth parallel to the direction of the pulse.

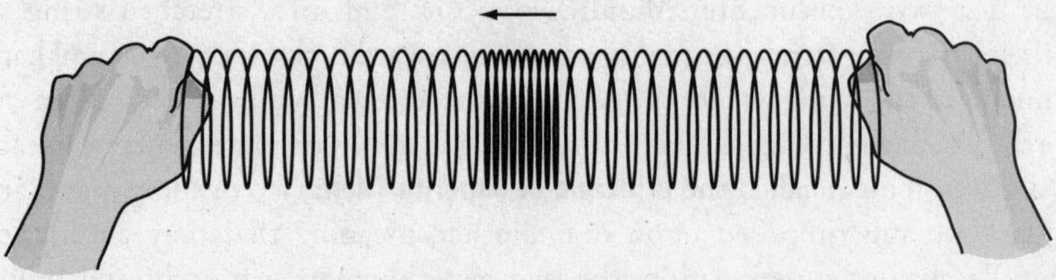

When the string on a violin, the surface of a bell, or the paper cone in a stereo speaker oscillates rapidly, it creates pulses of high air pressure, or compressions, with low pressure spaces in between, called rarefactions. These compressions and rarefactions are the equivalent of crests and troughs in transverse waves: the distance between two compressions or two rarefactions is a wavelength.

Pulses of high pressure propagate through the air much like the pulses of the slinky illustrated above, and when they reach our ears we perceive them as sound. Air acts as the medium for sound waves, just as string is the medium for waves of displacement on a string. The figure below is an approximation of sound waves in a flute—each dot represents something in the order of 10^{24} air molecules.

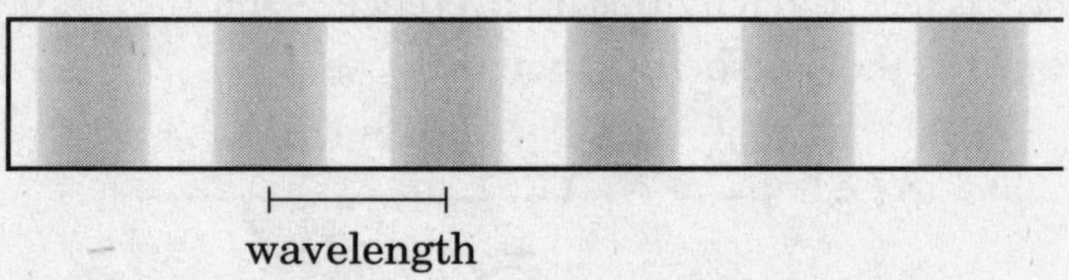

Loudness, Frequency, Wavelength, and Wave Speed

Many of the concepts describing waves are related to more familiar terms describing sound. For example, the square of the amplitude of a sound wave is called its **loudness**, or **volume**. Loudness is usually measured in **decibels**. The decibel is a peculiar unit measured on a logarithmic scale. You won't need to know how to calculate decibels, but it may be useful to know what they are.

The frequency of a sound wave is often called its **pitch**. Humans can hear sounds with frequencies as low as about 90 Hz and up to about 15,000 Hz, but many animals can hear sounds with much higher frequencies. The term wavelength remains the same for sound waves. Just as in a stretched string, sound waves in air travel at a certain speed. This speed is around 343 m/s under normal circumstances, but it varies with the temperature and pressure of the air. You don't need to memorize this number: if a question involving the speed of sound comes up on the SAT II, that quantity will be given to you.

Superposition

Suppose that two experimenters, holding opposite ends of a stretched string, each shake their end of the string, sending wave crests toward each other. What will happen in the middle of the string, where the two waves meet? Mathematically, you can calculate the displacement in the center by simply adding up the displacements from each of the two waves. This is called the **principle of superposition**: two or more waves in the same place are superimposed upon one another, meaning that they are all added together. Because of superposition, the two experimenters can each send traveling waves down the string, and each wave will arrive at the opposite end of the string undistorted by the other. The principle of superposition tells us that waves cannot affect one another: one wave cannot alter the direction, frequency, wavelength, or amplitude of another wave.

Destructive Interference

Suppose one of the experimenters yanks the string downward, while the other pulls up by exactly the same amount. In this case, the total displacement when the pulses meet will be zero: this is called **destructive interference**. Don't be fooled by the name, though: neither wave is destroyed by this interference. After they pass by one another, they will continue just as they did before they met.

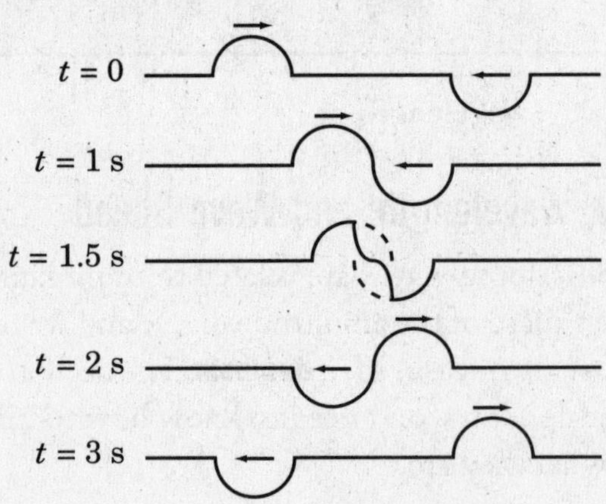

Constructive Interference

On the other hand, if both experimenters send upward pulses down the string, the total displacement when they meet will be a pulse that's twice as big. This is called **constructive interference**.

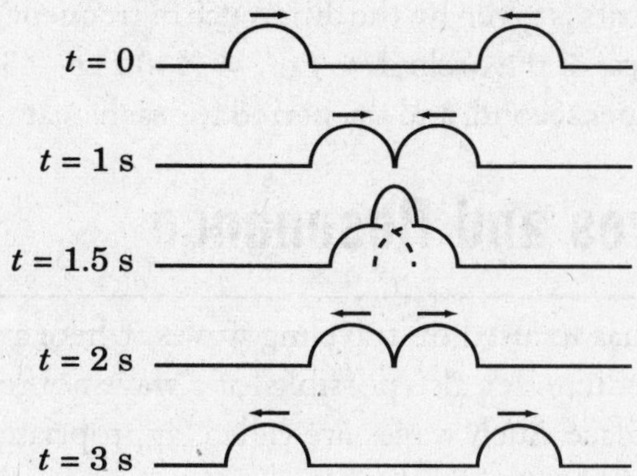

Beats

You may have noticed the phenomenon of interference when hearing two musical notes of slightly different pitch played simultaneously. You will hear a sort of "wa-wa-wa" sound, which results from repeated cycles of constructive interference, followed by destructive interference between the two waves. Each "wa" sound is called a **beat**, and the number of beats per second is given by the difference in frequency between the two interfering sound waves:

$$f_{\text{beat}} = |f_1 - f_2|$$

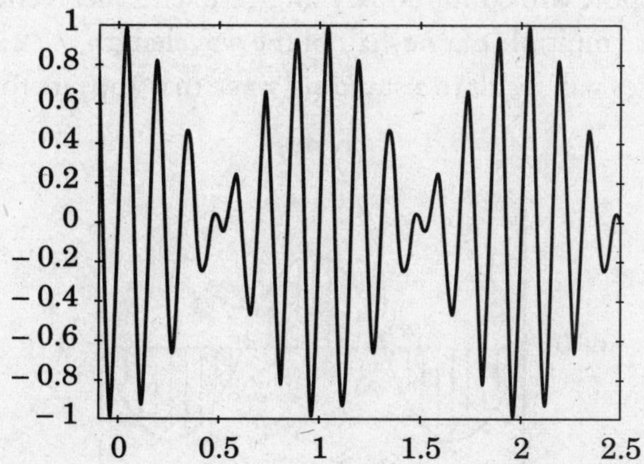

Example

Modern orchestras generally tune their instruments so that the note "A" sounds at 440 Hz. If one violinist is slightly out of tune, so that his "A" sounds at 438 Hz, what will be the time between the beats perceived by someone sitting in the audience?

The frequency of the beats is given by the difference in frequency between the out-of-tune violinist and the rest of the orchestra: $f_{\text{beat}} = |440 \text{ Hz} - 438 \text{ Hz}| = 2$ Hz. Thus, there will be two beats per second, and the period for each beat will be $T = 1/f = 0.5$ s.

Standing Waves and Resonance

So far, our discussion has focused on traveling waves, where a wave travels a certain distance through its medium. It's also possible for a wave not to travel anywhere, but simply to oscillate in place. Such waves are called, appropriately, **standing waves**. A great deal of the vocabulary and mathematics we've used to discuss traveling waves applies equally to standing waves, but there are a few peculiarities of which you should be aware.

Reflection

If a stretched string is tied to a pole at one end, waves traveling down the string will **reflect** from the pole and travel back toward their source. A reflected wave is the mirror image of its original—a pulse in the upward direction will reflect back in the downward direction—and it will interfere with any waves it encounters on its way back to the source. In particular, if one end of a stretched string is forced to oscillate—by tying it to a mass on a spring, for example—while the other end is tied to a pole, the waves traveling toward the pole will continuously interfere with their reflected copies. If the length of the string is a multiple of one-half of the wavelength, $\lambda/2$, then the superposition of the two waves will result in a standing wave that appears to be still.

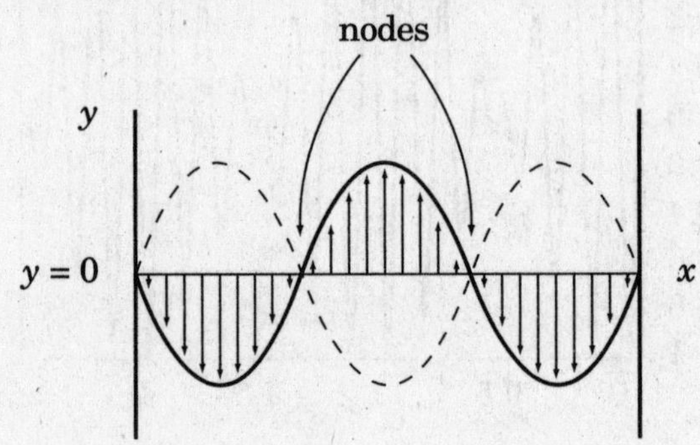

Nodes

The crests and troughs of a standing wave do not travel, or propagate, down the string. Instead, a standing wave has certain points, called **nodes**, that remain fixed at the equilibrium position. These are points where the original wave undergoes complete destructive interference with its reflection. In between the nodes, the points that oscillate with the greatest amplitude—where the interference is completely constructive—are called **antinodes**. The distance between successive nodes or antinodes is one-half of the wavelength, $\lambda/2$.

Resonance and Harmonic Series

The strings on musical instruments vibrate as standing waves. A string is tied down at both ends, so it can only support standing waves that have nodes at both ends, and thus can only vibrate at certain given frequencies. The longest such wave, called the **fundamental**, or **resonance**, has two nodes at the ends and one antinode at the center. Since the two nodes are separated by the length of the string, L, we see that the fundamental wavelength is $\lambda_1 = 2L$. The string can also support standing waves with one, two, three, or any integral number of nodes in between the two ends. This series of standing waves is called the **harmonic series** for the string, and the wavelengths in the series satisfy the equation $L = n(\lambda/2)$, or:

$$\lambda_n = \frac{2L}{n}$$

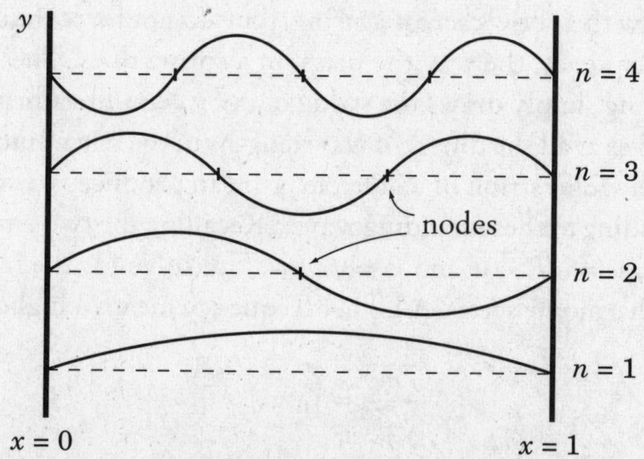

In the figure above, the fundamental is at the bottom, the first member of the harmonic series, with $n = 1$. Each successive member has one more node and a correspondingly shorter wavelength.

Example

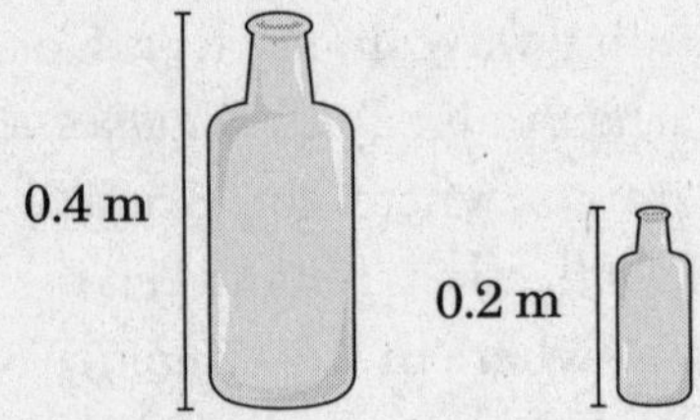

An empty bottle of height 0.2 m and a second empty bottle of height 0.4 m are placed next to each other. One person blows into the tall bottle and one blows into the shorter bottle. What is the difference in the pitch of the two sounds? What could you do to make them sound at the same pitch?

Sound comes out of bottles when you blow on them because your breath creates a series of standing waves inside the bottle. The pitch of the sound is inversely proportional to the wavelength, according to the equation $v = f\lambda$. We know that the wavelength is directly proportional to the length of the standing wave: the longer the standing wave, the greater the wavelength and the lower the frequency. The tall bottle is twice as long as the short bottle, so it vibrates at twice the wavelength and one-half the frequency of the shorter bottle. To make both bottles sound at the same pitch, you would have to alter the wavelength inside the bottles to produce the same frequency. If the tall bottle were half-filled with water, the wavelength of the standing wave would decrease to the same as the small bottle, producing the same pitch.

Pitch of Stringed Instruments

When violinists draw their bows across a string, they do not force the string to oscillate at any particular frequency, the way the mass on a spring does. The friction between the bow and the string simply draws the string out of its equilibrium position, and this causes standing waves at all the different wavelengths in the harmonic series. To determine what pitches a violin string of a given length can produce, we must find the frequencies corresponding to these standing waves. Recalling the two equations we know for the wave speed, $v = \sqrt{F_T/\mu}$ and $v = f\lambda$, we can solve for the frequency, f_n, for any term, n, in the harmonic series. A higher frequency means a higher pitch.

$$\frac{\sqrt{F_T}}{\sqrt{\mu}} = f_n\lambda_n$$

$$f_n = \frac{\sqrt{F_T}}{\lambda_n\sqrt{\mu}}$$

You won't need to memorize this equation, but you should understand the gist of it. This equation tells you that a higher frequency is produced by (1) a taut string, (2) a string with low mass density, and (3) a string with a short wavelength. Anyone who plays a stringed instrument knows this instinctively. If you tighten a string, the pitch goes up (1); the strings that play higher pitches are much thinner than the fat strings for low notes (2); and by placing your finger on a string somewhere along the neck of the instrument, you shorten the wavelength and raise the pitch (3).

The Doppler Effect

So far we have only discussed cases where the source of waves is at rest. Often, waves are emitted by a source that moves with respect to the medium that carries the waves, like when a speeding cop car blares its siren to alert onlookers to stand aside. The speed of the waves, v, depends only on the properties of the medium, like air temperature in the case of sound waves, and not on the motion of the source: the waves will travel at the speed of sound (343 m/s) no matter how fast the cop drives. However, the frequency and wavelength of the waves *will* depend on the motion of the wave's source. This change in frequency is called a **Doppler shift**. Think of the cop car's siren, traveling at speed v_s, and emitting waves with frequency f and period $T = 1/f$. The wave crests travel outward from the car in perfect circles (spheres actually, but we're only interested in the effects at ground level). At time T after the first wave crest is emitted, the next one leaves the siren. By this time, the first crest has advanced one wavelength, λ, but the car has also traveled a distance of v_sT. As a result, the two wave crests are closer together than if the cop car had been stationary.

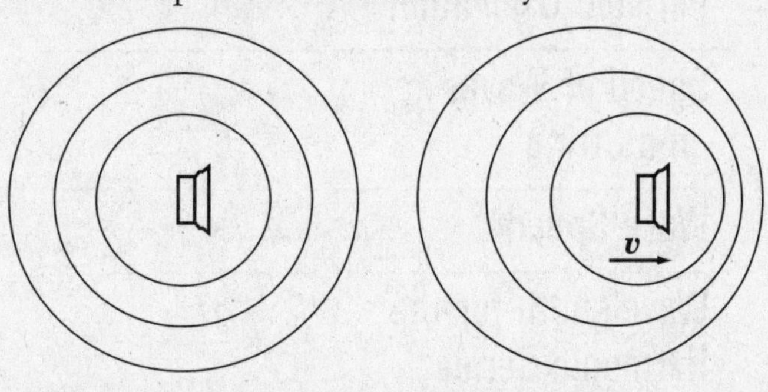

The shorter wavelength is called the Doppler-shifted wavelength, given by the formula $\lambda_D = \lambda - v_sT = \lambda(v - v_s)/v$. The Doppler-shifted frequency is given by the formula:

$$f_D = f\frac{v}{v - v_s}$$

Similarly, someone standing behind the speeding siren will hear a sound with a longer wavelength, $\lambda_D = \lambda + v_sT = \lambda(v + v_s)v$, and a lower frequency, $f_D = fv/(v + v_s)$.

You've probably noticed the Doppler effect with passing sirens. It's even noticeable with normal cars: the swish of a passing car goes from a higher hissing sound to a lower hissing sound as it speeds by. The Doppler effect has also been put to valuable use in astronomy, measuring the speed with which different celestial objects are moving away from the Earth.

Example

A cop car drives at 30 m/s toward the scene of a crime, with its siren blaring at a frequency of 2000 Hz. At what frequency do people hear the siren as it approaches? At what frequency do they hear it as it passes? The speed of sound in the air is 343 m/s.

As the car approaches, the sound waves will have shorter wavelengths and higher frequencies, and as it goes by, the sound waves will have longer wavelengths and lower frequencies. More precisely, the frequency as the cop car approaches is:

$$f_D = f\frac{v}{v - v_s} = (2000 \text{ Hz})\frac{343 \text{ m/s}}{343 \text{ m/s} - 30 \text{ m/s}} = 2192 \text{ Hz}$$

The frequency as the cop car drives by is:

$$f_D = f\frac{v}{v + v_s} = (2000 \text{ Hz})\frac{343 \text{ m/s}}{343 \text{ m/s} + 30 \text{ m/s}} = 1839 \text{ Hz}$$

Key Formulas

Frequency of Periodic Oscillation	$f = \frac{1}{T}$
Speed of Waves on a String	$v = \sqrt{\frac{F_T}{\mu}}$
Wave Speed	$v = f\lambda$
Wavelength for the Harmonic Series	$\lambda_n = \frac{2L}{n}$
Frequency for the Harmonic Series	$f_n = \frac{v}{\lambda_n} = \frac{nv}{2L}$
Beat Frequency	$f_{\text{beat}} = \lvert f_1 - f_2 \rvert$
Doppler Shift	$f_D = f\frac{v}{v - v_s}$

Practice Questions

1. Which of the following exhibit simple harmonic motion?

   I. A pendulum
   II. A mass attached to a spring
   III. A ball bouncing up and down, in the absence of friction

   (A) I only
   (B) II only
   (C) III only
   (D) I and II only
   (E) I, II, and III

2. If a wave has frequency $f = 10^{14}$ Hz and speed $v = 100$ m/s, what is its wavelength?

   (A) 10^{-16} m
   (B) 10^{-12} m
   (C) 10^{10} m
   (D) 10^{12} m
   (E) 10^{16} m

3. Two strings of equal length are stretched out with equal tension. The second string is twice as massive as the first string. If a wave travels down the first string with velocity v, how fast does a wave travel down the second string?

   (A) $\frac{1}{4}v$ (B) $\frac{1}{2}v$ (C) v (D) $2v$ (E) $4v$

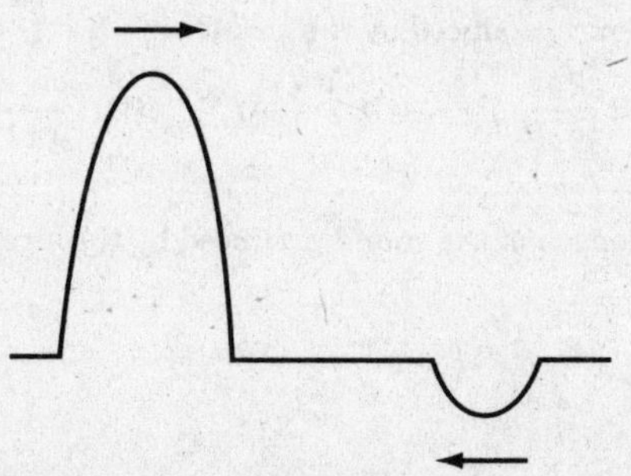

4. Two pulses travel along a string toward each other, as depicted above. Which of the following diagrams represents the pulses on the string at a later time?

   (A)
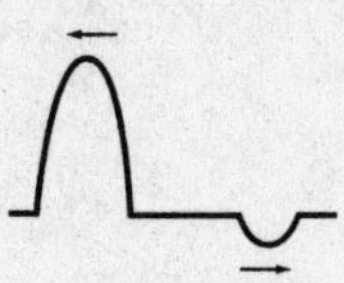
   (B)
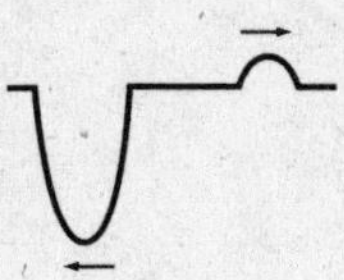
   (C)
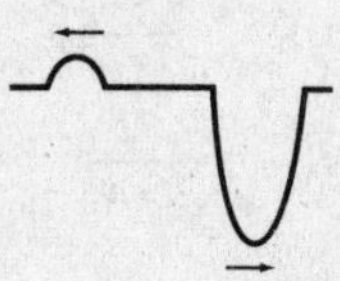
   (D)

   (E)
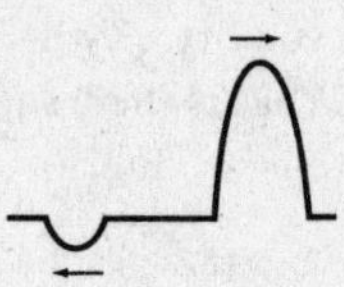

5. A piano tuner has a tuning fork that sounds with a frequency of 250 Hz. The tuner strikes the fork and plays a key that sounds with a frequency of 200 Hz. What is the frequency of the beats that the piano tuner hears?

(A) 0 Hz
(B) 0.8 Hz
(C) 1.25 Hz
(D) 50 Hz
(E) 450 Hz

6. How is the lowest resonant frequency, f_c, for a tube with one closed end related to the lowest resonant frequency, f_0, for a tube with no closed ends?

(A) $f_c = \frac{1}{4}f_0$ (B) $f_c = \frac{1}{2}f_0$ (C) $f_c = f_0$ (D) $f_c = 2f_0$ (E) $f_c = 4f_0$

7. What should a piano tuner do to correct the sound of a string that is flat; that is, it plays at a lower pitch than it should?

(A) Tighten the string to make the fundamental frequency higher
(B) Tighten the string to make the fundamental frequency lower
(C) Loosen the string to make the fundamental frequency higher
(D) Loosen the string to make the fundamental frequency lower
(E) Find a harmonic closer to the desired pitch

Questions 8 and 9 refer to a police car with its siren on, traveling at a velocity v_s toward a person standing on a street corner. As the car approaches, the person hears the sound at a frequency of f_d. Take the speed of sound to be v.

8. What is the frequency produced by the siren?

(A) $\frac{v - v_s}{v}f_d$ (B) $\frac{v}{v - v_s}f_d$ (C) $\frac{v + v_s}{v}f_d$ (D) $\frac{v - v_s}{v + v_s}f_d$ (E) $\frac{v}{v + v_s}f_d$

9. What is the wavelength of the sound produced by the siren?

(A) $\frac{v - v_s}{vf_d}$
(B) $\frac{v^2}{(v + v_s)f_d}$
(C) $\frac{v^2}{(v - v_s)f_d}$
(D) $\frac{v + v_s}{vf_d}$
(E) $\frac{v^2 f_d}{v - v_s}$

10. An ambulance driving with velocity $\frac{v_s}{8}$ where v_s is the speed of sound, emits a siren with a frequency of f. What is the frequency heard by a stationary observer toward whom the ambulance is driving?

(A) $\frac{f}{8}$ (B) $\frac{7f}{8}$ (C) f (D) $\frac{8f}{7}$ (E) $8f$

Explanations

1. (B)

Simple harmonic motion is defined as the oscillation of an object about an equilibrium position where the restoring force acting on the object is directly proportional to its displacement from the equilibrium position.

Though we often treat pendulum motion as simple harmonic motion, this is in fact a simplification. The restoring force acting on a pendulum is $mg \sin\theta$, where θ is the angle of displacement from the equilibrium position. The restoring force, then, is directly proportional to $\sin\theta$, and not to the pendulum bob's displacement, θ. At small angles, $\theta \approx \sin\theta$, so we can approximate the motion of a pendulum as simple harmonic motion, but the truth is more complicated.

The motion of a mass attached to a spring is given by Hooke's Law, $F = -kx$. Since the restoring force, F, is directly proportional to the mass's displacement, x, a mass on a spring does indeed exhibit simple harmonic motion.

There are two forces acting on a bouncy ball: the constant downward force of mg, and the occasional elastic force that sends the ball back into the air. Neither of these forces is proportional to the ball's displacement from any point, so, despite the fact that a bouncy ball oscillates up and down, it does not exhibit simple harmonic motion.

Of the three examples given above, only a mass on a spring exhibits simple harmonic motion, so the correct answer is (B).

2. (B)

The frequency, speed, and wavelength of a wave are related by the formula $v = \lambda f$. Solving for λ, we find:

$$\lambda = \frac{v}{f} = \frac{100 \text{ m/s}}{10^{14} \text{ Hz}}$$
$$= 10^{-12} \text{ m}$$

3. (B)

The formula relating wave speed, v, to the mass, m, length, ℓ, and tension, T, in a string is:

$$v = \frac{T\ell}{m}$$

Since v and m are inversely proportional, doubling m will halve v.

4. (E)

When two waves move toward one another, they pass through each other without one affecting the other. While both waves are in the same place, they will superimpose to form a single wave that is the sum of the two waves, but once they have passed one another, they will continue on their trajectory unaffected.

5. (B)

The frequency of the beats produced by two dissonant sounds is simply the difference between the two frequencies. In this case, the piano tuner will hear beats with a frequency of 250 Hz – 200 Hz = 50 Hz.

6. (B)

A tube closed at one end can support a standing wave with a node at the closed end and an antinode at the open end. A tube open at both ends can support a standing wave with antinodes at both ends.

$\lambda = 4L$ $\lambda = 2L$

As the figure shows, the wavelength for a standing wave in a tube closed at one end is twice the wavelength for a standing wave in a tube open at both ends. Since frequency is inversely proportional to wavelength, the frequency for a standing wave in a tube closed at one end is half the frequency of a standing wave in a tube open at both ends.

7. (A)

The easiest way to solve this problem is through simple intuition. When you tighten a string, it plays at a higher pitch, and when you loosen a string, it plays at a lower pitch. Pitch and frequency are the same thing, so in order to raise the pitch of the piano string, the tuner has to tighten the string, thereby raising its fundamental frequency.

8. (A)

In general, the frequency heard by the person is given by the formula:

$$f_d = \frac{v \pm v_d}{v \pm v_s} f_s$$

where f_d and v_d are the frequency heard by the person and the velocity of the person, respectively, and f_s and v_s are the frequency and the velocity of the police siren, respectively. Since the police car is traveling toward the person, the person will hear a higher frequency than that which the siren actually produces, so $f_d > f_s$. We also know that $v_d = 0$. If $f_d > f_s$, then the fraction in the equation above must be greater than one, so the denominator should read $v - v_s$, and not $v + v_s$. The resulting formula is:

$$f_d = \frac{v}{v - v_s} f_s$$
$$f_s = \frac{v - v_s}{v} f_d$$

9. (C)

Wavelength is related to velocity and frequency by the formula $\lambda = v/f$. In the previous question, we determined the frequency produced by the siren, so we can simply plug this formula into the formula for wavelength:

$$\begin{aligned}\lambda &= \frac{v}{f_s} \\ &= \frac{v}{((v - v_s)/v) f_d} \\ &= \frac{v^2}{(v - v_s) f_d}\end{aligned}$$

10. (D)

Generally speaking, the frequency heard by an observer is the frequency emitted at the source, multiplied by a factor of $(v_s - v_0)/(v_s - v)$, where v_s is the speed of sound, v_0 is the velocity of the observer, and v is the velocity of the source of the sound. Solving for f_0, the frequency heard by the observer, is just a matter of plugging the appropriate numbers into the formula:

$$\begin{aligned}f_0 &= f\frac{v_s}{v_s - v_s/8} = f\frac{v_s}{7/8 v_s} \\ &= \frac{8}{7} f\end{aligned}$$

Common intuition should save you from answering (A), (B), or (C): when an ambulance moves toward you, its siren sounds higher than it actually is.

Optics

Optics

Chapter Contents

The Electromagnetic Spectrum 321
Classical Optics 322
Optical Instruments 327
Wave Optics 335
Key Formulas 341
Practice Questions 342
Explanations 344

Having studied wave phenomena generally, let's take a look at the special case of **electromagnetic waves**. EM waves are transverse traveling waves produced by the oscillations of an electric field and a magnetic field. Because they are not transmitted by any material medium, as sound waves are through air molecules, EM waves can travel through the vacuum of space and give us valuable information about the universe beyond the Earth's atmosphere. Electromagnetic waves play a great many roles in our lives: we use EM waves of different wavelengths to microwave our dinner, to transmit radio signals, and to x-ray for broken bones. Most important, we are only able to see because our eyes can detect the EM waves that make up the spectrum of visible light.

Optics is the study of visible light, and how light can be manipulated to produce visual images. The distinction between the spectrum of visible light studied by optics and the rest of the electromagnetic spectrum is an arbitrary one based on the peculiarities of our eyes. All electromagnetic waves behave in the same way, and even if we can't see a radio wave or an X ray, they still respond to reflection and refraction, mirrors and lenses, in the same way as visible light.

The Electromagnetic Spectrum

Electromagnetic waves travel through a vacuum at the speed of light, $c = 3.00 \times 10^8$ m/s. As we'll see in the next chapter, this is the fastest speed there is: anything faster resides at present only in the realm of theoretical speculation. Because the speed of EM waves is constant, we can calculate a wave's frequency if we know its wavelength, and vice versa:

$$f = \frac{c}{\lambda} = \frac{3 \times 10^8 \text{ m/s}}{6 \times 10^{-7}\text{m}}$$
$$= 5 \times 10^{14} \text{ Hz}$$

Wavelength and frequency are the only qualities that distinguish one kind of EM wave from another. As a result, we can list all the kinds of EM waves on a one-dimensional graph called the **electromagnetic spectrum**.

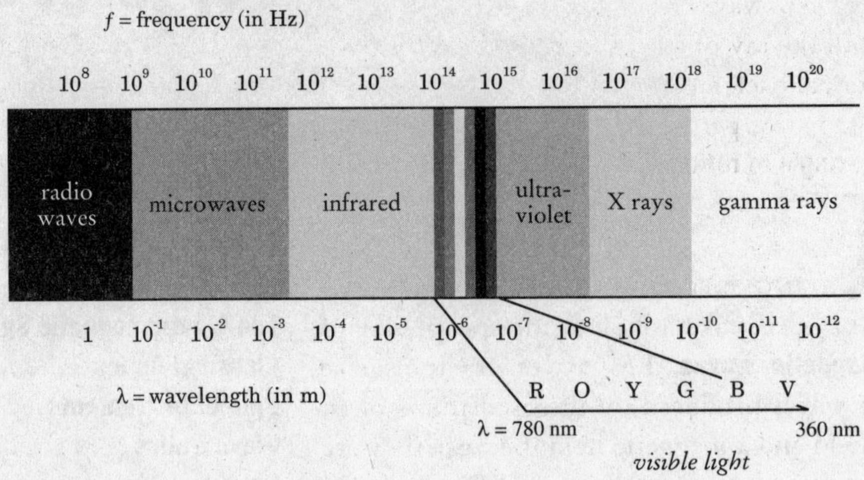

A higher frequency—and thus a shorter wavelength—corresponds to a wave with more energy. Though all waves travel at the same speed, those with a higher frequency oscillate faster, and a wave's oscillations are associated with its energy.

Visible light is the part of the electromagnetic spectrum between roughly 360 and 780 nanometers (1 nm = 10^{-9} m). When EM waves with these wavelengths—emitted by the sun, light bulbs, and television screens, among other things—strike the retina at the back of our eye, the retina sends an electrical signal to our brain that we perceive as color.

Classical Optics

"Classical" optics refers to those facts about optics that were known before the adoption of the wave model of light in the nineteenth century. In Newton's time, light was studied as if it had only particle properties—it moves in a straight line, rebounds off objects it bumps into, and passes through objects that offer minimal resistance. While this approximation of light as a particle can't explain some of the phenomena we will look at later in this chapter, it's perfectly adequate for dealing with most commonplace phenomena, and will serve as the basis for our examination of mirrors and lenses.

Reflection

When people think "**reflection**," they generally think of mirrors. However, everything that we see is capable of reflecting light: if an object couldn't reflect light, we wouldn't be able to see it. Mirrors do present a special case, however. Most objects absorb some light, reflecting back only certain frequencies, which explains why certain objects are of certain colors. Further, most objects have a rough surface—even paper is very rough on a molecular level—and so the light reflected off them deflects in all different directions. Mirrors are so smooth that they reflect all the light that strikes them in a very predictable and convenient way.

We call the ray of light that strikes a reflective surface an **incident ray**, and the ray that bounces back a **reflected ray**. The **angle of incidence**, θ, is the angle between the **normal**—the line perpendicular to the reflective surface—and the incident ray. Similarly, the **angle of reflection**, θ', is the angle between the normal and the reflected ray.

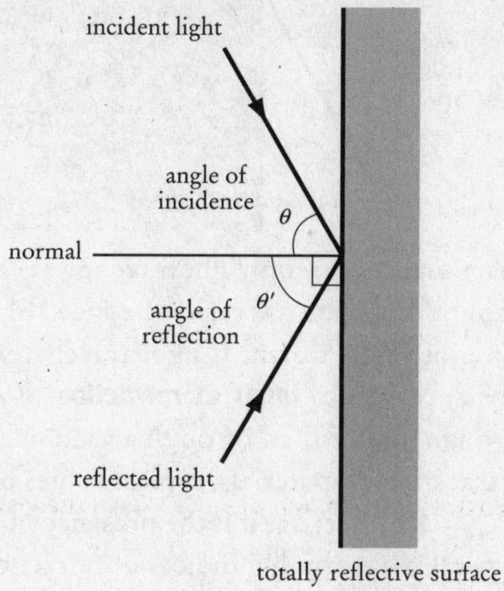

The **law of reflection** tells us that angle of incidence and angle of reflection are equal:

$$\theta_{\text{incidence}} = \theta_{\text{reflection}}$$

The reflection of a ray of light works in just the same way as a ball bouncing off a wall, except gravity has no noticeable effect on light rays.

Refraction

In addition to reflecting light, many surfaces also **refract** light: rather than bouncing off the surface, some of the incident ray travels through the surface, but at a new angle. We are able to see through glass and water because much of the light striking these substances is refracted and passes right through them.

Light passing from one substance into another will almost always reflect partially, so there is still an incident ray and a reflected ray, and they both have the same angle to the normal. However, there is also a third ray, the **refracted ray**, which lies in the same plane as the incident and reflected rays. The angle of the refracted ray will not be the same as the angle of the incident and reflected rays. As a result, objects that we see in a different medium—a straw in a glass of water, for instance—appear distorted because the light bends when it passes from one medium to another.

Optics

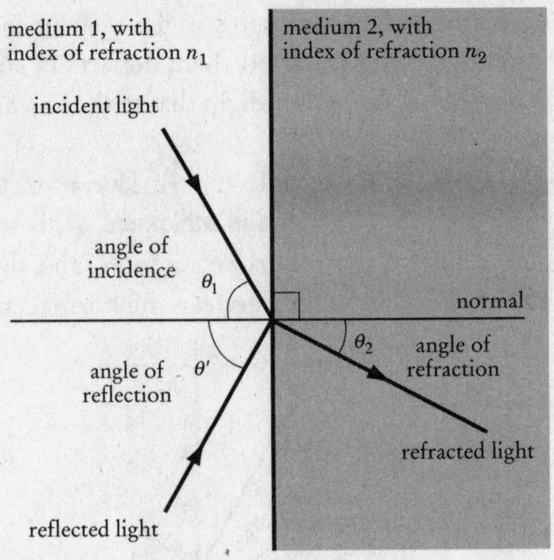

The phenomenon of refraction results from light traveling at different speeds in different media. The "speed of light" constant c is really the speed of light in a vacuum: when light passes through matter, it slows down. If light travels through a substance with velocity v, then that substance has an **index of refraction** of $n = c/v$. Because light always travels slower through matter than through a vacuum, v is always less than or equal to c, so $n \geq 1$. For transparent materials, typical values of n are quite low: $n_{air} = 1.0$, $n_{water} = 1.3$, and $n_{glass} = 1.6$. Because it is the presence of matter that slows down light, denser materials generally have higher indices of refraction.

A light ray passing from a less dense medium into a denser medium will be refracted toward the normal, and a light ray passing from a denser medium into a less dense medium will be refracted away from the normal. For example, water is denser than air, so the light traveling out of water toward our eyes is refracted away from the normal. When we look at a straw in a glass of water, we see the straw where it would be if the light had traveled in a straight line.

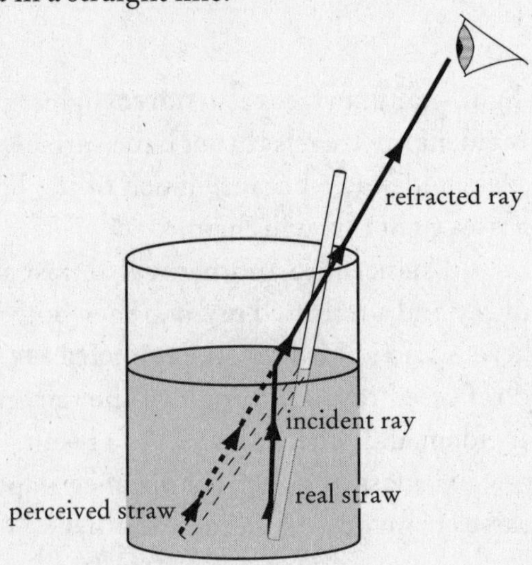

Given a ray traveling from a medium with index of refraction n_1 into a medium with index of refraction n_2, **Snell's Law** governs the relationship between the angle of incidence and the **angle of refraction**:

$$n_1 \sin\theta_1 = n_2 \sin\theta_2$$

Example

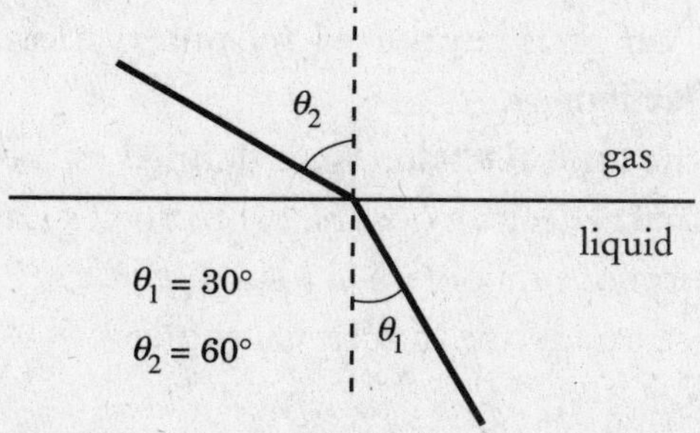

A ray of light passes from a liquid medium into a gas medium. The incident ray has an angle of 30° with the normal, and the refracted ray has an angle of 60° with the normal. If light travels through the gas at a speed of 2.5×10^8 m/s, what is the speed of light through the liquid medium? sin 30° = 0.500 and sin 60° = 0.866.

We know that the index of refraction for a substance, n, gives the ratio of the speed of light in a vacuum to the speed of light in that substance. Therefore, the index of refraction, n_1, in the liquid medium is related to the speed of light, v_1, in that medium by the equation $n_1 = c/v_1$; similarly, the index of refraction, n_2, in the gas medium is related to the speed of light, v_2, in that medium by the equation $n_2 = c/v_2$. The ratio between n_1 and n_2 is:

$$\frac{n_1}{n_2} = \frac{c/v_1}{c/v_2} = \frac{v_2}{v_1}$$

We can calculate the ratio between n_1 and n_2 using Snell's Law:

$$n_1 \sin\theta_1 = n_2 \sin\theta_2$$
$$n_1(0.500) = n_2(0.866)$$
$$\frac{n_1}{n_2} = 1.73$$

Since we know that the ratio of n_1/n_2 is equal to the ration of v_2/v_1, and since we know the value for v_2, we can now calculate the value for v_1:

$$\frac{v_2}{v_1} = 1.73$$
$$v_1 = \frac{2.5 \times 10^8 \text{ m/s}}{1.73}$$
$$= 1.45 \times 10^8 \text{ m/s}$$

Given $c = 3.0 \times 10^8$ m/s, we can also calculate that the index of refraction for the liquid substance is 2.1, while the index of refraction for the gas substance is 1.2.

Total Internal Reflection

The sine of an angle is always a value between –1 and 1, so for certain values of n_1, n_2, and θ_1, Snell's Law admits no solution for θ_2. For example, suppose medium 1 is glass, medium 2 is air and $\theta_1 = 87°$. Then the angle of refraction is given by sin θ_2 = 1.6, for which there is no solution. Mathematicians have not yet invented a physical angle with this property, so physicists just shrug their shoulders and conclude that there is no refracted ray, which is supported by observation. This phenomenon is known as **total internal reflection.**

For two given media, the **critical angle,** θ_c, is defined as the smallest angle of incidence for which total internal reflection occurs. From Snell's Law, we know that $\sin \theta_2 = n_1 \sin \theta_1 / n_2$, so refraction occurs only if $n_1 \sin \theta_1 / n_2 \leq 1$. Setting the left side of that equation to equal 1, we can derive the critical angle:

$$\theta_c = \arcsin \frac{n_2}{n_1}$$

Example

> The index of refraction for water is 1.3 and the index of refraction for air is 1.0. What is the maximum angle of incidence at which a ray of light can pass from water into the air?

If the angle of incidence is greater than the critical angle, then the ray of light will not be refracted into the air. The maximum angle of incidence, then, is the critical angle.

$$\begin{aligned}\theta_c &= \arcsin \frac{n_{\text{air}}}{n_{\text{water}}} \\ &= \arcsin \frac{1.0}{1.3} \\ &= 50.3°\end{aligned}$$

Dispersion

There is one subtlety of refraction that we've overlooked: the index of refraction depends slightly on the wavelength of the incident light. When a mixture of waves of different wavelength refract, each constituent color refracts differently—the different constituents **disperse.** Generally speaking, light of a longer wavelength and shorter frequency refracts less than light of a shorter wavelength and longer frequency, so $n_{\text{violet}} = n_{\text{red}}$.

The phenomenon of dispersion explains why we see a rainbow when sunlight refracts off water droplets in the air. The white light of the sun is actually a mixture of a multitude of different wavelengths. When this white light passes through water droplets in the air, the different wavelengths of light are refracted differently. The violet light is refracted at a steeper angle than the red light, so the violet light that reaches our eyes appears to be coming from higher in the sky than the red light, even though they both come from the same ray of sunlight. Because each color of light is refracted at a slightly different angle, these colors arrange themselves, one on top of the other, in the sky.

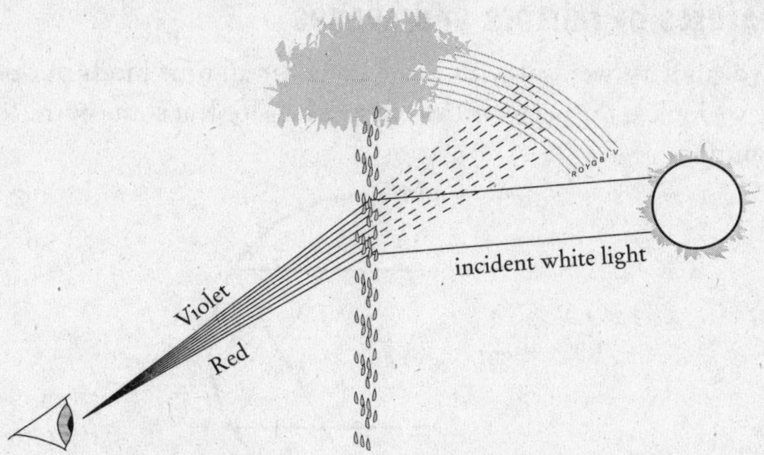

We find the same phenomenon with light shone into a glass prism.

Optical Instruments

The reflection and refraction we've dealt with so far have focused only on light interacting with flat surfaces. Lenses and curved mirrors are optical instruments designed to focus light in predictable ways. While light striking a curved surface is more complicated than the flat surfaces we've looked at already, the principle is the same. Any given light ray only strikes an infinitesimally small portion of the lens or mirror, and this small portion taken by itself is roughly flat. As a result, we can still think of the normal as the line perpendicular to the tangent plane.

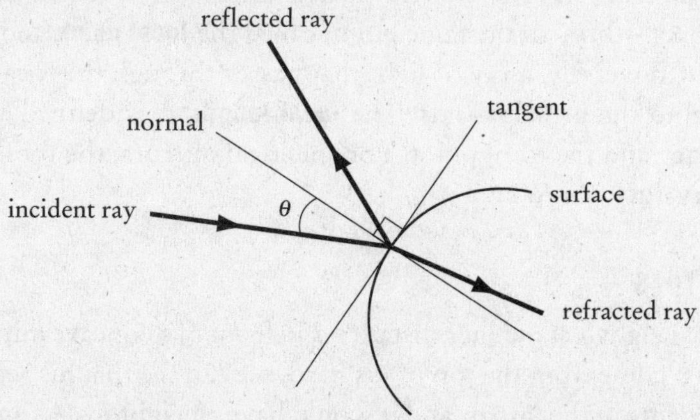

The four basic kinds of optical instruments—the only instruments that will be tested on the SAT II—are concave mirrors, convex mirrors, convex (or converging) lenses, and concave (or diverging) lenses. If you have trouble remembering the difference between concave and convex, remember that, like caves, concave mirrors and lenses curve inward. Convex lenses and mirrors bulge outward.

General Features of Mirrors and Lenses

Much of the vocabulary we deal with is the same for all four kinds of optical instruments. Before we look at the peculiarities of each, let's look at some of the features they all share in common.

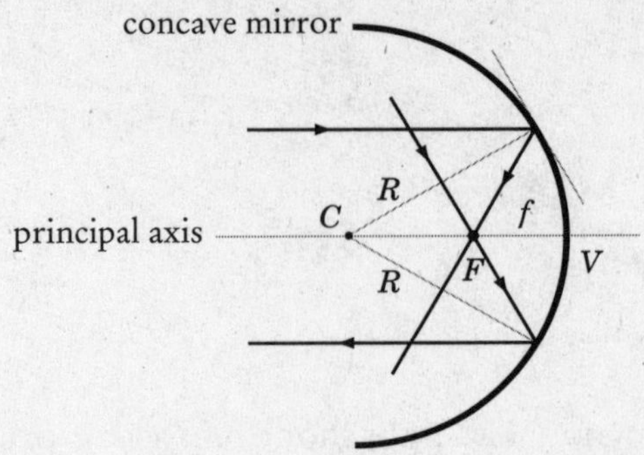

The diagram above shows a "ray tracing" image of a concave mirror, showing how a sample ray of light bounces off it. Though we will take this image as an example, the same principles and vocabulary apply to convex mirrors and to lenses as well.

The **principal axis** of a mirror or lens is a normal that typically runs through the center of the mirror or lens. The **vertex**, represented by V in the diagram, is the point where the principal axis intersects the mirror or lens.

The only kind of curved mirrors that appear on the SAT II are spherical mirrors, meaning they look like someone sliced off a piece of a sphere. Spherical mirrors have a **center of curvature**, represented by C in the diagram, which is the center of the sphere of which they are a slice. The radius of that sphere is called the **radius of curvature**, R.

All rays of light that run parallel to the principal axis will be reflected—or refracted in the case of lenses—through the same point, called the **focal point**, and denoted by F on the diagram. Conversely, a ray of light that passes through the focal point will be reflected parallel to the principal axis. The **focal length**, f, is defined as the distance between the vertex and the focal point. For spherical mirrors, the focal length is half the radius of curvature, $f = R/2$.

Concave Mirrors

Suppose a boy of height h stands at a distance d in front of a concave mirror. By tracing the light rays that come from the top of his head, we can see that his reflection would be at a distance d' from the mirror and it would have a height h'. As anyone who has looked into a spoon will have guessed, the image appears upside down.

The image at d' is a **real image**: as we can see from the ray diagram, the image is formed by actual rays of light. That means that, if you were to hold up a screen at position d', the image of the boy would be projected onto it. You may have noticed the way that the concave side of a spoon can cast light as you turn it at certain angles. That's because concave mirrors project real images.

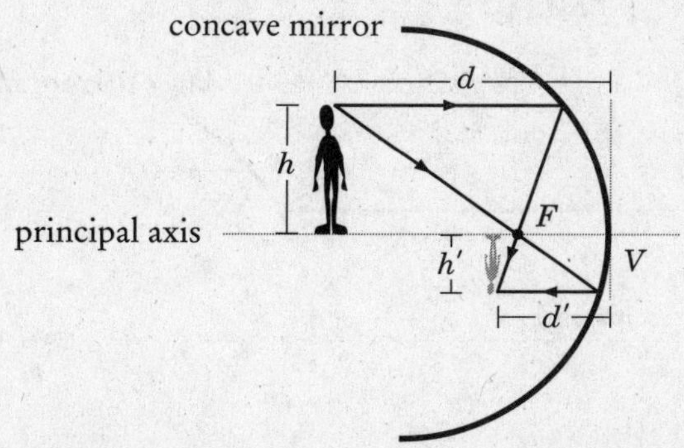

You'll notice, though, that we were able to create a real image only by placing the boy behind the focal point of the mirror. What happens if he stands in front of the focal point?

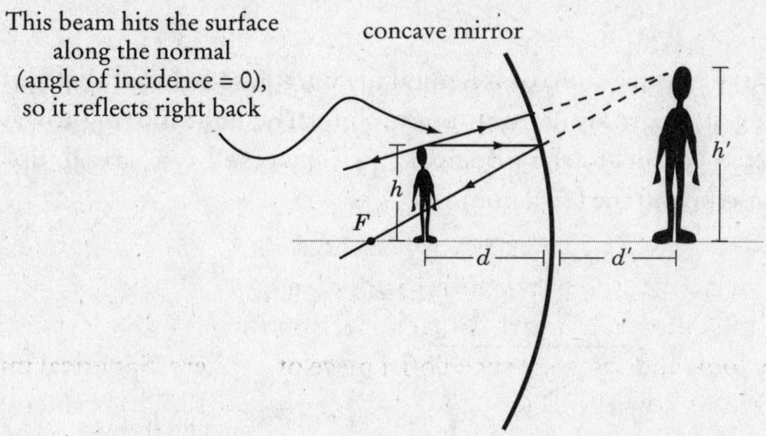

The lines of the ray diagram do not converge at any point in front of the mirror, which means that no real image is formed: a concave mirror can only project real images of objects that are behind its focal point. However, we can trace the diverging lines back behind the mirror to determine the position and size of a **virtual image**. Like an ordinary flat mirror, the image appears to be standing behind the mirror, but no light is focused on that point behind the mirror. With mirrors generally, an image is real if it is in front of the mirror and virtual if it is behind the mirror. The virtual image is right side up, at a distance d' from the vertex, and stands at a height h'.

You can test all this yourself with the right kind of spoon. As you hold it at a distance from your face, you see your reflection upside down. As you slowly bring it closer, the upside-down reflection becomes blurred and a much larger reflection of yourself emerges, this time right side up. The image changes from upside down to right side up as your face crosses the spoon's focal point.

Convex Mirrors

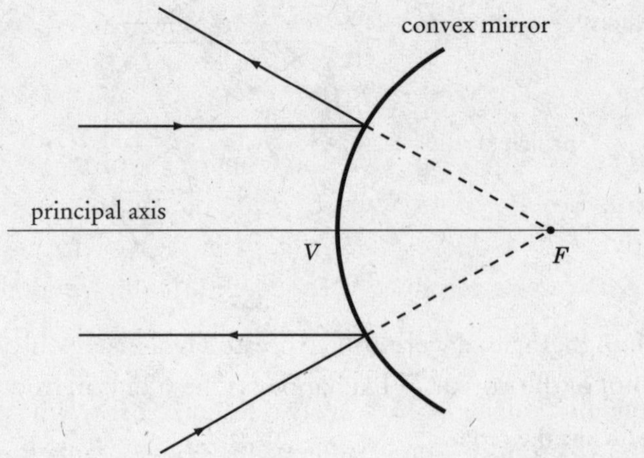

The focal point of a convex mirror is behind the mirror, so light parallel to the principal axis is reflected away from the focal point. Similarly, light moving toward the focal point is reflected parallel to the principal axis. The result is a virtual, upright image, between the mirror and the focal point.

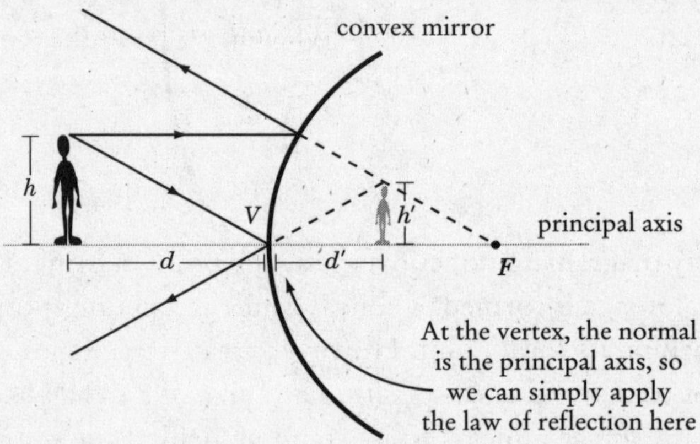

You've experienced the virtual image projected by a convex mirror if you've ever looked into a polished doorknob. Put your face close to the knob and the image is grotesquely enlarged, but as you draw your face away, the size of the image diminishes rapidly.

The Two Equations for Mirrors and Lenses

So far we've talked about whether images are real or virtual, upright or upside down. We've also talked about images in terms of a focal length f, distances d and d', and heights h and h'. There are two formulas that relate these variables to one another, and that, when used properly, can tell whether an image is real or virtual, upright or upside down, without our having to draw any ray diagrams. These two formulas are all the math you'll need to know for problems dealing with mirrors and lenses.

First Equation: Focal Length

The first equation relates focal length, distance of an object, and distance of an image:

$$\frac{1}{d} + \frac{1}{d'} = \frac{1}{f}$$

Values of d, d', and f are positive if they are in front of the mirror and negative if they are behind the mirror. An object can't be reflected unless it's in front of a mirror, so d will always be positive. However, as we've seen, f is negative with convex mirrors, and d' is negative with convex mirrors and with concave mirrors where the object is closer to the mirror than the focal point. A negative value of d' signifies a virtual image, while a positive value of d' signifies a real image.

Note that a normal, flat mirror is effectively a convex mirror whose focal point is an infinite distance from the mirror, since the light rays never converge. Setting $1/f = 0$, we get the expected result that the virtual image is the same distance behind the mirror as the real image is in front.

Second Equation: Magnification

The second equation tells us about the **magnification**, m, of an image:

$$m = \frac{h'}{h} = \frac{-d'}{d}$$

Values of h' are positive if the image is upright and negative if the image is upside down. The value of m will always be positive because the object itself is always upright.

The magnification tells us how large the image is with respect to the object: if $|m| > 1$, then the image is larger; if $|m| < 1$, the image is smaller; and if $m = 1$, as is the case in an ordinary flat mirror, the image is the same size as the object.

Because rays move in straight lines, the closer an image is to the mirror, the larger that image will appear. Note that d'/d will have a positive value with virtual images and a negative value with real images. Accordingly, the image appears upright with virtual images where m is positive, and the image appears upside down with real images where m is negative.

Example

A woman stands 40 cm from a concave mirror with a focal length of 30 cm. How far from the mirror should she set up a screen in order for her image to be projected onto it? If the woman is 1.5 m tall, how tall will her image be on the screen?

How far from the mirror should she set up a screen in order for her image to be projected onto it? The question tells us that $d = 40$ cm and $f = 30$ cm. We can simply plug these numbers into the first of the two equations and solve for d', the distance of the image from the mirror:

$$\frac{1}{40\text{ cm}} + \frac{1}{d'} = \frac{1}{30\text{ cm}}$$
$$\frac{1}{d'} = \frac{1}{120\text{ cm}}$$
$$d' = 120\text{ cm}$$

Because d' is a positive number, we know that the image will be real. Of course, we could also have inferred this from the fact that the woman sets up a screen onto which to project the image.

How tall will her image be on the screen? We know that $d = 40$ cm, and we now know that $d' = 120$ cm, so we can plug these two values into the magnification equation and solve for m:

$$m = \frac{-d'}{d} = \frac{-120\text{ cm}}{40\text{ cm}}$$
$$= -3$$

The image will be three times the height of the woman, or $1.5 \times 3 = 4.5$ m tall. Because the value of m is negative, we know that the image will be real, and projected upside down.

Convex Lenses

Lenses behave much like mirrors, except they use the principle of refraction, not reflection, to manipulate light. You can still apply the two equations above, but this difference between mirrors and lenses means that the values of d' and f for lenses are positive for distances behind the lens and negative for distances in front of the lens. As you might expect, d is still always positive.

Because lenses—both concave and convex—rely on refraction to focus light, the principle of dispersion tells us that there is a natural limit to how accurately the lens can focus light. For example, if you design the curvature of a convex lens so that red light is focused perfectly into the focal point, then violet light won't be as accurately focused, since it refracts differently.

A **convex lens** is typically made of transparent material with a bulge in the center. Convex lenses are designed to focus light into the focal point. Because they focus light into a single point, they are sometimes called "converging" lenses. All the terminology regarding lenses is the same as the terminology we discussed with regard to mirrors—the lens has a vertex, a principal axis, a focal point, and so on.

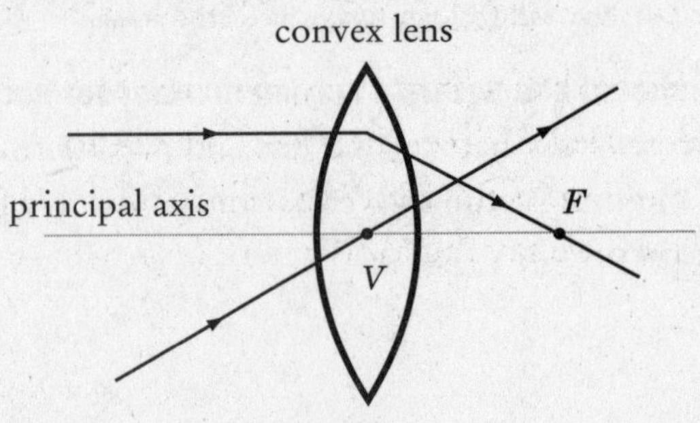

Convex lenses differ from concave mirrors in that their focal point lies on the opposite side of the lens from the object. However, for a lens, this means that $f > 0$, so the two equations discussed earlier apply to both mirrors and lenses. Note also that a ray of light that passes through the vertex of a lens passes straight through without being refracted at an angle.

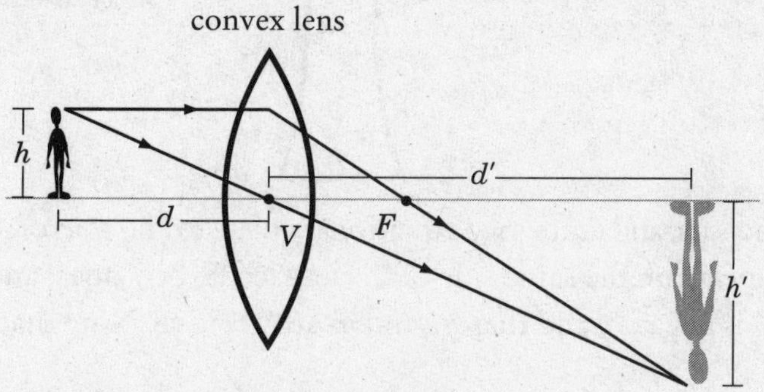

In this diagram, the boy is standing far enough from the lens that $d > f$. As we can see, the image is real and on the opposite side of the lens, meaning that d' is positive. Consequently, the image appears upside down, so h' and m are negative. If the boy were now to step forward so that $d < f$, the image would change dramatically:

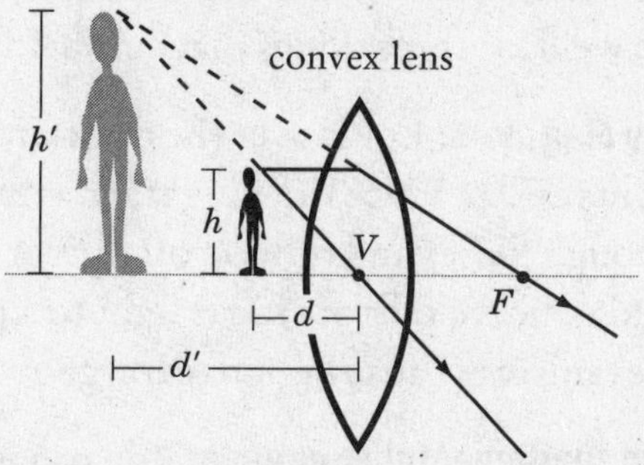

Now the image is virtual and behind the boy on the same side of the lens, meaning that d' is negative. Consequently, the image appears upright, so h' and m are positive.

Concave Lenses

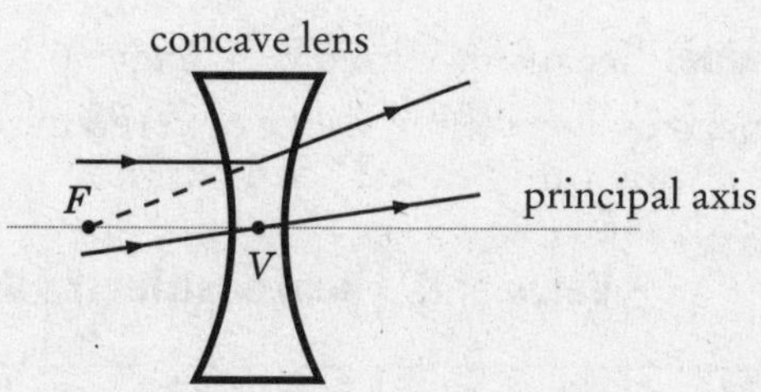

A **concave lens** is designed to divert light away from the focal point, as in the diagram. For this reason, it is often called a "diverging" lens. As with the convex lens, light passing through the vertex does not bend. Note that since the focal point F is on the same side of the lens as the object, we say the focal length f is negative.

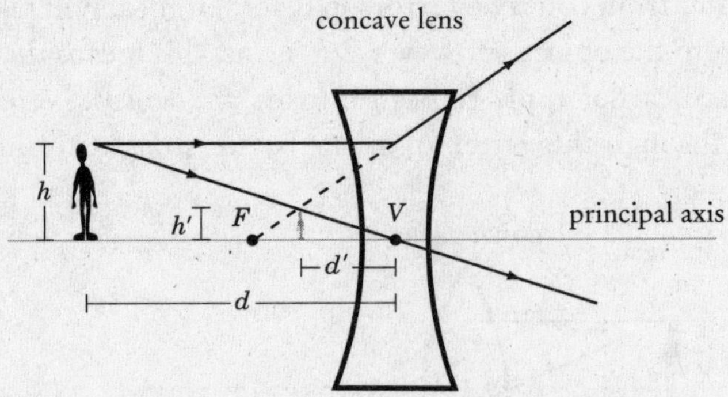

As the diagram shows us, and as the two equations for lenses and mirrors will confirm, the image is virtual, appears on the same side of the lens as the boy does, and stands upright. This means that d' is negative and that h' and m are positive. Note that $h > h'$, so $m < 1$.

Summary

There's a lot of information to absorb about mirrors and lenses, and remembering which rules apply to which kinds of mirrors and lenses can be quite difficult. However, this information is all very systematic, so once you grasp the big picture, it's quite easy to sort out the details. In summary, we'll list three things that may help you grasp the big picture:

1. **Learn to draw ray diagrams:** Look over the diagrams of the four kinds of optical instruments and practice drawing them yourself. Remember that light refracts through lenses and reflects off mirrors. And remember that convex lenses and concave mirrors focus light to a point, while concave lenses and convex mirrors cause light to diverge away from a point.
2. **Memorize the two fundamental equations:** You can walk into the SAT II knowing only the two equations for lenses and mirrors and still get a perfect score on the optical instruments questions, so long as you know how to apply these equations. Remember that f is positive for concave mirrors and convex lenses, and negative for convex mirrors and concave lenses.
3. **Memorize this table:** Because we love you, we've put together a handy table that summarizes everything we've covered in this section of the text.

Optical Instrument		Value of d'	Real or virtual?	Value of f	Upright or upside down?
Mirrors (d' and f are positive in front of mirror)	Concave $d > f$	Positive	Real	Positive	Upside down
	Concave $d < f$	Negative	Virtual	Positive	Upright
	Convex	Negative	Virtual	Negative	Upright
Lenses (d' and f are positive on far side of lens)	Convex $d > f$	Positive	Real	Positive	Upside down
	Convex $d < f$	Negative	Virtual	Positive	Upright
	Concave	Negative	Virtual	Negative	Upright

Note that when d' is positive, the image is always real and upside down, and when d' is negative, the image is always virtual and upright.

SAT II questions on optical instruments are generally of two kinds. Either there will be a quantitative question that will expect you to apply one of the two equations we've learned, or there will be a qualitative question asking you to determine where light gets focused, whether an image is real or virtual, upright or upside down, etc.

Wave Optics

As you may know, one of the weird things about light is that some of its properties can be explained only by treating it as a wave, while others can be explained only by treating it as a particle. The classical physics that we have applied until now deals only with the particle properties of light. We will now take a look at some phenomena that can only be explained with a wave model of light.

Young's Double-Slit Experiment

The wave theory of light came to prominence with Thomas Young's double-slit experiment, performed in 1801. We mention this because it is often called "Young's double-slit experiment," and you'd best know what the SAT II means if it refers to this experiment. The double-slit experiment proves that light has wave properties because it relies on the principles of **constructive interference** and **destructive interference**, which are unique to waves.

The double-slit experiment involves light being shone on a screen with—you guessed it—two very narrow slits in it, separated by a distance d. A second screen is set up a distance L from the first screen, upon which the light passing through the two slits shines.

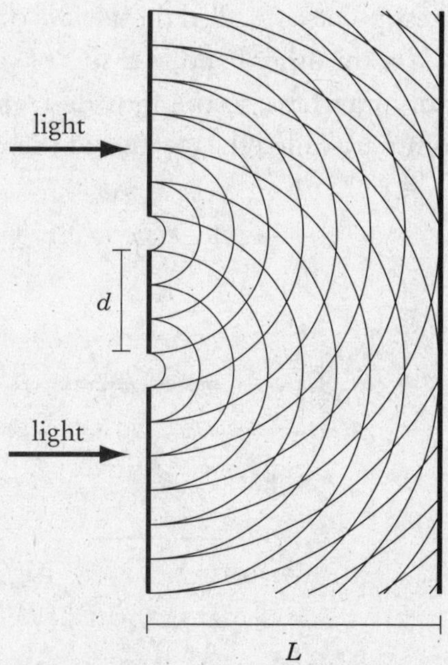

Suppose we have **coherent light**—that is, light of a single wavelength λ, which is all traveling in phase. This light hits the first screen with the two parallel narrow slits, both of which are narrower than λ. Since the slits are narrower than the wavelength, the light spreads out and distributes itself across the far screen.

At any point P on the back screen, there is light from two different sources: the two slits. The line joining P to the point exactly between the two slits intersects the front screen at an angle θ.

We will assume that the two screens are very far apart—somewhat more precisely, that L is much bigger than d. For this reason, this analysis is often referred to as the "far-field approximation." This approximation allows us to assume that angles α and β, formed by the lines connecting each of the slits to P, are both roughly equal to θ. The light from the right slit—the bottom slit in our diagram—travels a distance of $l = d \sin\theta$ more than the light from the other slit before it reaches the screen at the point P.

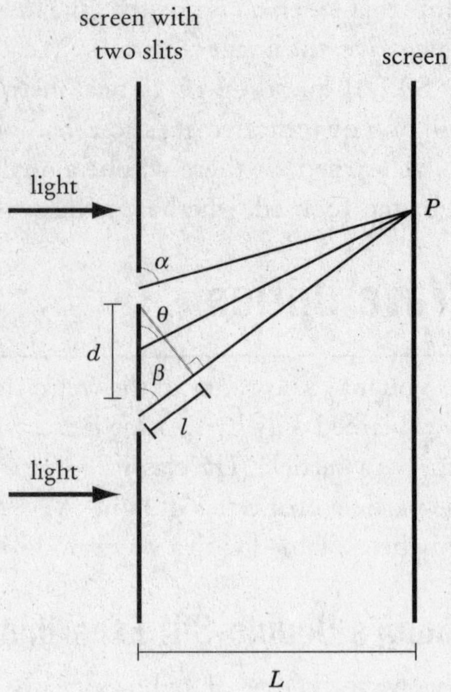

As a result, the two beams of light arrive at P out of phase by $d \sin\theta$. If $d \sin\theta = (n + 1/2)\lambda$, where n is an integer, then the two waves are half a wavelength out of phase and will destructively interfere. In other words, the two waves cancel each other out, so no light hits the screen at P. These points are called the **minima** of the pattern.

On the other hand, if $d \sin\theta = n\lambda$, then the two waves are in phase and constructively interfere, so the most light hits the screen at these points. Accordingly, these points are called the **maxima** of the pattern.

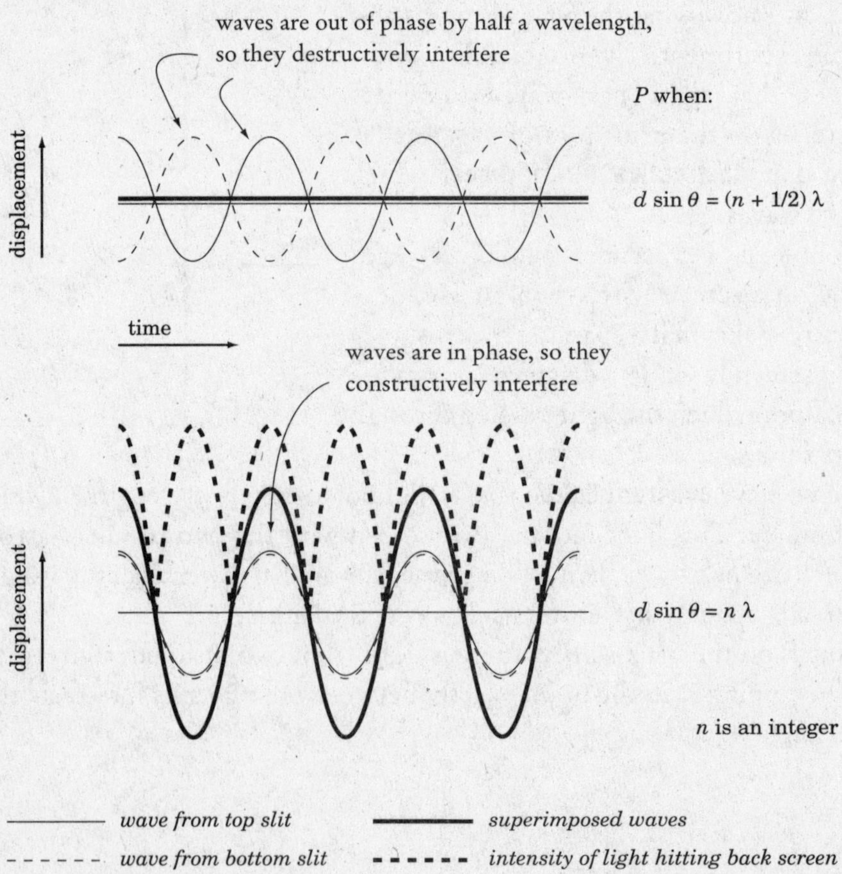

Because the far screen alternates between patches of constructive and destructive interference, the light shining through the two slits will look something like this:

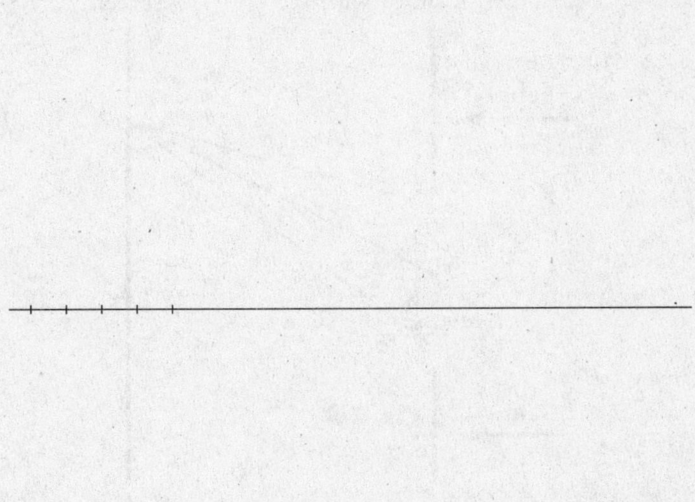

Note that the pattern is brightest in the middle, where $\theta = 0$, and moving outward the maxima become progressively less bright. Although the light spreads out when it passes through a slit, it is still brightest directly behind the slit, and it fades out on the sides as it spreads out.

Diffraction

Diffraction is an interference pattern caused by a lot of light sources, rather than just the two we saw with Young's double-slit experiment. For example, a **diffraction grating** is a screen with a bunch of parallel slits, each spaced a distance d apart. The analysis is exactly the same as in the double-slit case: there are still maxima at $d \sin \theta = n\lambda$ and minima at $d \sin \lambda = (n + 1/2)\lambda$. The only difference is that the pattern doesn't fade out as quickly on the sides.

Single-Slit Diffraction

You may also find single-slit diffraction on the SAT II. The setup is the same as with the double-slit experiment, only with just one slit. This time, we define d as the width of the slit and θ as the angle between the middle of the slit and a point P. To figure out where the maxima and minima are, consider light coming from the two sides of the slit.

Then, using the same reasoning as before, we can find the maxima at d sin $\lambda = n\lambda$ and the minima at d sin $\lambda = (n + 1/2)\lambda$, where n is an integer.

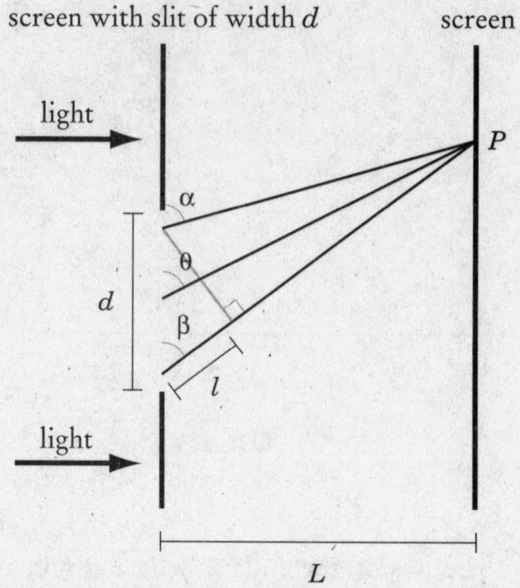

Actually, there are a lot of different paths that light can take to P—there is a path from any point in the slit. So really, the diffraction pattern is caused by the superposition of an infinite number of waves. However, paths coming from the two edges of the slit, since they are the farthest apart, have the biggest difference in phase, so we only have to consider these points to find the maxima and the minima.

Single-slit diffraction is nowhere near as noticeable as double-slit interference. The maximum at $n = 0$ is very bright, but all of the other maxima are barely noticeable. For this reason, we didn't have to worry about the diffraction caused by both slits individually when considering Young's experiment.

Polarization

Light is a transverse wave, meaning that it oscillates in a direction perpendicular to the direction in which it is traveling. However, a wave is free to oscillate right and left or up and down or at any angle between the vertical and horizontal.

Some kinds of crystals have a special property of **polarizing** light, meaning that they force light to oscillate only in the direction in which the crystals are aligned. We find this property in the crystals in Polaroid disks.

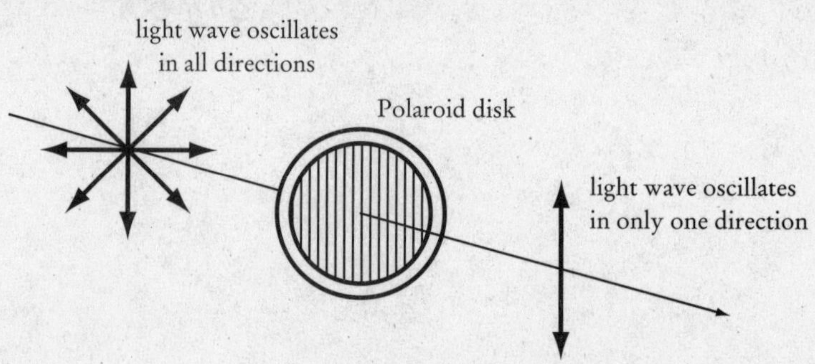

The human eye can't tell the difference between a polarized beam of light and one that has not been polarized. However, if polarized light passes through a second Polaroid disk, the light will be dimmed the more that second disk is out of alignment with the first. For instance, if the first disk is aligned vertically and the second disk is aligned horizontally, no light will pass through. If the second disk is aligned at a 45° angle to the vertical, half the light will pass through. If the second disk is also aligned vertically, all the light will pass through.

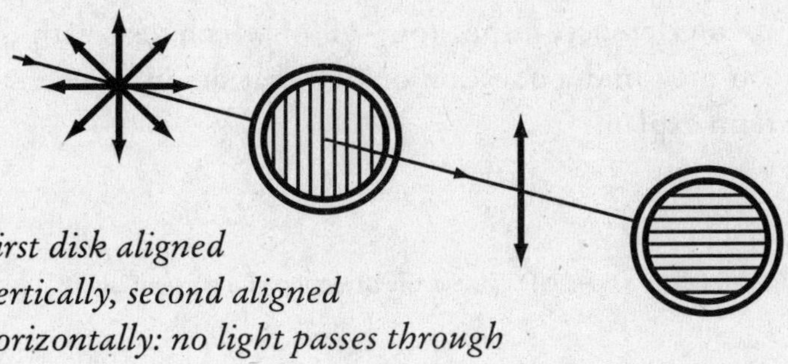

First disk aligned vertically, second aligned horizontally: no light passes through

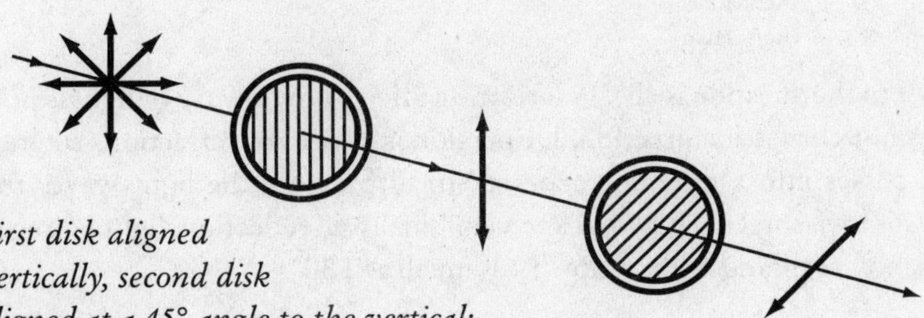

First disk aligned vertically, second disk aligned at a 45° angle to the vertical: half the light passes through at a 45° angle to the vertical

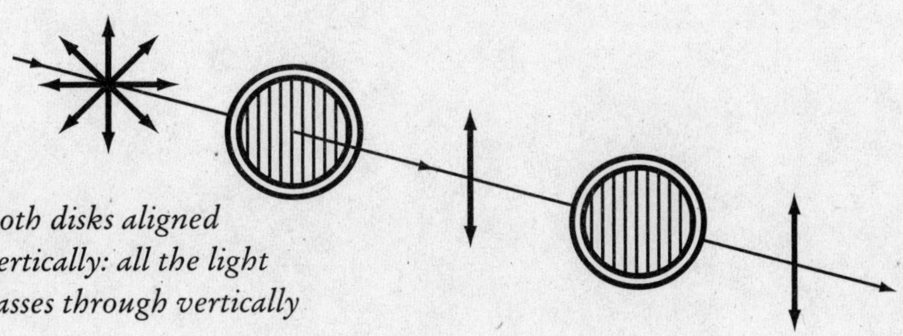

Both disks aligned vertically: all the light passes through vertically

Wave Optics on the SAT II

The SAT II will most likely test your knowledge of wave optics qualitatively. That makes it doubly important that you understand the physics going on here. It won't do you a lot of good if you memorize equations involving $d \sin \theta$ but don't understand when and why interference patterns occur.

One of the more common ways of testing wave optics is by testing your familiarity with different terms. We have encountered a number of terms—diffraction, polarization, reflection, refraction, interference, dispersion—all of which deal with different manipulations of light. You may find a question or two that describe a certain phenomenon and ask which term explains it.

Example

Which of the following phenomena does NOT affect the direction of a wave of light?

(A) Dispersion
(B) Polarization
(C) Diffraction
(D) Reflection
(E) Refraction

The answer to the question is (B). Polarization affects how a wave of light is polarized, but it does not change its direction. Dispersion is a form of refraction, where light is bent as it passes into a different material. In diffraction, the light waves that pass through a slit then spread out across a screen. Finally, in reflection, light bounces off an object, thereby changing its direction by as much as 180°.

Key Formulas

Frequency of an Electromagnetic Wave	$f = \frac{c}{\lambda} = \frac{3 \times 10^8 \text{ m/s}}{6 \times 10^{-7}\text{m}}$ $= 5 \times 10^{14} \text{ Hz}$
Law of Reflection	$\theta_{\text{incidence}} = \theta_{\text{reflection}}$
Index of Refraction	$n = \frac{c}{v}$
Snell's Law	$n_1 \sin \theta_1 = n_2 \sin \theta_2$
Critical Angle	$\theta_c = \arcsin \frac{n_2}{n_1}$
Focal Length for a Spherical Concave Mirror	$f = \frac{R}{2}$
Mirror and Lens Equation	$\frac{1}{d} + \frac{1}{d'} = \frac{1}{f}$
Magnification	$m = \frac{h'}{h} = \frac{-d'}{d}$
Minima in an Interference or Diffraction Pattern	$d \sin \theta = (n + \frac{1}{2})\lambda$, where n is an integer
Maxima in an Interference or Diffraction Pattern	$d \sin \theta = n\lambda$, where n is an integer

Optics

Practice Questions

1. Which of the following has the shortest wavelength?

   (A) Red light
   (B) Blue light
   (C) Gamma rays
   (D) X rays
   (E) Radio waves

2. Orange light has a wavelength of 6×10^{-7} m. What is its frequency? The speed of light is 3×10^{8} m/s.

   (A) 2×10^{15} Hz
   (B) 2×10^{-15} Hz
   (C) 5×1014 Hz
   (D) 5×10^{-14} Hz
   (E) 2×10^{14} Hz

3. When the orange light passes from air ($n = 1$) into glass ($n = 1.5$), what is its new wavelength?

   (A) 4×10^{-7} m
   (B) 4×10^{-6} m
   (C) 2.5×10^{-7} m
   (D) 6×10^{-7} m
   (E) 9×10^{-7} m

4. When a ray of light is refracted, the refracted ray does not have the same wavelength as the incident ray. Which of the following explain this phenomenon?

   I. Some of the energy of the incident ray is carried away by the reflected ray
   II. The boundary surface absorbs some of the energy of the incident ray
   III. The incident and refracted rays do not travel with the same velocity

   (A) I only
   (B) II only
   (C) III only
   (D) I and II only
   (E) I, II, and III

Questions 5 and 6 refer to a beam of light that passes through a sheet of plastic and out into the air. The angle the beam of light makes with the normal as it passes through the plastic is θ_1, and the angle the beam of light makes with the normal as it passes into the air is θ_2. The index of refraction for air is 1 and the index of refraction for plastic is 2.

5. What is the value of $\sin\theta_2$, in terms of θ_1?

   (A) $\frac{1}{2}\sin\theta_1$ (B) $2\sin\theta_1$ (C) $\sin 2\theta_1$ (D) $\sin\frac{1}{2}\theta_1$ (E) $4\sin\theta_1$

6. What is the minimum incident angle for which the light will undergo total internal reflection in the plastic?

   (A) $\sin^{-1}\frac{1}{4}$ (B) $\sin^{-1}\frac{1}{2}$ (C) $\sin^{-1}2$ (D) 0° (E) 90°

7. A person's image appears on the far side of an optical instrument, upside down. What is the optical instrument?

(A) Concave mirror
(B) Convex mirror
(C) Plane mirror
(D) Concave lens
(E) Convex lens

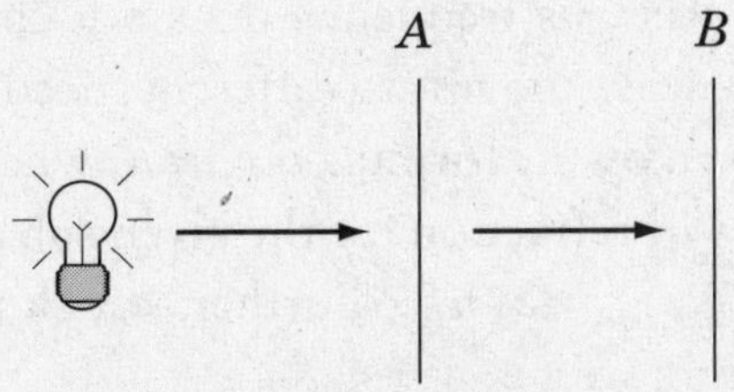

8. A physicist shines coherent light through an object, *A*, which produces a pattern of concentric rings on a screen, *B*. *A* is most likely:

(A) A polarization filter
(B) A single-slit diffraction grating
(C) A multiple-slit diffraction grating
(D) A prism
(E) A sheet with a pinhole

9. Sound waves do not exhibit polarization because, unlike light waves, they are not

(A) Longitudinal
(B) Coherent
(C) Dispersive
(D) Transverse
(E) Refractive

10. The solar glare of sunlight bouncing off water or snow can be a real problem for drivers. The reflecting sunlight is horizontally polarized, meaning that the light waves oscillate at an angle of 90° to a normal line drawn perpendicular to the Earth. At what angle relative to this normal line should sunglasses be polarized if they are to be effective against solar glare?

(A) 0° (B) 30° (C) 45° (D) 60° (E) 90°

Optics

Optics

Explanations

1. (C)

Gamma rays have wavelengths shorter than 10^{-12} m. Don't confuse wavelength and frequency: gamma waves have a very high frequency, thus they have a *short* wavelength.

2. (C)

Wavelength and frequency are related by the formula $v = f\lambda$. In the case of light, $v = c = 3 \times 10^8$ m/s, so we can solve for f with the following calculations:

$$f = \frac{c}{\lambda} = \frac{3 \times 10^8 \text{ m/s}}{6 \times 10^{-7}\text{m}} \\ = 5 \times 10^{14} \text{ Hz}$$

3. (A)

When the wave enters the glass, its frequency does not change; otherwise, its color would change. However, the wave moves at a different speed, since the speed of light, v, in different substances is given by the formula $v = c/n$, where c is the speed of light in a vacuum, and n is the index of refraction for the given substance. Since $\lambda_{glass} = v/f$, we can also reason that $\lambda_{glass} = c/nf$. Further, we know that $c/n = \lambda_{air}$, so substituting these equations in, we get:

$$\lambda_{\text{glass}} = \frac{\lambda_{\text{air}}}{n} = \frac{6 \times 10^{-7}\text{m}}{1.5} \\ = 4 \times 10^{-7}\text{m}$$

4. (C)

Statement I is true, but it doesn't explain why a refracted ray should have a different wavelength. The fact that some of the incident ray is reflected means that the refracted ray will have a different amplitude, but it will not affect the frequency.

Statement II is false, and even if it were true, a change in energy would affect the frequency of the wave, not its wavelength.

Statement III correctly explains why refracted rays have different wavelengths from their incident rays. A light ray will maintain the same frequency, and hence color, when it is refracted. However, since the speed of light differs in different substances, and since the wavelength is related to the speed of light, v, by the formula $\lambda = v/f$, a change in the speed of light will mean a change in the wavelength as well.

5. (A)

Snell's Law gives us the relationship between the indices of refraction and the angles of refraction of two different substances: $n_1 \sin \theta_1 = n_2 \sin \theta_2$. We know that n_1, the index of refraction for air, is 1, and we know that n_2, the index of refraction for plastic, is 2. That means we can solve for $\sin \theta_2$:

$$\sin \theta_2 = \frac{n_1}{n_2} \sin \theta_1 = \frac{1}{2} \sin \theta_1$$

6. (C)

Total internal reflection occurs when the refracted ray is at an angle of 90° or greater, so that, effectively, the refracted ray doesn't escape into the air. If $\theta_2 = 90°$, then $\sin\theta_2 = 1$, so by Snell's Law:

$$\sin\theta_1 = \frac{n_2}{n_1}\sin\theta_2 = \frac{n_2}{n_1}$$
$$= 2$$

If $\sin\theta_1 = 2$, then $\theta_1 = \sin^{-1} 2$.

7. (E)

Only concave mirrors and convex lenses can produce images that appear upside down. However, concave mirrors produce these images on the same side of the mirror as the object, while convex lenses produce these images on the opposite side of the mirror from the object.

8. (E)

Whenever we see a pattern of maxima and minima, we know we are dealing with the phenomenon of diffraction, which rules out the possibility that *A* is a polarization filter or a prism. Both single- and multiple-slit diffraction gratings tend to produce bands of light, but not concentric circles. The correct answer is (E), the pinhole: light passing through the pinhole will spread out in concentric circles and will alternate between bright and dark patches to produce concentric rings.

9. (D)

Visible light can be polarized because it travels as a transverse wave, meaning that it oscillates perpendicular to the direction of its motion. Polarization affects the oscillation of transverse waves by forcing them to oscillate in one particular direction perpendicular to their motion. Sound waves, on the other hand, are longitudinal, meaning that they oscillate parallel to the direction of their motion. Since there is no component of a sound wave's oscillation that is perpendicular to its motion, sound waves cannot be polarized.

10. (A)

The idea behind polarized sunglasses is to eliminate the glare. If the solar glare is all at a 90° angle to the normal line, sunglasses polarized at a 0° angle to this normal will not allow any of the glare to pass. Most other light is not polarized, so it will still be possible to see the road and other cars, but the distracting glare will cease to be a problem.

Modern Physics

Chapter Contents

Special Relativity 348
The Discovery of the Atom. 354
Quantum Physics 356
Nuclear Physics 365
Key Formulas 374
Review Questions 375
Explanations. 377

ALMOST EVERYTHING WE'VE COVERED in the previous 15 chapters was known by the year 1900. Taken as a whole, these 15 chapters present a comprehensive view of physics. The principles we've examined, with a few elaborations, are remarkably accurate in their predictions and explanations for the behavior of pretty much every element of our experience, from a bouncy ball to a radio wave to a thunderstorm. No surprise, then, that the physicist Albert Michelson should have claimed in 1894 that all that remained for physics was the filling in of the sixth decimal place for certain constants.

But as it turns out, the discoveries of the past 100 years show us that most of our assumptions about the fundamental nature of time, space, matter, and energy are mistaken. The "modern" physics of the past century focuses on phenomena so far beyond the scope of ordinary experience that Newton and friends can hardly be blamed for failing to notice them. Modern physics looks at the fastest-moving things in the universe, and at the smallest things in the universe. One of the remarkable facts about the technological advances of the past century is that they have brought these outer limits of nature in touch with palpable experience in very real ways, from the microchip to the atomic bomb.

One of the tricky things about modern physics questions on the SAT II is that your common sense won't be of very much use: one of the defining characteristics of modern physics is that it goes against all common intuition. There are a few formulas you are likely to be tested on—$E = hf$ in particular—but the modern physics questions generally test concepts rather than math. Doing well on this part of the test requires quite simply that you know a lot of facts and vocabulary.

Modern Physics

Special Relativity

Special relativity is the theory developed by Albert Einstein in 1905 to explain the observed fact that the speed of light is a constant regardless of the direction or velocity of one's motion. Einstein laid down two simple postulates to explain this strange fact, and, in the process, derived a number of results that are even stranger. According to his theory, time slows down for objects moving at near light speeds, and the objects themselves become shorter and heavier. The wild feat of imagination that is special relativity has since been confirmed by experiment and now plays an important role in astronomical observation.

The Michelson-Morley Experiment

As we discussed in the chapter on waves, all waves travel through a medium: sound travels through air, ripples travel across water, etc. Near the end of the nineteenth century, physicists were still perplexed as to what sort of medium light travels through. The most popular answer at the time was that there is some sort of invisible ether through which light travels. In 1879, Albert Michelson and Edward Morley made a very precise measurement to determine at what speed the Earth is moving relative to the ether. If the Earth is moving through the ether, they reasoned, the speed of light should be slightly different when hitting the Earth head-on than when hitting the Earth perpendicularly. To their surprise, the speed of light was the same in both directions.

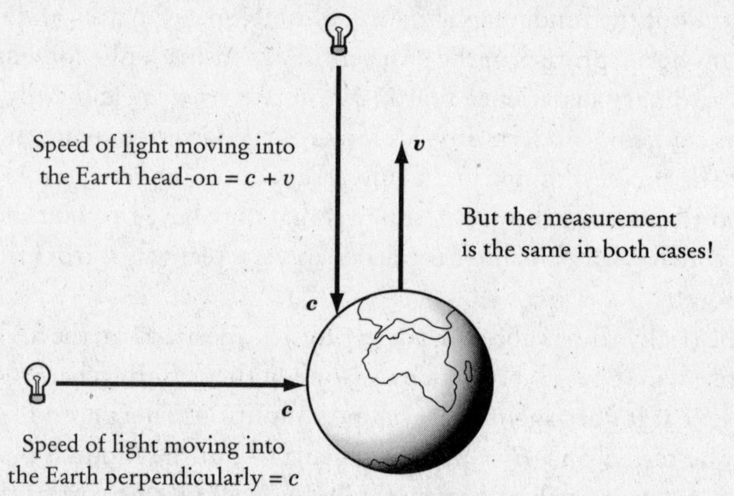

For people who believed that light must travel through an ether, the result of the **Michelson-Morley experiment** was like taking a ride in a boat and discovering that the boat crossed the wave crests at the same rate when it was driving against the waves as when it was driving in the same direction as the waves.

No one was sure what to make of the Michelson-Morley experiment until 1905, when Albert Einstein offered the two basic postulates of special relativity and changed forever the way we think about space and time. He asked all sorts of unconventional questions, such as, "what would I see if I were traveling at the speed of light?" and came up with all sorts of unconventional answers that experiment has since more or less confirmed.

The Basic Postulates of Special Relativity

Special relativity is founded upon two basic postulates, one a holdover from Newtonian mechanics and the other a seeming consequence of the Michelson-Morley experiment. As we shall see, these two postulates combined lead to some pretty counterintuitive results.

First Postulate

The laws of physics are the same in all inertial reference frames.

An **inertial reference frame** is one where Newton's First Law, the law of inertia, holds. That means that if two reference frames are moving relative to one another at a constant velocity, the laws of physics in one are the same as in the other. You may have experienced this at a train station when the train begins to move. Because the train is moving at a slow, steady velocity, it looks from a passenger's point of view that the station is moving backward, whereas for someone standing on the platform, it looks as if the train is moving forward.

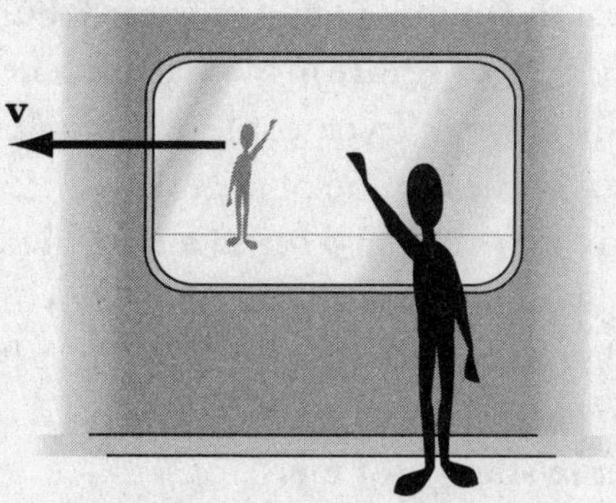

What you see from the train

What your friend sees from the platform

Einstein's first postulate tells us that neither the passenger on the train nor the person on the platform is wrong. It's just as correct to say that the train is still and the Earth is moving as it is to say that the Earth is still and the train is moving. Any inertial reference frame is as good as any other.

Second Postulate

> *The speed of light in a vacuum is a constant— 3.0×10^8 m/s—in every reference frame, regardless of the motion of the observer or the source of the light.*

This postulate goes against everything we've learned about vector addition. According to the principles of vector addition, if I am in a car moving at 20 m/s and collide with a wall, the wall will be moving at 20 m/s relative to me. If I am in a car moving at 20 m/s and collide with a car coming at me at 30 m/s, the other car will be moving at 50 m/s relative to me.

By contrast, the second postulate says that, if I'm standing still, I will measure light to be moving at 3.0×10^8 m/s, or c, relative to me, and if I'm moving toward the source of light at one half of the speed of light, I will still observe the light to be moving at c relative to me.

By following out the consequences of this postulate—a postulate supported by the Michelson-Morley experiment—we can derive all the peculiar results of special relativity.

Time Dilation

One of the most famous consequences of relativity is time dilation: time slows down at high speeds. However, it's important to understand exactly what this means. One of the consequences of the first postulate of special relativity is that there is no such thing as absolute speed: a person on a train is just as correct in saying that the platform is moving backward as a person on the platform is in saying that the train is moving forward. Further, both the person on the train and the person on the platform are in inertial reference frames, meaning that all the laws of physics are totally normal. Two people on a moving train can play table tennis without having to account for the motion of the train.

The point of time dilation is that, if you are moving *relative to me* in a very high-speed train at one-half the speed of light, it will appear to me that time is moving slower on board the train. On board the train, you will feel like time is moving at its normal speed. Further, because you will observe me moving at one-half the speed of light *relative to you*, you will think time is going more slowly for me.

What does this all mean? *Time is relative*. There is no absolute clock to say whether I am right or you are right. All the observations I make in my reference frame will be totally consistent, and so will yours.

We can express time dilation mathematically. If I were carrying a stopwatch and measured a time interval, t_0, you would get a different measure, t, for the amount of time I had the stopwatch running. The relation between these measures is:

$$t = \frac{t_0}{\sqrt{1 - (v^2/c^2)}}$$

So suppose I am moving at one-half the speed of light relative to you. If I measure 10 seconds on my stopwatch, you will measure the same time interval to be:

$$t = \frac{10\text{ s}}{\sqrt{1-(\frac{1}{2}c^2/c^2)}} \approx 14.1\text{ s}$$

This equation has noticeable effects only at near light speeds. The difference between t and t_0 is only a factor of $1/\sqrt{1-(v^2/c^2)}$. This factor—which comes up so frequently in special relativity that it has been given its own symbol, λ —is very close to 1 unless v is a significant fraction of c. You don't observe things on a train moving at a slower rate, since even on the fastest trains in the world, time slows down by only about 0.00005%.

Time Dilation and Simultaneity

Normally, we would think that if two events occur at the same time, they occur at the same time for all observers, regardless of where they are. However, because time can speed up or slow down depending on your reference frame, two events that may appear simultaneous to one observer may not appear simultaneous to another. In other words, special relativity challenges the idea of absolute simultaneity of events.

Example

> A spaceship of alien sports enthusiasts passes by the Earth at a speed of 0.8c, watching the final minute of a basketball game as they zoom by. Though the clock on Earth measures a minute left of play, how long do the aliens think the game lasts?

Because the Earth is moving at such a high speed relative to the alien spaceship, time appears to move slower on Earth from the aliens' vantage point. To be precise, a minute of Earth time seems to last:

$$\begin{aligned} t &= \frac{t_0}{\sqrt{1-v^2/c^2}} \\ &= \frac{60\text{ s}}{\sqrt{1-(0.8)^2}} \\ &= \frac{60\text{ s}}{\sqrt{1-0.64}} \\ &= \frac{60\text{ s}}{0.6} \\ &= 100\text{ s} \end{aligned}$$

Length Contraction

Not only would you observe time moving more slowly on a train moving relative to you at half the speed of light, you would also observe the train itself becoming shorter. The length of an object, l_0, contracts *in the direction of motion* to a length l when observed from a reference frame moving relative to that object at a speed v.

$$l = l_0\sqrt{1-(v^2/c^2)}$$

Modern Physics

Example

You measure a train at rest to have a length of 100 m and width of 5 m. When you observe this train traveling at 0.6*c* (it's a very fast train), what is its length? What is its width?

What is its length? We can determine the length of the train using the equation above:

$$l = (100\text{ m})\sqrt{1 - 0.36} = 80\text{ m}$$

What is its width? The width of the train remains at 5 m, since length contraction only works in the direction of motion.

Addition of Velocities

If you observe a person traveling in a car at 20 m/s, and throwing a baseball out the window in the direction of the car's motion at a speed of 10 m/s, you will observe the baseball to be moving at 30 m/s. However, things don't quite work this way at relativistic speeds. If a spaceship moving toward you at speed u ejects something in the direction of its motion at speed v_0 relative to the spaceship, you will observe that object to be moving at a speed v:

$$v = \frac{u + v_0}{1 + uv_0/c^2}$$

Example

A spaceship flying toward the Earth at a speed of 0.5*c* fires a rocket at the Earth that moves at a speed of 0.8*c* relative to the spaceship. What is the best approximation for the speed, v, of the rocket relative to the Earth?

(A) $v > c$
(B) $v = c$
(C) $0.8c < v < c$
(D) $0.5c < v < 0.8c$
(E) $v < 0.5c$

The most precise way to solve this problem is simply to do the math. If we let the speed of the spaceship be $u = 0.5c$ and the speed of the rocket relative to the spaceship be $v_0 = 0.8c$, then the speed, v, of the rocket relative to the Earth is

$$\begin{aligned} v &= \frac{u + v_0}{1 + uv_0/c^2} = \frac{1.3}{1 + 0.4} \\ &= \frac{13}{14} \end{aligned}$$

As we can see, the answer is (C). However, we could also have solved the problem by reason alone, without the help of equations. Relative to the Earth, the rocket would be moving faster than $0.8c$, since that is the rocket's speed relative to a spaceship that is speeding toward the Earth. The rocket cannot move faster than the speed of light, so we can safely infer that the speed of the rocket relative to the Earth must be somewhere between $0.8c$ and c.

Mass and Energy

Mass and energy are also affected by relativistic speeds. As things get faster, they also get heavier. An object with mass m_0 at rest will have a mass m when observed to be traveling at speed v:

$$m = \frac{m_0}{\sqrt{1 - (v^2/c^2)}}$$

"I'm not overweight, I'm just moving very fast."

Kinetic Energy

Because the mass increases, the kinetic energy of objects at high velocities also increases. Kinetic energy is given by the equation:

$$KE = (\frac{1}{\sqrt{1 - (v^2/c^2)}} - 1)mc^2$$

You'll notice that as v approaches c, kinetic energy approaches infinity. That means it would take an infinite amount of energy to accelerate a massive object to the speed of light. That's why physicists doubt that anything will ever be able to travel faster than the speed of light.

Mass-Energy Equivalence

Einstein also derived his most famous equation from the principles of relativity. Mass and energy can be converted into one another. An object with a rest mass of m_0 can be converted into an amount of energy, given by:

$$E = m_0c^2$$

We will put this equation to work when we look at nuclear physics.

Relativity and Graphs

One of the most common ways the SAT II tests your knowledge of special relativity is by using graphs. The key to remember is that, if there is a dotted line representing the speed of light, nothing can cross that line. For instance, here are two graphs of kinetic energy vs. velocity: the first deals with normal speeds and the second deals with relativistic speeds:

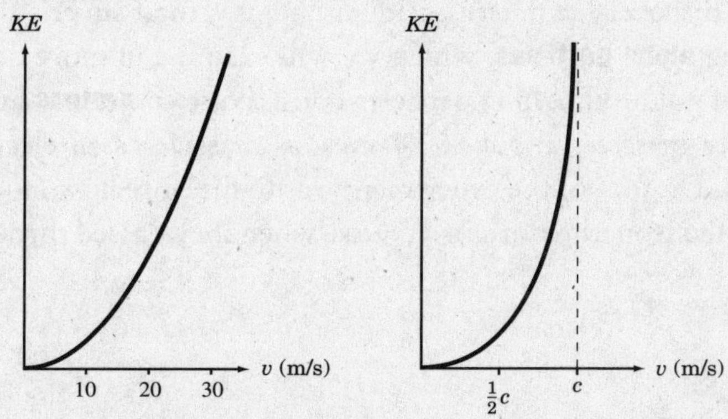

Modern Physics

In the first graph, we get a perfect parabola. The second graph begins as a parabola, but as it approaches the dotted line representing c, it bends so that it constantly approaches c but never quite touches it, much like a $y = 1/x$ graph will constantly approach the x-axis but never quite touch it.

The Discovery of the Atom

The idea that matter is made up of infinitely small, absolutely simple, indivisible pieces is hardly new. The Greek thinkers Leucippus and Democritus suggested the idea a good 100 years before Aristotle declared it was nonsense. However, the idea has only carried scientific weight for the past 200 years, and it only really took off in the past century.

Thompson's "Plum Pudding" Model

The first major discovery that set off modern atomic theory was that **atoms** aren't in fact the smallest things that exist. J. J. Thompson discovered the **electron** in 1897, which led him to posit a **"plum pudding" model** (a.k.a. the "raisin pudding" model) for the atom. Electrons are small negative charges, and Thompson suggested that these negative charges are distributed about a positively charged medium like plums in a plum pudding. The negatively charged electrons would balance out the positively charged medium so that each atom would be of neutral charge.

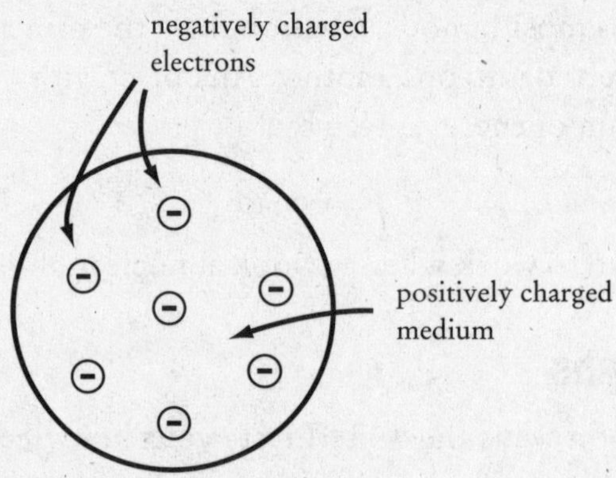

Rutherford's Gold Foil Experiment

In a series of experiments from 1909 to 1911, Ernest Rutherford established that atoms have nuclei. His discovery came by accident and as a total surprise. His experiment consisted of firing **alpha particles**, which we will examine in more detail shortly, at a very thin sheet of gold foil. Alpha particles consist of two **protons** and two **neutrons**: they are relatively massive (about 8000 times as massive as an electron), positively charged particles. The idea of the experiment was to measure how much the alpha particles were deflected from their original course when they passed through the gold foil.

Because alpha particles are positively charged and electrons are negatively charged, the electrons were expected to alter slightly the trajectory of the alpha particles. The experiment would be like rolling a basketball across a court full of marbles: when the basketball hits a marble, it might deflect a bit to the side, but, because it is much bigger than the marbles, its overall trajectory will not be affected very much. Rutherford expected the deflection to be relatively small, but sufficient to indicate how electrons are distributed throughout the "plum pudding" atom.

To Rutherford's surprise, most of the alpha particles were hardly deflected at all: they passed through the gold foil as if it were just empty space. Even more surprising was that a small number of the alpha particles were deflected at 180°, right back in the direction they came from.

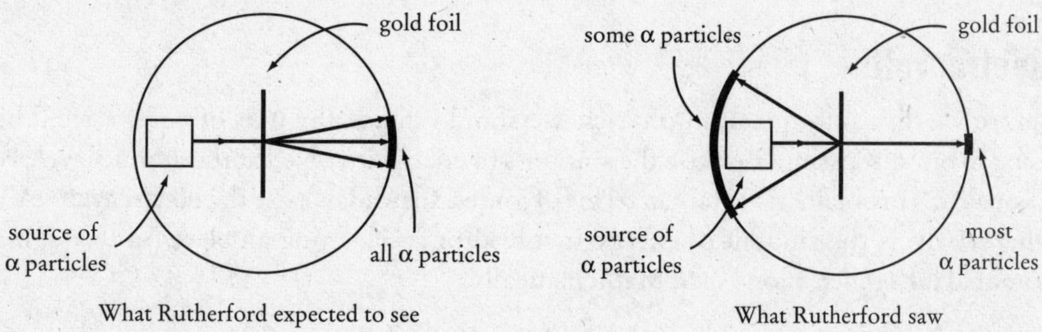

This unexpected result shows that the mass of an atom is not as evenly distributed as Thompson and others had formerly assumed. Rutherford's conclusion, known as the **Rutherford nuclear model**, was that the mass of an atom is mostly concentrated in a **nucleus** made up of tightly bonded protons and neutrons, which are then orbited by electrons. The electromagnetic force pulls the electrons into orbit around the nucleus in just the way that the gravitational force pulls planets into orbit around the sun.

The radius of an atom's nucleus is about 1/10,000 the radius of the atom itself. As a result, most of the alpha particles in Rutherford's **gold foil experiment** passed right through the sheet of gold foil without making contact with anything. A small number, however, bumped into the nucleus of one of the gold atoms and bounced right back.

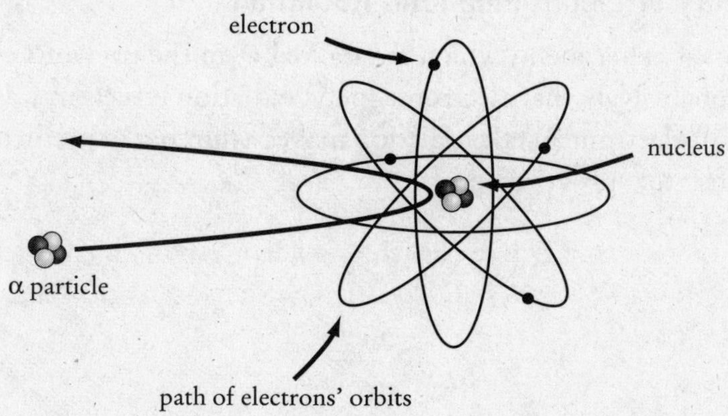

Quantum Physics

As physicists began to probe the mysteries of the atom, they came across a number of unexpected results along the lines of Rutherford's gold foil experiment. Increasingly, it became clear that things at the atomic level are totally unlike anything we find on the level of everyday objects. Physicists had to develop a whole new set of mechanical equations, called "quantum mechanics," to explain the movement of elementary particles. The physics of this "quantum" world demands that we upset many basic assumptions—that light travels in waves, that observation has no effect on experiments, etc.—but the results, from transistor radios to microchips, are undeniable. Quantum physics is very strange, but it works.

Electronvolts

Before we dive into quantum physics, we should define the unit of energy we'll be using in our discussion. Because the amounts of energy involved at the atomic level are so small, it's problematic to talk in terms of joules. Instead, we use the **electronvolt** (eV), where 1 eV is the amount of energy involved in accelerating an electron through a potential difference of one volt. Mathematically,

$$1 \text{ eV} = 1.6 \times 10^{-19} \text{ J}$$

The Photoelectric Effect

Electromagnetic radiation transmits energy, so when visible light, ultraviolet light, X rays, or any other form of electromagnetic radiation shines on a piece of metal, the surface of that metal absorbs some of the radiated energy. Some of the electrons in the atoms at the surface of the metal may absorb enough energy to liberate them from their orbits, and they will fly off. These electrons are called **photoelectrons**, and this phenomenon, first noticed in 1887, is called the **photoelectric effect**.

The Wave Theory of Electromagnetic Radiation

Young's double-slit experiment, which we looked at in the previous chapter, would seem to prove conclusively that electromagnetic radiation travels in waves. However, the wave theory of electromagnetic radiation makes a number of predictions about the photoelectric effect that prove to be false:

	Predictions of the wave theory	Observed result
Time lapse	Electrons need to absorb a certain amount of wave energy before they can be liberated, so there should be some lapse of time between the light hitting the surface of the metal and the first electrons flying off.	Electrons begin flying off the surface of the metal almost instantly after light shines on it.
Intensity	The intensity of the beam of light should determine the kinetic energy of the electrons that fly off the surface of the metal. The greater the intensity of light, the greater the energy of the electrons.	The intensity of the beam of light has no effect on the kinetic energy of the electrons. The greater the intensity, the greater the number of electrons that fly off, but even a very intense low-frequency beam liberates no electrons.
Frequency	The frequency of the beam of light should have no effect on the number or energy of the electrons that are liberated.	Frequency is key: the kinetic energy of the liberated electrons is directly proportional to the frequency of the light beam, and no electrons are liberated if the frequency is below a certain threshold.
Material	The material the light shines upon should not release more or fewer electrons depending on the frequency of the light.	Each material has a certain **threshold frequency**: light with a lower frequency will release no electrons.

Einstein Saves the Day

The young Albert Einstein accounted for these discrepancies between the wave theory and observed results by suggesting that electromagnetic radiation exhibits a number of particle properties. It was his work with the photoelectric effect, and not his work on relativity, that won him his Nobel Prize in 1921.

Rather than assuming that light travels as a continuous wave, Einstein drew on Planck's work, suggesting that light travels in small bundles, called **photons**, and that each photon has a certain amount of energy associated with it, called a **quantum**. Planck's formula determines the amount of energy in a given quantum:

$$E = hf$$

where h is a very small number, 6.63×10^{-34} J · s to be precise, called **Planck's constant**, and f is the frequency of the beam of light.

Work Function and Threshold Frequency

As the wave theory correctly assumes, an electron needs to absorb a certain amount of energy before it can fly off the sheet of metal. That this energy arrives all at once, as a photon, rather than gradually, as a wave, explains why there is no time lapse between the shining of the light and the liberation of electrons.

We say that every material has a given **work function**, ϕ, which tells us how much energy an electron must absorb to be liberated. For a beam of light to liberate electrons, the photons in the beam of light must have a higher energy than the work function of the material. Because the energy of a photon depends on its frequency, low-frequency light will not be able to liberate electrons. A liberated photoelectron flies off the surface of the metal with a kinetic energy of:

$$KE = hf - \phi$$

Example

> **Two beams of light, one blue and one red, shine upon a metal with a work function of 5.0 eV. The frequency of the blue light is 6.5×10^{14} Hz, and the frequency of the red light is 4.0×10^{14} Hz. What is the energy of the electrons liberated by the two beams of light?**

In order to solve this problem, we should translate h from units of J · s into units of eV · s:

$$h = (6.63 \times 10^{-34}\ \text{J} \cdot \text{s})\left(\frac{1\ \text{eV}}{1.6 \times 10^{-19}\ \text{J}}\right) = 4.14 \times 10^{-15}\ \text{eV} \cdot \text{s}$$

We know the frequencies of the beams of light, the work function of the metal, and the value of Planck's constant, h. Let's see how much energy the electrons liberated by the blue light have:

$$KE = hf - \phi = (4.14 \times 10^{-15}\ \text{eV} \cdot \text{s})(6.5 \times 10^{14}\ \text{Hz}) - 5.0\ \text{eV}$$
$$= 2.7\ \text{eV}$$

For the electrons struck by the red light:

$$KE = hf - \phi = (4.14 \times 10^{-15}\ \text{eV} \cdot \text{s})(4.0 \times 10^{14}\ \text{Hz}) - 5.0\ \text{eV}$$
$$= -3.3\ \text{eV}$$

The negative value in the sum means that $\phi > hf$, so the frequency of the red light is too low to liberate electrons. Only electrons struck by the blue light are liberated.

The Bohr Model of the Atom

Let's now return to our discussion of the atom. In 1913, the Danish physicist Niels Bohr proposed a model of the atom that married Planck's and Einstein's development of quantum theory with Rutherford's discovery of the atomic nucleus, thereby bringing quantum physics permanently into the mainstream of the physical sciences.

The Problem With Rutherford's Model

Light and other electromagnetic waves are emitted by accelerating charged particles. In particular, the electrons being accelerated in orbit about the nucleus of an atom release a certain amount of energy in the form of electromagnetic radiation. If we recall the chapter on gravity, the radius of an object in orbit is a function of its potential energy. If an electron gives off energy, then its potential energy, and hence the radius of its orbit about the nucleus, should decrease. But according to Rutherford's model, any radiating electron would give off all its potential energy in a fraction of a second, and the electron would collide with the nucleus. The fact that most of the atoms in the universe have not yet collapsed suggests a fundamental flaw in Rutherford's model of electrons orbiting nuclei.

The Mystery of Atomic Spectra

Another puzzling phenomenon unexplained by Rutherford's model, or anything else before 1913, is the spectral lines we see when looking through a **spectroscope**. A spectroscope breaks up the visible light emitted from a light source into a spectrum, so that we can see exactly which frequencies of light are being emitted.

The puzzling thing about atomic spectra is that light seems to travel only in certain distinct frequencies. For instance, we might expect the white light of the sun to transmit light in an even range of all different frequencies. In fact, however, most sunlight travels in a handful of particular frequencies, while very little or no light at all travels at many other frequencies.

Bohr's Hydrogen Atom

Niels Bohr drew on Rutherford's discovery of the nucleus and Einstein's suggestion that energy travels only in distinct quanta to develop an atomic theory that accounts for why electrons do not collapse into nuclei and why there are only particular frequencies for visible light.

Bohr's model was based on the hydrogen atom, since, with just one proton and one electron, it makes for the simplest model. As it turns out, Bohr's model is still mostly accurate for the hydrogen atom, but it doesn't account for some of the complexities of more massive atoms.

According to Bohr, the electron of a hydrogen atom can only orbit the proton at certain distinct radii. The closest orbital radius is called the electron's **ground state**. When an electron absorbs a certain amount of energy, it will jump to a greater orbital radius. After a while, it will drop spontaneously back down to its ground state, or some other lesser radius, giving off a photon as it does so.

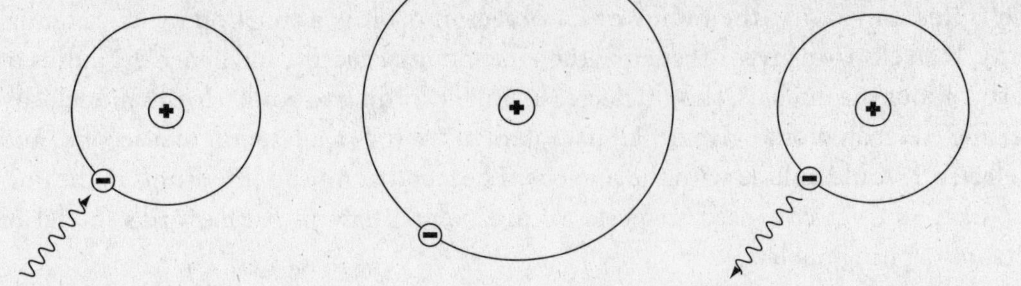

Because the electron can only make certain jumps in its energy level, it can only emit photons of certain frequencies. Because it makes these jumps, and does not emit a steady flow of energy, the electron will never spiral into the proton, as Rutherford's model suggests.

Also, because an atom can only emit photons of certain frequencies, a spectroscopic image of the light emanating from a particular element will only carry the frequencies of photon that element can emit. For instance, the sun is mostly made of hydrogen, so most of the light we see coming from the sun is in one of the allowed frequencies for energy jumps in hydrogen atoms.

Analogies With the Planetary Model

Because the electron of a hydrogen atom orbits the proton, there are some analogies between the nature of this orbit and the nature of planetary orbits. The first is that the centripetal force in both cases is $F = mv^2/r$. That means that the centripetal force on the electron is directly proportional to its mass and to the square of its orbital velocity, and is inversely proportional to the radius of its orbit.

The second is that this centripetal force is related to the electric force in the same way that the centripetal force on planets is related to the gravitational force:

$$F = \frac{mv^2}{r} = k_0\frac{(e)(Ze)}{r^2}$$

where e is the electric charge of the electron, and Ze is the electric charge of the nucleus. Z is a variable for the number of protons in the nucleus, so in the hydrogen atom, $Z = 1$.

The third analogy is that of potential energy. If we recall, the gravitational potential energy of a body in orbit is $U = -Gm_1m_2/r$. Analogously, the potential energy of an electron in orbit is:

$$U = -k_0\frac{(e)(Ze)}{r}$$

Differences from the Planetary Model

However, the planetary model places no restriction on the radius at which planets may orbit the sun. One of Bohr's fundamental insights was that the angular momentum of the electron, L, must be an integer multiple of $h/2\pi$. The constant $h/2\pi$ is so common in quantum physics that it has its own symbol, $\hbar$. If we take n to be an integer, we get:

$$L = mvr = n\hbar$$

Consequently, $v = n\hbar/mr$. By equating the formula for centripetal force and the formula for electric force, we can now solve for r:

$$F = k_0 \frac{(e)(Ze)}{r^2} = \frac{m(n\hbar/mr)^2}{r}$$

$$r = \frac{n^2\hbar^2}{mk_0 Ze^2}$$

Don't worry: you don't need to memorize this equation. What's worth noting for the purposes of the SAT II is that there are certain constant values for r, for different integer values of n. Note also that r is proportional to n^2, so that each successive radius is farther from the nucleus than the one before.

Electron Potential Energy

The importance of the complicated equation above for the radius of an orbiting electron is that, when we know the radius of an electron, we can calculate its potential energy. Remember that the potential energy of an electron is $U = -k_0(e)(Ze)/r$. If you plug in the above values for r, you'll find that the energy of an electron in a hydrogen atom at its ground state (where $n = 1$ and $Z = 1$) is -13.6 eV. This is a negative number because we're dealing with *potential* energy: this is the amount of energy it would take to free the electron from its orbit.

When the electron jumps from its ground state to a higher energy level, it jumps by multiples of n. The potential energy of an electron in a hydrogen atom for any value of n is:

$$U = \frac{1}{n^2}(-13.6 \text{ eV})$$

Modern Physics

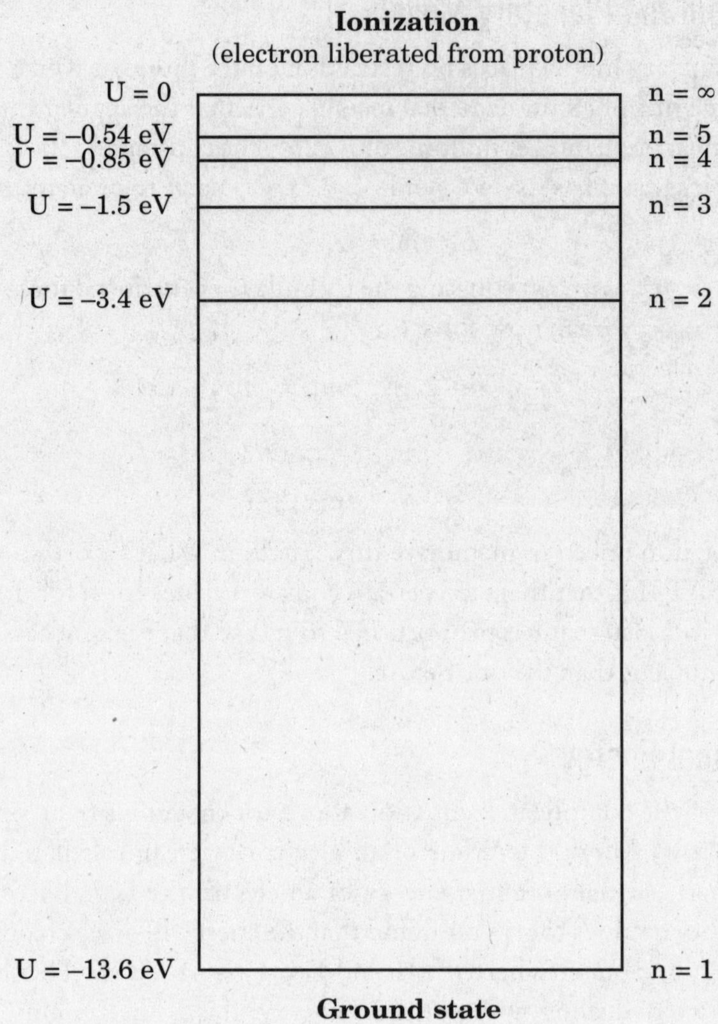

Frequency and Wavelength of Emitted Photons

As we said earlier, an excited hydrogen atom emits photons when the electron jumps to a lower energy state. For instance, a photon at $n = 2$ returning to the ground state of $n = 1$ will emit a photon with energy $E = (-3.4\text{ eV}) - (-13.6\text{ eV}) = 10.2\text{ eV}$. Using Planck's formula, which relates energy and frequency, we can determine the frequency of the emitted photon:

$$f = \frac{E}{h} = \frac{10.2\text{ eV}}{4.14 \times 10^{-15}\text{ eV} \cdot \text{s}}$$
$$= 2.46 \times 10^{-15}\text{ Hz}$$

Knowing the frequency means we can also determine the wavelength:

$$\lambda = \frac{c}{f} = 1.22 \times 10^{-7}\text{m}$$

As it turns out, this photon is of slightly higher frequency than the spectrum of visible light: we won't see it, but it will come across to us as ultraviolet radiation. Whenever an electron in a hydrogen atom returns from an excited energy state to its ground state it lets off an ultraviolet photon. When we consider that the sun is mostly made of hydrogen, we can understand why there is a great danger of ultraviolet radiation when we go to the beach without sunscreen.

Example

A hydrogen atom is energized so that its electron is excited to the $n = 3$ energy state. How many different frequencies of electromagnetic radiation could it emit in returning to its ground state?

Electromagnetic radiation is emitted whenever an electron drops to a lower energy state, and the frequency of that radiation depends on the amount of energy the electron emits while dropping to this lower energy state. An electron in the $n = 3$ energy state can either drop to $n = 2$ or drop immediately to $n = 1$. If it drops to $n = 2$, it can then drop once more to $n = 1$. There is a different amount of energy associated with the drop from $n = 3$ to $n = 2$, the drop from $n = 3$ to $n = 1$, and the drop from $n = 2$ to $n = 1$, so there is a different frequency of radiation emitted with each drop. Therefore, there are three different possible frequencies at which this hydrogen atom can emit electromagnetic radiation.

Modern Physics

Wave-Particle Duality

The photoelectric effect shows that electromagnetic waves exhibit particle properties when they are absorbed or emitted as photons. In 1923, a French graduate student named Louis de Broglie (pronounced "duh BRO-lee") suggested that the converse is also true: particles can exhibit wave properties. The formula for the so-called **de Broglie wavelength** applies to all matter, whether an electron or a planet:

$$\lambda = \frac{h}{mv}$$

De Broglie's hypothesis is an odd one, to say the least. What on earth is a wavelength when associated with matter? How can we possibly talk about planets or humans having a wavelength? The second question, at least, can be easily answered. Imagine a person of mass 60 kg, running at a speed of 5 m/s. That person's de Broglie wavelength would be:

$$\lambda = \frac{6.63 \times 10^{-34} \text{ J} \cdot \text{s}}{(60 \text{ kg})(5 \text{ m/s})} = 2.21 \times 10^{-36} \text{ m}$$

We cannot detect any "wavelength" associated with human beings because this wavelength has such an infinitesimally small value. Because h is so small, only objects with a very small mass will have a de Broglie wavelength that is at all noticeable.

Modern Physics

De Broglie Wavelength and Electrons

The de Broglie wavelength is more evident on the atomic level. If we recall, the angular momentum of an electron is $L = mvr = n\hbar$. According to de Broglie's formula, $mv = h/\lambda$. Therefore,

$$\frac{hr}{\lambda} = \frac{nh}{2\pi}$$
$$n\lambda = 2\pi r$$

The de Broglie wavelength of an electron is an integer multiple of $2\pi r$, which is the length of a single orbit. In other words, an electron can only orbit the nucleus at a radius where it will complete a whole number of wavelengths. The electron in the figure below completes four cycles in its orbit around the nucleus, and so represents an electron in the $n = 4$ energy state.

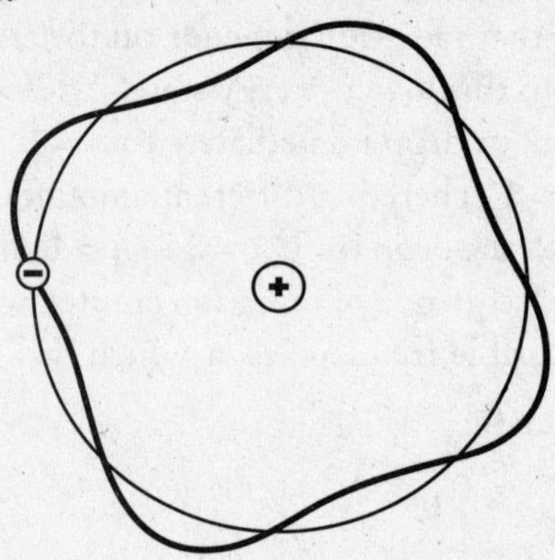

The de Broglie wavelength, then, serves to explain why electrons can orbit the nucleus only at certain radii.

Example

Which of the following explains why no one has ever managed to observe and measure a de Broglie wavelength of the Earth?

(A) The Earth is traveling too slowly. It would only have an observable de Broglie wavelength if it were moving at near light speed.

(B) The Earth is too massive. Only objects of very small mass have noticeable wavelengths.

(C) The Earth has no de Broglie wavelength. Only objects on the atomic level have wavelengths associated with them.

(D) "Wavelength" is only a theoretical term in reference to matter. There is no observable effect associated with wavelength.

(E) The individual atoms that constitute the Earth all have different wavelengths that destructively interfere and cancel each other out. As a result, the net wavelength of the Earth is zero.

This is the sort of question you're most likely to find regarding quantum physics on the SAT II: the test writers want to make sure you understand the theoretical principles that underlie the difficult concepts in this area. The answer to this question is (B). As we discussed above, the wavelength of an object is given by the formula $\lambda = h / mv$. Since h is such a small number, mv must also be very small if an object is going to have a noticeable wavelength. Contrary to (A), the object must be moving relatively slowly, and must have a very small mass. The Earth weighs 5.96×10^{24} kg, which is anything but a small mass. In fact, the de Broglie wavelength for the Earth is 3.71×10^{-64} m, which is about as small a value as you will find in this book.

Heisenberg's Uncertainty Principle

In 1927, a young physicist named Werner Heisenberg proposed a counterintuitive and startling theory: the more precisely we measure the position of a particle, the less precisely we can measure the momentum of that particle. This principle can be expressed mathematically as:

$$\Delta x \Delta p \geq \frac{h}{4\pi}$$

where Δx is the uncertainty in a particle's position and Δp is the uncertainty in its momentum.

According to the **uncertainty principle**, if you know exactly where a particle is, you have no idea how fast it is moving, and if you know exactly how fast it is moving, you have no idea where it is. This principle has profound effects on the way we can think about the world. It casts a shadow of doubt on many long held assumptions: that every cause has a clearly defined effect, that observation has no influence upon experimental results, and so on. For the SAT II, however, you needn't be aware of the philosophical conundrum Heisenberg posed—you just need to know the name of the principle, its meaning, and the formula associated with it.

Nuclear Physics

Until now, we've taken it for granted that you know what protons, neutrons, and electrons are. Within the past century, these objects have gone from being part of vaguely conjectured theories by advanced physicists to common knowledge. Unfortunately, the SAT II is going to test you on matters that go far beyond common knowledge. That's where we come in.

Basic Vocabulary and Notation

As you surely know, atoms are made up of a nucleus of protons and neutrons orbited by electrons. Protons have a positive electric charge, electrons have a negative electric charge, and neutrons have a neutral charge. An electrically stable atom will have as many electrons as protons.

Atomic Mass Unit

Because objects on the atomic level are so tiny, it can be a bit unwieldy to talk about their mass in terms of kilograms. Rather, we will often use the **atomic mass unit** (amu, or sometimes just u), which is defined as one-twelfth of the mass of a carbon-12 atom. That means that 1 amu = 1.6605×10^{-27} kg. We can express the mass of the elementary particles either in kilograms or atomic mass units

$$1\ amu = 1.6605 \times 10^{-27}\ kg$$

Particle	Mass (kg)	Mass (amu)
Proton	1.6725×10^{-27}	1.0073
Neutron	1.6747×10^{-27}	1.0086
Electron	9.11×10^{-31}	5.4863×10^{-4}

As you can see, the mass of electrons is pretty much negligible when calculating the mass of an atom.

Atomic Number, Neutron Number, and Mass Number

You're probably somewhat familiar with the periodic table and know that there are over 100 different chemical elements. An element is defined by the number of protons in the atomic nucleus. For instance, a nucleus with just one proton is hydrogen, a nucleus with two protons is helium, and a nucleus with 92 protons is uranium, the heaviest naturally occurring element. The number of protons in an atomic nucleus determines the **atomic number**, Z. In an electrically neutral atom of atomic number Z, there will be Z protons and Z electrons.

The number of neutrons in an atomic nucleus determines the **neutron number**, N. Different nuclei of the same atomic number—that is, atoms of the same element—may have different numbers of neutrons. For instance, the nuclei of most carbon atoms have six protons and six neutrons, but some have six protons and eight neutrons. Atoms of the same element but with different numbers of neutrons are called **isotopes**.

As we saw above, electrons weigh very little in comparison to protons and neutrons, which have almost identical masses. The sum of the atomic number and the neutron number, $Z + N$, gives us an atom's **mass number,** A.

Chemical Notation

The standard form for writing the chemical symbol of an element, X, is:

$$^{A}_{Z}X$$

The element's mass number is written in superscript, and the atomic number is written in subscript. You can infer the neutron number by subtracting $A - Z$. For instance, we would write the chemical symbol for the two carbon isotopes, called carbon-12 and carbon-14, as follows:

$$\text{carbon-12: } ^{12}_{6}\text{C}$$
$$\text{carbon-14: } ^{14}_{6}\text{C}$$

The same sort of system can be used to represent protons, neutrons, and electrons individually. Because a proton is the same thing as a hydrogen atom without an electron, we can represent protons by writing:

$$^{1}_{1}\text{H}^{+}$$

where the + sign shows that the hydrogen ion has a positive charge due to the absence of the electron. Neutrons are represented by the letter "n" as follows:

$$^{1}_{0}\text{n}$$

Electrons and **positrons**, which are positively charged electrons, are represented, respectively, as follows:

$$\text{electron: } ^{0}_{-1}\text{e}$$
$$\text{positron: } ^{0}_{+1}\text{e}$$

The number in subscript gives the charge of the particle—0 in the case of the neutron and –1 in the case of the electron. The number in superscript gives the mass. Though electrons have mass, it is so negligible in comparison to that of protons and neutrons that it is given a mass number of 0.

Some Other Elementary Particles

On the SAT II, you will not need to apply your knowledge of any elementary particles aside from the proton, the neutron, and the electron. However, the names of some other particles may come up, and you will at least need to know what they are.

No one entirely agrees on what **quarks** are, but they seem to be the building blocks of the protons, neutrons, and electrons that we once took to be the most elementary of particles. They generally have one-third charges, so that it takes three quarks combined to form a proton, neutron, or electron.

Neutrinos are neutrally charged particles of nearly zero mass that are released in certain kinds of radioactive decay.

Mesons are particles of mass somewhere between an electron and a proton. Common kinds of mesons are pi mesons, or **pions**, and mu mesons, or **muons**. Mesons can have a positive, negative, or neutral charge.

Radioactive Decay

Some configurations of protons and neutrons are more stable in a nucleus than others. For instance, the carbon-12 atom is more stable than the carbon-14 atom. While carbon-12 will remain stable, carbon-14 will spontaneously transform into a more stable isotope of nitrogen, releasing particles and energy in the process. Because these transformations take place at a very steady rate, archaeologists can date carbon-based artifacts by measuring how many of the carbon-14 atoms have decayed into nitrogen. These transformations are called **radioactive decay**, and isotopes and elements like carbon-14 that undergo such decay are called **radioactive**. There are three major kinds of radioactive decay.

Alpha Decay

When an atom undergoes **alpha decay**, it sheds an **alpha particle**, α, which consists of two protons and two neutrons. Through alpha decay, an atom transforms into a smaller atom with a lower atomic number. For instance, uranium-238 undergoes a very slow process of alpha decay, transforming into thorium:

$$^{238}_{92}\text{U} \rightarrow ^{234}_{90}\text{Th} + ^{4}_{2}\alpha$$

Notice that the combined mass number and atomic number of the two particles on the right adds up to the mass number and atomic number of the uranium atom on the left.

Beta decay

There are actually three different kinds of **beta decay**—β^- decay, β^+ decay, and electron capture—but the SAT II will only deal with β^- decay, the most common form of beta decay. In β^- decay, one of the protons in the nucleus transforms into a neutron, and an electron and a neutrino, υ, are ejected. A neutrino is a neutrally charged particle with very little mass. The ejected electron is called a **beta particle**, β.

The decay of carbon-14 into nitrogen is an example of β^- decay:

$$^{14}_{6}\text{C} \rightarrow ^{14}_{7}\text{N} + ^{0}_{-1}\text{e} + \nu$$

Note that the mass number of the carbon on the left is equal to the sum of the mass numbers of the nitrogen and the electron on the right: 14 = 14 + 0. Similarly, the atomic number of the carbon is equal to the sum of the atomic number of the nitrogen and the electron: 6 = 7 – 1. Because the neutrino has no charge and negligible mass, its presence has no effect on any aspect of beta decay that we will study. Still, it's important that you know the neutrino's there.

Gamma decay

Gamma decay is the most straightforward kind of decay. An element in a high-energy state can return to a lower-energy state by emitting a **gamma ray,** γ, which is an electromagnetic photon of very high frequency. No major particles are ejected and the nucleus doesn't transform from one element to another. All we get is an ejected gamma ray, as in this example with technetium:

$$^{99}_{43}\text{Tc} \rightarrow ^{99}_{43}\text{Tc} + \gamma$$

Example

$$^{24}_{11}\text{Na} \rightarrow ^{A}_{Z}X + ^{0}_{-1}\text{e}$$

The reaction schematized above is an example of what form of radioactive decay? What are the values for *A*, *Z*, and *X*?

What form of radioactive decay? In the above reaction, a sodium nucleus transforms into some other element and gives off an electron. Electrons are only released in beta decay. A neutrino is also released but, because its effects are negligible, it is often left out of the equation.

What are the values for *A*, *Z*, and *X*? We can calculate A and Z because the sum of the atomic numbers and the mass numbers on the right must add up to the atomic number and the mass number on the left. We can solve for A and Z with the following equations:

$$24 = A + 0$$
$$11 = Z - 1$$

So $A = 24$ and $Z = 12$. The resulting element is determined by the atomic number, Z. Consult a periodic table, and you will find that the element with an atomic number of 12 is magnesium, so X stands in for the chemical symbol for magnesium, Mg.

Binding Energy

Atomic nuclei undergo radioactive decay so as to go from a state of high energy to a state of low energy. Imagine standing on your hands while balancing a box on your feet. It takes a lot of energy, not to mention balance, to hold yourself in this position. Just as you may spontaneously decide to let the box drop to the floor and come out of your handstand, atomic nuclei in high-energy states may spontaneously rearrange themselves to arrive at more stable low-energy states.

Nuclear Forces

So far, all the physical interactions we have looked at in this book result either from the gravitational force or the electromagnetic force. Even the collisions we studied in the chapters on mechanics are the result of electromagnetic repulsion between the atoms in the objects that collide with one another. However, neither of these forces explains why the protons in an atomic nucleus cling together. In fact, the electromagnetic force should act to make the protons push away from one another, not cling together. Explaining how things work on the atomic level requires two additional forces that don't act beyond the atomic level: the **strong** and **weak nuclear forces**. The strong nuclear force binds the protons and neutrons together in the nucleus. The weak nuclear force governs beta decay. You don't need to know any of the math associated with these forces, but you should know what they are.

Mass Defect

As we have discussed, the mass of a proton is 1.0073 amu and the mass of a neutron is 1.0086 amu. Curiously, though, the mass of an alpha particle, which consists of two protons and two neutrons, is not 2(1.0073) + 2(1.0086) = 4.0318 amu, as one might expect, but rather 4.0015 amu. In general, neutrons and protons that are bound in a nucleus weigh less than the sum of their masses. We call this difference in mass the **mass defect**, Δm, which in the case of the alpha particle is 4.0318 – 4.0015 = 0.0202 amu.

Einstein's Famous Equation

The reason for this mass defect is given by the most famous equation in the world:

$$E = mc^2$$

As we discussed in the section on relativity, this equation shows us that mass and energy can be converted into one another.

The strong nuclear force binds the nucleus together with a certain amount of energy. A small amount of the matter pulled into the nucleus of an atom is converted into a tremendous amount of energy, the **binding energy**, which holds the nucleus together. In order to break the hold of the strong nuclear force, an amount of energy equal to or greater than the binding energy must be exerted on the nucleus. For instance, the binding energy of the alpha particle is:

$$\begin{aligned} E = mc^2 &= (3.354 \times 10^{-29}\ \text{kg})(3.0 \times 10^8\ \text{m/s})^2 \\ &= 3.0 \times 10^{-12}\ \text{J} \end{aligned}$$

Note that you have to convert the mass from atomic mass units to kilograms in order to get the value in joules. Often we express binding energy in terms of millions of electronvolts, MeV, per nucleon. In this case, 3.0×10^{-12} J = 18.7 MeV. Because there are four nucleons in the alpha particle, the binding energy per nucleon is 18.7/4 = 4.7 MeV/nucleon.

Example

A deuteron, a particle consisting of a proton and a neutron, has a binding energy of 1.12 MeV per nucleon. What is the mass of the deuteron?

Since there are two nucleons in a deuteron, the binding energy for the deuteron as a whole is $1.12 \times 2 = 2.24$ MeV. That energy, converted into mass, is:

$$m = \frac{E}{c^2} = \frac{3.59 \times 10^{-13}\ \text{J}}{(3.0 \times 10^8\ \text{m/s})^2}$$
$$= 3.99 \times 10^{-30}\ \text{kg} = 0.0024\ \text{amu}$$

The mass of a free proton plus a free neutron is 1.0073 + 1.0086 = 2.0159 amu. The mass of the deuteron will be 0.0024 amu less than this amount, since that is the amount of mass converted into energy that binds the proton and the neutron together. So the deuteron will weigh 2.0159 – 0.0024 = 2.0135 amu.

Decay Rates

On the SAT II, you probably won't be expected to calculate how long it takes a radioactive nucleus to decay, but you will be expected to know how the rate of decay works. If we take a sample of a certain radioactive element, we say that its **activity**, A, is the number of nuclei that decay per second. Obviously, in a large sample, A will be greater than in a small sample. However, there is a constant, called the **decay constant**, λ, that holds for a given isotope regardless of the sample size. We can use the decay constant to calculate, at a given time, t, the number of disintegrations per second, A; the number of radioactive nuclei, N; or the mass of the radioactive sample, m:

$$A = A_0 e^{-\lambda t}$$
$$N = N_0 e^{-\lambda t}$$
$$m = m_0 e^{-\lambda t}$$

A_0, N_0, and m_0 are the values at time $t = 0$. The mathematical constant e is approximately 2.718.

The decay constant for uranium-238 is about $5 \times 10^{-18}\ \text{s}^{-1}$. After one million years, a 1.00 kg sample of uranium-238 will weigh

$$m = m_0 e^{-\lambda t} = (1.00\ \text{kg})e^{-(5\times10^{-18}\ \text{s}^{-1})(3.16\times10^{9}\ \text{s})}$$
$$= 0.98\ \text{kg}$$

Uranium-238 is one of the slower decaying radioactive elements.

Half-Life

We generally measure the radioactivity of a certain element in terms of its **half-life**, $T_{1/2}$, the amount of time it takes for half of a given sample to decay. The equation for half-life, which can be derived from the equations above, is:

$$T_{1/2} = \frac{\ln 2}{\lambda}$$

You won't need to calculate the natural logarithm of 2—remember, no calculators are allowed on the test. What you will need to know is that, at time $t = T_{1/2}$, one-half of a given radioactive sample will have decayed. At time $t = 2T_{1/2}$, one-half of the remaining half will have decayed, leaving only one-quarter of the original sample. You may encounter a graph that looks something like this:

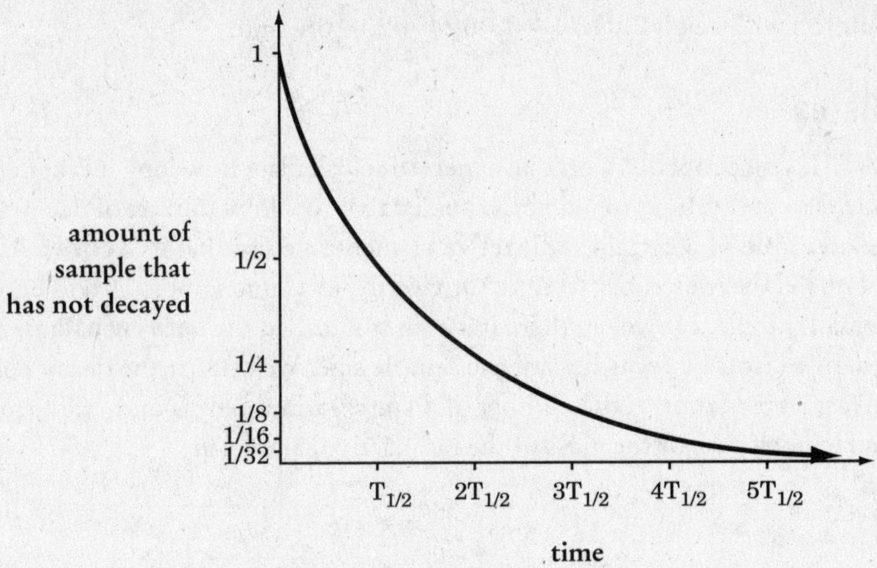

The graph of decay vs. time will get steadily closer to the x-axis, but will never actually reach it. The fewer atoms that remain undecayed, the less activity there will be.

Nuclear Reactions

Nuclear reactions are effectively the same thing as radioactivity: new particles are formed out of old particles, and the binding energy released in these transitions can be determined by the equation $E = mc^2$. The difference is that nuclear reactions that are artificially induced by humans take place very rapidly and involve huge releases of energy in a very short time. There are two kinds of nuclear reaction with which you should be familiar for the SAT II.

Nuclear Fission

Nuclear fission was used in the original atomic bomb, and is the kind of reaction harnessed in nuclear power plants. To produce nuclear fission, neutrons are made to bombard the nuclei of heavy elements—often uranium—and thus to split the heavy nucleus in two, releasing energy in the process. In the fission reactions used in power plants and atomic bombs, two or more neutrons are freed from the disintegrating nucleus. The free neutrons then collide at high speed with other atomic nuclei, starting what is called a **chain reaction**. By starting fission in just one atomic nucleus, it is possible to set off a chain reaction that will cause the fission of millions of other atomic nuclei, producing enough energy to power, or destroy, a city.

Nuclear Fusion

Nuclear fusion is ultimately the source of all energy on Earth: fusion reactions within the sun are the source of all the heat that reaches the Earth. These reactions fuse two or more light elements—often hydrogen—together to form a heavier element. As with fission, this fusion releases a tremendous amount of energy.

Fusion reactions can only occur under intense heat. Humans have only been able to produce a fusion reaction in the hydrogen bomb, or H-bomb, by first detonating an atomic bomb whose fission produced heat sufficient to trigger the fusion reaction. Scientists hope one day to produce a controllable fusion reaction, since the abundance of hydrogen found in this planet's water supply would make nuclear fusion a very cheap and nonpolluting source of energy.

Modern Physics

Key Formulas

Time Dilation	$t = \frac{t_0}{\sqrt{1-(v^2/c^2)}}$
Length Contraction	$l = l_0\sqrt{1-(v^2/c^2)}$
Addition of Relativistic Velocities	$v = \frac{u+v_0}{1+uv_0/c^2}$
Relativistic Mass	$m = \frac{m_0}{\sqrt{1-(v^2/c^2)}}$
Relativistic Kinetic Energy	$KE = (\frac{1}{\sqrt{1-(v^2/c^2)}} - 1)mc^2$
Mass-Energy Equivalence	$E = m_0c^2$
Electron-Volts Related to Joules	$1 \text{ eV} = 1.6 \times 10^{-19} \text{ J}$
Energy as a function of frequency	$E = hf$
Kinetic Energy of Liberated Photoelectron	$KE = hf - \phi$
Radius of Electron Orbit	$r = \frac{n^2\hbar^2}{mk_0Ze^2}$
Electron Potential Energy in a Hydrogen Atom	$U = \frac{1}{n^2}(-13.6 \text{ eV})$
De Broglie Wavelength	$\lambda = \frac{h}{mv}$
De Broglie Wavelength for Electron	$n\lambda = 2\pi r$
Heisenberg Uncertainty Principle	$\Delta x \Delta p \geq \frac{h}{4\pi}$
Atomic Mass Units in Kilograms	$1 \text{ amu} = 1.6605 \times 10{-}27 \text{ kg}$
Rate of Radioactive Decay	$A = A_0e^{-\lambda t}$ $N = N_0e^{-\lambda t}$ $m = m_0e^{-\lambda t}$
Half-Life of Radioactive Material	$T_{1/2} = \frac{\ln 2}{\lambda}$

Review Questions

1. A train at rest has a length of 100 m. At what speed must it approach a tunnel of length 80 m so that an observer at rest with respect to the tunnel will see that the entire train is in the tunnel at one time?

   (A) $1.25c$
   (B) $0.8c$
   (C) $0.64c$
   (D) $0.6c$
   (E) $0.36c$

2. A photon has 6.6×10^{-18} J of energy. Planck's constant, h, is 6.6×10^{-34} J · s. The frequency of the photon is most nearly:

   (A) 1.0×10^{-52} Hz
   (B) 1.0×10^{-16} Hz
   (C) 1.0 Hz
   (D) 1.0×10^{16} Hz
   (E) 1.0×10^{52} Hz

3. What happens to a stream of alpha particles that is shot at a thin sheet of gold foil?

   (A) All of the particles pass straight through
   (B) A few of the particles bounce back at 180°
   (C) All of the particles bounce back at 180°
   (D) Most of the particles are absorbed by the foil
   (E) None of the particles are deflected by more than 45°

4. According to Bohr's model of the atom, why do atoms emit or absorb radiation only at certain wavelengths?

   (A) Because the protons and electrons are distributed evenly throughout the atom
   (B) Because electrons can orbit the nucleus at any radius
   (C) Because electrons orbit the nucleus only at certain discrete radii
   (D) Because protons orbit the nucleus only at certain discrete radii
   (E) Because photons can only have discrete wavelengths

5. An electron is accelerated through a particle accelerator and then ejected through a diffraction grating. By means of the diffraction experiment, it is determined that the electron's de Broglie wavelength is $6{,}6 \times 10^{-10}$ m. What is the electron's linear momentum? Use Planck's constant, $h = 6.6 \times 10^{-34}$ J · s.

   (A) 1.0×10^{-44} kg · m/s
   (B) 1.0×10^{-24} kg · m/s
   (C) 1.0×10^{24} kg · m/s
   (D) 2.0×10^{24} kg · m/s
   (E) 1.0×10^{44} kg · m/s

6. Which of the following is the best definition of the uncertainty principle?

   (A) We cannot know for certain when any given radioactive particle will undergo decay
   (B) We cannot know both the momentum and the position of a particle at the same time
   (C) The laws of physics are the same in all intertial reference frames
   (D) Light exhibits both wave and particle properties
   (E) An unobserved particle can be in two places at the same time

7. Which of the following particles is most massive?

   (A) A proton
   (B) A neutron
   (C) An electron
   (D) A beta particle
   (E) An alpha particle

$$^{241}_{95}\text{Am} \rightarrow ^{237}_{93}\text{Np} + X$$

8. In this nuclear reaction, what particle is represented by *X*?

   (A) A proton
   (B) An electron
   (C) An alpha particle
   (D) A gamma ray
   (E) A beta particle

Questions 9 and 10 relate to the following graphs.

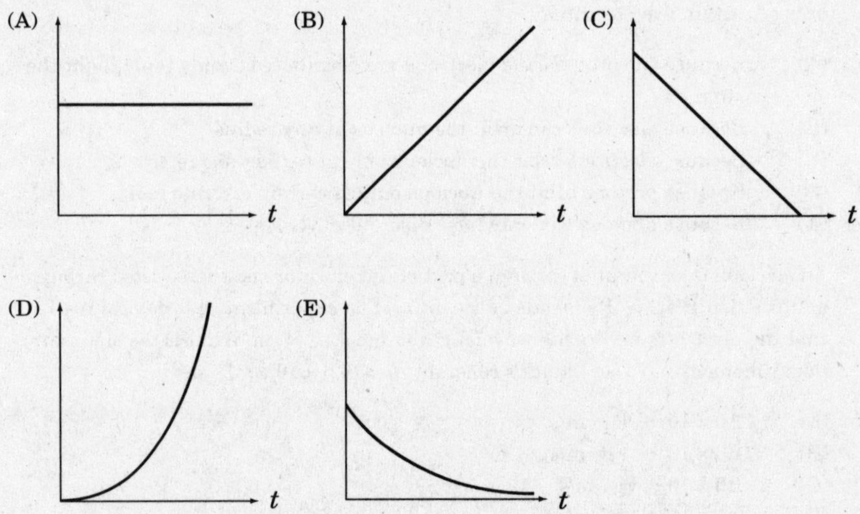

9. Which graph plots the activity of a radioactive substance as a function of time?

10. Which graph shows the half-life of a radioactive substance as a function of time?

Explanations

1. (D)

For an observer to see that the entire train is in the tunnel at one time, that observer must see that the train is only 80 m long. At relativistic speeds, the length of objects contracts in the direction of their motion according to the formula $l = l_0/\sqrt{1-(v^2/c^2)}$, where l is the relativistic length of the train, l_0 is the rest length of the train, and v is the speed of the train relative to the tunnel. Knowing that $l_0 = 100$ m and $l = 80$ m, we can solve for v:

$$\begin{aligned} 80\text{ m} &= 100\text{ m}\sqrt{1-v^2/c^2} \\ \sqrt{1-v^2/c^2} &= 0.8 \\ 1-v^2/c^2 &= 0.64 \\ v^2/c^2 &= 0.36 \\ v &= 0.6c \end{aligned}$$

2. (D)

Energy, frequency, and Planck's constant are related by the formula $E = hf$. Solving this problem is a matter of plugging numbers into this formula:

$$\begin{aligned} f &= \frac{E}{h} = \frac{6.6\times10^{-18}\text{ J}}{6.6\times10^{-34}\text{ J}\cdot\text{s}} \\ &= 1.0\times10^{16}\text{ Hz} \end{aligned}$$

3. (B)

Most of the particles will pass through with little deflection. However, some of the particles will hit the nucleus of one of the gold atoms and bounce back in the direction they came.

4. (C)

Answering this question is simply a matter of recalling what Bohr's atomic model shows us. According to Bohr's atomic model, electrons orbit the nucleus only at certain discrete radii, so (C) is the correct answer.

5. (B)

This problem asks that you apply the formula relating de Broglie wavelength to linear momentum, $\lambda = h/p$:

$$\begin{aligned} p &= \frac{h}{\lambda} = \frac{6.6\times10^{-34}\text{ J}\cdot\text{s}}{6.6\times10^{-10}\text{ m}} \\ &= 6.6\times10^{-24}\text{ kg}\cdot\text{m/s} \end{aligned}$$

6. (B)
Heisenberg's uncertainty principle tells us that we can never know both the momentum and the position of a particle at the same time, since the act of measuring one will necessarily affect the other.

7. (E)
An alpha particle is made up of two protons and two neutrons, so it is four times as massive as either a proton or a neutron. Further, protons and neutrons are nearly 2000 times as massive as an electron. A beta particle is the same thing as an electron.

8. (C)
Both atomic number and mass number are conserved in nuclear reactions. Since the mass number is 241 and the atomic number is 95 on the left side of the equation, the mass number must add up to 241 and the atomic number to 95 on the right side. Since the mass number of the Np atom is 237 and its atomic number is 93, the X atom must have a mass number of 4 and an atomic number of 2, which is the case with an alpha particle.

9. (E)
The activity of a radioactive sample, A, at time t is given by the formula $A = A_0e^{\lambda t}$, where A_0 is the activity at time $t = 0$, e is the natural constant, and λ is the decay constant. This formula tells us that the activity of a radioactive sample decreases exponentially over time, as expressed in graph (E).

10. (A)
The half-life of a radioactive substance is the constant that determines how long it will take for half of a radioactive sample to decay. Since half-life is a constant, its value does not change, as represented in graph (A).

PRACTICE TESTS

Practice Tests are Your Best Friends

In this crazy world, there is one thing that you can always take for granted: the SAT II Physics test will stay the same. From year to year and test to test, of the 75 questions on the SAT II Physics, about 5 will deal with kinematics, about 8 will deal with thermal physics, about 6 will deal with optics, and so on. Obviously, different versions of the test aren't *exactly* the same: individual questions will never repeat from test to test. But the subjects you'll be tested on, and the way in which you'll be tested, *will* remain constant. This constancy can be of great benefit to you as you study for the test.

Taking Advantage of the Test's Regularity

One day, an eleventh grader named Molly Bloom sits down at the desk in her room and takes a practice test for the SAT II Physics. She's a very bright young woman and gets only one question wrong. Molly checks her answers and then jumps from her chair and does a little dance that would be embarrassing if anyone else were around to see her.

After her euphoria passes, she begins to wonder which question she got wrong and returns to her chair. She discovers that the question dealt with optics. Looking over the question, Molly at first thinks the test made a mistake and that she was actually right, but then she realizes that she answered the question wrong because she had assumed the focal point of a diverging lens would have a positive value, when in fact it has a negative value. In thinking about the question, Molly realizes she didn't have a good grasp on which kinds of mirrors and lenses have which kinds of focal points. She studies up on her optics, sorts out *why* the focal point of a diverging lens must have a negative value, and memorizes what kinds of optical instruments have what kinds of focal points. All this takes her about ten minutes, after which she vows never to make a mistake on a question involving optics.

Analyzing Molly Bloom

Molly's actions here seem minor. All she did was study a question she got wrong until she understood why she got it wrong and what she should have done to get it right. But think about the implications. Molly answered the question incorrectly because she didn't understand the topic it was testing, and the practice test pointed out her mistaken understanding in the most noticeable way possible: she got the question wrong.

After doing her admittedly goofy little dance, Molly wasn't content simply to see what the correct answer was and get on with her day; she wanted to see *how* and *why* she got the question wrong and what she should have done, or needed to know, in order to get it right. So, with a look of determination, telling herself, "I will figure out why I got this question wrong, yes I will, yes," she spent a little while studying the question, discovered her mistaken understanding of diverging lenses, and nailed down the principles behind the situation. If Molly were to take that same test again, she definitely would not get that question wrong.

"But she never will take that test again, so she's never going to see that particular question again," some poor sap who hasn't read this guide might sputter. "She wasted her time. What a dork!"

Why that Poor Sap Really is a Poor Sap

In some sense, that poor sap is correct: Molly never will take that exact practice test again. But the poor sap is wrong to call Molly derogatory names, because, as we know, the SAT II Physics is remarkably similar from year to year—both in the topics it covers and in the way it poses questions about those topics. Therefore, when Molly taught herself about optics, she actually learned how to answer the similar questions dealing with converging lenses and concave and convex mirrors that will undoubtedly appear on every future practice test, and on the real SAT II Physics.

In studying the results of her practice test, in figuring out exactly why she got her one question wrong and what she should have known and done to get it right, Molly has targeted a weakness and overcome it.

Molly and You

Molly has it easy. She took a practice test and only got one question wrong. Less than 1% of all people who take the SAT II Physics will be so lucky. Of course, the only reason Molly got that many right was so that we could use her as an easy example.

So, what if you take a practice test and get fifteen questions wrong, and your errors span all the major topics in physics? Well, you should do exactly what Molly did: take your test and study it. Identify every question you got wrong, figure out why you got it wrong, and then teach yourself what you should have done to get the question right. If you can't figure out your error, find someone who can.

Think about it. What does an incorrect answer mean? That wrong answer identifies a weakness in your test taking, whether that weakness is an unfamiliarity with a particular topic or a tendency to be careless. If you got fifteen questions wrong on a practice test, then each of those fifteen questions identifies a weakness in your ability to take the SAT II Physics or your knowledge about the topics the SAT II Physics tests. But as you study each question and figure out why you got that question wrong, you are actually learning how to answer the very questions that will appear in similar form on the real SAT II Physics. You are discovering exactly where your weaknesses in physics lie, and addressing them. You are learning not just to understand the principles you're being tested on, but also the way that ETS will test you. Practice tests do for you what simply studying physics cannot: beyond helping you understand physics, they also prepare you specifically for the test you're going to take.

True, if you got fifteen questions wrong, the first time you study your test will take quite a bit of time. But if you invest that time and study your practice test properly, you will be eliminating future mistakes. Each successive practice test you take should have fewer errors, meaning less time spent studying those errors. Also, and more importantly, you'll be pinpointing what you need to study for the real SAT II Physics, identifying and overcoming your weaknesses, and learning to answer an increasing variety of questions on the specific topics covered by the test. Taking practice tests and studying them will allow you to teach yourself how to recognize and handle whatever the SAT II Physics has to throw at you.

Taking a Practice Test

Through the example of Miss Molly Bloom, we've shown you why studying practice tests is an extremely powerful study tool. Now we're going to explain the particulars of using that tool.

Controlling Your Environment

Although a practice test is practice, and no one but you ever needs to see your scores, you should do everything in your power to make the practice test feel like the real SAT II Physics. The closer your practice resembles the real thing, the more helpful it will be. When taking a practice test, follow these rules:

- **Time Yourself:** Don't give yourself any extra time. Be stricter with yourself than the meanest proctor you can think of. Don't give yourself time off for bathroom breaks. If you have to go to the bathroom, let the clock keep running; that's what will happen on the real SAT II.
- **Take the Test in a Single Sitting:** Training yourself to endure an hour of test taking is part of your preparation.

- **Eliminate Distractions:** Don't take the practice test in a room with lots of people walking through it. Go to a library, your bedroom, a well-lit closet—anywhere quiet.

Following these guidelines will help you to concentrate better and speed you toward your target score. However, don't be discouraged if you find these rules too strict; you can always bend a few. Preparing for the SAT II should not be torturous! Do whatever you have to do in order to make sure your studying is interesting and painless enough that you will actually do it.

Ultimately, if you can follow all of the above rules to the letter, you will probably be better off. But if following those rules makes studying excruciating, find little ways to bend them that won't interfere too much with your concentration.

Practice Test Strategy

You should take the test as if it were the real deal: go for the highest score you can get. This doesn't mean you should be more daring than you would be on the actual test, guessing blindly even when you can't eliminate an answer. It doesn't mean that you should speed through the test carelessly. The more closely your attitude and strategies during the practice test reflect those you'll employ during the actual test, the more accurately the practice test will reflect your strengths and weaknesses: you'll learn what areas you should study and how to pace yourself during the test.

Scoring Your Practice Test

After you take your practice test, you'll no doubt want to score it and see how you did. But don't just tally up your raw score. As a part of your scoring, you should keep a precise list of every question you got wrong and every question you skipped. This list will be your guide when you study your test.

Studying Your... No, Wait, Go Take a Break

You know how to have fun. Go do that for a while. Then come back when you're refreshed.

Studying Your Practice Test

After grading your test, you should have a list of the questions you answered incorrectly or skipped. Studying your test involves going down this list and examining each question you answered incorrectly. Make sure not just to learn the right answer but also to understand why you got the question wrong, and what you could have done to get the question right.

Why Did You Get the Question Wrong?

There are three main reasons why you might have gotten an individual question wrong.

1. You thought you knew the answer, but, actually, you didn't.
2. You couldn't answer the question directly, but you knew the general principles involved. Using this knowledge, you managed to eliminate some answer choices and then guessed among the remaining answers; sadly, you guessed incorrectly.
3. You knew the answer but somehow made a careless mistake.

You should know which of these reasons applies to every question you got wrong.

What You Could Have Done to Get the Question Right

If You Got a Question Wrong for Reason (1) or (2): Lack of Knowledge

Reasons (1) and (2) are variants of one another, and there is a pretty smooth continuum that runs between the two. Both result from a lack of knowledge of some of the principles of physics. Discovering a wrong answer in this domain gives you an opportunity to target your weakness. When addressing that weakness, make sure that you don't just look at the facts. For example, if you got a question wrong that dealt with resistors in parallel, don't just memorize the rule for calculating the total resistance of a set of resistors in parallel. First, go over how that rule can be derived from the simpler formulas that you're familiar with: understand *why* that rule is the way it is. And don't stop once you've grasped the principle of resistors in parallel. Take this opportunity to look over resistors in series and review DC circuits in general. Make sure you're comfortable with Kirchhoff's Rules: they're very handy in sorting out how current and voltage work in a circuit. There's no use in making sure you won't get that exact question wrong a second time, because you'll never have to answer that exact question again. Instead, cut to the heart of the problem, and understand the principles that would lead you to a correct answer on this question and any other related question.

You can also use these wrong answers to review your guessing strategy. Did you guess smartly? Could you have eliminated more answers? If yes, why didn't you? By thinking in this critical way about the decisions you made while taking the test, you can train yourself to make better, more informed decisions in a more decisive fashion.

Because you can better remember what you were thinking shortly after taking the test, it's important that you don't wait several days before coming back to the test and studying your wrong answers. Take a break after you finish writing and grading the test, but be sure to return to it while the test-taking experience is still fresh in your mind.

If You Got a Question Wrong for Reason (3): Carelessness

If you discover you got a question wrong because you were careless, it might be tempting to say to yourself, "Oh I made a careless error," and assure yourself you won't do that again. That is not enough. You made that careless mistake for a reason, and you should try to figure out why. While getting a question wrong because you didn't know the answer constitutes a weakness in your knowledge about the test, making a careless mistake represents a weakness in your *method of taking the test.*

To overcome this weakness, you need to approach it in the same critical way you would approach a lack of knowledge. Study your mistake. Reenact your thought process on the problem and see where and how your carelessness came about. Were you rushing? Did you jump at the first answer that seemed right instead of reading all the answers? Do you have trouble telling the difference between the letters C and D? Know your error, and look it in the eye. If you learn precisely what your mistake was, you are much less likely to make that mistake again.

If You Left a Question Blank

It is also a good idea to study the questions you left blank on the test, since those questions constitute a reservoir of lost points. A blank answer is a result either of (1) a total inability to answer a question, or (2) a lack of time.

If You Left a Question Blank For Reason (1): Lack of Knowledge

Look to see if there was some way you might have been able to eliminate an answer choice or two and put yourself in a better position to guess. You should also make a particular point to study up on that topic in physics, since you clearly have a good deal of trouble with it.

If You Left a Question Blank For Reason (2): Lack of Time

Look over the question and see whether you think you could have answered it. If you definitely could, then you know that you are throwing away points and probably working too slowly. If you couldn't, then carry out the steps above: study the relevant material and review your guessing strategy.

The Secret Weapon: Talking to Yourself

Yes, it's embarrassing. Yes, you'll look silly. But physical violence aside, talking to yourself is perhaps the best way to pound something into your brain. As you go through the steps of studying a question, you should talk them out. When you verbalize something, it's much harder to delude yourself into thinking you're working if you're really not.

We are serious about this advice, but, of course, it is just a suggestion. We can't enforce it and, anyway, we don't want to. But it is a nice little study trick. And it will help you.

SAT II Physics Practice Test I

PHYSICS TEST

Part A

Directions: Each set of lettered choices below refers to the numbered questions or statements immediately following it. Select the one lettered choice that best answers each question or best fits each statement, and then fill in the corresponding oval on the answer sheet. A choice may be used once, more than once, or not at all in each set.

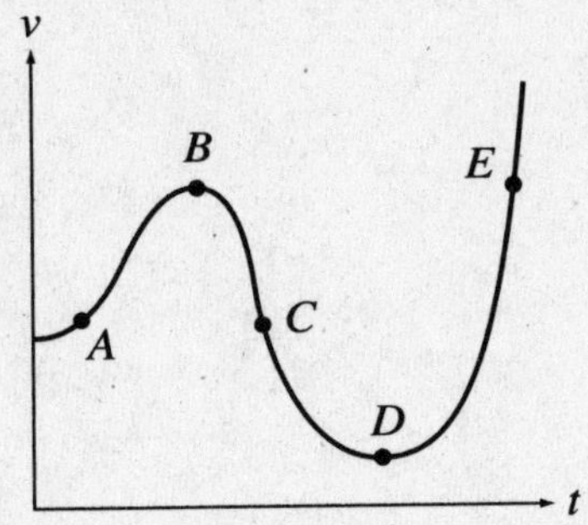

Questions 1–2

The graph of an object's velocity versus time is given above. For the following two questions, refer to the five points, labeled A through E, on the graph.

1. At what point is the acceleration greatest?

2. At what point is the acceleration the same as it is at point B?

Questions 3–5 refer to the following particles in an electric field.

(A) proton
(B) electron
(C) neutron
(D) alpha particle
(E) positron

3. Which particle does not experience an electric force from the field?

4. Which particle experiences the greatest force?

5. Which particle is accelerated in a direction opposite to the field?

Questions 6–7 relate to the following graphs of momentum vs. time.

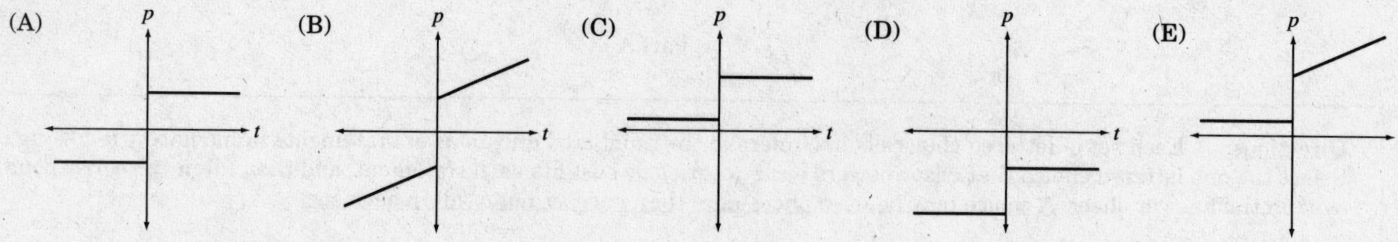

Select the graph that best expresses each of the following scenarios.

6. A hockey puck is moving with a constant velocity across the ice. A player hits the puck at time $t = 0$, and delivers an impulse to the puck in the same direction as the motion of the puck.

7. A rocket is travelling through space with a constant velocity. At time $t = 0$, the engines switch on, delivering an impulse to the rocket. The engines continue to function at a steady rate, causing the rocket to accelerate uniformly.

Questions 8–9 relate to the following graphs.

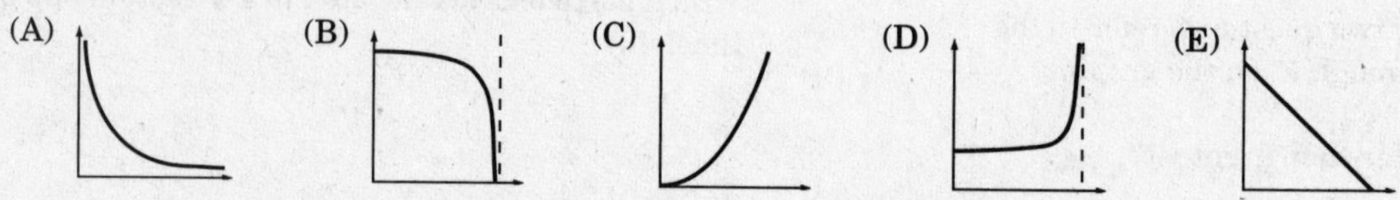

8. Which graph represents the relativistic length of an aluminum bar speeding through space vs. the speed of the bar?

9. Which graph represents the number of nuclei in a radioactive sample vs. time?

GO ON TO THE NEXT PAGE

<u>Questions 10–12</u>

Light rays from an illuminated object, O, enter a box, which contains an optical device. The resulting image is labeled I. The object is a distance d from the device in the box.

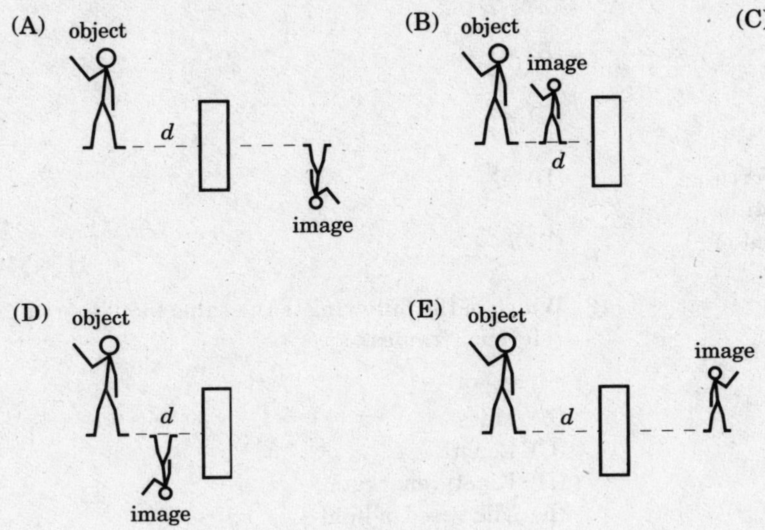

10. Which figure shows the image produced by a concave mirror whose focal length is less than d?

11. Which figure shows the image produced by a plane mirror?

12. Which figure shows a real image on the opposite side of the device from the object?

GO ON TO THE NEXT PAGE

13. A man of mass 80 kg sits in a chair. What is the magnitude of the normal force exerted by the chair on the man?

(A) 8.1 N
(B) 70.2 N
(C) 80 N
(D) 89.8 N
(E) 784 N

14. A force **F** acts on object A of mass m_1. An identical force acts on object B of mass m_2. A experiences acceleration a_1 and B experiences acceleration a_2. If $a_2 = 2a_1$, what is m_2 in terms of m_1?

(A) $\frac{1}{4}\ m_1$
(B) $\frac{1}{2}\ m_1$
(C) m_1
(D) $2m_1$
(E) $4m_1$

15. A chemist uses a mass spectrometer to find the mass ratio of an isotope in a substance. If the instrument returns the value 0.65314 for the ratio, but the chemist knows the instrument to be accurate only up to 2 percent, which of the following is the best report of the result of the measurement?

(A) 0.6
(B) 0.7
(C) 0.65
(D) 0.653
(E) 0.6531

16. The magnitude of the electric force between two charges depends on which of the following?

I. The distance between the charges
II. The magnitudes of the charges
III. The signs of the charges

(A) I only
(B) II only
(C) I and II only
(D) I and III only
(E) I, II, and III

17. Which formula describes the electric potential at a distance r from a charge q, if the potential is zero at an infinite distance?

(A) $k\frac{q}{r^2}$
(B) $k\frac{r}{q}$
(C) $k\frac{q}{r}$
(D) $k\frac{q^2}{r}$
(E) $k\frac{r}{q^2}$

18. Which of the following is the same in all inertial reference frames?

(A) Time
(B) Mass
(C) Length
(D) Kinetic energy
(E) The speed of light

19. A train traveling close to the speed of light passes by a platform. Which of the following observations might an observer on the platform make about the train?

I. The clocks inside the train move very slowly.
II. The seats in the train are shortened in the direction of the train's motion.
III. The train is very narrow from side to side.

(A) I only
(B) I and II only
(C) I and III only
(D) II and III only
(E) I, II, and III

GO ON TO THE NEXT PAGE

Questions 20–21 refer to a ball of mass 5 kg that is thrown straight up in the air to a height of 10 m, and then falls back to the ground. Assume the acceleration due to gravity is $g = 10\ m/s^2$.

20. What is the net work done by the force of gravity on the ball from the time it is thrown up to the time it returns to the ground?

(A) 0 J
(B) 5 J
(C) 50 J
(D) 500 J
(E) 1000 J

21. What is the ball's kinetic energy when it hits the ground?

(A) 0.2 J
(B) 5 J
(C) 20 J
(D) 50 J
(E) 500 J

22. A tennis player throws a ball of mass 0.500 kg straight up in the air. It takes 2.00 s for the ball to fall back to its original position. If a physicist calculates the total energy given to the ball by the player using the approximate value $g = 1.0 \times 10\ m/s^2$ for the acceleration due to gravity, which of the following is the most accurate answer he can give?

(A) 3×10 J
(B) 2.5×10 J
(C) 2.50×10 J
(D) 2.500×10 J
(E) 2.50000×10 J

Questions 23–24 refer to the diagram below.

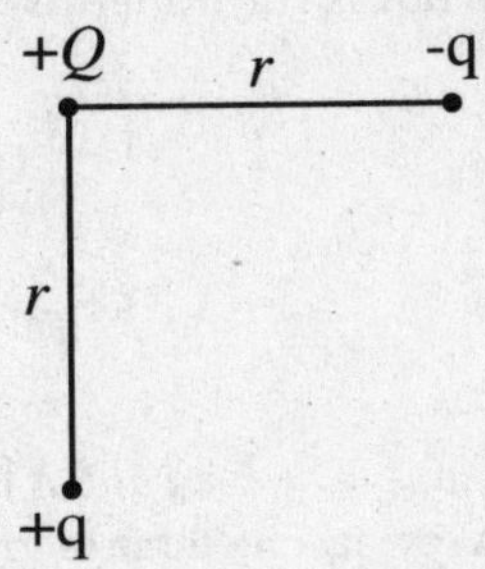

23. What is the direction of the force on the charge $+Q$?

(A) 
(B) 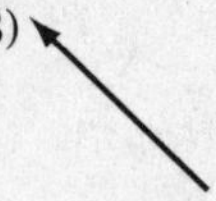
(C) 
(D) 
(E)

24. What is the magnitude of the force on the charge $+Q$?

(A) $\dfrac{2kqQ}{r^2}$
(B) $\dfrac{\sqrt{2}kqQ}{r^2}$
(C) $\dfrac{kqQ}{r^2}$
(D) $\dfrac{kqQ}{\sqrt{2}r^2}$
(E) $\dfrac{kqQ}{2r^2}$

GO ON TO THE NEXT PAGE

25. An athlete does a lap of a circular track at a speed of 10 m/s in 40 s. What is the athlete's displacement?

    (A) –400 m
    (B) –200 m
    (C) 0 m
    (D) 200 m
    (E) 400 m

26. A ball with mass $m_1 = 5$ kg and a ball with $m_2 = 10$ kg fall off a 20m tower. The force of gravity on the first ball is $\mathbf{F}_1$ and the force of gravity on the second ball is $\mathbf{F}_2$. What is the ratio $\mathbf{F}_2$ to $\mathbf{F}_1$?

    (A) 1:4
    (B) 1:2
    (C) 1:1
    (D) 2:1
    (E) 4:1

Questions 27–28 refer to the following graph, which shows the net force, **F**, in Newtons, exerted on a 5-kilogram ball as a function of time, *t*, in seconds. Assume that **F** acts in a fixed direction.

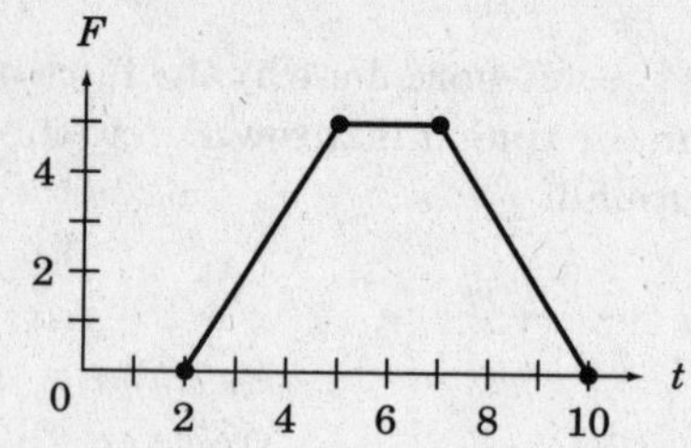

27. What is the acceleration of the ball at $t = 6$?

    (A) 0 m/s²
    (B) 1 m/s²
    (C) 5 m/s²
    (D) 10 m/s²
    (E) 25 m/s²

28. During which of the following time intervals is the speed of the ball constant?

    (A) $t = 0$ to $t = 2$
    (B) $t = 2$ to $t = 5$
    (C) $t = 5$ to $t = 7$
    (D) $t = 7$ to $t = 10$
    (E) Both $t = 0$ to $t = 2$ and $t = 5$ to $t = 7$

GO ON TO THE NEXT PAGE

29. Two vectors ***A*** and ***B*** have cross product, $\boldsymbol{C} = \boldsymbol{A} \times \boldsymbol{B}$. What angle between ***A*** and ***B*** will give the maximum magnitude for ***C***?

(A) 0°
(B) 30°
(C) 45°
(D) 90°
(E) 180°

30. A piano tuner tunes a piano by ringing a tuning fork and playing a key on the piano simultaneously. The tuning fork rings with a frequency of 100 Hz and the piano string vibrates with a frequency of 150 Hz. What is the frequency of the beats the piano tuner hears between the two dissonant sounds?

(A) –50 Hz
(B) 0.67 Hz
(C) 1.5 Hz
(D) 50 Hz
(E) 250 Hz

31. A person blows into the open end of a tube of length L that is open at one end and closed at the other. What is the wavelength of the first harmonic of the tube?

(A) $\frac{L}{4}$
(B) $\frac{L}{2}$
(C) L
(D) $2L$
(E) $4L$

32. The frequency of a mass on a spring that passes the equilibrium position twice every three seconds is most nearly

(A) 0.33 Hz
(B) 0.67 Hz
(C) 0.75 Hz
(D) 1.5 Hz
(E) 2.0 Hz

33. When light from the air enters a glass prism, which of the follwing changes?

I. The speed of the light
II. The frequency of the light
III. The wavelength of the light

(A) I only
(B) II only
(C) I and II only
(D) I and III only
(E) I, II, and III

34. Which of the following is true of an image reflected in a plane mirror?

I. It is real.
II. It is upright.
III. It is magnified ($|m| > 1$).

(A) II only
(B) I and II only
(C) II and III only
(D) I and III only
(E) I, II and III

35. Which of the following phenomena can be explained by Newton's Law of Universal Gravitation?

I. An astronaut in orbit experiences weightlessness.
II. A sandwich left on the beach is washed away by the tide.
III. A parachute slows down the descent of a parajumper.

(A) I only
(B) II only
(C) I and II only
(D) I and III only
(E) I, II, and II

36. A wire of resistance 2 Ω carries a current of 5 A. What is the emf of the circuit?

(A) 0.4 V
(B) 1 V
(C) 2.5 V
(D) 10 V
(E) 20 V

GO ON TO THE NEXT PAGE

37. A 5 kg object slides down a frictionless inclined plane of length 4 m. While it slides down, a force of 20 N perpendicular to the inclined plane acts on the object. How much work does this force do on the object?

(A) –100 J
(B) –80 J
(C) 0 J
(D) 20 J
(E) 80 J

38. A baseball of mass 0.15 kg is thrown straight up in the air from the ground with an initial speed of 10 m/s. Neglecting air resistance, what is the maximum height the baseball will reach?

(A) 0.1 m
(B) 0.2 m
(C) 1.5 m
(D) 5 m
(E) 10 m

39. A box of mass 10 kg slides on a surface at 30 m/s. The coefficient of kinetic friction between the box and the surface is 0.5. The amount of force needed to maintain the box's velocity is most nearly

(A) 0.05 N
(B) 0.5 N
(C) 5 N
(D) 25 N
(E) 50 N

40. A person stands 10 m away from a convex lens that has a focal length of 5 m. What is the magnification of the image produced by the lens?

(A) 1 (B) 2 (C) 5 (D) 10 (E) 50

41. Light of wavelength λ is incident on a barrier with two narrow slits separated by a distance d. The light forms a diffraction pattern on a screen a distance L from the barrier. How far apart are the bright fringes on the screen?

(A) $\frac{2d}{L\lambda}$

(B) $\frac{d}{2L\lambda}$

(C) $\frac{2L\lambda}{d}$

(D) $\frac{d}{L\lambda}$

(E) $\frac{L\lambda}{d}$

42. Which figures show appropriate magnetic field lines between two magnetic poles?

I.

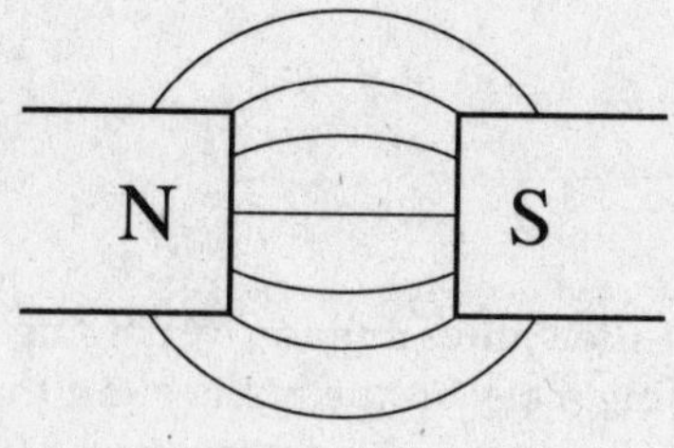

II.

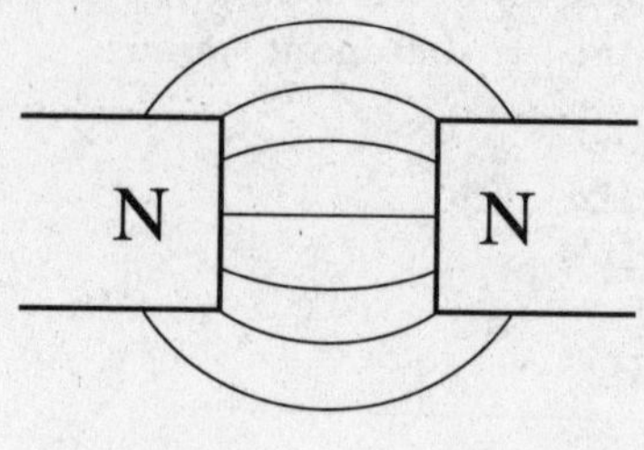

III.

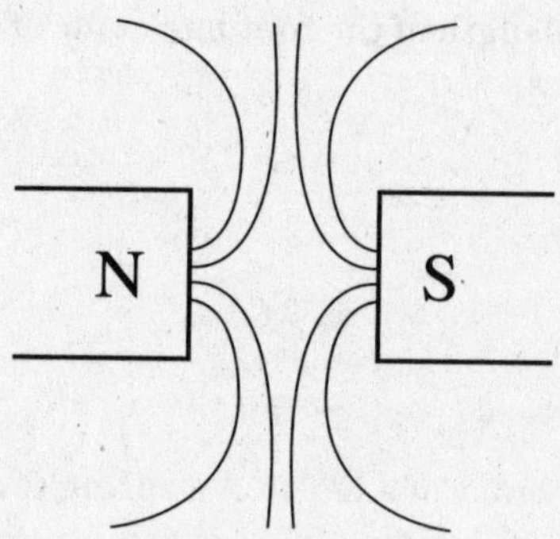

(A) I only
(B) II only
(C) I and II only
(D) II and III only
(E) I, II, and III

GO ON TO THE NEXT PAGE

43. An electron is moving in a direction perpendicular to the direction of a magnetic field. If its velocity doubles, what happens to the magnitude of the magnetic force on the electron?

(A) It is quartered.
(B) It is halved.
(C) It is unchanged.
(D) It is doubled.
(E) It is quadrupled.

44. A particle of charge 0.2 C moves at 400 m/s in a magnetic field of strength 0.2 T. If the particle's velocity is initially perpendicular to the magnetic field, what is the magnitude of the magnetic force?

(A) 1 N
(B) 4 N
(C) 16 N
(D) 40 N
(E) 160 N

45. A book that weighs 30 N is placed on a table 0.5m high. What is the book's potential energy relative to the floor?

(A) 1.5 J
(B) 15 J
(C) 60 J
(D) 150 J
(E) 600 J

46. A forklift lifts a 500 kg object up from the ground at a constant speed of 2 m/s. With what power does the forklift lift the object?

(A) 10 W
(B) 500 W
(C) 1000 W
(D) 5000 W
(E) 10,000 W

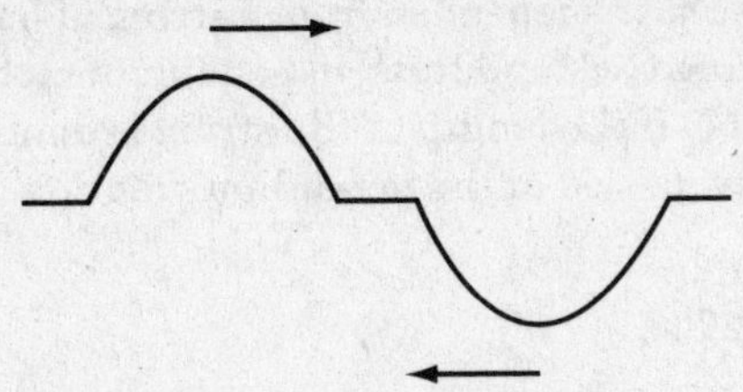

47. Two pulses with the same amplitude are moving toward each other on a rope. Which of the following statements describes their behavior?

(A) They will reflect off of each other and reverse their directions.
(B) They will cancel out, leaving no pulses in the rope.
(C) They will combine to form a pulse of twice the amplitude.
(D) They will pass through each other.
(E) They will combine to form a standing wave.

48. A violinist puts her finger onto one of the strings of her violin, reducing its length but not changing the tension. Which of the following statements about the fundamental standing wave in the string is true?

(A) The frequency increases and the wavelength decreases.
(B) The frequency decreases and the wavelength increases.
(C) The frequency and the wavelength both increase.
(D) The frequency and the wavelength both decrease.
(E) The frequency is unchanged, but the wavelength decreases.

49. Which of the following statements about nodes and antinodes in a standing wave are true?

I. There is always an antinode between two nodes.
II. Nodes and antinodes are evenly spaced.
III. Two nodes or two antinodes are separated by a distance equal to one half the wavelength of the sound wave.

(A) I only
(B) II only
(C) I and II only
(D) I and III only
(E) I, II, and III

GO ON TO THE NEXT PAGE

50. A mass m is suspended from a string of length L, swinging back and forth in pendulum motion, with a period T. If the length of the string is doubled, what is the new period of the pendulum's motion?

(A) $2T$
(B) $\sqrt{2}T$
(C) T
(D) $\frac{T}{\sqrt{2}}$
(E) $\frac{T}{2}$

51. If the radius of the earth is R, and we go down a very long mine shaft until we get to a radius of $^{R}/_{2}$, what would be the acceleration due to gravity, in terms of g, at the surface of the earth?

(A) $\frac{1}{8}g$
(B) $\frac{1}{4}g$
(C) $\frac{1}{2}g$
(D) $2g$
(E) $4g$

52. A metal rod of initial length L_0 is heated until its temperature increases by ΔT. Its length after heating is L. Which of the following is a correct expression for the rod's coefficient of linear expansion, α?

(A) $\frac{L}{L_o \Delta T}$
(B) $\Delta T(L - L_o)$
(C) $\frac{L - L_o}{L_o \Delta T}$
(D) $\Delta T\sqrt{\frac{L}{L_o}}$
(E) $\Delta T(L - L_0)$

53. It takes 1 calorie of heat to raise the temperature of 1 g of water by 1°C. The latent heat of vaporization for water is 540 cal/g. How much heat is needed to transform 1kg of water at 90°C into steam at 100°C?

(A) 550 cal
(B) 5.5×10^5 cal
(C) 640 cal
(D) 6.4×10^5 cal
(E) 541 cal

54. A heat engine absorbs 500 J from a heat source and discharges 250 J of exhaust in each heat cycle. What is the efficiency of the heat engine?

(A) 25%
(B) 50%
(C) 75%
(D) 33%
(E) 66%

55. Which of the following is the correct expression for the resistance R of wire of length L and circular cross-sectional area A, made of a material with resistivity ρ?

(A) $R = \rho\frac{A}{L}$
(B) $R = \rho\frac{L^2}{A}$
(C) $R = \rho\frac{L}{A}$
(D) $R = \rho\frac{A}{L^2}$
(E) $R = \rho\frac{L}{2A}$

56. In a circuit with a resistance of 5 Ω, what is the maximum voltage that can be applied before a 3 A fuse blows?

(A) $\frac{3}{5}$ V
(B) $\frac{5}{3}$ V
(C) 3 V
(D) 5 V
(E) 15 V

57. A cannon fires a ball at an angle of 30° with the horizontal and with an initial velocity of 100 m/s. If we take the acceleration due to gravity to be 10 m/s², how long does it take the cannonball to hit the ground? sin 30° = cos 60° = 0.500, sin 60° = cos 30° = 0.866

(A) 5 s
(B) 17.32 s
(C) 10 s
(D) 34.64 s
(E) 20 s

GO ON TO THE NEXT PAGE

58. A 10kg object, initially at rest, is subject to a uniform force with magnitude **F** = 50N. What is the velocity of the object after 4 m/s?

(A) 20 m/s
(B) 40 m/s
(C) 75 m/s
(D) 125 m/s
(E) 200 m/s

59. 5 g of water at 10°C is added to a beaker which contains 20 g of water at 60°C. What is the temperature of the water in the beaker when it reaches thermal equilibrium?

(A) 24°C
(B) 25°C
(C) 50°C
(D) 70°C
(E) 103°C

60. Which of the following are methods of heat transfer?

I. Conduction
II. Radiation
III. Diffraction

(A) I only
(B) II only
(C) I and II only
(D) I and III only
(E) I, II, and III

61. A hot-air balloonist heats the air in his balloon to make it rise higher. Why does this work?

(A) Increasing the temperature in the balloon lowers the air pressure.
(B) Increasing the temperature lowers the entropy of the gas in the balloon.
(C) Increasing the temperature lowers the density of the air in the balloon.
(D) Increasing the temperature lowers the mass of the air in the balloon.
(E) Increasing the temperature lowers the kinetic energy of the air in the balloon.

62. Which of the following particles has a mass number of two and a charge of +1?

(A) proton
(B) electron
(C) neutron
(D) alpha particle
(E) deuteron

63. When ${}^{239}_{94}\text{Pt}$ decays to ${}^{235}_{92}\text{U}$, which of the following particles is emitted?

(A) alpha particle
(B) electron
(C) beta particle
(D) positron
(E) gamma ray

GO ON TO THE NEXT PAGE

Questions 64–66 refer to a ball with mass m = 10 kg attached to one end of a massless string of length r = 5 m. The other end of the string is fixed. The ball rotates at a frequency f = 1 Hz.

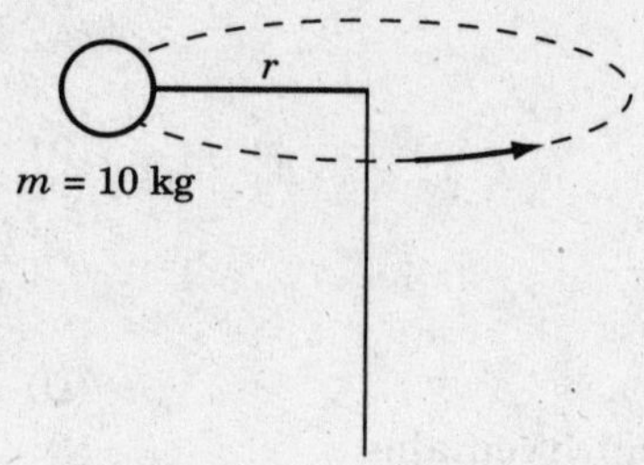

64. What is the tension on the string caused by the ball?

(A) 0.2 N
(B) 2π N
(C) $40\pi^2$ N
(D) 50 N
(E) $200\pi^2$ N

65. If the string is cut while the ball is rotating, at what speed does the ball fly away?

(A) 1 m/s
(B) 5 m/s
(C) 5π m/s
(D) 10π m/s
(E) $50\pi^2$ m/s

66. While the ball is rotating, the string is reeled in until its length is r = 3 m. If the system is not otherwise disrupted in any way, what is the new frequency of rotation?

(A) $\frac{25}{9}$ Hz
(B) $\frac{5}{3}$ Hz
(C) 1 Hz
(D) $\frac{3}{5}$ Hz
(E) $\frac{9}{25}$ Hz

GO ON TO THE NEXT PAGE

67. Which of the following uses the principle of electromagnetic induction?

(A) Transformer
(B) Electric generator
(C) Electric motor
(D) Capacitor
(E) Voltmeter

Questions 68–70 refer to the diagram below. A block of mass m is pressed against a spring with spring constant k. It is displaced a distance Δx from the equilibrium position, point B, to point A. The block is stationed on a frictionless table a distance h above the ground. When it is released, the spring pushes the block off the table and it falls to the ground. A ball of mass M is suspended by a string, also a distance h above the ground.

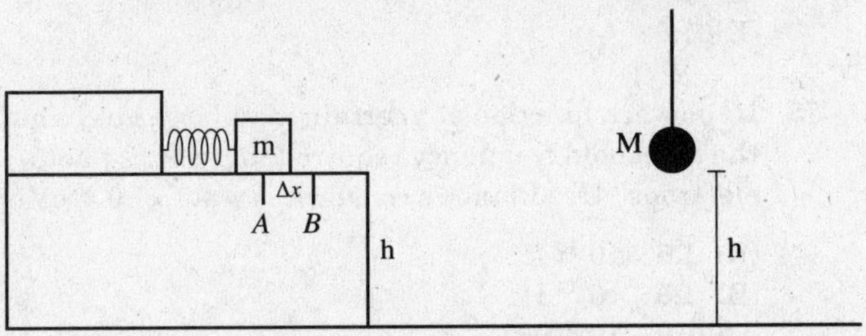

68. The block and the spring are released in such a way that the spring pushes the block off the table. As the block shoots off the table, the string attached to the ball is cut, and the ball falls to the ground. Which object will hit the ground first, and how long will it take for the other object to hit the ground after it?

(A) The block will hit the ground $\frac{m}{M}$ seconds before the ball.
(B) The block will hit the ground $\frac{M}{m}$ seconds before the ball.
(C) The ball will hit the ground $\frac{m}{M}$ seconds before the block.
(D) The ball will hit the ground $\frac{M}{m}$ seconds before the block.
(E) They will hit the ground at exactly the same time.

69. The potential energy of the block before it is released is U_{block}, and the potential energy of the ball before it is released is U_{ball}. What is the difference in potential energy between the block and the ball, $\Delta U = U_{block} - U_{ball}$?

(A) $\frac{1}{2}k\Delta x^2 - (M - m)gh$
(B) mgh
(C) $\frac{1}{2}k\Delta x^2$
(D) $mgh - \frac{1}{2}k\Delta x^2$
(E) Zero

70. What is the magnitude of the velocity of the block when it hits the ground?

(A) $\sqrt{\frac{k}{m}}\Delta x + \sqrt{2gh}$
(B) $\sqrt{\frac{k}{m}\Delta x^2 + 2gh}$
(C) $\sqrt{\frac{k}{m}\Delta x^2 2gh}$
(D) $\sqrt{2gh}$
(E) $\sqrt{\frac{k}{m}}\Delta x$

GO ON TO THE NEXT PAGE

71. A light bulb is attached to a DC circuit with a 2 A current. The voltage drop across the light bulb is 5 V. Every second, 8J of energy is dissipated as heat; the rest is dissipated as light. What is the efficiency of the light bulb?

(A) 100%
(B) 80%
(C) 50%
(D) 30%
(E) 20%

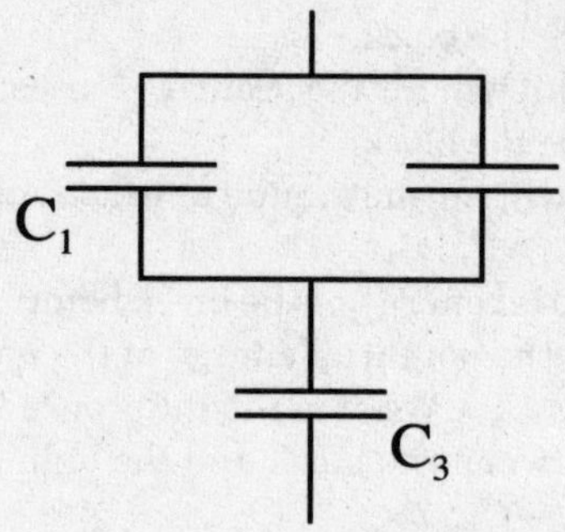

72. The capacitances of the three capacitors in the diagram above are $C_1 = 3\ \mu F$, $C_2 = 6\ \mu F$, and $C_3 = 3\ \mu F$. What is the equivalent capacitance?

(A) $\frac{4}{9}$ μF
(B) $\frac{2}{3}$ μF
(C) $\frac{3}{2}$ μF
(D) $\frac{9}{4}$ μF
(E) $\frac{7}{2}$ μF

73. A dielectric is inserted into a capacitor. What happens to the potential difference between the conductors?

(A) It increases.
(B) It decreases.
(C) First it increases, then it decreases.
(D) First it decreases, then it increases.
(E) It remains the constant.

74. Radioactive substance A has a half-life of λ, and radioactive substance B has a half-life of 2λ. If an equal amount of each substance is left to decay, what is the ratio of substance A to substance B at time $t = 6\lambda$?

(A) 1:3
(B) 3:1
(C) 1:8
(D) 8:1
(E) 1:6

75. If the work function of a certain metal is 4.1 eV, what is the threshold frequency required to produce photo-electrons? Use Planck's constant, $h \approx 4.1 \times 10^{-15}$ eV·s.

(A) 1.6×10^{-16} Hz
(B) 1.0×10^{-15} Hz
(C) 1.0×10^{15} Hz
(D) 2.0×10^{15} Hz
(E) 1.6×10^{-16} Hz

S T O P

IF YOU FINISH BEFORE TIME IS CALLED, YOU MAY CHECK YOUR WORK ON THIS TEST ONLY.
DO NOT TURN TO ANY OTHER TEST IN THIS BOOK.

SAT II Physics Practice Test I Explanations

Test I Explanations

Calculating Your Score

Your raw score for the SAT II Physics test is calculated from the number of questions you answer correctly and incorrectly. Once you have determined your composite score, use the conversion table on page 6 of this book to calculate your scaled score.

To Calculate Your Raw Score

Count the number of questions you answered correctly: ________
A

Count the number of questions you answered incorrectly, and multiply that number by $\frac{1}{4}$:

________ X $\frac{1}{4}$ = ________
B C

Subtract the value in field C from the value in field A: ________
D

Round the number in field D to the nearest whole number. This is your raw score: ________
E

Answers to SAT II Physics Practice Test I

Question Number	Correct Answer	Right	Wrong	Question Number	Correct Answer	Right	Wrong
1.	E	___	___	39.	E	___	___
2.	D	___	___	40.	A	___	___
3.	C	___	___	41.	E	___	___
4.	D	___	___	42.	B	___	___
5.	B	___	___	43.	D	___	___
6.	C	___	___	44.	C	___	___
7.	E	___	___	45.	B	___	___
8.	B	___	___	46.	E	___	___
9.	A	___	___	47.	D	___	___
10.	D	___	___	48.	A	___	___
11.	C	___	___	49.	E	___	___
12.	A	___	___	50.	B	___	___
13.	E	___	___	51.	C	___	___
14.	B	___	___	52.	C	___	___
15.	C	___	___	53.	B	___	___
16.	C	___	___	54.	B	___	___
17.	C	___	___	55.	C	___	___
18.	E	___	___	56.	E	___	___
19.	B	___	___	57.	C	___	___
20.	A	___	___	58.	A	___	___
21.	E	___	___	59.	C	___	___
22.	B	___	___	60.	C	___	___
23.	C	___	___	61.	C	___	___
24.	B	___	___	62.	E	___	___
25.	C	___	___	63.	C	___	___
26.	D	___	___	64.	E	___	___
27.	B	___	___	65.	D	___	___
28.	A	___	___	66.	C	___	___
29.	D	___	___	67.	B	___	___
30.	D	___	___	68.	E	___	___
31.	E	___	___	69.	C	___	___
32.	A	___	___	70.	B	___	___
33.	D	___	___	71.	E	___	___
34.	A	___	___	72.	D	___	___
35.	C	___	___	73.	B	___	___
36.	D	___	___	74.	C	___	___
37.	C	___	___	75.	C	___	___
38.	D	___	___	TOTAL			

Physics Shared Answer Questions

1. **(E)** Kinematics *Recall*
The slope of the graph gives us the acceleration, so the point where the slope is greatest is the point with the greatest acceleration.

2. **(D)** Kinematics *Recall*
Since acceleration is given by the slope of the graph, we need to find a point with the same slope as *B*. Both points *B* and *D* have a slope of zero. Don't be deceived by (E): at point *E*, the velocity is equal to the velocity at point *B*, but the acceleration at these two points is considerably different.

3. **(C)** Electric Forces, Fields, and Potential *Recall*
An electric field only exerts a force on charged particles. A neutron is uncharged, so it feels no electric force.

4. **(D)** Electric Forces, Fields, and Potential *Single-Concept Problem*
The electric force on a particle is given by $\mathbf{F} = q\mathbf{E}$; that is, the force due to an electric field is directly proportional to the magnitude of the charge. The alpha particle has two protons, so it has twice the electric charge of the other particles, and so it feels the greatest force.

5. **(B)** Electric Forces, Fields, and Potential *Single-Concept Problem*
Electric field lines point in the direction that a positively charged particle would move in that field. The proton, positron, and alpha particle are all positively charged, so they would move in the direction of the field lines, while the neutron would not move at all. Only the negatively charged electron would move in the opposite direction of the field.

6. **(C)** Linear Momentum *Single-Concept Problem*
Before the puck is hit, it is traveling with constant velocity—which means constant momentum. After the hit is over, no horizontal force will be acting on the puck, and it will also be traveling with constant momentum. The impulse delivered to the puck by the hockey player increases the magnitude of the momentum in the direction of motion. So we need to find a graph with constant momentum on the negative side of the *p*-axis, and constant momentum of the same sign but greater magnitude on the positive side of the *p*-axis.

7. **(E)** Linear Momentum *Single-Concept Problem*
Remember that momentum is proportional to velocity. Since the rocket has an initial constant velocity, it should also have an initial constant momentum. When the rocket begins to accelerate, its velocity increases at a steady rate, and so should its momentum. We need to find a graph that has a constant value for *p* on the negative side of the *p*-axis and a steadily increasing value for *p* on the positive side of the *p*-axis.

8. **(B)** Modern Physics *Single-Concept Problem*
An object of rest length l_o traveling at speed v will have a relativistic length, l, given by the formula $l = l_o\sqrt{1 - v^2/c^2}$. According to this formula, l will decrease as v increases, approaching 0 as v gets close to c. This is depicted in (B).

Be careful not to fall for answer (D). This choice may be tempting because it shows characteristic relativistic behavior. Remember, though, that while mass, kinetic energy, and time are extended at relativistic speeds, length contracts.

9. **(A)** Modern Physics *Single-Concept Problem*

The number of nuclei, N, in a given sample is given by the formula $N = N_0 e^{-\gamma t}$, where N_0 is the number of nuclei at time t = 0, e is the natural constant, and γ is the decay constant. This formula suggests that N decreases at t increases, but N never becomes less than zero. This relationship is represented in graph (A).

10. **(D)** Optics *Multi-Concept Problem*

You don't need to do a calculation so long as you remember that an object reflected in a concave mirror at a distance greater than the focal point will produce a real image. A real image is inverted and on the same side of the mirror as the object. Only (D) fits these criteria.

11. **(C)** Optics *Recall*

A plane mirror creates an upright image on the opposite side of the mirror from the object that is the same size and scale as the object. A plane mirror is the kind of mirror we normally use in bathrooms and bedrooms.

12. **(A)** Optics *Single-Concept Problem*

Real images are inverted, so the answer must be (A). Incidentally, this image must be produced by a lens, since mirrors only produce real images on the same side of the mirror as the object.

Physics Solitary Multiple Choice Questions

13. **(E)** Dynamics *Single-Concept Problem*

Newton's Third Law tells us that for every action, there is an equal and opposite reaction. When one object exerts a force on another object, that other object exerts a reaction force on the first object. By sitting in the chair, the man exerts a gravitational force, equal to his weight, on the chair. In response, the chair exerts a normal force equal to the man's weight back on the man. The man's weight, and hence the normal force exerted by the chair on the man, is

$$W = mg = (80 \text{ kg})(9.8 \text{ m/s}^2)$$
$$= 784 \text{ N}$$

14. **(B)** Dynamics *Single-Concept Problem*

Newton's Second Law tells us that $F = ma$. Since F is the same for both objects, and we know that $a_2 = 2a_1$, we can solve for m_2 in terms of m_1:

$$F = m_2 a_2 = m_1 a_1$$

$$2m_2 a_1 = m_1 a_1$$

$$m_2 = \frac{1}{2} m_1$$

15. **(C)** Miscellaneous (Equation Manipulation, Graph Analysis, Significant Digits) *Single-Concept Problem*

When reporting a measurement with a given accuracy, one should never report more digits than the procedure can confidently measure. The last digit reported in an account of a measurement is taken to be the first uncertain digit. Thus, for example, a mass reported as 0.10 kg is understood to lie between 0.15 and 0.05 kg. In this case, 2% of 0.653414 is 0.013, so that the true mass ratio could lie anywhere between 0.666 and 0.640. The hundredths' place is thus the first uncertain digit, so the best report of the experimental result is (C).

16. **(C)** Electric Forces, Fields, and Potential *Single-Concept Problem*

The magnitude of the electric force is given by Coulomb's Law, $F = kq_1q_2r^2$. The electric force is directly proportional to the magnitudes of the charges and inversely proportional to the square of the distance between the charges. The signs of the charges affect the direction of the force, but not its magnitude.

17. **(C)** Electric Forces, Fields, and Potential *Single-Concept Problem*

The electric potential of a point charge is directly proportional to the magnitude of the charge and inversely proportional to the distance from the charge.

18. **(E)** Modern Physics *Recall*

The theory of special relativity is based on two basic postulates. The first is that the laws of physics apply equally to all inertial reference frames. The second is that the speed of light is a constant in all inertial reference frames. From these two postulates we can derive results that show that time, mass, length, and kinetic energy all vary depending on one's reference frame.

19. **(B)** Modern Physics *Recall*

Einstein's special theory of relativity predicts that time will appear to pass more slowly in an inertial reference frame that is moving very fast relative to an observer. Therefore, the theory predicts that an observer on the platform will think that time is moving very slowly on the train, so statement I is true.

The special theory also predicts that the length of an object will contract in the direction of its motion. That means that the seats will appear shortened, and statement II is true. However, length contraction only takes place in the direction of the train's motion, so while the train will appear to be shorter, it won't appear to be any narrower, so statement III is false.

Since only statements I and II are true, we can conclude that (B) is the correct answer.

20. **(A)** Work, Energy, and Power *Single-Concept Problem*

Work is the dot product of force and displacement. The displacement vector for the ball's trajectory is zero, since it ends up at ground level, just where it started. As a result, the net work done on the ball is zero.

21. **(E)** Work, Energy, and Power *Single-Concept Problem*

The law of conservation of energy tells us that the ball's kinetic energy when it hits the ground is equal to its gravitational potential energy at its highest point, $U_g = mgh$. By calculating the gravitational potential energy at its highest point, we will know the ball's kinetic energy when it hits the ground.

$$KE = U_g = mgh = (5\text{ kg})(10\text{ m/s}^2)(10\text{ m})$$
$$= 500\text{ J}$$

22. **(B)** Miscellaneous (Equation Manipulation, Graph Analysis, Significant Digits) *Single-Concept Problem*

If you look at the answer choices, you can see that you don't actually need to calculate the energy given to the ball. All the answer choices give the same value, but they give different numbers of significant digits. When making a measurement that involves several different quantities, the resulting measurement can only be accurate to the number of significant digits of the quantity that has the smallest number of significant digits. Even though the mass of the ball and the time of the ball's flight are given to three significant digits, the value of g is given only to two significant digits, so the answer can only be given in terms of two significant digits, as in choice (B).

23. **(C)** Electric Forces, Fields, and Potential *Single-Concept Problem*

The positive point charge beneath +Q exerts a repulsive force, pushing +Q upward. The negative point charge to the right of +Q exerts an attractive force, pulling +Q to the right. The vector sum of these two forces is a force diagonally up and to the right.

24. **(B)** Electric Forces, Fields, and Potential *Single-Concept Problem*

Coulomb's law tells us that, for two charges of magnitude Q and q separated by a distance r, the force exerted by each on each is $F = kqQ/r^2$. Because the two forces acting on +Q act at right angles to one another, we can calculate the magnitude of the vector sum of these two forces using the Pythagorean Theorem. If each force has a magnitude of F, then the net force acting on +Q has a magnitude of

$$\sqrt{F^2 + F^2} = \sqrt{2}F$$

Since both forces have a magnitude of $F = kqQ/r^2$, the net force must be $\sqrt{2}F = \sqrt{2}kqQ/r^2$.

25. **(C)** Kinematics *Single-Concept Problem*

Displacement is a vector quantity, measuring the straight line distance between a starting point and an ending point. At the end of the lap, the athlete is back at the starting point, so the total displacement is zero.

26. **(D)** Dynamics *Single-Concept Problem*

The force on the balls due to gravity is $F = mg$. Since m_2 is twice m_1, that means that F_2 is twice F_1.

If you answered (C), you were probably thinking of the acceleration due to gravity. All objects have the same acceleration under the influence of gravity, but the force acting upon them differs. Remember, the term for the force of gravity acting on an object is its weight, and not all objects weigh the same.

27. **(B)** Dynamics *Single-Concept Problem*

According to the graph, the force acting on the ball at $t = 6$ is 5 N. Applying Newton's Second Law, we can solve for the acceleration of the ball:

$$a = \frac{F}{M} = \frac{5 \text{ N}}{5 \text{ kg}}$$
$$= 1 \text{ m/s}^2$$

28. **(A)** Dynamics *Single-Concept Problem*

The speed of the ball is constant when its acceleration is zero. According to Newton's Second Law, $F = ma$, so if $a = 0$, then $F = 0$ as well. According to the graph, $F = 0$ only between t = 0 and t = 2.

Remember that, though the force is constant between $t = 5$ and $t = 7$, the speed is not. When there is a constant, non-zero force acting on an object, it experiences acceleration, which changes the speed of the ball.

29. **(D)** Vectors *Single-Concept Problem*

The magnitude of a vector, **C**, formed by the cross product between two vectors, **A** and **B**, is given by

$$\mathbf{C} = \mathbf{AB} \sin \theta$$

where θ is the angle between the two vectors. Since the sine function is maximum at 90°, the answer is (D).

Don't be confused by (A) or (E). These angles maximize the dot product, not the cross product. The magnitude of the cross product is maximized when the vectors are perpendicular. The magnitude of the dot product is maximized when the vectors are parallel.

30. **(D)** Waves *Single-Concept Problem*

When two sounds of different frequencies play at the same time, they produce beats with a frequency that is the difference between the frequencies of the two sounds. In this case, the beats have a frequency of 150 Hz – 100 Hz = 50 Hz.

Don't be fooled by answer (A): frequency, as a measure of beats per second, can only have a positive value, so this answer choice is nonsensical.

31. **(E)** Waves *Recall*

The open tube of the question looks like this:

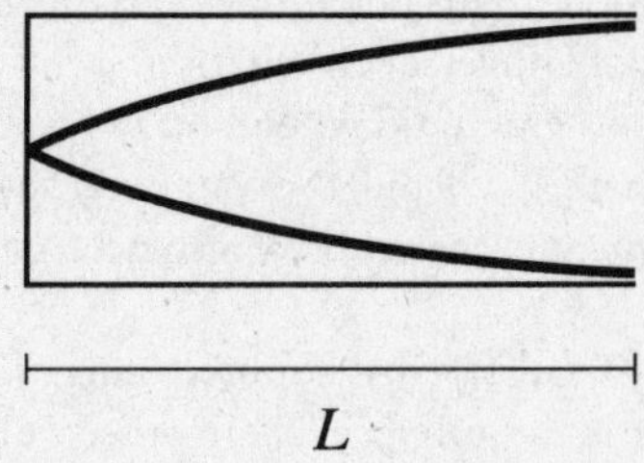

A wavelength is the distance it takes a wave to complete a cycle. From this figure, we can see that the length of the tube is only one quarter of a wavelength. That means a whole wavelength is four times the length of the tube.

32. **(A)** Waves *Single-Concept Problem*

In a given cycle, a mass on a spring will pass the equilibrium position twice. That means that the mass has one cycle every three seconds. Frequency is defined as the number of cycles per second, so in this case, if the mass undergoes one cycle every three seconds, that means it undergoes one third of a cycle in a second, and so has a frequency of 0.33 Hz.

33. **(D)** Optics *Recall*

The speed of light depends on the medium it's in: the denser the medium, the slower the speed of light. When light passes from air into glass, it slows down, so I is true.

The frequency of the light does not change. You might be tempted to think that II is true, since white light that passes into a prism breaks into many different colors of many different frequencies. However, this is not because the white light is changing frequency. White light is made up of light of all different frequencies. When it passes into a prism, the phenomenon of dispersion causes the white light to break up into its constituent frequencies, but these frequencies themselves do not change, so II is false.

The velocity, frequency, and wavelength of a beam of light are related by the formula $v = f\lambda$. If v changes when light passes into glass and f doesn't, then λ must change in direct proportion to the change in v. Therefore, III is true.

Since I and III are true but II is not, we can safely conclude that (D) is the correct answer.

34. **(A)** Optics *Multi-Concept Problem*

A plane mirror is a mirror without a focal point, or rather, a mirror for which $f = \infty$, and so $1/f = 0$. Applying this result to the lens and mirror equation, we find:

$$\frac{1}{d} + \frac{1}{d'} = \frac{1}{f}$$

$$\frac{1}{d} = -\frac{1}{d'}$$

$$d = -d'$$

This formula tells us a number of things. First, since d' is negative, we know that the image appears behind the mirror and is upright. Since it is behind the mirror, it is virtual, so I is false, and since it is upright, II is true. Further, since magnification depends on the relative distance of the object and image from the mirror, and since both object and image are the same distance from the mirror, the magnification of the image is $m = 1$, so III is false. We can conclude that the answer is (A).

35. **(C)** Gravity *Recall*

A satellite orbits the earth because it is in continual free-fall around the earth. An astronaut in such a satellite is in free-fall along with everything else in the satellite. Because they are all falling at the same rate, they are weightless relative to one another. This phenomenon is due to the gravitational pull of the earth, which is explained by Newton's Law of Universal Gravitation.

The tides are affected by the moon's gravitational pull on the waters of the earth. When the moon is directly above a certain point on the earth, it will pull the water ever so slightly in that direction, causing the tide to rise. As a gravitational phenomenon, the tides are also determined by Newton's Law of Universal Gravitation.

A parachute slows a parajumper's fall not by counteracting the effects of gravity, but rather by increasing the amount of air resistance in the jumper's fall. In other words, the parachute sets up a force that opposes the force of gravity, but it doesn't change the effects of the gravitational force.

Since I and II are gravitational phenomena, and III is due to air resistance, the correct answer is (C).

36. **(D)** Circuits and Circuit Elements *Single-Concept Problem*

Solving this problem is a simple matter of applying Ohm's Law:

$$\begin{aligned} V &= IR = (5\text{ A})(2\ \Omega) \\ &= 10\text{ V} \end{aligned}$$

37. **(C)** Work, Energy, and Power *Single-Concept Problem*

The work done on an object by a force is equal to the product of the object's displacement and the component of the force in the direction of the object's displacement. In this question, the object's displacement takes place parallel to the inclined plane. Since the force exerted is perpendicular to the inclined plane, it is perpendicular to the object's displacement. As a result, there is no component of the force that is in the direction of the object's displacement, so the force does no work on the object.

38. **(D)** Work, Energy, and Power *Multi-Concept Problem*

This is a conservation of energy question: as the baseball rises, it gains gravitational potential energy, but its total mechanical energy must remain constant, so it must compensate for that gain in potential energy with a corresponding loss in kinetic energy. It will reach its maximum height when all of its kinetic energy has been converted to potential energy and its velocity is zero. Therefore, the ball's kinetic energy at ground level must equal its gravitational potential energy at the peak of its flight.

$$\begin{aligned} mgh &= \frac{1}{2}mv^2 \\ h &= \frac{v^2}{2g} \\ &= \frac{(10\text{ m/s})^2}{2(10\text{ m/s}^2)} \\ &= 5\text{ m} \end{aligned}$$

39. **(E)** Dynamics *Multi-Concept Problem*

Newton's First Law tells us that the box will maintain a constant velocity if the net fore acting on it is zero. Since there is a force of friction opposing the box's motion, the box will only maintain a constant velocity if a force is applied to the box that is equal and opposite to the force of friction. Since we know the coefficient of kinetic friction, μ_k, and the mass of the box, m, we can calculate the magnitude of the force of friction, F_f:

$$\begin{aligned} F_f &= \mu_k mg \\ &= (0.5)(10\text{ kg})(10\text{ m/s}^2) \\ &= 50\text{ N} \end{aligned}$$

Since the force of friction is 50 N, an equal and opposite force must be applied to maintain the box's velocity.

40. **(A)** Optics *Multi-Concept Problem*

The equation for magnification tells us that the magnification, m, of an object is directly proportional to the distance of the image from the lens, d', and inversely proportional to the distance of the object from the lens, d:

$$m = d'/d$$

We know the value of d, but we need to use the lens and mirror equation to calculate the value of d':

$$\frac{1}{d} + \frac{1}{d'} = \frac{1}{f}$$

$$\frac{1}{d} = \frac{1}{10}$$

$$d' = 10\text{ m}$$

Since both d and d' have the same value, namely 10 m, the magnification of the image will be $m = d'/d = 1$.

41. **(E)** Optics *Recall*

This question describes Young's double-slit experiment. Rather than recap it now, we suggest you refer back to the chapter on Optics. In that chapter's discussion of the double-slit experiment, we derive a formula that tells us that bright fringes are separated by a distance of $L\lambda/d$.

42. **(B)** Magnetic Fields and Forces *Single-Concept Problem*

Magnetic field lines go from one pole to the opposite pole, as in II, and not as in III. When two north poles are placed together, as in I, the lines push away from one another due to the repulsive force between two like poles.

43. **(D)** Magnetic Fields and Forces *Single-Concept Problem*

The magnitude of the magnetic force on a particle of charge q moving with velocity v perpendicular to the field lines of magnetic field **B** is given by $\mathbf{F} = qv\mathbf{B}$. Since force and velocity are directly proportional, doubling the velocity means doubling the force.

44. **(C)** Magnetic Fields and Forces *Single-Concept Problem*

Solving this problem is a simple matter of plugging numbers into the formula for magnetic force:

$$\begin{aligned} F = qvB &= (0.2\text{ C})(400\text{ m/s})(0.2\text{ T}) \\ &= 16\text{ N} \end{aligned}$$

45. **(B)** Work, Energy, and Power *Single-Concept Problem*

The potential energy of the book is given by the formula $U = mgh$. Since the weight of the book is $W = mg$, the potential energy is simply the book's weight multiplied by its height from the ground:

$$U = Wh = (30\text{ N})(0.5\text{ m}) = 15\text{ J}$$

46. **(E)** Work, Energy, and Power *Single-Concept Problem*

Power is given by the formula $P = W/t$. Since work is given by the formula $W = Fd$, we can equally express power as $P = Fd/t = Fv$. When a force is exerted at a constant velocity, power is the product of force and velocity.

Because the forklift is lifting the object with a constant velocity, the net force acting on the object is zero. That means that the force exerted by the forklift on the object is equal and opposite to the force of gravity on the object, which is $F = mg$. With this information, we can solve for P:

$$P = Fv = mgv = (500\text{ kg})(10\text{ m/s}^2)(2\text{ m/s}) = 10{,}000\text{ W}$$

47. **(D)** Waves *Recall*

As the pulses meet, they will undergo superposition, but they will then separate. After they meet, their direction and shape are unchanged.

48. **(A)** Waves *Multi-Concept Problem*

The frequency of a fundamental standing wave in a string is given by the formula $v/2L$, where v is the wave speed and L is the length of the string. The wave speed depends only on the type of string and the tension in the string; it does not change if the length changes. Thus if L is halved, then the frequency is doubled. Further, since frequency is inversely proportional to wavelength, an increase in frequency means a decrease in wavelength.

49. **(E)** Waves *Recall*

A standing wave is essentially a sinusoidal wave, where nodes are the places where the amplitude of the standing wave is zero and antinodes are the places where the amplitude is largest.

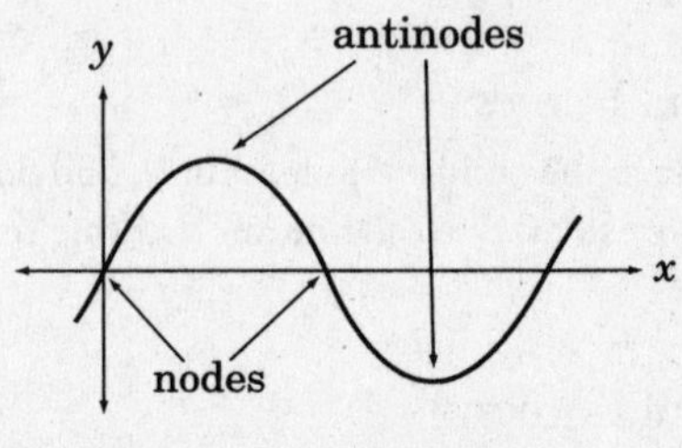

From this diagram, we can see that nodes and antinodes must be evenly spaced, and that there must be an antinode between each node. Further, we can see that there are two nodes and two antinodes in a wavelength. Therefore, all three statements are true, and so (E) is the correect answer.

50. **(B)** Special Problems in Mechanics (Pulleys, Springs, Pendulums, Inclined Planes) *Single-Concept Problem*

The period for a pendulum in motion is given by the formula:

$$T = 2\pi\sqrt{\frac{L}{g}}$$

This equation tells us that the period is directly proportional to the square root of the length of the pendulum string. If L is doubled, then T is increased by a multiple of $\sqrt{2}$.

51. (C) Gravity *Single-Concept Problem*

The gravitational constant, g, is given by $g = GM_{earth}/R_{earth}^2$. As we get closer to the center of the earth, both R and M decrease in value. R decreases because we're closer to the center of the earth, and M decreases because the only mass that's relevant to the gravitational pull is the mass that's closer to the center of the earth than we are. It turns out that there is a linear relationship between the value of g and the value of R between the surface of the earth and the center of the earth, so that if we go halfway into the earth, the value of g is halved. This can be proved with geometry, but it's easier just to memorize this fact.

52. (C) Thermal Physics *Single-Concept Problem*

The formula for linear expansion of a heated object of initial length L_o is $L - L_o = \alpha L_o\Delta T$, where α is the coefficient of linear expansion. Dividing both sides of the equation by $L_o\Delta T$ yields the coefficient $\alpha = (L - L_o)/L_o\Delta T$.

53. (B) Thermal Physics *Multi-Concept Problem*

To answer this question, we first need to calculate how much heat is needed to raise the temperature of the water to 100°C, and then calculate how much heat is needed to transform that water into steam.

1 kg of water is the same thing as 1000 g. If we want to raise 1000 g of water by 10°C it will require 10,000 cal of heat, or 1.0×10^4 cal. When it comes to turning this water into steam, it takes 540 cal per gram, so it must take 540,000 cal, or 5.4×10^5 cal, to boil 1000 g. If we add the amount of heat required to bring the water up to 100°C to the amount of heat required to boil the water, we find that the total heat needed is $1.0 \times 10^4 + 5.4 \times 10^5 = 5.5 \times 10^5$ cal.

54. (B) Thermal Physics *Single-Concept Problem*

Efficiency is given by the formula $e = 1 - Q_{out}/Q_{in}$. Substituting 250 J for Q_{out} and 500 J for Q_{in}, we find that $e = 0.5$ or 50%.

55. (C) Circuits and Circuit Elements *Recall*

The correct expression is $R = \rho L/A$. The resistance of a wire is directly proportional to its resistivity and length and inversely proportional to its cross-sectional area.

56. (E) Circuits and Circuit Elements *Single-Concept Problem*

The fuse will blow when more than 3 A of current flows through it. The voltage that would produce a current of 3 A is

$$V = IR = (3\text{ A})(5\ \Omega)$$
$$= 15\text{ V}$$

57. **(C)** Kinematics *Multi-Concept Problem*

This is a projectile motion problem involving gravity. We can break the cannonball's velocity into an x-component, which determines how far it will fly, and a y-component, which determines how high it will fly.

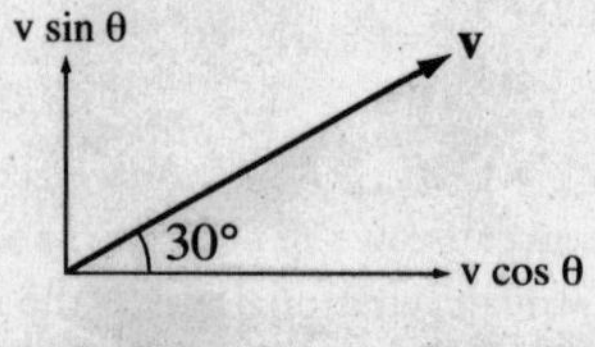

The time of the cannonball's flight is entirely dependent on the y-component of its velocity. It will spend the same amount of time in the air as a cannonball that was shot directly up into the air with a velocity $v \sin \theta$. Effectively, we are faced with a kinematics problem where we know the initial velocity, $v \sin \theta = (100 \text{ m/s})(0.500) = 50 \text{ m/s}$, the acceleration, -10 m/s^2, and the displacement, 0, and need to calculate the time, t. Displacement is zero, since when the cannonball hits the ground it will be at the same height as when it was initially fired. So let's apply the kinematic equation that disregards final velocity:

$$x = v_0 t + \frac{1}{2} a t^2$$

$$0 = (50 \text{ m/s})t + \frac{1}{2}(-10 \text{ m/s}^2)t^2$$

$$50\text{t} - 5t^2 = 0$$

$$t(10 - t) = 0$$

$$t = 0 \text{ or } 10$$

This equation tells us that the displacement of the cannonball is zero at the instant it is fired, and then 10 seconds later.

58. **(A)** Kinematics *Multi-Concept Problem*

The force gives the object an acceleration of

$$a = F/m = 50 \text{ N}/10 \text{ kg}$$

$$= 5 \text{ m/s}^2$$

Since the object starts at rest and has a constant acceleration, its velocity is the product of acceleration and time. We know the acceleration, and we know the time, so the velocity is $v = (5 \text{ m/s}^2)(4 \text{ s}) = 20 \text{ m/s}$.

59. **(C)** Thermal Physics *Multi-Concept Problem*

The heat lost by the hot water is equal to the heat gained by the cold water. The amount of heat gained or lost is given by the formula $Q = mc\Delta T$. Setting the heat change for the two substances equal to one another, we can solve for the final temperature, T, as follows:

$$m_{\text{hot}}c(T_{\text{hot}} - T) = m_{\text{cold}}c(T - T_{\text{cold}})$$

$$(20 \text{ g})c(60°\text{C} - T) = (5 \text{ g})c(T - 10°\text{C})$$

$$1200c - 20cT = 5cT - 50c$$

$$25T = 1250$$

$$T = 50°\text{C}$$

60. **(C)** Thermal Physics *Recall*

Heat can flow by conduction, convection, or radiation. Diffraction refers to the bending of waves (including electromagnetic radiation) around barriers.

61. **(C)** Thermal Physics *Single-Concept Problem*

To answer this question it is useful to recall the ideal gas law, $PV = nRT$. From this law we see that an increase in temperature at a constant volume is accompanied by an increase in pressure, so (A) is out. Similarly, an increase in temperature is accompanied by an increase in entropy, so (B) is wrong. And temperature is defined as a measure of the average kinetic energy of the gas molecules in the balloon, so (E) cannot be right. Finally, the mass of the gas isn't affected by a change in temperature, so (D) is also no good.

This leaves only (C). Heating a hot-air balloon causes the balloon to expand, and hence lowers the density of the gas inside the balloon. Because the density of the gas is lower than the density of the surrounding air, the balloon rises.

62. **(E)** Modern Physics *Recall*

The proton and neutron have a mass number of one, the electron has a mass number of zero and the alpha particle, since it consists of two protons and two neutrons, has a mass number of four. Even if you don't know what a deuteron is, you should be able to eliminate the other four answer choices. A deuteron is a hydrogen nucleus with a neutron; that is, it consists of a proton and a neutron, so it has a mass number of two and a charge of +1.

63. **(C)** Modern Physics *Single-Concept Problem*

In all radioactive decays, the mass number and atomic number must be conserved. $^{235}_{92}\text{U}$ has a mass number four lower than $^{239}_{94}\text{Pt}$ and an atomic number two lower, so the particle that is emitted must have atomic number two and mass number four. An α-particle, consisting of two protons and two neutrons, fits this description. So the correct answer is (A).

All the other listed particles have no mass number. In fact, a γ-ray, has no atomic number, either. You can also eliminate choices (B) and (C) because a β-particle is an electron. They can't both be the right answer, so they must both be the wrong answer.

64. **(E)** Circular Motion and Rotation *Multi-Concept Problem*

The tension in the string provides the centripetal force that keeps the ball in circular motion. The equation for centripetal force is $F = mv^2/r$. We know the values for m and r, and must calculate the value for v.

The average speed of the ball on the string in a given rotation is the total distance traveled divided by the total time. The distance traveled is the circumference of the circle, or $2\pi r$. The time of a single rotation is the period, T. Period is related to frequency by the formula $T = 1/f$, so the period is $T = 1/1\text{ Hz} = 1\text{ s}$. That means the average speed of the ball is $v = 2\pi r/1\text{ s} = 2\pi r/\text{s}$.

Now that we know the mass, velocity, and radius of the ball's rotation, we can calculate the force of tension:

$$F = mv^2/r = m(2\pi r\text{ s}^{-1})^2/r = 4m\pi^2r^2/r\text{ s}^2 = 4m\pi^2a = 4\pi^2(10\text{ kg})(5\text{ m/s}^2)$$
$$= 200\pi^2\text{ kg}\cdot\text{m/s}^2$$

65. **(D)** Circular Motion and Rotation *Single-Concept Problem*

When the string is cut, the ball will fly away with the velocity it had at the instant the string was cut. We calculated in the previous question that the instantaneous velocity of the ball is $v = 2\pi r/\text{s}$. Since $r = 5$ m, the instantaneous velocity of the ball is $v = 10\pi$ m/s.

66. **(C)** Circular Motion and Rotation *Multi-Concept Problem*

We know that angular momentum, $L = mvr$, is a conserved quantity. We calculated earlier that $v = 2\pi rf$. If r_1 and f_1 are the initial radius and frequency, and r_2 and f_2 are the final radius and frequency, then we can solve for f_2:

$$m(2\pi\, r_1 f_1)r_1 = m(2\pi\, r_2 f_2)r_2$$

$$f_2 = \frac{f_1 r_1^2}{r_2^2} = \frac{(1\text{ Hz})(5\text{ m})^2}{(3\text{ m})^2}$$

$$= \frac{25}{9}\text{ Hz}$$

67. **(B)** Electromagnetic Induction *Recall*

An electric generator uses induction to turn mechanical energy into electrical energy.

68. **(E)** Special Problems in Mechanics (Pulleys, Springs, Pendulums, Inclined Planes) *Multi-Concept Problem*

To calculate how long it takes for an object to hit the ground from a certain height, we need only know the height from which the object falls, and the acceleration due to gravity, *g*, which is the same for all objects. Since both these objects fall from the same height, they will hit the ground at the same time. The difference in mass, and the fact that the block is also moving horizontally, are immaterial.

69. **(C)** Special Problems in Mechanics (Pulleys, Springs, Pendulums, Inclined Planes) *Multi-Concept Problem*

The potential energy of the block is the sum of the stored spring energy and the gravitational potential energy. The potential energy of the spring is given by $U_s = {}^1\!/_2\, k\Delta x^2$, and the gravitational potential energy is given by $U_g = mgh$, so the total potential energy for the block is $U_{block} = {}^1\!/_2\, k\Delta x^2 + mgh$.

The gravitational potential energy of the ball is simply $U_{ball} = Mgh$, so the difference the two potential energies is:

$$\begin{aligned}\Delta U &= U_{block} - U_{ball} \\ &= \frac{1}{2} k\,\Delta x^2 + mgh - Mgh \\ &= \frac{1}{2} k\,\Delta x^2 - (M-m)gh\end{aligned}$$

70. **(B)** Special Problems in Mechanics (Pulleys, Springs, Pendulums, Inclined Planes) *Multi-Concept Problem*

We can solve this problem by applying the law of conservation of energy. The potential energy of the block before it is released is equal to the kinetic energy of the block when it hits the ground. We know that the kinetic energy is ${}^1\!/_2\, mv^2$. The potential energy of the block is the sum of its spring potential energy and its gravitational potential energy, $U = {}^1\!/_2\, k\Delta x^2 + mgh$. Equating potential and kinetic energies, we can solve for v:

$$\begin{aligned}\frac{1}{2} mv^2 &= \frac{1}{2} k\,\Delta x^2 + mgh \\ v^2 &= (k/m)\,\Delta x^2 + 2gh \\ v &= \sqrt{\frac{k}{m}\Delta x^2 + 2gh}\end{aligned}$$

71. **(E)** Thermal Physics *Multi-Concept Problem*

The amount of energy dissipated by the light bulb in a given time period is the power of that light bulb. The power dissipated at a point in a circuit is given by the formula $P = IV$, where I is the current and V is the voltage drop at that point in the circuit. Therefore, the power dissipated through the light bulb is $P = (2\text{ A})(5\text{ V}) = 10\text{ W}$. The unit of power, the watt, is equal to one joule per second. In other words, the light bulb dissipates 10 joules of energy per second. If 8 J of that energy are dissipated as heat, only 2 J are dissipated as light.

Efficiency is given by the formula $e = 1 - \Delta Q_{out}/\Delta Q_{in}$, where ΔQ_{out} is the amount of heat that is wasted, and ΔQ_{in} is the amount of heat that is initially put into the light bulb. Since we know that each second the light bulb dissipates 10 J of heat, 8 of which are wasted, we can calculate the efficiency:

$$e = 1 - 8\text{ J}/10\text{ J} = 0.2$$

An efficiency of 0.2 is equivalent to 20%.

72. **(D)** Circuits and Circuit Elements *Multi-Concept Problem*

First we need to determine the equivalent capacitance of the two capacitors in parallel. Adding capacitors in parallel is like adding resistors in series:

$$C_{1+2} = C_1 + C_2 = 9\ \mu\text{F}$$

Now we need to add this equivalent capacitance to the third capacitor, which is in series with the other two. We add capacitors in series as we add resistors in parallel:

$$\frac{1}{C_{1+2+3}} = \frac{1}{C_{1+2}} + \frac{1}{C_3} = \frac{1}{9}\ \mu\text{F} + \frac{1}{3}\ \mu\text{F}$$

$$C_{1+2+3} = \frac{9}{4}\ \mu\text{F}$$

73. **(B)** Circuits and Circuit Elements *Single-Concept Problem*

Inserting a dielectric into a capacitor reduces the electric field between the plates. The electrons in the dielectric are attracted to the positive plate, and create their own electric field in the dielectric. This field opposes the original field in the capacitor, and the overall electric field is reduced. Since potential difference is directly proportional to the strength of the electric field, it also decreases.

74. **(C)** Modern Physics *Single-Concept Problem*

If the half-life of substance A is λ, that means that at time $t = \lambda$ there will be $\frac{1}{2}$ of the initial sample of A, at time $t = 2\lambda$ there will be $(\frac{1}{2})^2$ of the initial sample of A, and at time $t = 6\lambda$ there will be $(\frac{1}{2})^6$ of the initial sample of A. If the half-life of substance B is 2λ, that means that at time $t = 2\lambda$ there will be $\frac{1}{2}$ of the initial sample of B, at time $t = 4\lambda$ there will be $(\frac{1}{2})^2$ of the initial sample of B, and at time $t = 6\lambda$ there will be $(\frac{1}{2})^3$ of the initial sample of B. The ratio of A to B, then, is $(\frac{1}{2})^6 : (\frac{1}{2})^3$, or $\frac{1}{64} : \frac{1}{8}$. If we multiply both sides of this ratio by 64, we get 1:8.

75. **(C)** Modern Physics *Single-Concept Problem*

If the metal has a work function of 4.1 eV, that means that a photoelectron must carry an energy of 4.1 eV in order to free an electron from the metal. Energy is related to frequency by the formula $E = hf$. Since we know that the energy required is 4.1 eV, we can use this formula to solve for f:

$$f = \frac{E}{h} = \frac{4.1\ \text{eV}}{4.1 \times 10^{-15}\ \text{eV} \cdot \text{s}}$$

$$= 1.0 \times 10^{15}\ \text{Hz}$$

SAT II Physics Practice Test II

PHYSICS TEST

Part A

Directions: Each set of lettered choices below refers to the numbered questions or statements immediately following it. Select the one lettered choice that best answers each question or best fits each statement, and then fill in the corresponding oval on the answer sheet. A choice may be used once, more than once, or not at all in each set.

Questions 1–2 refer to two point charges, q_1 and q_2, separated by a distance r.

(A) It is doubled.
(B) It is halved.
(C) It is quadrupled.
(D) It is quartered.
(E) It is unchanged.

1. If the charge of q_2 is doubled and q_1 remains the same, what happens to the magnitude of the force on q_1?

2. If r is halved, what happens to the magnitude of the force on q_1?

Questions 3–4

A mover pushes a box weighing 200 N along a floor. The coefficient of static friction between the box and the floor is 0.8, while the coefficient of kinetic friction between the box and the floor is 0.2. The mover pushes the box from rest to a constant speed of 10 m/s.

(A) 0 N
(B) 40 N
(C) 120 N
(D) 160 N
(E) 200 N

3. What is the net force acting on the box when it is moving at 10 m/s, in Newtons?

4. How much harder must the mover push initially to get the box going than to maintain the box's speed at 10 m/s?

Questions 5–6

The following particles are moving in a direction perpendicular to the field lines of a magnetic field.

(A) Proton
(B) Electron
(C) Neutron
(D) Alpha particle
(E) Positron

5. Which particle is not deflected?

6. If the particles all have the same velocity, which is deflected the most?

GO ON TO THE NEXT PAGE

Questions 7–8 refer to the following graphs.

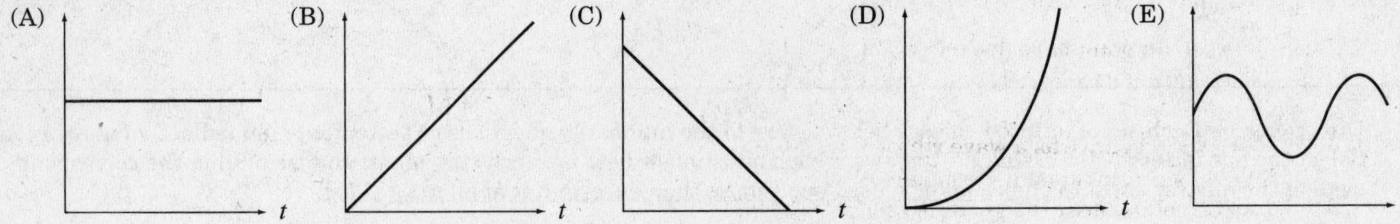

7. Which graph shows the power dissipated in a resistor carrying a constant current as a function of time?

8. Which graph shows the total heat dissipated in a resistor carrying a constant current as a function of time?

GO ON TO THE NEXT PAGE

Questions 9–11

Match each observation or concept with the phenomenon or theory that best explains it.

(A) Electrons in an atom have discrete energy levels.
(B) The decay rate of a radioactive nucleus is constant in time.
(C) All matter can behave as a wave whose wavelength is inversely proportional to its momentum.
(D) Objects travelling near the speed of light appear to be shortened in the direction of their motion.
(E) Energy from light is absorbed by matter in discrete bundles called photons; their size depends on the frequency of the light.

9. When a beam of electrons is directed against a screen with a very narrow opening, the pattern of electrons emerging from the hole exhibits a pattern of high-intensity bands separated by regions with no electrons.

10. The photoelectric effect.

11. Bohr's model of the atom.

Questions 12–13

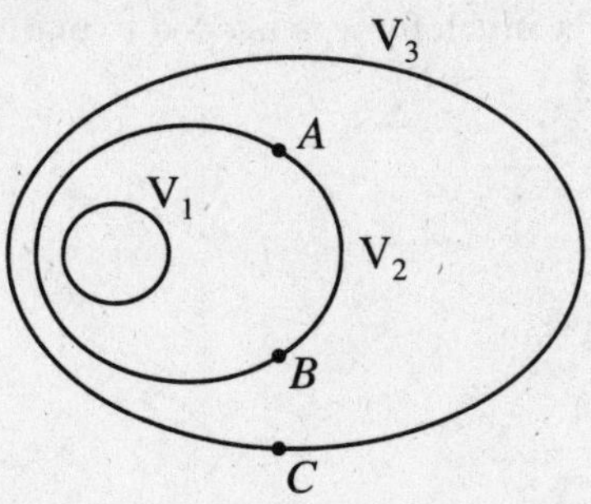

The diagram above shows equipotential lines in a region of space. A charge q is at point A.

(A) zero

(B) $q(V_3 - V_2)$

(C) $-q(V_3 - V_2)$

(D) $(V_3 - V_2)$

(E) $\frac{-q}{V_3 - V_2}$

12. What is the work done in moving the charge from point A to point B?

13. What is the work done in moving the charge from point A to point C?

GO ON TO THE NEXT PAGE

14. A box of mass 10kg slides on a frictionless surface at 30 m/s. How much force is needed to maintain this velocity?

(A) 0 N

(B) $\frac{1}{3}$ N

(C) 3 N

(D) 30 N

(E) 300 N

15. A 15-kilogram ball is suspended by a rope from the ceiling. The magnitude of the force exerted on the ball by the rope is most nearly

(A) zero
(B) 1.5 N
(C) 15 N
(D) 150 N
(E) 200 N

16. If Body 2 is twice as massive as Body 1, and the net force acting on Body 1 is four times as great as the net force acting on Body 2, what is the ratio between the acceleration of Body 1 and the acceleration of Body 2?

(A) 1 : 8
(B) 1 : 2
(C) 1 : 1
(D) 2 : 1
(E) 8 : 1

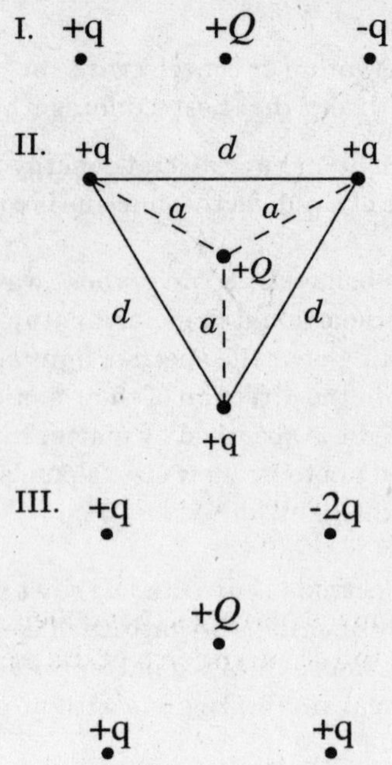

17. In which of these diagrams is the point charge +*Q* in static equilibrium?

(A) I only
(B) II only
(C) III only
(D) I and II only
(E) I, II, and III

GO ON TO THE NEXT PAGE

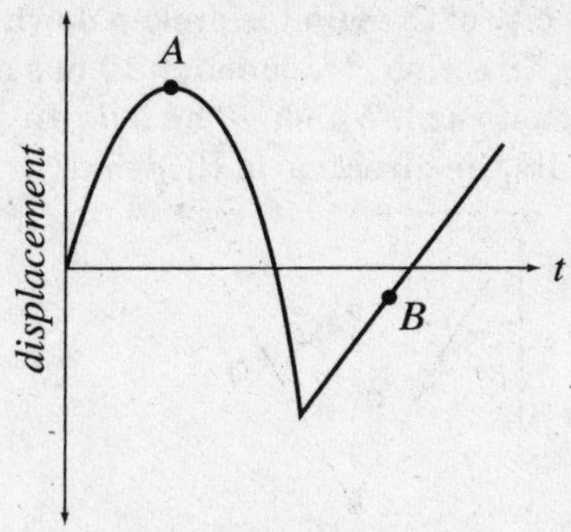

Questions 18–19 refer to the displacement vs. time graph of a particle, pictured above.

18. What of the following gives correct statements about the velocity v and the acceleration a of the particle at point A?

(A) $v > 0, a = 0$
(B) $v = 0, a > 0$
(C) $v = 0, a = 0$
(D) $v = 0, a < 0$
(E) $v < 0, a = 0$

19. What are the values of the velocity, v, and acceleration, a, of the particle at point B?

(A) $v > 0, a > 0$
(B) $v > 0, a = 0$
(C) $v = 0, a = 0$
(D) $v < 0, a > 0$
(E) $v < 0, a = 0$

20. What is the resulting image of an object placed in front of a concave mirror, where the object is closer to the mirror than the mirror's focal point?

(A) real, upright, and larger
(B) virtual, upright, and larger
(C) real, inverted, and smaller
(D) virtual, inverted, and larger
(E) virtual, upright, and smaller

21. Eyeglasses work by which of the following principles?

(A) Reflection
(B) Refraction
(C) Diffraction
(D) Interference
(E) Dispersion

22. 30 J of heat are added to a system, then the system does 50 J of work. What is the change in the internal energy of the system?

(A) +80 J
(B) +20 J
(C) It does not change.
(D) –20 J
(E) –80 J

23. In a certain car, for every 100 J produced in the engine, 80J is dissipated as heat in the engine, 10 J work the car's internal electronics, 4 J is dissipated because of air resistance, and 6 J is dissipated because of friction with the road. What is the efficiency of the car's engine in its capacity to propel the car forward?

(A) 80%
(B) 14%
(C) 10%
(D) 6%
(E) 4%

GO ON TO THE NEXT PAGE

24. A force **F** = 10 N acts on a 5 kg object while the object moves a distance of 20 m. If **F** is parallel to direction of motion, how much work does it do?

(A) 0 J
(B) 0.5 J
(C) 2 J
(D) 20 J
(E) 200 J

25. Two motors lift identical loads to a platform at constant speeds. The first motor takes 3 minutes and the second takes 6 minutes. What is the power output of the second motor, P_2, in terms of that of the first, P_1?

(A) $P_2 = \frac{1}{4}P_1$

(B) $P_2 = \frac{1}{2}P_1$

(C) $P_2 = P_1$

(D) $P_2 = 2P_1$

(E) $P_2 = 4P_1$

26. A man drags a box 10 m along a floor, against a frictional force of 20 N. How much work does the frictional force do in moving the box forward?

(A) –200 J
(B) –20 J
(C) 0 J
(D) 20 J
(E) 200 J

27. The velocity of the wind is broken down into its compass direction components: 10 m/s northward and 20 m/s eastward. Which of the following vectors best represents the direction of the wind?

(A) 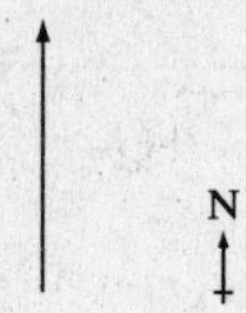

(B) 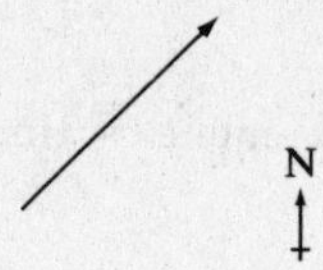

(C) 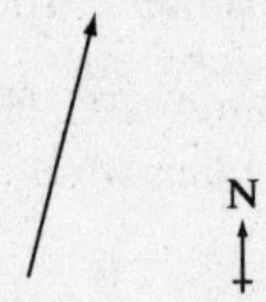

(D) 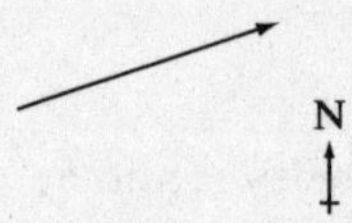

(E) 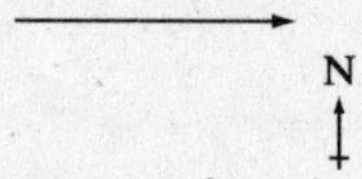

GO ON TO THE NEXT PAGE

Questions 28–29 refer to the diagram below.

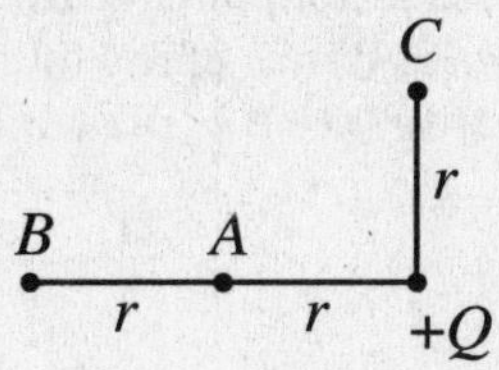

28. If a point charge is moved from point A to point B, then how does its electric potential from the $+Q$ charge source change?

(A) It is halved.
(B) It is doubled.
(C) It is quartered.
(D) It is quadrupled.
(E) It is unchanged.

29. A point charge $-q$ is moved from A to C. How much work is done by the electric field created by the $+Q$ charge?

(A) $\frac{kQq}{2r}$

(B) $-\frac{kQq}{2r}$

(C) $\frac{kQq}{4r}$

(D) $-\frac{kQq}{4r}$

(E) Zero

30. Which of the following are examples of transverse waves?

I. Radio Waves
II. Sound waves
III. Vibrations on a violin string

(A) I only
(B) II only
(C) III only
(D) I and III only
(E) I, II, and III

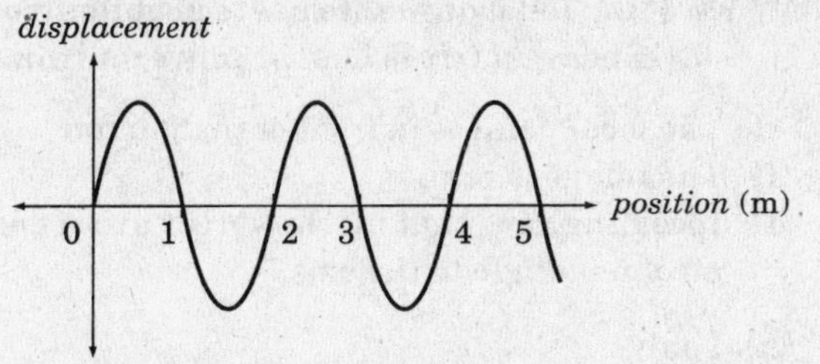

31. The diagram above depicts a wave at an instant in time. If the wave travels at a speed of 10 m/s, what is its frequency?

(A) 1 Hz
(B) 2 Hz
(C) 5 Hz
(D) 10 Hz
(E) 20 Hz

32. A wave on a string propagates with velocity v. If the tension in the string is doubled, what is the new velocity of the wave?

(A) $\frac{v}{2}$

(B) $\frac{v}{\sqrt{2}}$

(C) v

(D) $\sqrt{2}v$

(E) $2v$

33. Which of the following properties is the same for all visible light in a vacuum?

(A) Polarization
(B) Wavelength
(C) Frequency
(D) Amplitude
(E) Wave speed

GO ON TO THE NEXT PAGE

34. Which of the following statements about mirrors are true for concave mirrors but not convex mirrors?

    I. The focal point is in front of the mirror.
    II. The image is real.
    III. Incoming rays that hit the vertex are reflected at an equal angle to the axis.

    (A) I only
    (B) II only
    (C) I and II only
    (D) I and III only
    (E) I, II, and III

35. A scientist measures the mass of a sample to an accuracy of 1%. Which of the following is most likely the mass he measured?

    (A) 44.4 g
    (B) 44 g
    (C) 4.4 g
    (D) 4 g
    (E) 0.4 g

36. A 1 kg book is being pushed across a horizontal table with a force of 5 N. The coefficient of kinetic friction between the book and the table is 0.3. What is the acceleration of the book?

    (A) 0 m/s^2
    (B) 2 m/s^2
    (C) 5 m/s^2
    (D) 8 m/s^2
    (E) 10 m/s^2

37. A construction worker pushes a box 10 m up an inclined plane, exerting a force of 300 N along the plane. If it takes the worker 15 seconds to get the box to the top of the incline, what is the average power with which he pushes?

    (A) 7 W
    (B) 20 W
    (C) 70 W
    (D) 100 W
    (E) 200 W

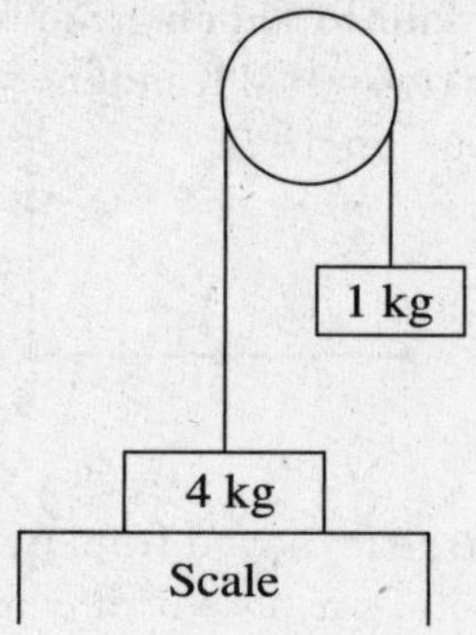

38. The figure above shows a pulley system in equilibrium. What is the weight read by the scale? Assume that $g = 10\ m/s^2$.

    (A) 0 N
    (B) 10 N
    (C) 30 N
    (D) 40 N
    (E) 50 N

39. Which statements describe properties of longitudinal waves?

    I. The waves oscillate in a direction parallel to the direction of wave propagation.
    II. The wave speed is the product of the wavelength and the frequency.
    III. They cannot undergo diffraction.

    (A) I only
    (B) II only
    (C) I and II only
    (D) I and III only
    (E) I, II, and III

40. Coherent sound waves are produced in one medium and travel into a second medium. Which of the following characteristics of the waves must remain constant?

    I. Wavelength
    II. Amplitude
    III. Frequency

    (A) I only
    (B) II only
    (C) III only
    (D) I and III only
    (E) I, II, and III

GO ON TO THE NEXT PAGE

41. Which of the following gives a formula for f_n, the nth harmonic (or resonant) frequency of a wave with speed v on a string of length L?

(A) $f_n = \frac{2nv}{L}$

(B) $f_n = \frac{nv}{L}$

(C) $f_n = \frac{2L}{nv}$

(D) $f_n = \frac{L}{nv}$

(E) $f_n = \frac{nv}{2L}$

42. Which of the following is a correct statements about the structure of the atom?

(A) The number of neutrons in an atom must be greater than the number of protons.
(B) An atom's mass number must be greater than its atomic number.
(C) Two atoms with the same atomic number must have the same mass number.
(D) The number of protons in an atom must be greater than the number of neutrons.
(E) The atomic number of an atom that emits an alpha particle must decrease by two.

43. Which of the following successfully accounts for the photoelectric effect?

(A) The Bohr model of the hydrogen atom
(B) The Rutherford model of the atom
(C) The Heisenberg uncertainty principle
(D) de Broglie wavelength
(E) Young's double-slit experiment

44. Which of the following has the same mass as a beta particle?

(A) A proton
(B) An electron
(C) A neutrino
(D) An alpha particle
(E) A nucleon

45. Which of the following is not a formulation of the Second Law of Thermodynamics?

(A) Heat flows spontaneously from a hotter body to a colder body.
(B) The internal energy of a system doing work decreases.
(C) The entropy in the universe is increasing.
(D) There is no such thing as a heat engine with 100% efficiency.
(E) Time flows in the direction of entropy increase.

46. Which of the following phrases best describes convection?

(A) An exchange of particles that increases the disorder in a system
(B) An exchange of heat between a hot solid object and a cold solid object
(C) An exchange of heat between warmer and cooler regions in a gas or liquid
(D) An exchange of gas particles between higher-pressure and lower-pressure regions
(E) An exchange of momenta between a moving particle and a still particle in a collision

47. An ideal gas is enclosed in a sealed container. Upon heating, which property of the gas does not change?

(A) Volume
(B) Pressure
(C) The average speed of the molecules
(D) The rate of collisions of the molecules with each other
(E) The rate of collisions of the molecules with the walls of the container

GO ON TO THE NEXT PAGE

48. A ball is dropped from a window and hits the ground 2 seconds later. Assuming $g = 10$ m/s^2, how high is the window from the ground?

(A) 10 m
(B) 20 m
(C) 25 m
(D) 40 m
(E) 80 m

49. A car drives 150 km east in 3 hrs., then 50 km west in 2 hrs. What is the car's average velocity?

(A) 20 km/hr east
(B) 25 km/hr east
(C) 37.5 km/hr east
(D) 40 km/hr east
(E) 50 km/hr east

50. A ball of mass m drops off a tower of height h. The force of air resistance is $\mathbf{F}_r$. What is the ball's kinetic energy when it hits the ground?

(A) $\mathbf{F}_r h$
(B) mgh
(C) $mg\,\mathbf{F}_r h^2$
(D) $mgh - \mathbf{F}_r h$
(E) $mgh + \mathbf{F}_r h$

51. A 1kg object slides along a rough surface with an initial velocity of 20 m/s. If the coefficient of kinetic friction between the object and the surface it slides along is 0.5, how far does the object travel before it comes to a stop?

(A) 10 m
(B) 20 m
(C) 25 m
(D) 40 m
(E) 50 m

52. A man standing at rest on a frictionless surface fires a gun directly forward. A 10g bullet leaves the gun with velocity $v_{\text{bullet}} = 300$ m/s. The mass of the man and gun is 100kg. What is the velocity of the man after firing the gun?

(A) 3 m/s, in the direction of the bullet
(B) 0 m/s
(C) 0.03 m/s, opposite the direction of the bullet
(D) 3 m/s, opposite the direction of the bullet
(E) 300 m/s, opposite the direction of the bullet

53. The center of mass of a set of particles is directly proportional to the sum of the products of each particle's mass and position, and inversely proportional to the total mass. Which of the following is the correct expression for the center of mass of a system of n particles along the x-axis?

(A) $\dfrac{m_1x_1 + m_2x_2 + \ldots + m_nx_n}{m_1 + m_2 + \ldots + m_n}$

(B) $\dfrac{(m_1 + m_2 + \ldots + m_n)(x_1 + x_2 + \ldots + x_n)}{m_1 + m_2 + \ldots + m_n}$

(C) $(m_1x_1 + m_2x_2 + \ldots + m_nx_n)(m_1 + m_2 + \ldots + m_n)$

(D) $(m_1 + m_2 + \ldots + m_n)^2(x_1 + x_2 + \ldots + x_n)$

(E) $\dfrac{m_1x_1 + m_2x_2 + \ldots + m_nx_n}{(m_1 + m_2 + \ldots + m_n)^2}$

GO ON TO THE NEXT PAGE

54. Planet A is in an orbit of radius R around a star. Planet B orbits the same star with orbit radius $4R$. The mass of planet B is four times the mass of planet A. If the gravitational potential energy of A is U, what is the gravitational potential energy of B?

(A) $\frac{U}{16}$
(B) $\frac{U}{4}$
(C) $\frac{U}{2}$
(D) U
(E) $2U$

55. A speaker plays a loud, constant tone. How does a listener, riding a bicycle towards the speaker, perceive the sound?

(A) The listener hears a higher frequency of sound wave than that produced by the speaker.
(B) The listener hears a higher amplitude of sound wave than that produced by the speaker.
(C) The listener hears a longer wavelength of sound than that produced by the speaker.
(D) The listener hears a lower frequency of sound wave than that produced by the speaker.
(E) The listener hears a lower amplitude of sound wave than that produced by the speaker.

56. A string of length 10 m is fixed at both ends. Which of the following is a possible wavelength for a standing wave on this string?

(A) 8 m
(B) 20 m
(C) 20π m
(D) 40 m
(E) None of these

57. An experimenter adds thermal energy to a block of some solid substance, but the temperature of the substance does not change. Which of the following explains this effect?

I. The thermal energy is used in a phase transformation.
II. The thermal energy increases the entropy of the substance.
III. The thermal energy is lost because heat engines can never be perfectly efficient.

(A) I only
(B) II only
(C) III only
(D) I and II only
(E) I and III only

58. A bridge girder with a co-efficient of linear expansion $\alpha = 1.0 \times 10^{-5}\ ^\circ\text{C}^{-1}$ is 10m long on a 10°C day. On a 30°C day, how much longer is the girder?

(A) 1.2×10^{-3} m
(B) 2.0×10^{-5} m
(C) 1.2×10^{-6} m
(D) 1.0×10^{-2} m
(E) 2.0×10^{-3} m

GO ON TO THE NEXT PAGE

<u>Questions 59–60</u> refer to a mover pulling a 20-kilogram box along a frictionless surface using a rope that makes an angle of 30° with the horizontal. The mover pulls on the rope with a force of 200 N.
sin 30°= 0.500, cos 30°= 0.866

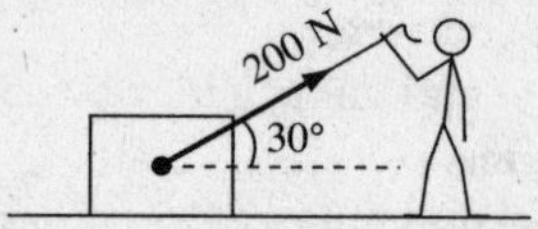

59. What is the acceleration of the box along the ground?

(A) 0 m/s^2
(B) 5.00 m/s^2
(C) 8.66 m/s^2
(D) 10.0 m/s^2
(E) 17.32 m/s^2

60. What is the normal force exerted by the ground on the box?

(A) 27 N
(B) 100 N
(C) 200 N
(D) 300 N
(E) 373 N

61. When a figure skater spinning with her arms outstretched brings her arms closer to her body, she begins to rotate at a much faster rate. When she stretches out her arms once again, she rotates more slowly. Which of the following best explains this phenomenon?

(A) When the skater's arms are out, she has more air resistance, so she moves more slowly.
(B) The act of bringing her arms closer to her body pushes her around more quickly.
(C) With her arms together, she has more weight pressing down on a smaller area, which allows her to move faster.
(D) With her arms outstreched, her moment of inertia is larger, so when she brings them in, her angular velocity must increase due to conservation of angular momentum.
(E) She must do work to bring her arms into her body, and so that work goes into increasing her angular velocity.

GO ON TO THE NEXT PAGE

Questions 62–64 refer to the following circuit. All four resistors have a resistance of 2 Ω.

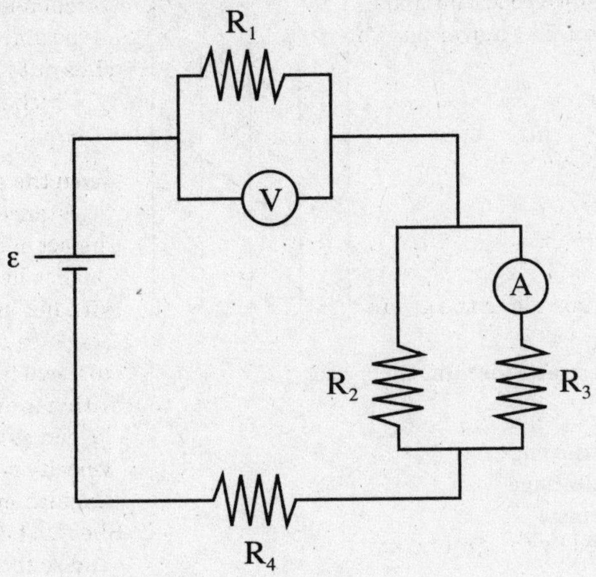

62. What is the equivalent resistance of the circuit?

(A) $\frac{1}{2}$ Ω
(B) 2 Ω
(C) 4 Ω
(D) 5 Ω
(E) 8 Ω

63. If the battery supplies 10 V, what is measured by the voltmeter?

(A) $\frac{1}{2}$ V
(B) 2 V
(C) 4 V
(D) 5 V
(E) 10 V

64. What is measured by the ammeter?

(A) $\frac{1}{2}$ A
(B) 1 A
(C) 2 A
(D) 4 A
(E) 10 A

GO ON TO THE NEXT PAGE

Questions 65–67 refer to the diagram of a positively charged particle moving with velocity v in a magnetic field **B**.

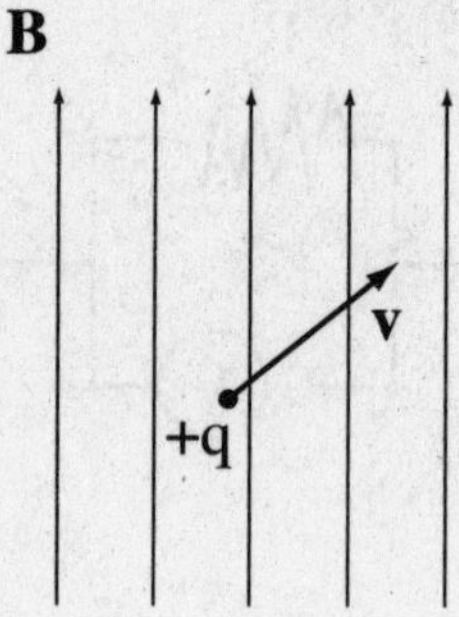

65. What is the direction of the magnetic force on the charged particle?

(A) To the right in the plane of the page
(B) Downward in the plane of the page
(C) Upward in the plane of the page
(D) Into the page
(E) Out of the page

66. What would the direction of the magnetic force be if the particle were negatively charged?

(A) To the right in the plane of the page
(B) Downward in the plane of the page
(C) Upward in the plane of the page
(D) Into the page
(E) Out of the page

67. Which of the following figures best represents the particle's trajectory through the magnetic field?

(A)

(B)
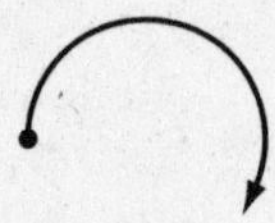

(C)

(D)

(E)

GO ON TO THE NEXT PAGE

Questions 68–69 refer to the diagram of a block of mass m on a plane inclined at angle θ. The block is at height h from the ground.

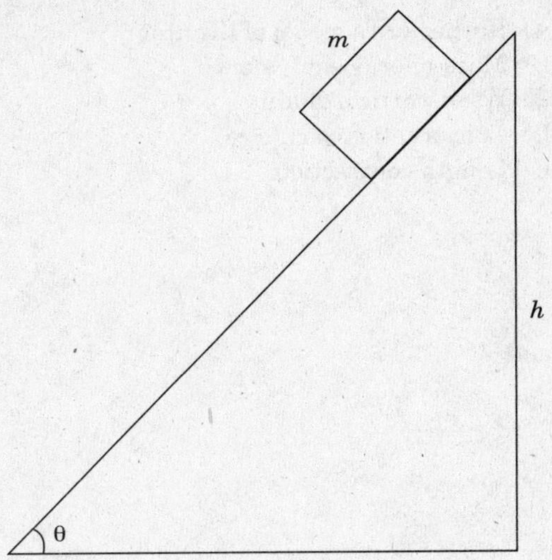

68. If the block is released from rest, and there is no friction between the block and the inclined plane, what is the velocity of the block when it reaches the ground?

(A) $\sqrt{2mgh}$
(B) $\sqrt{mgh}$
(C) $\sqrt{2gh}$
(D) $\sqrt{gh}$
(E) $2gh^2$

69. If the force of friction between the block and the incline is given by the equation $\mathbf{F}_{fr} = \mu\mathbf{N}$, where μ is the coefficient of kinetic friction, and $\mathbf{N}$ is the normal force, what is the acceleration of the block as it slides down the incline?

(A) $g(\cos\theta - \mu)$
(B) $mg - \mu$
(C) $\cos\theta(1 - \mu)$
(D) $g(\cos\theta - \mu\sin\theta)$
(E) $\mu\sqrt{2gh}$

70. Which of the following explains why white light striking a prism separates into different colors?

(A) Light of different wavelengths create different interference patterns.
(B) The index of refraction is slightly different for different wavelengths of incident light.
(C) Irregularities in the surface of the prism cause the emission of light of different wavelengths.
(D) The glass of the prism polarizes the light, causing the colors to separate.
(E) The prism acts as a lens, whose focal point differs for different wavelengths of light.

71. A double-slit diffraction pattern is set up against a screen. If the distance between the two slits is increased, how does the diffraction pattern change?

(A) The bright patches on the screen will be farther apart.
(B) The bright patches on the screen will be closer together.
(C) The bright patches will become brighter.
(D) The bright patches will become less bright.
(E) Changing the distance between the two slits has no effect on the diffraction pattern.

72. A bar magnet is approaching a loop of wire from above, south pole first. Which statement describes the current induced before the magnet reaches the coil?

(A) It flows clockwise as viewed from above.
(B) It flows counterclockwise as viewed from above.
(C) It first flows clockwise and then flows counterclockwise.
(D) It first flows counterclockwise and then flows clockwise.
(E) No current is induced before the magnet reaches the coil.

GO ON TO THE NEXT PAGE

Questions 73–74 refer to a spaceship passing by the earth at a speed of $0.6c$.

73. Relative to an observer on the earth, how long will it take for a clock on the spaceship to count one minute?

(A) 36 s
(B) 48 s
(C) 75 s
(D) 80 s
(E) 94 s

74. If the ship's crew measures the length of the ship as 200 m, what will an observer on earth measure the ship's length to be?

(A) 120 m
(B) 128 m
(C) 160 m
(D) 250 m
(E) 333 m

75. Which of the following best explains why two protons in the nucleus of an atom don't push away from one another in accordance with Coulomb's Law?

(A) Rutherford's model of the atom
(B) Mass-energy equivalence
(C) Wave-particle duality
(D) Conservation of charge
(E) Length contraction

S T O P

IF YOU FINISH BEFORE TIME IS CALLED, YOU MAY CHECK YOUR WORK ON THIS TEST ONLY. DO NOT TURN TO ANY OTHER TEST IN THIS BOOK.

SAT II Physics Practice Test II Explanations

Calculating Your Score

Your raw score for the SAT II Physics test is calculated from the number of questions you answer correctly and incorrectly. Once you have determined your composite score, use the conversion table on page 6 of this book to calculate your scaled score.

To Calculate Your Raw Score

Count the number of questions you answered correctly: ________
A

Count the number of questions you answered incorrectly, and multiply that number by $\frac{1}{4}$:

________ X $\frac{1}{4}$ = ________
B C

Subtract the value in field C from the value in field A: ________
D

Round the number in field D to the nearest whole number. This is your raw score: ________
E

Answers to SAT II Physics Practice Test II

Question Number	Correct Answer	Right	Wrong	Question Number	Correct Answer	Right	Wrong
1.	A	___	___	39.	C		___
2.	C	___	___	40.	C		___
3.	A	___	___	41.	E		___
4.	C	___	___	42.	E		___
5.	C	___	___	43.	A		___
6.	D	___	___	44.	B		___
7.	A	___	___	45.	B		___
8.	B	___	___	46.	C		___
9.	C	___	___	47.	A		___
10.	E	___	___	48.	B		___
11.	A	___	___	49.	A		___
12.	A	___	___	50.	D		___
13.	C	___	___	51.	D		___
14.	A	___	___	52.	C		___
15.	D	___	___	53.	A		___
16.	E	___	___	54.	D		___
17.	B	___	___	55.	A		___
18.	D	___	___	56.	B		___
19.	B	___	___	57.	D		___
20.	B	___	___	58.	E		___
21.	B	___	___	59.	C		___
22.	D	___	___	60.	B		___
23.	D	___	___	61.	D		___
24.	E	___	___	62.	D		___
25.	B	___	___	63.	C		___
26.	A	___	___	64.	B		___
27.	D	___	___	65.	E		___
28.	A	___	___	66.	D		___
29.	E	___	___	67.	D		___
30.	D	___	___	68.	C		___
31.	C	___	___	69.	D		___
32.	D	___	___	70.	B		___
33.	E	___	___	71.	B		___
34.	C	___	___	72.	A		___
35.	A	___	___	73.	C		___
36.	B	___	___	74.	A		___
37.	E	___	___	75.	B		___
38.	C	___	___	TOTAL			

Physics Shared Answer Questions

1. **(A)** Electric Forces, Fields, and Potential *Single-Concept Problem*

According to Coulomb's Law, the magnitude of an electric force between two particles is directly proportional to the charges of the particles and inversely proportional to the square of the distance between them. Doubling the charge of one of the particles doubles the force.

2. **(C)** Electric Forces, Fields, and Potential *Single-Concept Problem*

Coulomb's Law tells us that the magnitude of the electric force between two charged particles is directly proportional to the charge of the two particles and inversely proportional to the square of the distance between them. If the distance between them is multiplied by $\frac{1}{2}$, then the force is multiplied by $1/(\frac{1}{2})^2 = 4$.

3. **(A)** Dynamics *Single-Concept Problem*

Newton's First Law tells us that an object maintains a constant velocity unless a net force is acting on it. Since the box is moving with a constant velocity, the net force on the box is zero, so the answer is (A). The force exerted by the mover and the force of kinetic friction cancel each other out.

4. **(C)** Dynamics *Single-Concept Problem*

The box is at constant velocity both while it is at rest ($v = 0$) and after it has started moving. So in both situations, the box is not accelerating and the net force is zero. Therefore, the mover exerts a force that is equal and opposite to the force of friction, so that the two forces cancel each other out.

The frictional forces, both static and kinetic, are proportional to the normal force exerted by floor. Since the floor is flat, the normal force is just the weight of the box. Before the box starts moving, the mover's push is opposed by the force of static friction, which is given by $F_s = \mu_s W = (0.8)(200 \text{ N}) = 160 \text{ N}$. After the box is moving, the mover's push is opposed by the force of kinetic friction, given by $F_k = \mu_k W = (0.2)(200 \text{ N}) = 40 \text{ N}$. The difference between the two forces is $160 \text{ N} - 40 \text{ N} = 120 \text{ N}$, and the correct answer is (C).

5. **(C)** Magnetic Fields and Forces *Recall*

A magnetic field only exerts a force on a moving particle if the particle is charged. A neutron is uncharged, so it feels no magnetic force and is not deflected.

6. **(D)** Magnetic Fields and Forces *Single-Concept Problem*

The magnetic force on a charged particle moving in a magnetic field in a direction perpedicular to the field lines is given by the formula $F = qvB$. Here, q is the charge, v is the velocity of the particle, and B is the strength of the field. The alpha particle has the greatest charge, so it feels the largest force and is deflected the most.

7. **(A)** Circuits and Circuit Elements *Single-Concept Problem*

The power dissipated in a resistor is given by the formula $P = IV$. We are told that the current remains constant, and the voltage source cannot change spontaneously, so the value for P is a constant, as in graph (A).

8. **(B)** Circuits and Circuit Elements *Single-Concept Problem*

The total heat dissipated in a resistor is given by the formula $H = Pt$. In the previous question, we saw that P has a constant value. That means that H is proportional to time t: we get a linear relationship, as in (B).

9. **(C)** Modern Physics *Single-Concept Problem*
Electrons, like light, exhibit characteristic of both as waves and particles, depending on the experiment. The diffraction pattern observed when electrons are directed against a narrow opening is a manifestation of the wave-like nature of matter.

10. **(E)** Modern Physics *Recall*
The photoelectric effect is observed when light of a single frequency is shined on the surface of a metal, exciting electrons on the surface and causing them to fly off. The maximum energy of ejected electrons does not depend on the intensity of the light, but rather on its frequency. This is explained by the quantum theory of light, which regards light as a stream of photons each with energy proportional to its frequency.

11. **(A)** Modern Physics *Recall*
Bohr's model of the atom asserts that electrons can only occupy certain discrete energy levels and orbital radii. This explains why electrons don't fall into the nucleus of the atom when they emit radiation.

12. **(A)** Electric Forces, Fields, and Potential *Single-Concept Problem*
A and *B* are both on line V_2, which means they have the same potential. The work done on a charge is relative to the change in the charge's potential energy. Since the charge experiences no change in potential energy, no work is done on it.

13. **(C)** Electric Forces, Fields, and Potential *Single-Concept Problem*
The work done is $W_{AC} = -\Delta U = -q\Delta V = -q(V_3 - V_2)$.

14. **(A)** Dynamics *Single-Concept Problem*
Newton's First Law tell us that an object in motion will remain in motion unless a force acts upon it. Since the surface is frictionless, no force acts on the box and no force is needed to maintain its constant velocity.

15. **(D)** Dynamics *Single-Concept Problem*
Since the ball isn't moving, the net force acting upon it is zero. That means that the force of tension in the rope must be pulling it up with an equal and opposite force to the force of gravity pulling it downward. Since the force of tension is of equal in magnitude to the force of gravity, we simply need to calculate the force of gravity, or weight, of the ball, which is

$$W = mg = (15\text{ kg})(10\text{ m/s}^2)$$
$$= 150\text{ N}$$

Physics Solitary Multiple Choice Questions

16. **(E)** Dynamics *Single-Concept Problem*
We can denote the net force on Body 1, its mass, and its acceleration as F_1, m_1, and a_1, respectively, and we can denote the net force on Body 2, its mass, and its acceleration as F_2, m_2, and a_2, respectively. We are told that $m_2 = 2m_1$ and that $F_1 = 4F_2$. With these substitutions, we can solve for a_1 using Newton's Second Law:

$$a_1 = \frac{F_1}{m_1} = \frac{4F_2}{0.5m_2} = \frac{8F_2}{m_2}$$
$$= 8a_2$$

The ratio of a_1 to a_2 is, then,

$$a_1 : a_2 = 8a_1 : a_2 = 8 : 1$$

17. **(B)** Electric Forces, Fields, and Potential *Single-Concept Problem*

A point charge is in static equilibrium when the net force acting on it is zero. Because we are dealing with a positive point charge, the force exerted by a negative charge will be attractive, and the force exerted by a positive charge will be repulsive.

In the first diagram, the negative charge to the right of the point charge pulls the point charge to the right. The positive charge to the left of the point charge pushes the point charge to the right. As a result, there will be a net force in the rightward direction.

In the second diagram, all three charges will exert a repulsive force on the point charge. However, since they are at the three vertices of an equilateral triangle of which the point charge is the center, these three repulsive forces will cancel one another out, and the net force acting on the point charge will be zero. Hence, II represents a case of static equilibrium.

In the third diagram, the three $+q$ charges will combine to push the point charge in the direction of the $-2q$ charge, which will in turn pull the point charge in that direction. There will be a net force on the point charge in the diagonally up and right direction.

18. **(D)** Kinematics *Multi-Concept Problem*

The slope of the line at *A* is zero, so the velocity at that point will be zero. However, the slope of the line to the left of *A* is positive while the slope of the line to the right of *A* is negative. This means that the velocity is changing from positive to negative, so the acceleration must be negative.

19. **(B)** Kinematics *Multi-Concept Problem*

The graph at point *B* is a straight line with a positive slope. The slope of the graph tells us what the velocity is, so since the slope is positive, we know that the velocity is also positive. Since the line is straight, its slope is not changing; the velocity is constant and the acceleration must be zero.

20. **(B)** Optics *Single-Concept Problem*

If you can't remember the answer offhand, the best way to solve this problem is to draw a ray diagram:

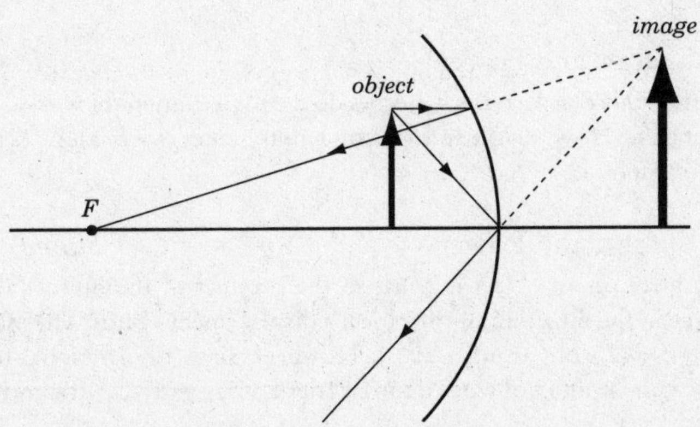

We can see immediately that the image is upright and larger that the object. We can conclude the image is virtual because it is behind the mirror: the object appears to be behind the mirror, but no light is actually focused on any point behind the mirror.

21. (B) Optics *Recall*

Eyeglasses are lenses that focus light in order to compensate for the defective lenses in the eye. Lenses work according to the principle of refraction: light passing from the open air into the medium of glass are bent as they enter the glass and bent once again as the pass out of the glass back into the open air.

22. (D) Thermal Physics *Single-Concept Problem*

According to the First Law of Thermodynamics, a system gains energy when heat is added to it, and loses energy when it does work. The net amount of energy change in the system is therefore $30 - 50 = -20$ J.

23. (D) Thermal Physics *Multi-Concept Problem*

A car propels itself forward by the force of friction between the car tires and the road. Energy dissipated as friction with road is thus energy used to propel the car forward. The other uses of energy, though useful in the case of powering the car's internal electronics, do nothing to propel the car forward.

The efficiency of a heat engine is given by the formula $e = 1 - \Delta Q_{out}/\Delta Q_{in}$, where ΔQ_{out} is the amount of heat that is wasted, and ΔQ_{in} is the amount of heat produced. The value of ΔQ_{out} is 94 J, since all but the 6 J of friction with the road are wasted. The value of ΔQ_{in} is 100 J, so the efficiency of the car is

$$e = 1 - 94\text{ J}/100\text{ J} = 0.06$$

An efficiency of 0.06 translates to 6%. Alternately, we could simply have reasoned that 6 out of every 100 J works to propel the car forward, suggesting that the efficiency should be 6%.

24. (E) Work, Energy, and Power *Single-Concept Problem*

Since the force is exerted parallel to the direction of motion, the work it does is given by

$$\begin{aligned} W &= Fd = (10\text{ N})(20\text{ m}) \\ &= 200\text{ J} \end{aligned}$$

Don't let the 5 kg mentioned in the question fool you. If you know the force exerted on the object, the object's mass plays no role in determining the work done.

25. (B) Work, Energy, and Power *Single-Concept Problem*

Power is given by the formula $P = W/t$. Both motors do the same amount of work, since they lift loads of equal weight equal distances. However, since the second motor takes twice as long as the first motor, it uses half the power of the first motor.

26. (A) Work, Energy, and Power *Single-Concept Problem*

The work done by a force on an object is equal to the product of the object's displacement and the component of the force in the direction of the object's displacement vector. The magnitude of the work done is simply $W = Fd = (20\text{ N})(10\text{ m}) = 200$ J. However, since the frictional force acts against the displacement of the box, the amount of work done by friction is negative, so the correct answer is (A).

27. (D) Vectors *Single-Concept Problem*

We know that the vector for the wind has both a northward and an eastward component. We also know that the eastward component is twice the northward component. That means the vector should be on a diagonal, pointing more to the east than to the north.

28. (A) Electric Forces, Fields, and Potential *Single-Concept Problem*

Electric potential is given by the equation $V = kQ/r$. In other words, electric potential and distance are inversely proportional. In moving from A to B, the distance of the point charge from $+Q$ is doubled, so the electric potential is halved.

29. **(E)** Electric Forces, Fields, and Potential *Single-Concept Problem*

Both *A* and *C* are a distance *r* from +Q. As a result, they both have the same electric potential, and no work is done in moving a point charge between these two points.

30. **(D)** Waves *Recall*

Transverse waves oscillate in a direction perpendicular to their propagation. For instance, the vibrations on a violin string are transverse because the waves travel up and down the string while vibrating (or oscillating) perpendicular to the length of the string. Radio waves, as a form of electromagnetic wave, are similarly transverse. On the other hand, sound waves are not transverse, but longitudinal, meaning that the air particles transmitting sound oscillate in a direction parallel to the propagation of the wave. Therefore, the correct answer is (D).

31. **(C)** Waves *Multi-Concept Problem*

The wavelength of a wave is the distance is covers in completing a single cycle. The wave depicted here has a wavelength of 2 m. Wavelength, λ, is related to wave speed, v, and frequency, f, by the formula $v = f\lambda$. Since we have been given the wave speed and have calculated the wavelength, we can solve for f:

$$f = \frac{v}{\lambda} = \frac{10 \text{ m/s}}{2 \text{ m}}$$

$$= 5 \text{ Hz}$$

32. **(D)** Waves *Single-Concept Problem*

The velocity of a wave on a string is given by the formula

$$v = \sqrt{\frac{T}{\mu}}$$

where T is the tension in the string and μ is the linear mass density of the string. Since velocity is directly proportional to the square root of the tension in the string, increasing the tension by a multiple of two will increase the velocity by a multiple of the square root of two.

33. **(E)** Optics *Recall*

All electromagnetic waves travel through a vacuum at the same speed, $c = 3 \times 10^8$ m/s. Every color of visible light has a particular frequency associated with it, so the frequency of visible light varies from color to color. Since wavelength is inversely proportional to frequency, it too varies as frequency varies. The intensity of a beam of light is directly proportional to the square of the amplitude of the wave, so brighter light will have a greater amplitude than dimmer light. Finally, visible light can be polarized in many different ways, which may be indistinguishable to the human eye.

34. **(C)** Optics *Multi-Concept Problem*

If you've memorized all the information about mirrors, great. If you haven't, it's still possible to reason your way through this one by drawing ray diagrams.

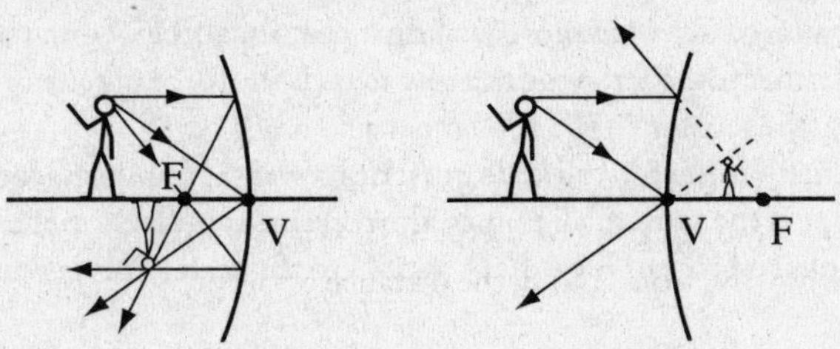

The focal length is the distance between the focal point, F, and the mirror. With mirrors, the focal length is positive if F is in front of the mirror and negative is F is behind the mirror. The focal point F is where the reflection of rays coming in parallel to the axis will intersect. As seen in the diagram, for a concave mirror, the focal point is in front of the mirror. For a convex mirror, the rays diverge as they bounce off the mirror, so the focal point is located behind the mirror. We can conclude that I is correct.

An image is real if it appears in front of the mirror, since then it is composed of real light rays, and it is virtual if it appears behind the mirror. Our ray diagrams show us that concave mirrors produce real images and convex mirrors produce virtual images, so II is also correct.

The vertex, V, is the point where the principal axis intersects the mirror, right in its middle. The ray diagrams show us that the incident angle of a ray striking the vertex is equal to the reflected angle. This is true of concave mirrors, but since it is also true of convex mirrors, III is false.

Since I and II are true and III is false, we can conclude that the answer to the above question is (C).

35. **(A)** Miscellaneous (Equation Manipulation, Graph Analysis, Significant Digits) *Single-Concept Problem*

If the measurement is accurate to 1%, the final significant digit must be one-hundredth the value of the first significant digit. In other words, the measurement must contain three significant digits.

36. **(B)** Dynamics *Multi-Concept Problem*

According to Newton's Second Law, the acceleration of an object is given according to its mass and the net force acting on it by the formula $a = F/m$. We know the mass of the book, but we need to calculate the net force acting on it. The net force is the 5 N push minus the force of kinetic friction resisting this push. The force of kinetic friction is:

$$F_f = m_k mg = (0.3)(1\text{ kg})(10\text{ m/s}^2)$$
$$= 3\text{ N}$$

Now that we know the force of friction, we can solve for the acceleration of the book:

$$a = \frac{F_{net}}{m} = \frac{F_{push} - F_f}{1\text{ kg}} = \frac{5\text{ N} - 3\text{ N}}{1\text{ kg}}$$
$$= 2\text{ m/s}^2$$

37. **(E)** Work, Energy, and Power *Single-Concept Problem*

Power is given by the formula $P = W/t$. We can calculate how much work the worker does quite easily: he exerts a force of 300 N over a distance of 10 m, so he does $F \cdot d = 3000$ J of work. If it takes him 15 seconds, then we know the worker pushes with a power of

$$P = \frac{W}{t} = \frac{3000\text{ J}}{15\text{ s}}$$
$$= 200\text{ W}$$

38. **(C)** Special Problems in Mechanics (Pulleys, Springs, Pendulums, Inclined Planes) *Single-Concept Problem*

Since the 4 kg mass is resting on the scale, the system is in static equilibrium, which means that the net force acting on both masses illustrated above is zero. We know that the force of gravity acting on the 1 kg mass is $F = mg = 10$ N. The only other force acting on the 1 kg mass is the force of tension in the pulley rope. Since the 1 kg mass isn't moving, the tension in the rope must also be 10 N.

The force that the 4 kg mass applies to the scale must be its weight minus the force of tension in the rope. Since the tension in the rope is the same at all points, the rope must pull up on the 4 kg mass with a force of 10 N. So the 4 kg mass applies a force of $mg - 10$ N = 30 N to the scale. This is what the scale reads.

39. **(C)** Waves *Multi-Concept Problem*

Statement I is a definition of longitudinal waves, while statement II is true of waves of all kinds. And while we normally study the phenomenon of diffraction with transverse light waves, longitudinal waves can also exhibit diffraction, so statement III is false. Therefore, the correct answer is (C).

40. **(C)** Waves *Recall*

Sound waves passing from one medium into another experience refraction, just like light waves. That means that the speed of sound in one medium is different from the speed of sound in another medium.

The frequency of the sound waves remains unchanged: a person snorkeling underwater will hear sounds from above water at the same pitch as someone hearing them in the open air. However, the speed of sound is different, and since speed, wavelength, and frequency are related by the formula $v = f\lambda$, a change in the speed of sound must also result a change in the wavelength.

The amplitude of the sound wave is related to how loud the sound is, and this is affected by the medium through which it travels. The sounds heard by a person snorkeling underwater are more muffled than the sounds heard by a person in the open air, since water transmits sound waves with a much lower amplitude than the open air.

Since only frequency remains unaffected when sound waves pass from one medium into another, (C) is the correct answer.

41. **(E)** Waves *Recall*

A standing wave can form when the string is a multiple of $\lambda/2$ long; that is, where $\lambda_n = 2L/n$. Since λ is related to frequency and wave speed by the formula $\lambda_n = v/f_n$, we just need to solve for f_n:

$$\frac{v}{f_n} = \frac{2L}{n}$$

$$f_n = \frac{nv}{2L}$$

42. **(E)** Modern Physics *Recall*

The atomic number of an atom tells us how many protons there are in that atom. An alpha particle contains two protons and two neutrons, so when an atom emits an alpha particle, it emits two protons and its atomic number decreases by two.

43. **(A)** Modern Physics *Multi-Concept Problem*

The photoelectric effect can only be explained by accepting that electromagnetic waves have particle properties; only certain discrete energies are possible. Bohr's model of the hydrogen atom shows that electrons only orbit the nucleus at certain radii because the electron can only have certain levels of potential energy. To be released, as in the photoelectric effect, photoelectrons must absorb the right frequency of electromagnetic radiation.

44. **(B)** Modern Physics *Recall*

A beta particle is the same thing as an electron, so, naturally, it has the same mass. Protons and alpha particles, and nucleons (protons or neutrons) are both much more massive than the beta particle. Neutrinos have practically no mass.

45. (B) Thermal Physics *Single-Concept Problem*

The most straightforward definition of the Second Law of Thermodynamics is that heat flows spontaneously from a hotter body to a colder body. However, this one law has all sorts of implications from which a number of different equivalent forms can be derived. All the answer choices given here are among these equivalent forms except (B). That the internal energy of a system doing work decreases is a formulation of the First Law of Thermodynamics, not the Second.

46. (C) Thermal Physics *Recall*

Convection is a type of motion found in a gas or liquid where there is a temperature difference between the regions. For example, in a boiling pot of water, the water closer to the flame is hotter than the water at the top of the pot. The hot water rises, displacing the colder water. The correct answer is (C).

Be careful not to fall for answer (B). Convection only takes place in liquid or gas media. The transfer of heat between solid objects occurs through conduction, without the particles being displaced.

47. (A) Thermal Physics *Single-Concept Problem*

Since the gas is in a closed container, its volume remains constant, so the correct answer is (A).

When the gas is heated, its temperature increases, meaning that the average speed of the gas molecules increase. An increase in temperature also means there are more collisions between molecules.

According to the Ideal Gas Law, when volume is constant and temperature is increased, then pressure will also increase. Pressure is determined by the rate of collisions of the gas molecules with the walls of the container.

48. (B) Kinematics *Single-Concept Problem*

To solve this problem, we can use the kinematic equation that doesn't include a value for final velocity v, and solve for x:

$$x = v_0t + \frac{1}{2}at^2$$

$$= (0)(2 \text{ s}) + \frac{1}{2}(10 \text{ m/s}^2)(2 \text{ s})^2$$

$$= 20 \text{ m}$$

49. (A) Kinematics *Single-Concept Problem*

Average velocity is a measure of total displacement over total time. The total displacement is 150 km east – 50 km west = 100 km east. Remember, since displacement and velocity are vector quantities, you can't simply add up the distance traveled. Since the total time is 5 hrs., the average velocity is simply 100 km east /5 hrs. = 20 km/hr east.

50. (D) Work, Energy, and Power *Multi-Concept Problem*

In cases where there is no air resistance, the law of conservation of energy states that the gravitational potential energy, $U_g = mgh$, of an object when it is released is equal to the kinetic energy of that object when it hits the ground. In the absence of air resistance, the ball's kinetic energy would be mgh. However, the force of air resistance does some work on the ball, and so some of the ball's kinetic energy is dissipated. The amount of work done by the air resistance on the ball is $W = -F_fh$. This is the product of the force exerted and the distance over which it is exerted. It is negative, since the ball travels in the opposite direction of the force. If we add the amount of work done on the ball by air resistance to the ball's kinetic energy in the absence of air resistance, we find that its kinetic energy at the bottom of its fall is KE = $mgh - F_fh$.

51. (D) Work, Energy, and Power *Multi-Concept Problem*

The object has an initial kinetic energy of

$$KE = \frac{1}{2}mv^2 = \frac{1}{2}(1\text{ kg})(20\text{ m/s})^2$$
$$= 200\text{ J}$$

By the time the object comes to a stop, it has no more kinetic energy. At this point, all the kinetic energy has been dissipated by the work done on the object by the force of friction. In other words, the force of friction must do 200 J of work on the object to bring it to a stop.

The force of friction is given by the formula $F = \mu_k mg$, where μ_k is the coefficient of kinetic friction. Since the work done by friction is the product of this force times the object's displacement, we can calculate the object's displacement by dividing the amount of work done by friction by the magnitude of the force of friction:

$$d = \frac{W}{F} = \frac{W}{\mu_k mg} = \frac{200\text{ J}}{(0.5)(1\text{ kg})(10\text{ m/s}^2)}$$
$$= 40\text{ m}$$

52. (C) Linear Momentum *Single-Concept Problem*

Initially, the system is at rest, the linear momentum of the system is zero. After firing the gun, the bullet has a linear momentum of $p = (300\text{ m/s})(0.01\text{ kg}) = 3\text{ kg}\cdot\text{m/s}$. To conserve momentum, the man and gun must have the same momentum in the opposite direction, and since they weigh 100 kg, their velocity must be

$$3\text{ kg}\cdot\text{m/s}/100\text{ kg} = 0.03\text{ m/s}$$

You cannot use conservation of energy to solve such a problem, since energy is added to the system when the gun is fired.

53. (A) Miscellaneous (Equation Manipulation, Graph Analysis, Significant Digits) *Single-Concept Problem*

Since the center of mass is directly proportional to the sum of the products of each particle's mass and position, $m_1x_1 + m_2x_2 + \ldots + m_nx_n$ must be in the numerator of the expression. Since the center of mass is inversely proportional to the total mass, the denominator of the expression is $m_1 + m_2 + \ldots + m_n$.

54. (D) Gravity *Single-Concept Problem*

The equation for the gravitational potential energy of a planet in orbit is

$$U = -G\frac{Mm}{R}$$

where M is the mass of the star, m is the mass of the planet, and R is the radius of the orbit. This equation tells us that potential energy is directly proportional to m and inversely proportional to R. If both m and R are multiplied by a factor of 4, both the numerator and the denominator of the fraction will be multiplied by 4, and the value of U will remain unchanged.

55. (A) Waves *Recall*

As the listener approaches the speaker, the perceived distance between the crests of sound waves becomes less than the distance between the crests as they leave the speaker. Since the crests are closer together, the frequency of the sound the listener hears is higher than the frequency coming from the speaker. This phenomenon is known as the Doppler Effect.

Since frequency and wavelength are inversely proportional, an increase in frequency does not result in an increase in wavelength. The amplitude of the sound wave is proportional to the volume of the sound, which remains unaffected by the velocity of the listener.

56. (B) Waves *Single-Concept Problem*

Since the string is fixed at both ends, the end points are nodes and do not move. That means that there must be a whole number of nodes on the standing wave. Since there are two nodes in a given wavelength, the possible wavelengths for this standing wave are $2(10\text{ m})/n$, where n is a positive integer. Where $n = 1$, we have a wavelength of 20 m, (B). None of the other answer choices are possible wavelengths of a standing wave of this length.

57. (D) Thermal Physics *Recall*

We can immediately eliminate III, since energy, including heat energy, is never lost. The second law of thermodynamics does imply that heat engines cannot convert all of the thermal energy they are given into work, but the excess energy is generally ejected as heat, never lost.

Heat energy can be used to change the phase of a substance, for example by heating a block of ice to turn it into liquid water. In this case the temperature remains constant even though heat is being added to the system. Therefore, I is a correct explanation.

The entropy of a solid is lower than the entropy of the liquid or gas phase of the same substance at the same temperature, and the latent heat of transformation that is added to a solid to melt it increases its entropy. Therefore, II is also correct.

58. (E) Thermal Physics *Single-Concept Problem*

The change in length of the girder is given by $\Delta L = \alpha L_o \Delta T$, where L_o is the girder's initial length, and ΔT is the change in temperature. With this information, we can easily solve:

$$\Delta L = (1.0 \times 10^{-5}\ ^\circ\text{C}^{-1})(10\text{ m})(20\ ^\circ\text{C}) = 2.0 \times 10^{-3}\text{ m}$$

59. (C) Dynamics *Multi-Concept Problem*

The force pulling on the box can be broken down into a horizontal component and a vertical component. The acceleration of the box along the ground is due only to the horizontal component of the force. If the total force is F, then the horizontal component of the force is $F \cos 30°$. Applying this knowledge and Newton's Second Law, we can solve for the acceleration of the box along the ground:

$$a = (F \cos 30°/m) = (200\text{ N}/20\text{ kg}) \cos 30° = (10\text{ m/s}^2)(0.866)$$
$$= 8.66\text{ m/s}^2$$

60. (B) Dynamics *Multi-Concept Problem*

The normal force is the upward force exerted by the ground on the box to counter the downward force exerted by the box on the ground. Normally, the downward force would be the box's weight, $mg = (20\text{ kg})(10\text{ m/s}^2) = 200\text{ N}$. However, since the mover is pulling slightly upward on the box, the net downward force is the box's weight minus the vertical component of the mover's pull. Taking the force exerted by the mover to be F, we can solve for the normal force, N:

$$N = mg - F \sin 30° = 200\text{ N} - (200\text{ N})(0.500)$$
$$= 100\text{ N}$$

61. **(D)** Circular Motion and Rotation *Recall*

The correct answer is (D). Angular momentum is the product of moment of inertia and angular velocity, $\mathbf{L} = I\boldsymbol{\omega}$. When the skater's arms are outstretched, her moment of inertia is greater than when they are close to her body. Because angular momentum is conserved, a decrease in the value of I calls for a corresponding increase in the magnitude of $\boldsymbol{\omega}$.

You may be tempted to answer (A). While this is actually true, it doesn't account for the tremendous difference between the two angular velocities. (D) is the better answer.

(C) may also be tempting. The skater does do work in bringing her arms closer to her body—she must expend calories to move her muscles—the energy mostly goes into heating her muscles.

62. **(D)** Circuits and Circuit Elements *Single-Concept Problem*

First, let's determine the equivalent resistance of the two resistors in parallel, R_2 and R_3:

$$1/R_{2+3} = 1/R_2 + 1/R_3 = 1/2\Omega + 1/2\Omega$$
$$R_{2+3} = 1\Omega$$

The resistors R_1 and R_4 are both in series with the two resistors in parallel, so we can now simply add up their resistance and the equivalent resistance of R_2 and R_3 to find the equivalent resistance of the circuit, R_t:

$$R_t = R_4 + R_{2+3} + R_4 = 2\,\Omega + 1\,\Omega + 2\,\Omega$$
$$= 5\,\Omega$$

63. **(C)** Circuits and Circuit Elements *Multi-Concept Problem*

The voltmeter measures the voltage drop across R_1, which is given by Ohm's Law as $V = I_tR_1$, where I_t is the total current in the circuit. I_t is, in turn given by Ohm's Law as

$$I = V/R = 10\text{ V}/5\,\Omega$$
$$= 2\text{ A}$$

With this value for I_t, we can now solve for the voltage drop across R_1:

$$V = I_tR_1 = (2\text{ A})(2\,\Omega)$$
$$= 4\text{ V}$$

64. **(B)** Circuits and Circuit Elements *Multi-Concept Problem*

As we saw in the previous question, the total current in the circuit is $I_t = 2$ A. However, the ammeter only measures the current across R_3. The total current is split between R_2 and R_3. Since both resistors are of equal strength, the current is split evenly between them, so that the current across R_3 is simply $I_t/2 = 1$ A.

65. **(E)** Magnetic Fields and Forces *Single-Concept Problem*

The magnetic force on a moving charged particle is given by the formula $\mathbf{F} = q(\mathbf{v} \times \mathbf{B})$. Since we are dealing with a positive charge, q is positive and $\mathbf{F}$ will have the same direction as $\mathbf{v} \times \mathbf{B}$. Using the right-hand rule, we find that $\mathbf{F}$ points out of the page.

66. **(D)** Magnetic Fields and Forces *Multi-Concept Problem*

For a negatively charged particle, the $\mathbf{F}$ vector points in the opposite direction of the $\mathbf{v} \times \mathbf{B}$ vector. The right-hand rule tells us that the $\mathbf{v} \times \mathbf{B}$ vector points out of the page, so the $\mathbf{F}$ vector must point into the page.

67. **(D)** Magnetic Fields and Forces *Single-Concept Problem*

Since the particle is moving at an angle with respect to the magnetic field, we can break its trajectory down into a component that's parallel with the magnetic field and a component that's perpendicular to the magnetic field. The component that's parallel will be unaffected by the magnetic field, and so will continue traveling upward in the plane of the page. The component that's perpendicular will be drawn into a circular orbit that's perpendicular to the magnetic field lines. Adding these two components together, we get a helix pattern, where the particle spirals about while moving upward in the plane of the page, as we see in (D).

68. **(C)** Special Problems in Mechanics (Pulleys, Springs, Pendulums, Inclined Planes) *Multi-Concept Problem*

We can most easily solve this problem by using the law of conservation of energy. At the top of the incline, the block has potential energy *mgh* and no kinetic energy. At the bottom of the incline, the block has kinetic energy $\frac{1}{2}mv^2$ and no potential energy. Since the energy at the top is equal to the energy at the bottom, we can set these two formulas to be equal and solve for v:

$$\frac{1}{2}mv^2 = mgh$$
$$v^2 = 2gh$$
$$v = \sqrt{2gh}$$

69. **(D)** Special Problems in Mechanics (Pulleys, Springs, Pendulums, Inclined Planes) *Multi-Concept Problem*

The acceleration of the block is equal to the net force acting on the block divided by the block's mass: $a = F/m$. The net force acting on the block, as shown in the free-body diagram below, is $mg \sin\theta - \mu N$.

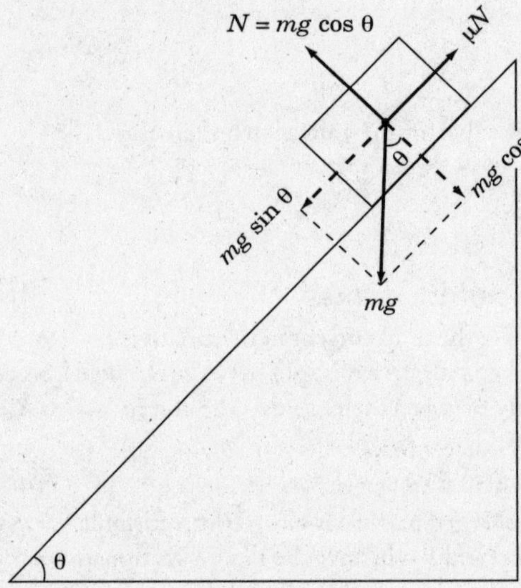

Since $N = mg\cos\theta$, we can solve for a:

$$a = F/m$$
$$= (mg\sin\theta - \mu\, mg\cos\theta)/m$$
$$= g(\sin\theta - \mu\cos\theta)$$

Remember, the normal force is not *mg* in this case! Since the block only exerts a force of $mg\cos\theta$ into the inclined plane, the plane only exerts a normal force of $mg\cos\theta$ in return.

70. **(B)** Optics *Recall*

The phenomenon of dispersion when light passes through a prism is due to the refraction of light as it passes through the glass. The index of refraction differs slightly for the different wavelengths of light, so the white light that passes into the prism is broken up into its spectral elements.

71. **(B)** Optics *Multi-Concept Problem*

Each point *P* on the screen can be identified by the angle θ between the line connecting the midpoint between the two slits and *P*, and the perpendicular between the slits and the screen. The bands of maximum intensity in a double-slit diffraction pattern occur at points for which $n= d \sin\theta \lambda$ is an integer. Here, *d* is the distance between the two slits and λ is the (constant) wavelegth. For a fixed *n*, then, increasing *d* decreases $\sin \theta$, which corresponds to a decrease in the distance between bright bands.

72. **(A)** Electromagnetic Induction *Single-Concept Problem*

As the magnet moves down, flux is decreasing in the downward direction. By the right-hand rule, the induced current must flow clockwise to create flux downward.

73. **(C)** Modern Physics *Single-Concept Problem*

Solving this problem involves the formula for time dilation, where t_o is the time measured on the spaceship:

$$\begin{aligned} t &= t_o/\sqrt{1 - v^2/c^2} \\ &= 60\text{ s}/\sqrt{1 - (0.6c)^2/c^2} \\ &= 60\text{ s}/\sqrt{0.64} \\ &= 60\text{ s}/0.8 \\ &= 75\text{ s} \end{aligned}$$

74. **(A)** Modern Physics *Single-Concept Problem*

Solving this problem involves plugging numbers into the formula for length contraction, where l_o is the length measured by the ship's crew.

$$\begin{aligned} l &= l_o\sqrt{1 - v^2/c^2} \\ &= (200\text{ m})\sqrt{1 - (0.6c)^2/c^2} \\ &= (200\text{ m})\sqrt{0.64} \\ &= (200\text{ m})(0.8) \\ &= 160\text{ m} \end{aligned}$$

75. **(B)** Modern Physics *Recall*

At very short distances, the strong nuclear force pulls protons together with a much stronger force than the repulsion between like charges exercised by the electromagnetic force. Because of the strong nuclear force, a nucleus has a smaller mass than the sum of the masses of the individual protons and neutrons that make it up. This discrepancy, or mass defect, is the amount of matter that is converted into energy when the nucleus is formed, according to Einstein's principle of mass-energy equivalence.

SAT II Physics Practice Test III

PHYSICS TEST

Part A

Directions: Each set of lettered choices below refers to the numbered questions or statements immediately following it. Select the one lettered choice that best answers each question or best fits each statement, and then fill in the corresponding oval on the answer sheet. A choice may be used once, more than once, or not at all in each set.

Questions 1–3

Which of the following are best applied to answer the questions given below?

(A) Newton's Second Law
(B) Newton's Third Law
(C) Law of conservation of energy
(D) Law of conservation of linear momentum
(E) Law of conservation of angular momentum

1. An ice skater pushes against a wall with a certain force. With what force does the skater recoil from the wall?

2. A box slides down a frictionless incline. What is the speed of the box when it reaches the bottom of the incline?

3. A box is pushed with a constant velocity along a surface that has a certain coefficient of kinetic friction. If the box is pushed onto a frictionless surface, and the same pushing force is still applied to it, what is the box's acceleration?

Questions 4–5

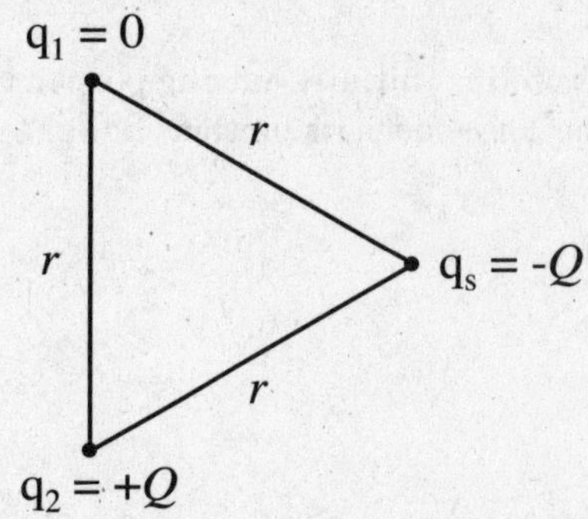

Three particles with charges $q_1 = 0$, $q_2 = +Q$, and $q_3 = -Q$ are arranged in an equilateral triangle with sides of length r. The following choices refer to the magnitude of the electric force on the particles.

(A) $k\frac{Q^2}{r^2}$

(B) $k\frac{Q^2}{r}$

(C) kQr

(D) $k\frac{Q}{r^2}$

(E) zero

4. What is the magnitude of the electric force on q_1?

5. What is the magnitude of the electric force on q_3?

GO ON TO THE NEXT PAGE

Questions 6–7 refer to the motion of a proton initially moving in a straight line after it enters a region with a magnetic field.

(A) It continues in a straight line.
(B) It stops.
(C) It reverses direction.
(D) It curves to one side.
(E) It oscillates back and forth.

6. If the proton is initially moving parallel to the field lines, how does its motion change?

7. If the proton is initially moving perpendicular to the field lines, how does its motion change?

Questions 8–10

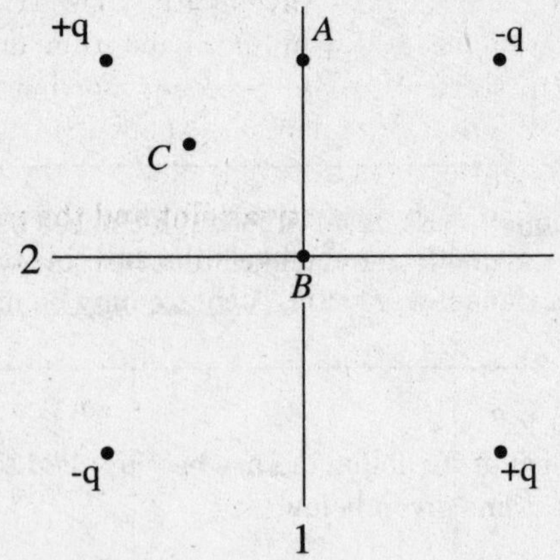

The diagram above shows four charges of equal magnitude arranged in a square. The following choices refer to the positions *A*, *B*, and *C*, and lines 1 and 2, marked on the diagram.

(A) *A* only
(B) *B* only
(C) *C* only
(D) Along lines 1 and 2
(E) Everywhere

8. Where is the electric field equal to zero?

9. Where is the electric potential equal to zero?

10. Where would a positive point charge feel no electric force?

GO ON TO THE NEXT PAGE

<u>Questions 11–13</u> relate to a double-slit experiment, where a beam of light with wavelength λ passes through two slits, shown on the left of the diagram, and sets up a diffraction pattern on a screen, shown on the right of the diagram. Five points are shown on the screen. In the diagram, θ is the angle made by the line drawn between each respective point and the point O, directly between the two slits, and d is the distance between the two slits.

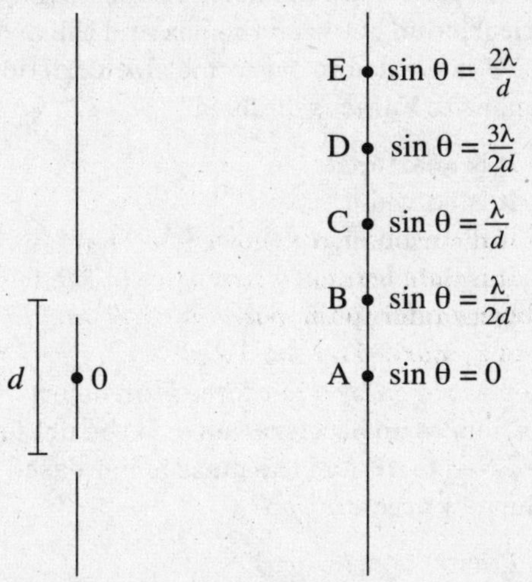

11. At what point on the screen is the light the brightest?

12. Which point on the screen will be lit as brightly as point B?

13. If the distance between the two slits is halved, which of the five points will receive the least amount of light?

14. The mass of an object that weighs 50 N is most nearly

    (A) 0.2 kg
    (B) 5 kg
    (C) 10 kg
    (D) 50 kg
    (E) 500 kg

15. A ball is dropped from the top of a building. Ignoring air resistance, which of the following remains constant for the ball?

    (A) Acceleration
    (B) Kinetic energy
    (C) Momentum
    (D) Potential energy
    (E) Velocity

GO ON TO THE NEXT PAGE

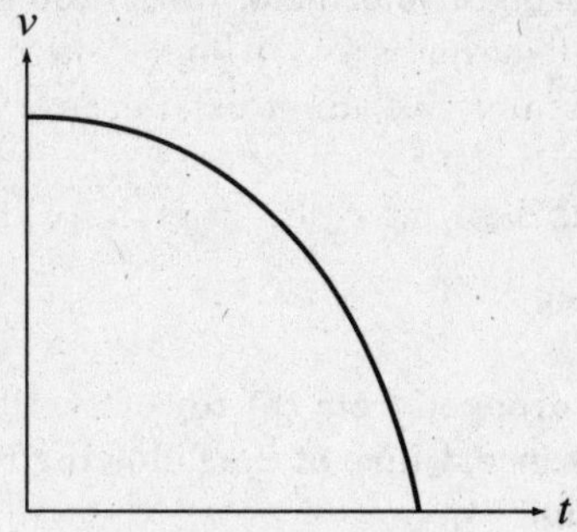

16. The graph of velocity versus time for an object moving along a straight line is given above. During the time shown on the graph, which of the following is a true statement about the position and the acceleration of the object?

(A) position is increasing; acceleration is increasing
(B) position is increasing; acceleration is decreasing
(C) position is decreasing; acceleration is increasing
(D) position is decreasing, acceleration is decreasing
(E) position is decreasing, acceleration is constant

17. If the speed of an object is doubled, what happens to its kinetic energy?

(A) It is quartered
(B) It is halved
(C) It is unchanged
(D) It is doubled
(E) It is quadrupled

18. A 100 kg aluminum box is at the top of a 100 m office building. The potential energy of the box with respect to the ground is most nearly

(A) 10 J
(B) 10^2 J
(C) 10^3 J
(D) 10^4 J
(E) 10^5 J

19. A 5 kg box is suspended from a rope. The force of tension in the rope is most nearly

(A) 0 N
(B) 0.5 N
(C) 5 N
(D) 25 N
(E) 50 N

20. A box is resting on the floor. The coefficient of maximal static friction between the box and the floor is μ_s. A force **F** is needed to overcome static friction. What happens to **F** if μ_s is doubled?

(A) It is quartered
(B) It is halved
(C) It is unchanged
(D) It is doubled
(E) It is quadrupled

21. When acted on by a net force **F** an object of mass M experiences an acceleration a. If the net force is increased to 4**F** and the mass is increased to $2M$ what is the new acceleration?

(A) $\frac{1}{8}a$
(B) $\frac{1}{2}a$
(C) a
(D) $2a$
(E) $8a$

GO ON TO THE NEXT PAGE

22. Which of the following graphs shows a function with constant slope?

(A)

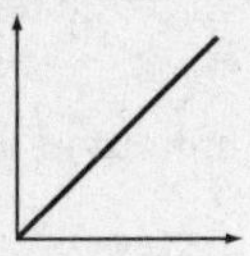

(B)

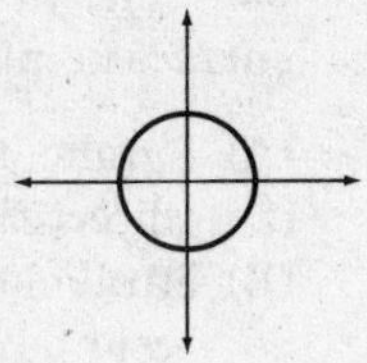

(C)

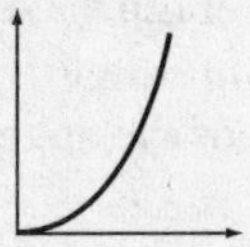

(D)

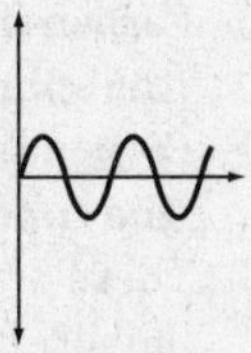

(E)

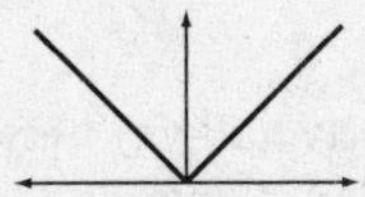

23. According to the Ideal Gas Law, which of the following may decrease when the volume of a gas is increased?

(A) The temperature of the gas
(B) The number of molecules in the gas
(C) The pressure of the gas
(D) The average kinetic energy of the gas molecules
(E) The atomic number of the atoms in the gas

24. After a process during which an ideal gas did 7 joules of work, the internal energy of the gas increased by 12 joules. Some heat may have been transferred between the gas and its surroundings. Which of the following is true about this heat flow?

(A) 19 joules of heat has been added to the gas.
(B) 5 joules of heat has been added to the gas
(C) Heat has neither been added to nor lost by the gas.
(D) 5 joules of heat has been lost by the gas.
(E) 19 joules of heat has been lost by the gas.

Questions 25–26 refer to a 2.5 g coin dropped from the top of a building that is 125 m high. You may use $g = 10\ m/s^2$ and assume that air resistance is negligible.

25. How long does the penny stay in free fall?

(A) 1 s
(B) 1.25 s
(C) 5 s
(D) 10 s
(E) 12.5 s

26. How far has the coin traveled when its velocity is 30 m/s?

(A) 9 m
(B) 30 m
(C) 36 m
(D) 45 m
(E) 90 m

GO ON TO THE NEXT PAGE

27. Which of the following has the highest frequency?

(A) Radio waves
(B) Microwaves
(C) Red light
(D) X Rays
(E) Gamma Rays

28. What is the frequency of the wave pictured below?

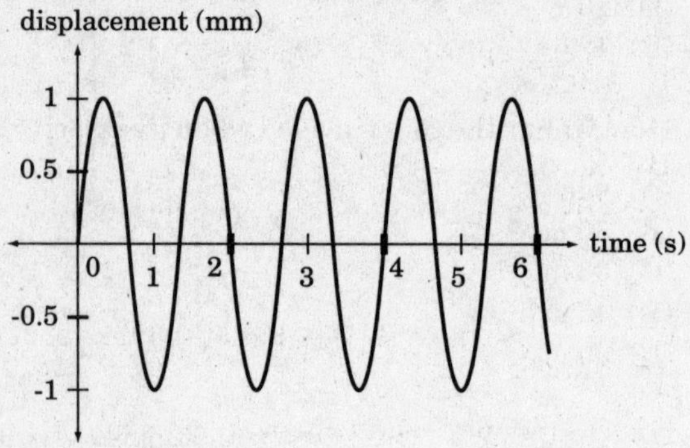

(A) 0.25 Hz
(B) 0.50 Hz
(C) 0.75 Hz
(D) 1.0 Hz
(E) 1.3 Hz

29. A spherical conductor has a charge $-Q$ distributed evenly throughout it. In what direction do the electric field lines just above the surface of the conductor surface point?

(A) Parallel to the surface of the sphere in a clockwise direction
(B) Parallel to the surface of the sphere in a counterclockwise direction
(C) Perpendicular to the surface of the sphere and out of the sphere
(D) Perpendicular to the surface of the sphere and into the sphere
(E) There are no electric field lines at the surface of the sphere.

30. A positive charge Q is exactly halfway between two unknown charges, q_1 and q_2. If the force on the positive charge is zero, what can be said about the unknown charges?

(A) q_1 and q_2 have opposite signs and the same magnitude.
(B) q_1 and q_2 have the same sign and the same magnitude.
(C) q_1 and q_2 are both positive.
(D) q_1 and q_2 are both negative.
(E) q_1 and q_2 have opposite signs and different magnitudes.

GO ON TO THE NEXT PAGE

Questions 31–32 refer to the following diagram of a 1 kg mass suspended from a spring whose spring constant is $k = 4$ N/m. The system is in equilibrium when the center of mass is at point A.

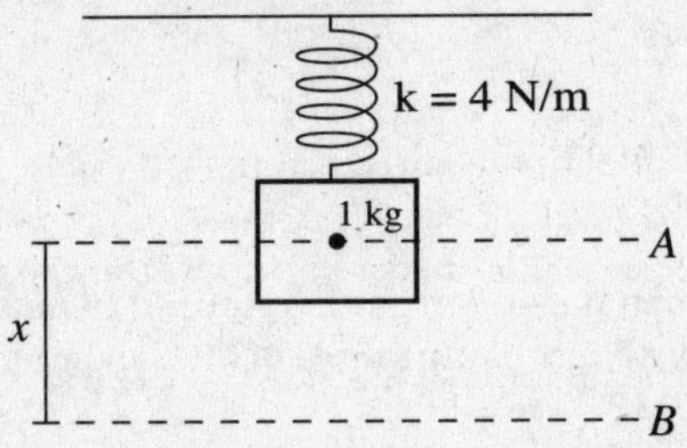

31. If the mass is pulled from point A to point B, a distance of x meters from A, what is the energy stored in the spring?

(A) $-kx$

(B) kx

(C) $-\frac{1}{2}kx^2$

(D) $\frac{1}{2}kx^2$

(E) 0

32. At point B, what is the force exerted on the mass by the spring, if we now assume that $x = 0.25$ m?

(A) 1 N, pointing down
(B) 1 N, pointing up
(C) 0.25 N, pointing down
(D) 0.25 N, pointing up
(E) zero

33. Which of the following is explained by Thompson's "raisin pudding" model of the atom?

(A) Atoms have nuclei
(B) Atoms contain electrons
(C) Atoms contain protons
(D) Electrons orbit only at certain radii
(E) Electrons exhibit wave properties

34. What is the same for every isotope of a given element?

(A) Decay rate
(B) Mass number
(C) Atomic number
(D) Neutron number
(E) Percent distribution in a random sample of the element

35. Which of the following is a true statement about atomic structure?

(A) The number of protons in an atom must equal the number of neutrons.
(B) The number of protons in an atom must equal the number of electrons.
(C) The number of protons in an atom plus the number of neutrons in that atom gives the mass number of that atom.
(D) Two atoms of one element may have different numbers of protons.
(E) Two atoms of one element must have the same number of neutrons.

GO ON TO THE NEXT PAGE

36. What happens to the pressure of an ideal gas when its volume is doubled and its temperature is halved?

(A) It is quartered.
(B) It is quadrupled.
(C) It is halved.
(D) It is doubled.
(E) It remains unchanged.

37. The heat needed to raise m kilograms of a substance with specific heat c by ΔT degrees is given by which of the following formulas?

(A) $Q = \frac{mc}{\Delta T}$

(B) $Q = \frac{m\Delta T}{c}$

(C) $Q = mc\Delta T$

(D) $Q = mc(\Delta T)^2$

(E) $Q = \frac{mc}{(\Delta T)^2}$

38. A heat engine goes through a cyclic process in which 300 J of work is done on the external environment. If the efficiency of the engine is 75%, how much heat does the engine consume during a given cycle?

(A) 66 J
(B) 225 J
(C) 300 J
(D) 400 J
(E) 1200 J

39. A box resting is guven a push up a frictionless ramp. If its initial speed is 10 m/s, how high above the ground will the box be when it stops moving?

(A) 1 m
(B) 5 m
(C) 10 m
(D) 50 m
(E) 100 m

40. A man pushes an 8 kg box, initially at rest, with a constant force of 10N a distance of 10m across a frictionless floor. What is the final velocity of the box?

(A) 4 m/s
(B) 5 m/s
(C) 7 m/s
(D) 8 m/s
(E) 10 m/s

41. How much power output is required to push a 10 kg object with a constant force of 30 N over a distance of 5m in 10 s?

(A) 0.15 W
(B) 1.5 W
(C) 15 W
(D) 150 W
(E) 1500 W

42. A quantity A is directly proportional to B and inversely proportional to C^2. Which of the following graphs is the best representation of B vs. C when A is held constant?

(A)

(B)
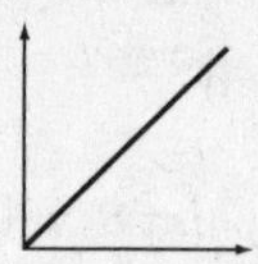

(C)
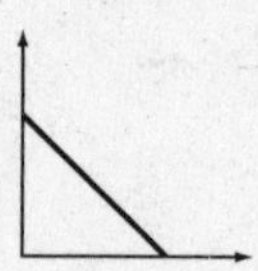

(D)
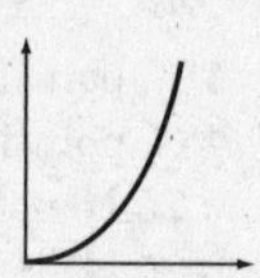

(E)
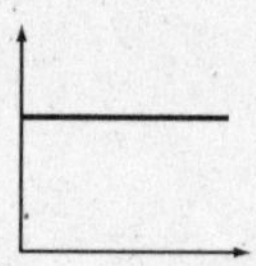

GO ON TO THE NEXT PAGE

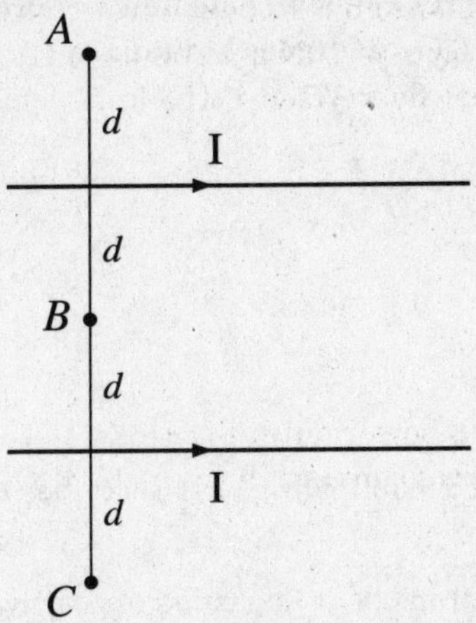

43. Two long, straight, parallel wires each carry current I in the same direction as pictured above. Where is the magnetic field created by the wires zero?

(A) A
(B) B
(C) C
(D) A and C
(E) Nowhere

<u>Questions 44–45</u> refer to a planet of mass m orbiting a star of mass M at speed v. The orbit is circular with radius R.

44. What would the planet's velocity be if it orbited at a radius of $2R$?

(A) $\frac{v}{2}$
(B) $\frac{v}{\sqrt{2}}$
(C) v
(D) $\sqrt{2}v$
(E) $2v$

45. What would the planet's velocity be if its mass were $2m$?

(A) $\frac{v}{2}$
(B) $\frac{v}{\sqrt{2}}$
(C) v
(D) $\sqrt{2}v$
(E) $2v$

GO ON TO THE NEXT PAGE

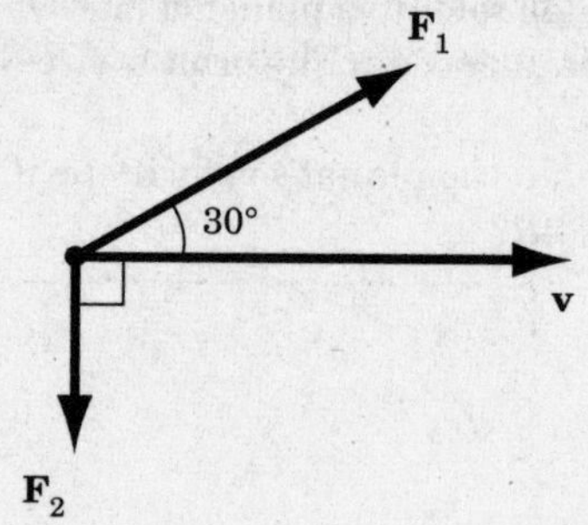

46. Two forces act on an object as pictured above. The object moves to the right. If $\mathbf{F}_1$ = 60 N, what is $\mathbf{F}_2$? sin 30° = 0.500, cos 30° = 0.866

(A) 30 N
(B) 52 N
(C) 60 N
(D) 69 N
(E) 120 N

47. A 2 kg box is pulled along the floor by a rope. The tension in the rope is a constant 10 N, and the box accelerates at a rate of 2 m/s^2. The coefficient of kinetic friction between the box and the floor is most nearly:

(A) 0.1
(B) –0.3
(C) 0.3
(D) 0.7
(E) 1

48. Which of the following elements is produced when an atom of $^{14}_{6}C$ undergoes beta-decay?

(A) $^{12}_{6}C$
(B) $^{11}_{5}B$
(C) $^{15}_{8}O$
(D) $^{15}_{7}N$
(E) $^{14}_{7}N$

49. Which of the following best describes the Bohr model of the atom?

(A) Electrons float in a cloud of positive charge.
(B) Electrons orbit the positively charged nucleus at discrete radii.
(C) Positive and negative charge is scattered randomly about the atom.
(D) Positive charge is concentrated at the center of a cloud of negative charge.
(E) Negative charge is concentrated at the center of a cloud of positive charge.

50. Two circular wires of the same material have the same length. The diameter of wire 1 is twice that of wire 2. What is the ratio of the resistance in 1 to the resistance in 2?

(A) 1 : 4
(B) 1 : 2
(C) 1 : 1
(D) 2 : 1
(E) 4 : 1

GO ON TO THE NEXT PAGE

Questions 51–53 refer to a particle of charge -2×10^{-10} C and mass 2×10^{-15} kg moving in a 0.5T magnetic field. The particle's initial velocity is 4×10^5 m/s, perpendicular to the field.

51. What is the magnetic force on the particle?

(A) 0 N
(B) 2×10^{-5} N
(C) 4×10^{-5} N
(D) 8×10^{-5} N
(E) 8×10^{-15} N

52. After 5 seconds, what is the magnitude of its velocity?

(A) 2×10^{10} m/s
(B) 3×10^{8} m/s
(C) 4×10^{5} m/s
(D) 4×10^{-5} m/s
(E) 2×10^{-10} m/s

53. If the particle's initial velocity were parallel to the magnetic field, what would be the magnetic force on the particle?

(A) 0 N
(B) 4×10^{-5} N
(C) 8×10^{-5} N
(D) 4×10^{-15} N
(E) 8×10^{-15} N

Questions 54–56 refer to a circuit of two resistors in series with a 3 V battery. The resistors are $R_1 = 3\Omega$ and $R_2 = 6\Omega$.

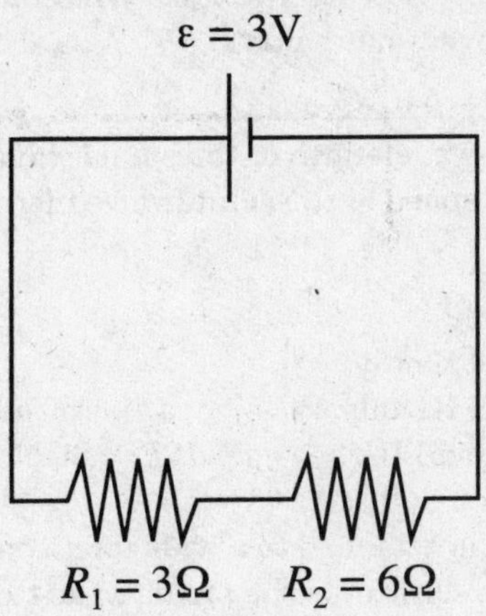

54. What is the total current in the wire?

(A) $\frac{1}{3}$ A
(B) $\frac{2}{3}$ A
(C) 2 A
(D) 3 A
(E) 18 A

55. What is the voltage drop across R_1?

(A) $\frac{1}{3}$ V
(B) $\frac{1}{2}$ V
(C) 1 V
(D) 2 V
(E) 3 V

56. What is the power dissipated in R_1?

(A) $\frac{1}{3}$ W
(B) $\frac{2}{3}$ W
(C) 1 W
(D) $\frac{3}{2}$ W
(E) 3 W

GO ON TO THE NEXT PAGE

57. A simple flute is a tube with one closed end. The player changes the pitch of the sound by covering and uncovering holes in the tube. What happens when the flutist covers more holes?

    I. The pitch increases.
    II. The wavelength of the sound wave increases.
    III. The speed of the sound wave increases.

    (A) I only
    (B) II only
    (C) I and II only
    (D) I and III only
    (E) I, II, and III

58. A string is attached to a wall at one end. At the other end a person yanks the string so as to send a wave pulse down toward the wall. Which of the following describes what happens to the wave after it bounces off the wall?

    (A) Not inverted, increased wavelength
    (B) Inverted, increased wavelength
    (C) Not inverted, same wavelength
    (D) Inverted, same wavelength
    (E) The wave is destroyed.

59. A science teacher measures the length of her car to be 3.46 m. What is the margin of error in her measurement?

    (A) ± 0.001 m
    (B) ± 0.005 m
    (C) ± 0.01 m
    (D) ± 0.05 m
    (E) ± 0.06 m

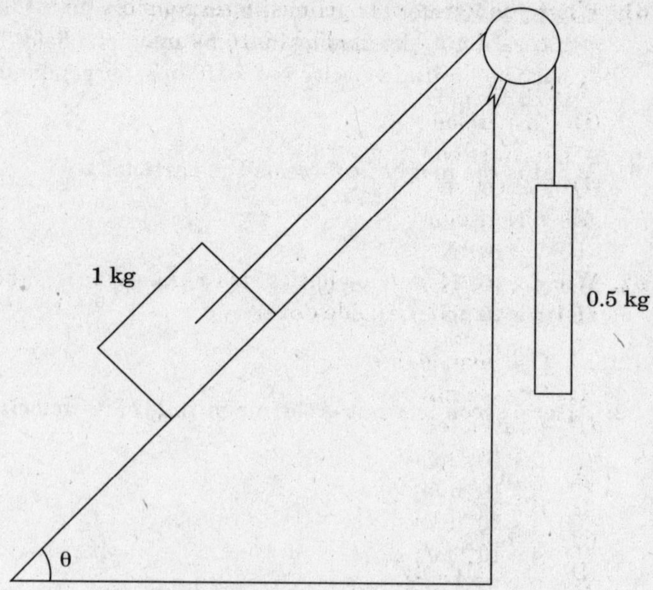

60. <let>A 1 kg mass on a frictionless inclined plane is connected by a pulley system to a suspended 0.5 kg mass, as in the diagram above. At what angle will the system be in equilibrium?
$\cos 30° = \sin 60° = \sqrt{3}/2$, $\cos 60° = \sin 30° = 1/2$,
$\cos 45° = \sin 45° = 1/\sqrt{2}$.

    (A) 0°
    (B) –30°
    (C) 30°
    (D) 45°
    (E) 60°

GO ON TO THE NEXT PAGE

61. What phenomenon is primarily responsible for a prism's splitting white light into colors?

(A) Diffraction
(B) Dispersion
(C) Refraction
(D) Reflection
(E) Absorption

62. Which of the following optical instruments can make an image appear upside down?

I. Concave mirror
II. Convex mirror
III. Convex lens

(A) I only
(B) II only
(C) III only
(D) I and III only
(E) II and III only

63. A person stands 12 m away from a convex mirror that has a focal length of 6 m. How far from the mirror is the reflected image, and is it real or virtual?

(A) A real image 4 m from the mirror
(B) A virtual image 4 m from the mirror
(C) A virtual image 6 m from the mirror
(D) A real image 12 m from the mirror
(E) A virtual image 12 m from the mirror

64. A heat engine's high-temperature source has a mass, m, and a specific heat, c. In a cycle of the engine, as it extracts energy from its high-temperature source, the temperature of that source drops by temperature ΔT (assume 100% efficiency for this heat extraction). The engine then does some work and then exhausts an amount of energy, Q, into a cold reservoir. What is the efficiency of the heat engine?

(A) $1 - \frac{Q}{mc\Delta T}$

(B) $1 - \frac{mc\Delta T}{Q}$

(C) $\frac{Q}{mc\Delta T}$

(D) $mc\Delta T - Q$

(E) $\frac{1}{mc\Delta T - Q}$

65. A gas initially at 200K is heated at constant volume, so that its pressure doubles. A bar of metal is in thermodynamic equilibrium with the gas, so that its temperature remains the same as that of the gas. If the metal's coefficient of volume expansion is $\beta = 2.0 \times 10^{-5}\ K^{-1}$, and its volume before being heated is 0.50 L, by how much will its volume change?

(A) 1.0×10^{-3} L
(B) 2.0×10^{-3} L
(C) 2.5×10^{-3} L
(D) 5.0×10^{-3} L
(E) The volume will not change

66. Which of the following is the correct formula for C, the capacitance of a capacitor with charge Q and potential difference V?

(A) $C = QV$

(B) $C = \frac{Q}{V}$

(C) $C = \frac{Q^2}{V}$

(D) $C = Q^2V$

(E) $C = QV^2$

67. A dielectric is inserted into a capacitor while the charge on the plates is held constant. Which of the following occurs?

I. The electric field increases.
II. The capacitance increases.
III. The stored energy increases.

(A) I only
(B) II only
(C) I and II only
(D) II and III only
(E) I, II, and III

GO ON TO THE NEXT PAGE

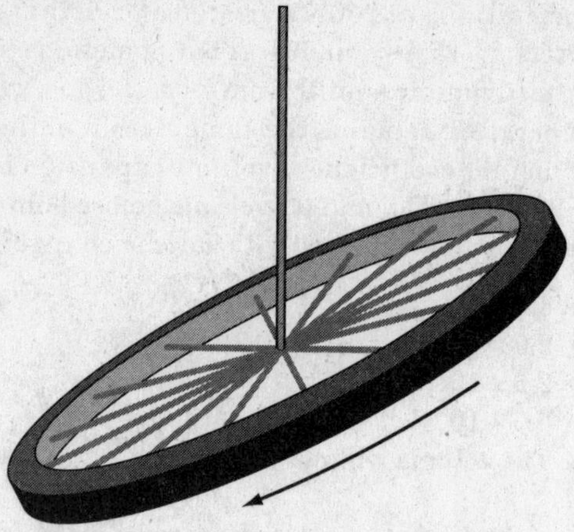

68. A bicycle wheel is suspended on a string tied to its hub, and the wheel is set in motion. Rather than falling into a vertical position, the wheel will precess around the axis of the string, maintaining a constant angle of tilt. What best explains this phenomenon?

(A) The law of conservation of momentum dictates that the wheel's angular momentum must remain constant, so that it will maintain a constant angle of tilt. The torque on the wheel due to gravity will cause the wheel to precess around the string.
(B) The wheel gains kinetic energy when it is set in motion, and it must continue to precess about according to the law of conservation of energy.
(C) By holding the string, we exert a force on it and cause it to rotate around.
(D) The magnetic field of the earth will cause the axis of the bicycle wheel to line up with it. In the Southern Hemisphere, the wheel will spin the opposite direction.
(E) The bicycle wheel does not precess, but rather falls immediately into a vertical position, under the influence of gravity.

Questions 69–70 refer to a violin string of mass m, length L, and tension T.

69. If we change the mass of the string to $2m$, which of the following properties will change?

I. The velocity of the wave on the string
II. The frequency of the first harmonic
III. The wavelength of the first harmonic

(A) I only
(B) II only
(C) I and II only
(D) II and III only
(E) I, II, and III

70. If we change the mass of the string to $2m$ and the length of the string to $2L$, which of the following properties will change?

I. The velocity of the wave on the string
II. The frequency of the first harmonic
III. The wavelength of the first harmonic

(A) I only
(B) II only
(C) I and II only
(D) II and III only
(E) I, II, and III

S T O P

IF YOU FINISH BEFORE TIME IS CALLED, YOU MAY CHECK YOUR WORK ON THIS TEST ONLY.
DO NOT TURN TO ANY OTHER TEST IN THIS BOOK.

SAT II Physics Practice Test III Explanations

Calculating Your Score

Your raw score for the SAT II Physics test is calculated from the number of questions you answer correctly and incorrectly. Once you have determined your composite score, use the conversion table on page 6 of this book to calculate your scaled score.

To Calculate Your Raw Score

Count the number of questions you answered correctly: ________
A

Count the number of questions you answered incorrectly, and multiply that number by $\frac{1}{4}$:

________ X $\frac{1}{4}$ = ________
B C

Subtract the value in field C from the value in field A: ________
D

Round the number in field D to the nearest whole number. This is your raw score: ________
E

Answers to SAT II Physics Practice Test III

Question Number	Correct Answer	Right	Wrong	Question Number	Correct Answer	Right	Wrong
1.	B			39.	B		
2.	D			40.	B		
3.	A			41.	C		
4.	E			42.	D		
5.	A			43.	B		
6.	A			44.	B		
7.	D			45.	C		
8.	B			46.	A		
9.	D			47.	C		
10.	B			48.	E		
11.	A			49.	B		
12.	D			50.	A		
13.	C			51.	C		
14.	B			52.	C		
15.	A			53.	A		
16.	B			54.	A		
17.	E			55.	C		
18.	E			56.	A		
19.	E			57.	B		
20.	D			58.	D		
21.	D			59.	D		
22.	A			60.	C		
23.	C			61.	B		
24.	A			62.	D		
25.	C			63.	B		
26.	D			64.	A		
27.	E			65.	B		
28.	C			66.	B		
29.	D			67.	B		
30.	B			68.	A		
31.	D			69.	C		
32.	B			70.	D		
33.	B			71.	C		
34.	C			72.	B		
35.	C			73.	A		
36.	A			74.	B		
37.	C			75.	E		
38.	D			TOTAL			

Physics Shared Answer Questions

1. **(B)** Dynamics *Recall*

Newton's Third Law tells us that for every action, there is an equal and opposite reaction. When the skater pushes on the wall, the wall pushes back on the skater. That means that the force with which the skater recoils is exactly equal and opposite to the force the skater applies to the wall.

2. **(D)** Work, Energy, and Power *Recall*

At the top of the incline, the box is motionless, so it has no kinetic energy, but a certain amount of gravitational potential energy. At the bottom of the incline, the box has no gravitational potential energy left: it has all been converted into kinetic energy. That means that the gravitational potential energy of the box at the top of the incline is equal to the kinetic energy of the box at the bottom of the incline. If we know the height of the incline, we can calculate the velocity of the box at the bottom.

3. **(A)** Dynamics *Recall*

While the box moves at a constant velocity, the net force acting on it is zero. This means that the force of friction opposing the box's motion is equal to the force of the push. When the box is pushed onto the frictionless surface, the same force is still applied to push it forward, but there is no longer a force of friction to resist this push, so the box accelerates. To calculate the degree to which it accelerates, we need to know its mass, and then apply Newton's Second Law, $F = ma$.

4. **(E)** Electric Forces, Fields, and Potential *Recall*

Since q_1 has no charge, it feels no electric force.

5. **(A)** Electric Forces, Fields, and Potential *Single-Concept Problem*

The negatively charged particle q_3 feels an attractive force toward the positively charged particle q_2. Coulomb's Law tells us that electric force between two particles is directly proportional to the charge of the two particles and inversely proportional to the square of the distance between them. Since both particles have a charge of magnitude Q, the force between them is proportional to Q^2/r^2, as in (A).

6. **(A)** Magnetic Fields and Forces *Recall*

The magnitude of the force on the proton is given by $F = qvB \sin \theta$, where θ is the angle between the velocity vector and the magnetic field lines. Since the proton is moving parallel to the field lines, $\sin \theta = 0$ and the field exerts no force on it, so it continues in a straight line.

7. **(D)** Magnetic Fields and Forces *Recall*

When the magnetic field and velocity vectors are perpendicular, a magnetic force is exerted in a direction perpendicular to both velocity and magnetic field vectors. A perpendicular force acting on the proton will pull it into a circular orbit, causing its path to curve.

8. **(B)** Electric Forces, Fields, and Potential *Single-Concept Problem*

The electric field is equal to zero at a point where the net force acting on a positive charge is zero. At any point in the figure, the electric field is the sum of the electric fields from each charge. You can solve this problem qualitatively by drawing electric field arrows in for each of the points.

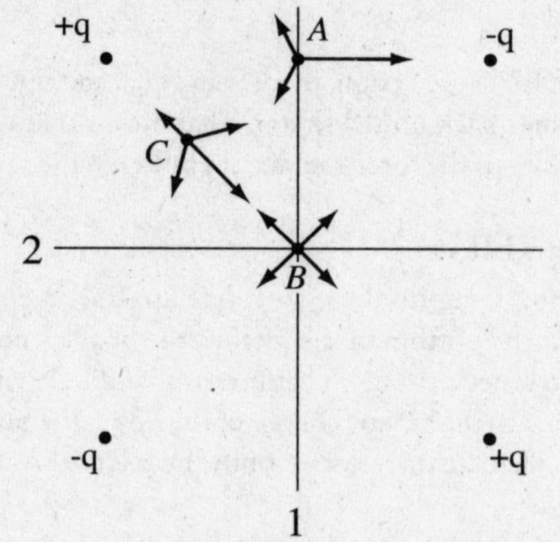

As you can see, only at B is the net force zero. You could also reason that B is the only point that is equidistant from both positive charges and equidistant from both negative charges, and so only a charge at B will feel not net push or pull toward or away from one of the negative or positive charges.

9. **(D)** Electric Forces, Fields, and Potential *Single-Concept Problem*

The potential of a charge is $V = kq/r$. If there is more than one charge involved, then the electric potential is simply the sum of the potentials from each individual charge. So, for instance, the electric potential at A is the sum of the electric potentials from all four charges at the corners of the square. The electric potential due to the positive charges will be kq/r, where r is the distance from those charges, and the electric potential due to the negative charges will be $-kq/r$. Because the positive and negative charges are an equal distance away, the sum of all the potentials will be zero. In fact, this will be the case at any point along lines 1 and 2, where there is an equal distance between the positive charges and the negative charges. The answer, then, must be (D).

10. **(B)** Electric Forces, Fields, and Potential *Single-Concept Problem*

Electric force, **F**, can be expressed in terms of electric field: $\mathbf{F} = q\mathbf{E}$. Since the field is zero only at point B, the force on a charge is also zero only at point B.

11. **(A)** Optics *Recall*

The brightest maximum in a double-slit diffraction pattern is the point directly between the two slits, where $d \sin \theta = 0$.

12. **(D)** Optics *Single-Concept Problem*

There will be minima in the diffraction pattern at every point that satisfies the equation $d \sin \theta = (n + \frac{1}{2}) \lambda$. Since both points B and D satisfy this equation, no light will shine on these points.

13. **(C)** Optics *Multi-Concept Problem*

Normally, we find minima at those points that satisfy the equation $d \sin \theta = (n + \frac{1}{2}) \lambda$. If the value of d is halved, then we have to substitute $\frac{1}{2} d$ for d:

$$\frac{d}{2}\sin\theta = (n + \frac{1}{2})\lambda$$

$$\sin\theta = (\frac{2}{d})(n + \frac{1}{2})\lambda$$
$$= \frac{2n+1}{d}\lambda$$
$$= \lambda/d, 3\lambda/d, 5\lambda/d, \ldots$$

The only one of the five points that satisfies this new equation for a minimum is C.

Physics Solitary Multiple Choice Questions

14. **(B)** Dynamics *Single-Concept Problem*

The formula relating weight and mass is $W = mg$. We simply need to rearrange that formula and solve for m:

$$m = \frac{W}{g} = \frac{50 \text{ N}}{10 \text{ m/s}^2}$$
$$= 5 \text{ kg}$$

15. **(A)** Kinematics *Recall*

You may recall that gravitational acceleration is given by $g = 9.8 \text{ m/s}^2$: in the absence of air resistance, objects in free-fall on earth always accelerate downward at this constant rate.

Although total energy is conserved, neither kinetic energy (B) nor potential energy (D) is conserved independently. Also, because the ball is accelerating, its velocity is constantly increasing. As the velocity increases, so does the momentum, since momentum is the product of an object's mass and its velocity.

16. **(B)** Kinematics *Single-Concept Problem*

Because the graph is above the x-axis over the time period shown, that means that the velocity is constantly positive. That in turn means that the position is increasing. On the other hand, the slope of the velocity graph is not only negative, but the slope is changing so that, over time, the slope decreases. Since the slope represents the acceleration, we know that the acceleration is decreasing.

17. **(E)** Work, Energy, and Power *Single-Concept Problem*

The formula for the kinetic energy of an object is $KE = \frac{1}{2}mv^2$. Since kinetic energy is directly proportional to the square of the velocity, doubling the kinetic energy would mean quadrupling the velocity. Since speed is just the magnitude of velocity, the speed is also quadrupled when kinetic energy is doubled.

18. **(E)** Work, Energy, and Power *Single-Concept Problem*

The formula for gravitational potential energy is $U = mgh$. As long as we're careful to add the right number of zeros, we will find that

$$U = mgh = (100 \text{ kg})(10 \text{ m/s}^2)(100 \text{ m})$$
$$= 10^5 \text{ J}$$

19. (E) Dynamics *Single-Concept Problem*

Newton's Third Law tells us that, for every action, there is an equal and opposite reaction. When one object exerts a force on another object, that other object exerts a reaction force on the first object. The box is under the influence of gravity, and so it will pull down on the rope. As a result, the rope will pull on the box in the upward direction, which is its force of tension. In other words, the force of tension in the rope is equal to the force of gravity acting on the box. The force of gravity acting on the box is the box's weight, $W = mg$. Approximating $g = 10\ \text{m/s}^2$, the weight of the box is

$$W = mg = (5\ \text{kg})(10\ \text{m/s}^2) \\ = 50\ \text{N}$$

20. (D) Dynamics *Single-Concept Problem*

The force needed to start the box moving, F, is equal and opposite to the maximum force of static friction, F_s, which is given by the formula $F_s = \mu_s N$, where N is the normal force. Since F_s is directly proportional to μ_s, so is F. That means that if μ_s is doubled, then F is doubled as well.

21. (D) Dynamics *Single-Concept Problem*

Solving this problem is simply a matter of applying Newton's Second Law:

$$a_{\text{new}} = \frac{4F}{2M} = \frac{2F}{M} \\ = 2a$$

22. (A) Miscellaneous (Equation Manipulation, Graph Analysis, Significant Digits) *Recall*

The graph of a function with constant slope is a single straight line. Only (A) fits this criterion. Answer choice (E) consists of straight lines, but there are two different slopes in the graph, one positive and one negative.

23. (C) Thermal Physics *Single-Concept Problem*

The Ideal Gas Law states that $PV = nRT$, where P is the pressure of a gas, V is the volume, n is the number of moles of gas molecules, R is the ideal gas constant, and T is the temperature of the gas. If the value of V increases, then, all else being equal, the value of P will decrease. The temperature of the gas would not decrease, but rather increase. The average kinetic energy of the gas molecules is the same thing as temperature. There is no reason the number of molecules in the gas or the atomic number of the atoms in the gas should change at all.

24. (A) Thermal Physics *Single-Concept Problem*

The First Law of Thermodynamics tells us that $\Delta U = \Delta Q + \Delta W$. In English, that means that the change in the internal energy of a system is equal to the heat added to the system and the work done on the system. In this example, the value of ΔU is 12 joules, and the value of ΔW is –7 joules. ΔW has a negative value, because work is done by the system, not on it. With this information, we can now solve for ΔQ:

$$\Delta Q = \Delta U - \Delta W = 12\ \text{J} - (-7\ \text{J}) \\ = 19\ \text{J}$$

The equation shows us that 19 joules have been added to the system.

25. **(C)** Kinematics *Single-Concept Problem*

To solve this problem, you need to recall the kinematic equation $x = v_o t + \frac{1}{2}at^2$. We have $v_o = 0$ and $a = g = 10\ \text{m/s}^2$. Plugging in, we can solve for t:

$$125\ \text{m} = (0)t + \frac{1}{2}(10\ \text{m/s}^2)t^2$$
$$t^2 = 2(125\ \text{m})/10\ \text{m/s}^2$$
$$t^2 = 25\ \text{s}^2$$
$$t = 5\ \text{s}$$

26. **(D)** Kinematics *Multi-Concept Problem*

One way to solve this problem is to recall the kinematic equation $v^2 = v_o^2 + 2ax$. Plugging the appropriate numbers into the equation, we can solve for x:

$$(30\ \text{m/s})^2 = (0\ \text{m/s})^2 + 2(10\ \text{m/s}^2)x$$
$$x = 900\ \text{m}^2/\text{s}^2/20\ \text{m/s}^2$$
$$= 45\ \text{m}$$

27. **(E)** Waves *Recall*

The only real way to solve this problem is to be familiar with the electromagnetic spectrum. You may be misled by the fact that radio waves have the longest wavelengths and gamma rays have the shortest wavelengths. However, frequency is inversely proportional to wavelength, so short wavelengths corresponds to high frequencies.

28. **(C)** Waves *Single-Concept Problem*

The frequency of a wave is the number of cycles it completes per second, where a cycle is a single "round trip" for the wave. On the diagram, we see that the first cycle is completed at some point between $t = 1$ and $t = 2$, but it's not clear exactly where. An easier way to judge is to notice that exactly three cycles are completed in four seconds. Since frequency, f, is a measure of the number of cycles per second, we can conclude that:

$$f = \frac{3\ \text{cycles}}{4\ \text{seconds}} = \frac{3}{4}\ \text{cycles/second}$$
$$= 0.75\ \text{Hz}$$

29. **(D)** Electric Forces, Fields, and Potential *Recall*

Electric field lines always point in the direction that a positive point charge would move if it were at that point. A positive point charge would be attracted to a conductor with a charge of $-Q$ in the same way a massive object is attracted to the surface of the earth. The field lines will point directly toward the center of the sphere, perpendicular to the surface of the sphere, (D).

30. **(B)** Electric Forces, Fields, and Potential *Single-Concept Problem*

Both q_1 and q_2 exert a force on Q, so if the net force on Q is zero, then the force exerted by the two charges must be equal and opposite. If q_1 is a negative charge, it will attract Q, so q_2 must have a negative charge of equal magnitude in order to balance out this attraction. Similarly, if q_1 is positive, it will repel Q, so q_2 must have a positive charge of equal magnitude in order to balance out this repulsion. q_1 and q_2 can be either positive or negative, but they must both be of the same sign, and they must both be of the same magnitude.

31. **(D)** Special Problems in Mechanics (Pulleys, Springs, Pendulums, Inclined Planes) *Single-Concept Problem*

The energy stored in any spring of spring constant k, displaced by a distance x, is given by (D).

If you answered (A) or (B) you were thinking of the force on the mass at B.

Answer (C) is incorrect because energy stored in a spring is always non-negative.

32. **(B)** Special Problems in Mechanics (Pulleys, Springs, Pendulums, Inclined Planes) *Single-Concept Problem*

The spring will exert a restoring force on the mass, according to Hooke's Law: $F = -kx$. Since $k = 4$ N/m and $x = 0.25$ m, that force is $F = -(4 \text{ N/m})(0.25 \text{ m}) = -1$ N. Because this force is a restoring force, we know that it will be directed upward, in the direction of the equilibrium position, A. Because we know the direction in which the force is acting, we needn't worry about the negative sign in our answer.

33. **(B)** Modern Physics *Recall*

Thompson's "raisin pudding" model of the atom was based on his 1897 discovery of the electron. According to the model, atoms consist of a positively charged medium with small, negatively charged electrons distributed throughout. The other four answer choices were all later discoveries: the existence of nuclei and protons were established by Rutherford, the fact that electrons orbit only at certain radii was discovered by Bohr, and the fact that electrons exhibit wave properties was discovered by de Broglie.

34. **(C)** Modern Physics *Single-Concept Problem*

The atomic number gives the number of protons in an atomic nucleus, which in turn defines what element that atom is. Different isotopes of an element have different numbers of neutrons, so their neutron numbers and mass numbers will vary. The decay rate and distribution in a random sample both vary from isotope to isotope as well.

35. **(C)** Modern Physics *Recall*

The mass number of an atom is the sum of the number of protons and neutrons. All atoms of the same element have the same number of protons, but may have different numbers of neutrons or electrons.

Don't be fooled by (B): the number of protons in an atom must equal the number of electrons in an atom only if that atom is electrically neutral. A positively or negatively charged ion is still an atom, even though there are more or fewer electrons than there are protons.

36. **(A)** Thermal Physics *Single-Concept Problem*

According to the Ideal Gas Law, $P = nRT/V$. Since volume is inversely proportional to pressure, doubling the volume will halve the pressure, and since temperature is directly proportional to pressure, halving the temperature will also halve the pressure. Since the pressure is halved twice over, the net result is that it is divided by four.

37. **(C)** Thermal Physics *Recall*

More heat is needed to raise the temperature of a larger mass or to raise the temperature by a larger amount. Also, the higher the specific heat of an object, the greater the heat needed to raise its temperature.

38. **(D)** Thermal Physics *Single-Concept Problem*

The efficiency of a heat engine is determined by the ratio of the amount of work it can do to the amount of heat that must be supplied: $e = W/Q_{in}$. Since we know that $e = 75\% = 0.75$ and $W = 300$ J, we should have no trouble solving for Q_{in}:

$$Q_{in} = \frac{W}{e} = \frac{300 \text{ J}}{0.75}$$
$$= 400 \text{ J}$$

39. **(B)** Work, Energy, and Power *Single-Concept Problem*

Because this ramp is frictionless, we can apply the law of conservation of energy: the kinetic energy of the box at the bottom of the ramp is equal to its gravitational potential energy when it reaches its highest point. By equating the formula for kinetic energy to the formula for gravitational potential energy, we can solve for the box's height, h:

$$mgh = \frac{1}{2}mv^2$$
$$h = \frac{v^2}{2g}$$
$$= \frac{(10 \text{ m/s})^2}{2(10 \text{ m/s}^2)}$$
$$= 5 \text{ m}$$

40. **(B)** Work, Energy, and Power *Multi-Concept Problem*

Because the man pushes the box along a frictionless surface, the law of conservation of energy applies, and all the work done on the box is converted into kinetic energy. The amount of work done on the box is

$$W = Fd = (10 \text{ N})(10 \text{ m})$$
$$= 100 \text{ J}$$

Since we know that the man imparts 100 J of kinetic energy to the box, we can plug this number into the formula for kinetic energy and solve for velocity, v:

$$\frac{1}{2}mv^2 = 100 \text{ J}$$
$$v^2 = 200 \text{ J}/m$$
$$v = \sqrt{200 \text{ J}/8 \text{ kg}}$$
$$= 5 \text{ m/s}$$

41. **(C)** Work, Energy, and Power *Single-Concept Problem*

Power is given by the formula $P = W/t$, and work is given by the formula $W = Fd$, so the power output is

$$P = \frac{W}{t} = \frac{Fd}{t} = \frac{(30 \text{ N})(5 \text{ m})}{10 \text{ s}}$$
$$= 15 \text{ W}$$

Since we know the amount of force exerted on the object, knowing its mass is irrelevant for the question.

42. **(D)** Miscellaneous (Equation Manipulation, Graph Analysis, Significant Digits) *Multi-Concept Problem*

We can express the relationship between A, B, and C according to the equation $A = k\,B/C^2$, where k is some constant. We can rearrange this equation to solve for B:

$$B = \frac{1}{k}AC^2$$

Since B is directly proportional to C^2 and A/k is held constant, the resulting graph will be a parabola that opens up from the origin. This is represented in choice (D).

43. **(B)** Magnetic Fields and Forces *Single-Concept Problem*

Since the current is in the same direction, and of the same magnitude, in both wires, the magnetic fields created by both wires are the same: the magnetic field points into the page to the right of the wire, and out of the page to the left of the wire. Point *B* is directly between the two wires, where the field created by the two wires is equal and opposite, thus canceling each other out.

44. **(B)** Gravity *Multi-Concept Problem*

We have two equations for the force acting on the planet. Newton's Law of Universal Gravitation tells us that the force is $F = GMm/R^2$, and the equation for centripetal force tells us that the force is $F = mv^2/R$. If we set these two equations equal to one another, we can solve for v:

$$\frac{mv^2}{R} = \frac{GMm}{R^2}$$

$$v^2 = \frac{GM}{R}$$

$$v = \sqrt{GM/R}$$

This equation shows us that v is inversely proportional to the square root of R. That means that if R is doubled, then v is multiplied by a factor of $1/\sqrt{2}$.

45. **(C)** Gravity *Multi-Concept Problem*

If we refer to the equation we derived in the previous question, we find that the planet's velocity is determined by the equation

$$v = \sqrt{GM/R}$$

The M in this equation is for the mass of the star, not the mass of the planet. The mass of the planet has no effect on its orbital velocity, just like the mass of an object above the surface of the earth has no effect on its acceleration. The gravitational force is proportional to the planet's mass, but the planet's acceleration is inversely proportional to its mass, so the factor for the planet's mass cancels out.

46. **(A)** Dynamics *Single-Concept Problem*

If the object moves to the right, the net vertical component of the force acting upon it is zero. That means that the upward component of $\mathbf{F}_1$ must be equal and opposite to $\mathbf{F}_2$, so that the two forces cancel each other out in the vertical direction. So in order to calculate the magnitude of $\mathbf{F}_2$, we need to calculate the magnitude of the upward component of $\mathbf{F}_1$:

$$F_{1y} = \mathbf{F}_1 \sin 30° = (60 \text{ N})(0.500)$$
$$= 30 \text{ N}$$

47. **(C)** Dynamics *Multi-Concept Problem*

Newton's Second Law tells us the net force acting on the box:

$$F_{net} = ma = (2 \text{ kg})(2 \text{ m/s}^2)$$
$$= 4 \text{ N}$$

The 10 N force of tension in the rope pulls the box forward, so there must be a frictional force of 6 N opposing the force of tension pulling the box forward so that the net force on the box can be $10 - 6 = 4$ N. The magnitude of the force of kinetic friction is given by the formula $F_f = \mu_k mg$, where μ_k is the coefficient of kinetic friction. Knowing the magnitude of the force of friction and the mass of the box, we can solve for μ_k:

$$6 \text{ N} = (2 \text{ kg})(10 \text{ m/s}^2)\mu_k$$

$$\mu_k = \frac{6 \text{ N}}{20 \text{ N}}$$

$$= 0.3$$

48. (E) Modern Physics *Single-Concept Problem*

Both the mass number and atomic number of atoms must be conserved in nuclear reactions. The initial particle, ${}^{14}_{6}\text{C}$, has a mass number of 14 and an atomic number of 6. When it undergoes beta-decay, it emits a beta particle, with a mass number of 0 and an atomic number of –1, as well as a neutrino, that has no mass number or atomic number. This reaction can be expressed algebraically as follows:

$${}^{14}_{6}\text{C} = {}^{a}_{z}X + {}^{0}_{-1}\beta + \nu$$

We can solve for the mass number, *a*, and the atomic number, *z*, of our mystery element, *X*, with two simple equations:

$$14 = a + 0 \quad\quad 6 = z - 1$$
$$a = 14 \quad \text{and} \quad z = 7$$

If we know that the element with atomic number 7 is nitrogen, we can correctly conclude that the result of this beta-decay is ${}^{14}_{7}\text{N}$.

49. (B) Modern Physics *Recall*

In the Bohr model of the atom, electrons orbit the positively charged nucleus at discrete radii.

50. (A) Circuits and Circuit Elements

The resistance, R, of a wire is given by the formula $R = \rho L/A$, where ρ is the resistivity of the wire material, L is the length of the wire, and A is the cross-sectional area of the wire. We can relate the area, A, of a circular wire to its diameter, d, by the formula $A = \pi d^2/4$. Since area is directly proportional to the square of the diameter, doubling the diameter will quadruple the area. And since resistance is inversely proportional to cross-sectional area, quadrupling the area will quarter the resistance.

51. (C) Magnetic Fields and Forces *Single-Concept Problem*

The magnitude of the magnetic force on a particle moving in a direction perpendicular to the lines of the field is given by the formula $F = qvB$. Here, q is the charge of the particle, v is its velocity, and B is the field. We can apply that formula to find

$$F = (2 \times 10^{-10}\ \text{C})(4 \times 10^{5}\ \text{m/s})(0.5\ \text{T})$$
$$= 4 \times 10^{-5}\ \text{N}.$$

52. (C) Magnetic Fields and Forces *Single-Concept Problem*

The magnetic force is perpendicular to the velocity, pulling it circular motion. The centripetal magnetic force can change the direction, but not the magnitude, of the velocity of a moving object.

53. (A) Magnetic Fields and Forces *Multi-Concept Problem*

The magnitude of the magnetic force is given by the formula $F = qvB \sin\theta$, where θ is the angle between the velocity vector and the magnetic field vector. If these two vectors are parallel, then $\theta = 0°$, and so $F = 0$.

54. (A) Circuits and Circuit Elements *Multi-Concept Problem*

The equivalent resistance of the two resistors in series is 3 Ω + 6 Ω = 9 Ω. We can then solve for the total current, using Ohm's Law:

$$I = \frac{V}{R} = \frac{3\ \text{V}}{9\ \text{W}}$$
$$= \frac{1}{3}\ \text{A}$$

55. **(C)** Circuits and Circuit Elements *Multi-Concept Problem*

Because the resistors are in series, the total current of $^1/_3$ A flows through each of them. Using Ohm's Law, then, we can calculate the voltage drop across R_1:

$$V = IR = (\frac{1}{3}\text{ A})(3\ \Omega)$$
$$= 1\text{ V}$$

56. **(A)** Circuits and Circuit Elements *Multi-Concept Problem*

We have calculated that the voltage drop across R_1 is 1 V and that the current across R_1 is the constant current for the circuit of $^1/_3$ A. With these two figures, we can calculate the power dissipated in R_1:

$$P = IV = (\frac{1}{3}\text{ A})(1\text{ A})$$
$$= \frac{1}{3}\text{ W}$$

57. **(B)** Waves *Recall*

Covering more holes makes the tube longer. This increases the wavelength and decreases the frequency and pitch. The length of a tube containing standing waves has no effect on the speed of the sound wave, so the answer is (B).

58. **(D)** Waves *Recall*

When a wave pulse hits a wall, it will bounce back inverted. The wavelength of the wave depends on the linear mass density and tension of the string, neither of which are affected by the wave bouncing off the wall, so the wavelength will remain constant. Therefore, the correct answer is (D).

59. **(D)** Miscellaneous (Equation Manipulation, Graph Analysis, Significant Digits) *Single-Concept Problem*

When reporting a measurement, the last digit is always uncertain, and can be off by as much as one half the value of the second to last digit. In this case, the science teacher guessed the final 6, meaning that her car could really have a length anywhere between 3.41 m and 3.51 m.

60. **(C)** Special Problems in Mechanics (Pulleys, Springs, Pendulums, Inclined Planes) *Multi-Concept Problem*

The system will be in equilibrium when the net force acting on the 1 kg mass is equal to zero. A free-body diagram of the forces acting on the 1 kg mass shows that it is in equilibrium when the force of tension in the pulley rope is equal to $mg \sin \theta$, where $m = 1$ kg and θ is the angle of the inclined plane.

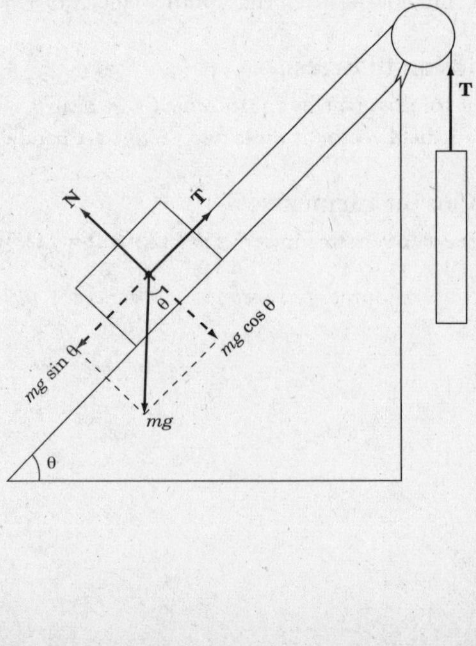

Since the system is in equilibrium, the tension in the rope must be equal and opposite to the force of gravity acting on the 0.5 kg mass. The force of gravity on the 0.5 kg mass, and hence the force of tension in the rope, has a magnitude of (0.5 kg)g. Knowing that the force of tension is equal to $mg \sin \theta$, we can now solve for θ:

$$mg \sin \theta = 0.5g$$
$$(1 \text{ kg}) \sin \theta = 0.5$$
$$\sin \theta = 0.5$$
$$\theta = 30°$$

61. **(B)** Optics *Recall*

Dispersion is a variation in wave speed with frequency that causes the index of refraction, $n = c/v$, to vary with the frequency of the light. When white light hits a glass prism, each wavelength is refracted by a slightly different angle, so the white light breaks up into its constituent frequencies, or colors.

62. **(D)** Optics *Recall*

If you don't know the answer to this question off the top of your head, the easiest way to solve it is to draw three ray diagrams, and see which diagrams produce inverted images.

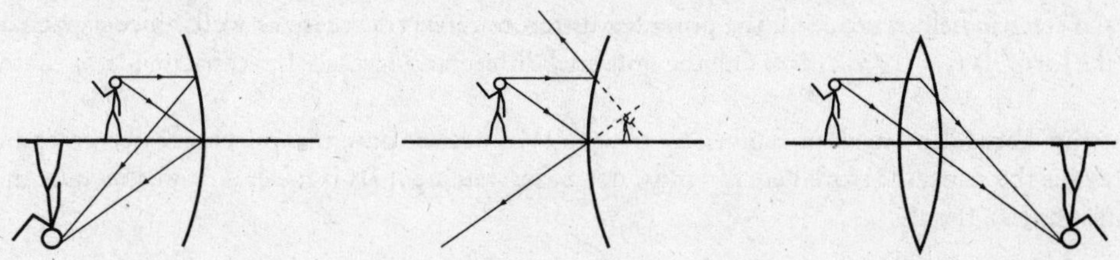

concave mirror convex mirror convex lens

From these three diagrams, you can see that only concave mirrors and convex lenses produce upside down images, so (D) is the correct answer.

63. **(B)** Optics *Multi-Concept Problem*

Solving this problem would be a simple application of the lens and mirror equation if not for a little catch: because we are dealing with a convex mirror, the focal point is behind the mirror, and so the value of f is negative. With that in mind, we can solve for d', the distance of the image from the mirror:

$$\frac{1}{d} + \frac{1}{d'} = \frac{1}{f}$$
$$\frac{1}{d'} = -\frac{1}{6} \text{ m} - \frac{1}{12} \text{ m}$$
$$d' = -4 \text{ m}$$

The image appears 4 m behind the mirror. Since the mirror is convex and d' is negative, we know that the image produced is virtual.

64. **(A)** Thermal Physics *Multi-Concept Problem*

The efficiency of a heat engine is given by the formula $e = 1 - Q_{out}$ over Q_{in}, where Q_{out} is the amount of heat exhausted into a cold reservoir, and Q_{in} is the amount of heat put into the engine. In this question, the value of Q_{out} is Q, since that is the energy exhausted into the cold reservoir. The value of Q_{in} is the amount of heat extracted from the high temperature source. The amount of heat taken from the high temperature source is given by the formula $Q = mc\Delta T$. Plugging these values for Q_{out} and Q_{in} into the formula for efficiency, we get (A).

65. **(B)** Thermal Physics *Multi-Concept Problem*

The law of thermal expansion tells us that, given an initial volume, V_o, a coefficient of volume expansion, β, and a change in temperature, ΔT, an object will expand by an amount ΔT, according to the formula $\Delta V = \beta V_o \Delta T$. We know the values of β and V_o, but we need to know the value of ΔT.

The ideal gas law states that, at a constant volume, temperature and pressure are directly proportional. That is, if the pressure is doubled, so it the temperature. If the temperature is initially at 200K, then it ends up at 400 K, with a change of 200K. Plugging this value into our formula for volume expansion, we find:

$$\Delta V = (2.0 \times 10^{-5}\ \text{K}^{-1})(0.50\ \text{L})(200\text{K}) = 2.0 \times 10^{-3}\ \text{L}$$

66. **(B)** Circuits and Circuit Elements *Recall*

The correct expression is $C = Q/V$. Capacitance is a measure of the ability to store charge. The larger the capacitance, the greater the amount of charge that can be stored at a constant potential difference.

67. **(B)** Circuits and Circuit Elements *Multi-Concept Problem*

The electrons in the dielectric are attracted to the positive plate, and create their own electric field in the dielectric. This field opposes the original field, and the overall electric field is reduced. Statement I is false.

Since the electric field is reduced, the potential difference must decrease as well. Since capacitance is given by the formula $C = Q/V$, decreasing the potential difference increases the capacitance. So statement II is true.

The energy stored in a capacitor is given by $U = \frac{1}{2}QV$. Since the potential difference decreases and the charge remains the same, the stored energy must decrease; statement III is false. We can conclude that the answer must be (B), II only.

68. **(A)** Circular Motion and Rotation *Recall*

If the wheel is spinning in a clockwise direction, its angular momentum vector points at an angle in a downward direction. The force of gravity working on the different points in the wheel serve to give the wheel a counterclockwise torque:

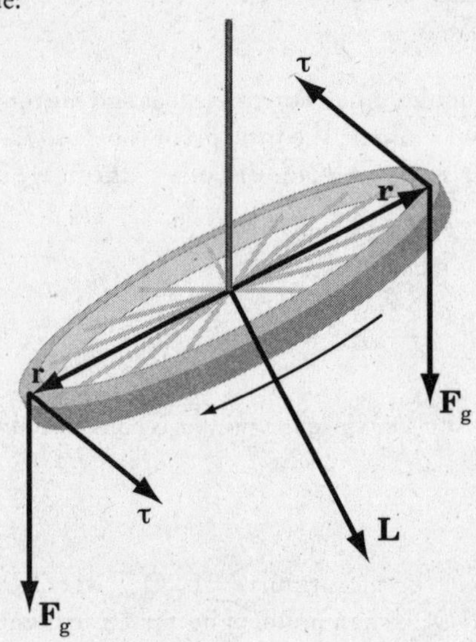

The torque caused by the force of gravity will cause the wheel to precess about its axis, but it will not cause the wheel to collapse into a vertical position, since the wheel's angular momentum must be conserved. So the wheel will continue to rotate in a clockwise direction while it precesses in a counterclockwise direction.

If you answered (B), you were on the right track. Energy is indeed conserved. However, energy is a scalar quantity, while angular momentum is a vector. A scalar quantity like kinetic energy can't place constraints upon the direction of an object's motion of rotation.

69. **(C)** Waves *Multi-Concept Problem*

The relationship between the velocity of a wave on a string and the mass of the string is given by the formula

$$v = \sqrt{\frac{TL}{m}}$$

From this formula, we can see that doubling the mass will decrease the velocity of the wave on the string.

The wavelength of the first harmonic is given by the formula $\lambda = 2L$: as we can see, the wavelength depends only on L, and not on m, so it will remain constant when the mass is doubled.

Wave speed, frequency, and velocity are related by the formula $v = f\lambda$. Since the wave speed is affected by a change in mass and the wavelength is held constant, the frequency must change in direct proportion to the change in wave speed.

Since wave speed and frequency are affected by a change in mass but wavelength is not, the correct answer is (C).

70. **(D)** Waves *Multi-Concept Problem*

The wave speed of a wave on a string is related to the mass, density, and length of the string by the formula:

$$v = \sqrt{\frac{TL}{m}}$$

If we double both L and m, the value of the fraction, and hence the value of v, remains constant.

The wavelength of the first harmonic is given by the formula $\lambda = 2L$. If L is doubled, then the wavelength of the first harmonic will also be doubled.

Wave speed, wavelength, and frequency are related by the formula $v = f\lambda$. If λ increases while v is kept constant, then f must decrease.

Since the frequency and wavelength of the first harmonic change but the wave speed does not, the correct answer is (D).

71. **(C)** Electromagnetic Induction *Recall*

The emf is induced by changing the magnetic flux through the loop. The induced emf is given by Faraday's Law: $\mathcal{E} = -\Delta\Phi / \Delta t$. The current flows so that it opposes the change in magnetic flux by creating its own magnetic field.

72. **(B)** Electromagnetic Induction *Single-Concept Problem*

Initially, the magnetic flux through the ring is zero. When we turn on the current, we create a magnetic field pointing out the top of the solenoid. By Lenz's Law, the ring will create a magnetic field to oppose the magnetic field created by the solenoid. To oppose the solenoid's magnetic field, the iron ring will induce an electric current that goes in the opposite direction of the solenoid's. Since opposite currents repel one another, the ring will fly off the solenoid, as described in (B).

73. **(A)** Linear Momentum *Single-Concept Problem*

According to the law of conservation of momentum, the momentum of the bullet before it hits the block, $p = mv$, is equal to the momentum of the bullet-block system immediately after impact, $p' = (m + M)v'$, where v' is the velocity of the bullet-block system. By applying the equation $p = p'$, we can solve for v':

$$(m + M)v' = mv$$

$$v' = \frac{m}{m + M}v$$

74. **(B)** Special Problems in Mechanics (Pulleys, Springs, Pendulums, Inclined Planes) *Multi-Concept Problem*

The law of conservation of energy tells us that the initial kinetic energy of the bullet-block system when it starts to move is equal to the gravitational potential energy of the bullet-block system when it reaches its highest point. In other words, $\frac{1}{2}mv^2 = mgh$. Since we solved for v' in the previous question, we can now solve for h:

$$mgh = \frac{1}{2}m\left(\frac{m}{m + M}v\right)^2$$

$$h = \frac{1}{2g}\left(\frac{m}{m + M}\right)^2 v^2$$

75. **(E)** Modern Physics *Multi-Concept Problem*

The particle's rest energy is given by the formula $E_o = mc^2$, where m is its mass. The total energy is $E = E_o + KE = mc^2/\sqrt{1 - v^2/c^2}$, where KE is the kinetic energy and v is the velocity. We can therefore solve for $KE = E - E_o = (1/\sqrt{1 - v^2/c^2} - 1)mc^2$. If we plug $v = 0.6c$ into the equation for relativistic kinetic energy, KE, we can solve for KE in terms of E_o:

$$\begin{aligned} KE &= (1/\sqrt{1 - v^2/c^2} - 1)mc^2 \\ &= (1/\sqrt{1 - (0.6c)^2/c^2} - 1)mc^2 \\ &= (1/\sqrt{0.64} - 1)mc^2 \\ &= (1/0.8 - 1)mc^2 \\ &= (1.25 - 1)mc^2 \\ &= (0.25)mc^2 \\ &= 0.25\,E_o \end{aligned}$$